Lecture Notes in Computer Science

Commenced Publication in 1973
Founding and Former Series Editors:
Gerhard Goos, Juris Hartmanis, and Jan van Leeuwen

Editorial Board

Oded Goldreich   Arnold L. Rosenberg
Alan L. Selman (Eds.)

Theoretical Computer Science

Essays in Memory of Shimon Even

 Springer

Volume Editors

Oded Goldreich
Weizmann Institute of Science
Department of Computer Science
Rehovot, Israel
E-mail: oded.goldreich@weizmann.ac.il

Arnold L. Rosenberg
University of Massachusetts Amherst
Department of Computer Science
Amherst, MA 01003, USA
E-mail: rsnbrg@cs.umass.edu

Alan L. Selman
University at Buffalo, The State University of New York
Department of Computer Science and Engineering
Buffalo, NY 14260-2000, USA
E-mail: selman@cse.buffalo.edu

The illustration appearing on the cover of this book is the work of Daniel Rozenberg (DADARA).

Library of Congress Control Number: 2006922002

CR Subject Classification (1998): F.2.2, G.1.2, G.2.2, C.2.4, E.3

LNCS Sublibrary: SL 1 – Theoretical Computer Science and General Issues

ISSN   0302-9743
ISBN-10   3-540-32880-7 Springer Berlin Heidelberg New York
ISBN-13   978-3-540-32880-3 Springer Berlin Heidelberg New York

Springer is a part of Springer Science+Business Media

springer.com

© Springer-Verlag Berlin Heidelberg 2006
Printed in Germany

Typesetting: Camera-ready by author, data conversion by Boller Mediendesign
Printed on acid-free paper   SPIN: 11685654   06/3142   5 4 3 2 1 0

Shimon Even (1935–2004)

Preface

On May 1, 2004, the world of theoretical computer science suffered a stunning loss: Shimon Even passed away. Few computer scientists have had as long, sustained, and influential a career as Shimon.

Shimon Even was born in Tel-Aviv in 1935. He received a B.Sc. in Electrical Engineering from the Technion in 1959, an M.A. in Mathematics from the University of Northern Carolina in 1961, and a Ph.D. in Applied Mathematics from Harvard University in 1963. He held positions at the Technion (1964–67 and 1974–2003), Harvard University (1967–69), the Weizmann Institute (1969–74), and the Tel-Aviv Academic College (2003-04). He visited many universities and research institutes, including Bell Laboratories, Boston University, Cornell, Duke, Lucent Technologies, MIT, Paderborn, Stanford, UC-Berkeley, USC and UT-Dallas.

Shimon Even played a major role in establishing computer science education in Israel and led the development of academic programs in two major institutions: the Weizmann Institute and the Technion. In 1969 he established at the Weizmann the first computer science education program in Israel, and led this program for five years. In 1974 he joined the newly formed computer science department at the Technion and shaped its academic development for several decades. These two academic programs turned out to have a lasting impact on the evolution of computer science in Israel.

Shimon Even was a superb teacher, and his courses deeply influenced many of the students attending them. His lectures, at numerous international workshops and schools, inspired a great number of students and researchers. His books, especially his celebrated *Graph Algorithms*, carried his educational message also to computer scientists who were not fortunate enough to meet him in person. As a mentor to aspiring researchers, Shimon was almost without peer, nurturing numerous junior researchers and advising many graduate students, who went on to have their own successful research careers.

Shimon Even was a pioneer in the areas of graph algorithms and cryptography, and his research contributions to these areas influenced the course of their development. Shimon was famous for not confining his interests to a few topics, but choosing rather to work in such diverse areas as switching and automata theory, coding theory, combinatorial algorithms, complexity theory, distributed computing, and circuit layout. In each of these areas, he produced high-quality, innovative research for more than four decades.

Shimon was the purest of pure theoreticians, following his nose toward research problems that were "the right" ones at the moment, not the faddish ones. His standards were impeccable, to the point where he would balk at employing any result whose proof he had not mastered himself. His integrity was unimpeachable: he would go to great lengths to defend any principle he believed in.

Shimon had a great passion for computer science as well as a great passion for truth. He valued simplicity, commitment to science, natural questions and carefully prepared expositions. By merely following his own way, Shimon influenced numerous researchers to adopt his passions and values. We hope that this is reflected in the current volume.

This volume contains research contributions and surveys by former students and close collaborators of Shimon. We are very pleased that Reuven Bar-Yehuda, Yefim Dinitz, Guy Even, Richard Karp, Ami Litman, Yehoshua Perl, Sergio Rajsbaum, Adi Shamir, and Yacov Yacobi agreed to send contributions. In accordance with Shimon's style and principles, the focus of these contributions is on addressing natural problems and being accessible to most researchers in theoretical computer science. The contributions are of three different types, reflecting three main scientific activities of Shimon: original research, technical surveys, and educational essays.

The Contributions

The contributions were written by former students and close collaborators of Shimon. In some cases the contributions are co-authored by researchers who were not fortunate enough to be close to Shimon or even to have met him in person. Below we comment on particular aspects of each contribution that we believe Shimon would have appreciated.

Original Research

Needless to say, everybody likes original research, and Shimon was no exception. We believe that Shimon would have been happy with the attempt to make these research contributions accessible to a wide range of researchers (rather than merely to experts in the area). In order to promote this goal, these contributions were reviewed both by experts and by non-experts.

- P. Fraigniaud, D. Ilcinkas, S. Rajsbaum and S. Tixeuil: *The Reduced Automata Technique for Graph Exploration Space Lower Bounds*. Shimon liked connections between areas, and the areas of graph algorithms and of automata theory were among his favorites.
- O. Goldreich: *Concurrent Zero-Knowledge with Timing, Revisited*. Shimon would have joked at Oded's tendency to write long papers.
- R.M. Karp: *Fair Bandwidth Allocation Without Per-Flow State*. Shimon would have like the fact that the starting point of this work is a practical problem, and that it proceeds by distilling a clear computational problem and resolving it optimally.
- R.M. Karp, T. Nierhoff and T. Tantau: *Optimal Flow Distribution Among Multiple Channels with Unknown Capacities*. This paper has the same flavor as the previous one, and Shimon would have liked it for the very same reason.

- A. Litman: *Parceling the Butterfly and the Batcher Sorting Network*. Shimon would have liked the attempt to present a new complexity measure that better reflects the actual cost of implementations.
- X. Zhou, J. Geller, Y. Perl, and M. Halper: *An Application Intersection Marketing Ontology*. Shimon would have liked the fact that simple insights of graph theory are used for a problem that is very remote from graph theory.
- R.L. Rivest, A. Shamir and Y. Tauman: *How to Leak a Secret: Theory and Applications of Ring Signatures*. Shimon would have like the natural ("daily") problem addressed in this paper as well as the elegant solution provided to it.
- O. Yacobi with Y. Yacobi: *A New Related Message Attack on RSA*. Shimon would have enjoyed seeing a father and son work together.

Technical Surveys

Shimon valued the willingness to take a step back, look at what was done (from a wider perspective), and provide a better perspective on it. We thus believe that he would have been happy to be commemorated by a volume that contains a fair number of surveys.

- R. Bar-Yehuda and D. Rawitz: *A Tale of Two Methods*. Shimon liked stories, and he also liked the techniques surveyed here. Furthermore, he would have been excited to learn that these two techniques are in some sense two sides of the same coin.
- Y. Dinitz: *Dinitz' Algorithm: The Original Version and Even's Version*. Shimon is reported to have tremendously enjoyed Dinitz's lecture that served as a skeleton to this survey.
- C. Glaßer, A.L. Selman, and L. Zhang: *Survey of Disjoint NP-pairs and Relations to Propositional Proof Systems*. This survey focuses on one of the applications of promise problems, which was certainly unexpected in 1984 when Shimon Even, together with Alan Selman and Yacov Yacobi, introduced this notion.
- O. Goldreich: *On Promise Problems*. This survey traces the numerous and diverse applications that the notion of promise problems found in the two decades that have elapsed since the invention of the notion.
- G. Malewicz and A.L. Rosenberg: *A Pebble Game for Internet-Based Computing*. Shimon liked elegant models, and would have been interested to see pebble games used to model an Internet-age problem.

Educational Essays

Shimon liked opinionated discussions and valued independent opinions that challenge traditional conventions. So we are sure he would have enjoyed reading these essays, and we regret that we cannot have his reaction to them.

- G. Even: *On Teaching Fast Adder Designs: Revisiting Ladner & Fischer.*
  Shimon would have been very proud of this insightful and opinionated exposition of hardware implementations of the most basic computational task.
- O. Goldreich: *On Teaching the Basics of Complexity Theory.* Shimon would
  have appreciated the attempt to present the basics of complexity theory in
  a way that appeals to the naive student.
- A.L. Rosenberg: *State.* Shimon would have supported the campaign, launched
  in this essay, in favor of the Myhill-Nerode Theorem.

December 2005   Oded Goldreich (Weizmann Institute of Science)
                       Arnold L. Rosenberg (University of Massachusetts Amherst)
                                  Alan L. Selman (University at Buffalo)

Table of Contents

The Reduced Automata Technique
for Graph Exploration Space Lower Bounds*

Pierre Fraigniaud[1] **, David Ilcinkas[2] **, Sergio Rajsbaum[3] ***, and
Sébastien Tixeuil[4] †

[1] CNRS, LRI, Université Paris-Sud, France
pierre@lri.fr
[2] LRI, Université Paris-Sud, France
ilcinkas@lri.fr
[3] Instituto de Matemáticas, Univ. Nacional Autónoma de México, Mexico
rajsbaum@math.unam.mx
[4] LRI & INRIA, Université Paris-Sud, France
tixeuil@lri.fr

Abstract. We consider the task of exploring graphs with anonymous
nodes by a team of non-cooperative robots, modeled as finite automata.
For exploration to be completed, each edge of the graph has to be traversed by at least one robot. In this paper, the robots have no a priori
knowledge of the topology of the graph, nor of its size, and we are interested in the amount of memory the robots need to accomplish exploration, We introduce the so-called *reduced automata technique*, and we
show how to use this technique for deriving several space lower bounds
for exploration. Informally speaking, the reduced automata technique
consists in reducing a robot to a simpler form that preserves its "core"
behavior on some graphs. Using this technique, we first show that any
set of $q \geq 1$ non-cooperative robots, requires $\Omega(\log(\frac{n}{q}))$ memory bits
to explore all n-node graphs. The proof implies that, for any set of q
K-state robots, there exists a graph of size $O(qK)$ that no robot of this
set can explore, which improves the $O(K^{O(q)})$ bound by Rollik (1980).
Our main result is an application of this latter result, concerning *terminating* graph exploration with one robot, i.e., in which the robot is
requested to stop after completing exploration. For this task, the robot
is provided with a pebble, that it can use to mark nodes (without such a
marker, even terminating exploration of cycles cannot be achieved). We
prove that terminating exploration requires $\Omega(\log n)$ bits of memory for
a robot achieving this task in all n-node graphs.

* A preliminary version of this paper appears in the proceedings of the 12th Inter-
national Colloquium on Structural Information and Communication Complexity
(SIROCCO), Mont Saint-Michel, France, May 24-26, 2005, as part of [13].
** Supported by the INRIA project "Grand Large", and the projects "PairAPair" of the
ACI "Masses de Données", and "FRAGILE" of the ACI "Sécurité et Informatique".
*** Supported by LAFMI and PAPIIT projects. Part of this work was done while visiting
LRI, Univ. Paris Sud, Orsay.
† Supported by the INRIA project "Grand Large". Additional support from the
project "FRAGILE" of the ACI "Sécurité et Informatique".

O. Goldreich et al. (Eds.): Shimon Even Festschrift, LNCS 3895, pp. 1–26, 2006.

1 Introduction

The problem of exploring an unknown environment occurs in a variety of situations, like robot navigation, network maintenance, resource discovery, and WWW search. In these situations the entities performing exploration can be either a physical mobile device or a software agent. In this paper, we restrict our attention to the case where the environment in which the mobile entities are moving is modeled as a graph. At an abstract level, graph exploration is the task where one or more mobile entities, called *robots* in this paper, are trying to collectively traverse every edge of a graph. In addition to the aforementioned applications, graph exploration is important due to its strong relation to complexity theory, and in particular to the undirected *st*-connectivity (USTCON) problem (cf., e.g., [6]). Given an undirected graph G and two vertices s and t, the USTCON problem is to decide whether s and t are in the same connected component of G. The directed version of the problem is denoted STCON. It is known that STCON is complete for NL, the class of non-deterministic log-space solvable problems. Whether USTCON is complete for L, the class of problems solvable by deterministic log-space algorithms, has been a challenging open problem for quite a long time, and it is only very recently that Reingold proved that USTCON is indeed complete for L [15]. Note that the existence of a finite set of finite-state automata able to explore all graphs would have put USTCON in L, and proving or disproving this existence had therefore motivated quite a long sequence of studies. Cook and Rackoff [6] eventually proved that even a more powerful machine, called JAG, for "Jumping Automaton for Graphs", cannot explore all graphs (a JAG is a finite set of globally cooperative finite-state automata enhanced with the ability, for every automaton, to "jump" from its current position to any node occupied by another automaton). Since this latter result, the exploration graph problem is focussing on determining the space complexity of robots able to explore all graphs.

As far as upper bounds in concerned, Reingold showed in [15] that his log-space algorithm for USTCON implies the existence of log-space constructible *universal exploration sequences* (UXS) of polynomial length. Roughly speaking, a UXS [14] is a sequence of integers that (1) tell a robot how to move from node to node in a graph (the exit port at the kth step of the traversal is obtained by adding the kth integer of the UXS to the entry port), and (2) guarantee to explore every node of a graph of appropriate size (a UXS is defined for a given size, and a given degree). Rephrasing this latter result, there is a $O(\log n)$-space robot that explores all the graphs of size n. The extend to which this bound can be decreased by using a set of $q > 1$ cooperative robots remains open. Also, the question of the existence of log-space constructible *universal traversal sequences* (UTS) [1] remains open (a UTS is a sequence of port-numbers so that the output port at the kth step of the traversal is the kth element of the sequence).

As far as lower bounds are concerned, most papers are dealing with the design of small *traps* for arbitrary teams of robots, i.e., small graphs that no robot of the team can explore. (Formally, a trap consists of a graph and a node from where the robots start the exploration.) The first trap for a finite state robot is generally attributed to Budach [5] (the trap is actually a planar graph). The trap constructed by Budach is however of large size, and a much smaller trap was described in [12] which proved that, for any K-state robot, there exists a trap of at most $K + 1$ nodes. In [16], Rollik proved that no finite set of finite locally-cooperative automata, i.e., automata that exchange information only when they meet at a node, can explore all graphs. In the proof of this result, the author uses as a tool a trap for a set of q non-cooperative K-state robots (such robots may have different transition functions, hence they will follow different paths in the explored graph). This latter trap is of size $O(K^{O(q)})$ nodes. Rollik's trap for cooperative robots is even larger: $\tilde{O}(K^{K^{\cdot^{\cdot^{\cdot^{K}}}}})$ nodes, with $2q + 1$ levels of exponentials where the $\tilde{O}$ notation hides logarithmic factors. In this paper, we present a new lower bound technique for graph exploration, called *reduced automata technique*. Roughly, this technique consists in reducing a robot to a simpler form that preserves its "core" behavior on some graphs: except for some easily described closed paths, the reduced robot follows the path of the original robot, on any such graph.

The interested reader can find other pointers to the literature in, e.g., [3–5, 7, 8, 12]. To complete the picture, and before describing our results in more details, let us point out that Shimon Even, whom this book is dedicated to, was interested in graph exploration problems early on in his career. In particular, in his 1976 seminal paper with Tarjan [11], he presented a way of numbering nodes during a DFS traversal that proved to be useful in many algorithms. In collaboration with A. Litman and P. Winkler [10], he then studied traversal in directed networks. With G. Itkis and S. Rajsbaum [9], he described a traversal strategy for undirected graphs that constructs a subgraph with good connectivity but few edges. And recently, in collaboration with S. Bhatt, D. Greenberg, and R. Tayard [2], he studied the problem of using a robot as simple as possible (with access to some local memory stored in the vertices) to find an Eulerian cycle in mazes and graphs.

1.1 Problem Settings

As in [6, 16], we are interested in exploration of undirected graphs where nodes are not uniquely labeled. Note that, besides the theoretical interest of understanding when or at what cost such graphs can be explored, the unlabeled-node setting can occur in practice, due to, e.g., privacy concerns, limited capabilities of the robots, or simply anonymous edge intersections. The robots, modeled as a deterministic automata, can however identify the edges incident to a node through unique port labels, from 1 to the degree of the node. We consider two types of exploration:

- Perpetual exploration, in which the task of the robots is to, eventually, traverse all edges.
- Terminating exploration, in which the robots, after completing exploration, must eventually stop.

In acyclic graphs, terminating exploration is strictly more difficult than perpetual exploration. In particular, it is shown in [7] that terminating exploration in n-node bounded degree trees requires a robot with memory size $\Omega(\log \log \log n)$, whereas perpetual exploration is possible with $O(1)$ bits. In arbitrary graphs, terminating exploration cannot be achieved. Indeed, it is easy to see that a robot can traverse all edges of some graphs, say a cycle, but that it cannot recognize when it has visited a node twice, because nodes are not uniquely labeled. That is, there are graphs that a robot can explore perpetually, but it can never stops. Thus, as in previous work in this setting, e.g., [3,4,8], we assume that, for terminating exploration, robots can mark nodes: a robot can drop a pebble in a node and later identify it and pick it up.

Following the common conventions in the literature, the robots aiming at performing perpetual exploration are not given pebbles, whereas robots aiming at performing terminating exploration are given one or more pebbles. As a consequence, the two problems becomes incomparable. Indeed, terminating exploration is more demanding than perpetual exploration, but the "machines" designed for these two tasks do not have the same power.

A *team* of robots is a set of deterministic automata with possibly different transition functions, all starting from the same starting point. When sets or teams of robots are considered, the robots of a team can communicate in various manners. Four cases are considered in the literature:

- Non-cooperative robots: the robots are oblivious of each other, and each of them acts independently from the others.
- Locally cooperative robots: robots meeting at a node can exchange information, including their identities and their current states.
- Globally cooperative robots: the robots are perpetually aware of the states of the others, of whether they meet and who they meet, and of the degrees of the nodes occupied by the robots.
- Jumping Automaton: the robots are globally cooperative, and any robot is able to jump from the node it is currently occupying to a node currently occupied by any other robot.

In this paper, we restrict our attention to the two weakest models: non-cooperative robots, and locally cooperative robots.

1.2   Our Results

In this paper, we present a new lower bound technique for graph exploration, called *reduced automata technique*. Based on this technique, the lower bounds presented in this paper are obtained as follows. Assume a set of q robots is given. Then construct the smallest possible graph, called a *trap* for this set of robots, such that if the robots are placed in some specified nodes of the graphs, then there is at least one edge that is not traversed by any of the robots. If the q robots have K states each, and the trap has $f_q(K)$ nodes, then the space lower bound for a set of q robots exploring all n-node graphs is $\Omega(\log f_q^{-1}(n))$ bits.

The reduced automata technique for the design of space lower bounds for graph exploration is described in Section 2. This lower bound technique allows us to concentrate on a subclass of graphs, called *homogeneous:* edge-colored and regular. For such graphs, a robot can be described by a very simple automaton, whose transition function consists of a graph formed by a directed path followed by a directed cycle. The reduced automata technique applies to homogeneous graphs. Roughly speaking, a *reduced robot* has the property that if it traverses an edge $\{u, v\}$ at some step of the exploration, say from u to v, then its next move will not be traversing the edge back to u. This property is achieved by transforming a robot into a reduced robot whose transition function never has two consecutive edges with the same label. We construct a *trap core* directly from the transition function of a reduced robot, which is then easily extended to a trap for the original robot.

In Section 3 we use the technique of reducing a robot to construct a degree 3 trap for a K-state robot, of size $K + 3$. The proof technique can be generalized to produce traps of any degree, but for illustrating the technique, it is sufficient to work with degree 3 graphs. Indeed, [12] presents a trap of size $K + 1$, planar and valid for graphs of any degree. The proof we present is somewhat simpler than the one of [12], and moreover, it illustrates the technique used to prove our results in the following sections.

In Section 4 we present our new results about traps for collective exploration by a set of *non* cooperative robots. The robots do not communicate at all, and every edge must be traversed by at least one robot. We show (cf. Theorem 4) that for any set of q non-cooperative K-state robots, there exists a 3-regular graph G, and two pairs $\{u, u'\}$ and $\{v, v'\}$ of neighboring nodes, such that any robot of the set, starting from u or u', fails to traverse the edge $\{v, v'\}$. The graph G has $O(qK)$ nodes. This improves the $O(K^{O(q)})$ bound of Rollik [16] (cf. Corollary 2).

By simply plugging this new trap for non-cooperative robots into Rollik's construction, we get (cf. Corollary 4) a new trap for locally-cooperative exploration of size $\tilde{O}(K^{K^{\cdot^{\cdot^{\cdot^{K}}}}})$ with $q + 1$ levels of exponential, to be compared with the $2q + 1$ levels of [16]. Our trap is thus smaller than the one in [16].

In Section 5 we show that Theorem 4 has a significant impact on the space complexity of terminating graph exploration by a single robot. As mentioned above, when terminating exploration is required, the robot is provided with a pebble. We prove (cf. Theorem 5) that terminating exploration requires a robot with $\Omega(\log n)$ bits for the family of graphs with at most n nodes. As mentioned before, in arbitrary graphs, perpetual exploration and terminating exploration are not comparable because even if perpetual exploration is a simpler task than terminating exploration, in the latter case the robot is given a pebble. Therefore, even if the existence of traps with at most $K + O(1)$ nodes for any K-state robot implies an $\Omega(\log n)$ bits lower bound for the memory size of a robot that performs perpetual exploration in all graphs with at most n nodes, the $\Omega(\log n)$ lower bound for terminating exploration is not a consequence of the first result about perpetual exploration.

2   Preliminaries

In Section 2.1 we define formally what we mean by a robot exploring a graph. In Section 2.2 we describe basic properties of a robot. In Section 2.3 we show how to simplify the structure of a robot, for the proofs of the following sections.

2.1   Graphs, Robots and Traps

A robot considered in this paper traverses a graph by moving from node to node along the edges of the graph. We first describe the basic model of a robot traversing a graph, and what we mean by a trap, namely a pair (G, u) where G is a graph and $u \in V(G)$ such that a robot starting from u cannot explore G. To construct a trap for a robot, we first design a graph that the robot cannot leave, called a trap core, and then we add to it edges that the robot does not explore. We explain how the description of a robot is simplified when traversing a more symmetric kind of graph, called homogeneous. The simpler description will be crucial in the rest of the paper.

The Basic Model of a Robot Traversing a Graph. In a graph where nodes have no identifiers, two nodes are indistinguishable to the robot, unless they have different degree. However, edges have local port numbers, so the robot can distinguish two different edges incident to a node. In more detail, each edge has two labels, each one associated to one of its two endpoints. The labels of the edges incident to a node v are arbitrary and pairwise distinct in the set $\{0, \dots, \delta_v - 1\}$, where δ_v denotes the degree of v. When a robot is in a node, it sees only the labels at the endpoints of the edges incident to the node. This allows the robot to distinguish the edges incident to the node through their unique labels, called *local port numbers*. Notice that an edge may have different port numbers in its two endpoints. We refer to those graphs as *port-labeled graphs*.

A robot is an automaton with a single initial state; at the beginning, it is placed on an arbitrary starting node of the graph in this state. When a robot is in a node u and *traverses* an edge $\{u, v\}$ to get to v, it learns the label at v's endpoint of the edge once it enters v. The robot decides which edge to take to leave v based on this label, as well as on the degree of v, and of course, based on its local state. We do not define formally such a robot because we will study its behavior only on a special class of graphs, called homogeneous, for which a very simple representation of a robot is possible, that we *will* define formally. In Section 5 we will consider an extended definition of a robot that can drop a pebble in a node and pick it up when it returns to the node to drop it somewhere else.

A *trap* for a set of robots is a pair (G, U), where G is a port-labeled graph and U is a set of nodes of G, such that if all the robots are placed in nodes $u \in U$, each in its initial state, then there will be an edge $\{u, v\}$ that is never traversed by the robots. To make our lower bound results stronger, sometimes we present a *simple trap*, namely with no parallel edges and self-loops.

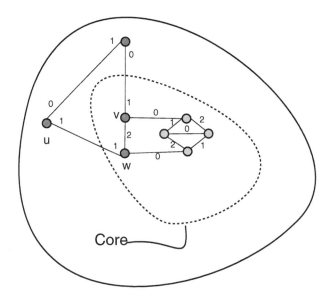

Fig. 1. A trap and its core

Homogeneous Graphs and Trap Cores. We will study the behavior of a robot in a graph where both port numbers of an edge coincide. In such a graph a robot can be described by a very simple automaton, as we shall see next. A δ-*homogeneous* undirected graph is a graph that is δ-regular and δ-edge-colored. A graph is δ-*regular* if each of its nodes has degree δ, and it is δ-*edge-colored* if each edge is labeled with one of the integers in the set $\Delta = \{0, 1, \ldots, \delta - 1\}$ in a way that no two edges incident to the same node have the same color. For the sake of clarity, we mainly focus on graphs with maximum degree three.

When a robot traverses a 3-homogeneous graph, each time it arrives to a node the local environment looks exactly the same as in any other node: all nodes are equal and in each node all local ports are 0, 1, or 2. Thus, the robot decides which edge to take to exit the node based only on its current state. Formally, a *robot* is an automaton $A = (\Delta, \mathcal{S}, f, \hat{s})$, with a finite set of states $\mathcal{S}$, an initial state $\hat{s} \in \mathcal{S}$, and a transition function $f : \mathcal{S} \to \mathcal{S} \times \Delta$. For a state $s \in \mathcal{S}$ with $f(s) = (s', i)$, denote $f_{st}(s) = s'$ and $f_{\ell}(s) = i$. The robot A moves on a 3-regular graph as follows. Initially A is placed on a node of the graph in state $\hat{s}$. If A is in a node v in state s then A moves to the node v' such that the edge $\{v, v'\}$ is labeled $f_{\ell}(s)$, and changes to state $f_{st}(s)$.

When considering the formal definition of a robot for homogeneous graphs, one can construct a trap by first defining a graph G that is edge-colored, but not necessarily 3-regular, and then adding some edges and nodes to obtain a trap in which the trap core looks homogeneous to the robot. We do not demand that a trap is 3-homogeneous as long as a robot never tries to take an edge that is not defined in the graph. Formally:

Definition 1 (Trap Core). *A* trap core *for a set of robots is a pair* (G, U), *where* G *is a 3-edge-colored graph and* U *is a set of nodes of* G, *such that if all the robots are placed in nodes* $u \in U$, *each in its initial state, then each time a robot* $A = (\Delta, \mathcal{S}, f, \hat{s})$ *is in some node* u *in some state* s, *if* $f_\ell(s) = i$ *then an edge* $\{u, v\}$ *labeled* i *must be in* G.

From a Trap Core to a Trap. Once we have built a trap core (G, U) it is not difficult to construct a trap (G', U), by adding to it some edges and a constant number of nodes. Notice that if (G, U) is a trap core for a set of robots, then (G', U) is a trap for the same set of robots, because G is a strict subgraph of G' that the robots never leave. We first show how to construct G' from a 3-edge colored graph G, by adding at most 2 nodes, and adding edges that guarantee that every node of G has degree exactly 3, and we define local port labels for the newly added edges. Thus, as in Figure 1, edges that were originally in G have the same port labels in both endpoints (e.g. $\{v, w\}$ in the figure), while newly added edges may have different port labels (e.g. $\{u, w\}$ in the figure). Afterward we show how to construct an homogeneous G' with at most 13 new nodes.

Definition 2 (Simple Trap Extension). *Given a 3-edge-colored graph* $G = (V, E)$, *the labeled simple graph* $G' = (V', E')$, $|V'| \leq |V| + 2$, *obtained from* G *in the following construction is called the* simple trap extension *of* G.

To construct G' first we can assume that there are at most 2 nodes of degree less than 3. Otherwise, there are two nodes of degree less than 3 that are not connected by an edge, and we may add an edge connecting them, with appropriate local port labels. Now, we add at most 2 new nodes. Each time we add a new node, we connect it to nodes with degree less than 3, with appropriate local port labels. If all nodes of G have now degree exactly 3, we are done, else we add a new node and repeat the procedure. At the end we obtain the desired 3-edge colored graph G', where all original nodes have exactly degree 3, while the new nodes have degree at most 3. Moreover, G' is a simple graph.

Remark. Using the same type of arguments as above, it is possible to construct a simple trap extension for arbitrary degree δ, by adding at most $\delta - 1$ nodes.

We can construct a trap extension G^{hom} from G that is homogeneous, by adding a few more nodes.

Definition 3 (Homogenous Extension). *Given a 3-edge colored graph* $G = (V, E)$, *the graph* $G^{hom} = (V', E')$, $|V'| \leq |V| + 13$, *obtained from* G *in the following construction is called the* homogeneous extension *of* G.

Add to each node of G of degree i less than 3, $3 - i$ pending "half-edges" colored differently from each other and from the colors of edges incident to the node. For $\ell = 0, 1, 2$, let $parity(\ell)$ be the parity of the number of pending half-edges labeled ℓ in the resulting graph[5] G'.

---

[5] We will use this notion of "graph" with "half-edges" several times in this paper.

Claim. For any $\ell, \ell' \in \{0, 1, 2\}$, $parity(\ell) = parity(\ell')$.

Proof. An edge of G' can be considered as two non-pending half-edges. For $\ell \in \{0, 1, 2\}$, let t_ℓ be the total number of half-edges of G' labeled ℓ, and p_ℓ, resp. np_ℓ, be the number of pending, resp. non-pending, half-edges of G' labeled ℓ. All nodes in G' are exactly of degree 3 and are incident to one half-edge of each label. Thus $t_0 = t_1 = t_2$, and this is equal to the number of nodes in G', $|G'|$. In G', if a half-edge is not pending, then it forms an edge with another non-pending edge with the same label. Therefore, all the np_ℓ's are even. Since $t_\ell = p_\ell + np_\ell$, t_ℓ and p_ℓ have the same parity, and thus all the p_ℓ's have the same parity. $\qquad \square$

We now construct the desired homogeneous graph G^{hom}. Let ϱ be the parity of the number of pending half-edges of a given label in G'. If ϱ is odd, then we add to G' a node connected to one of the half-edges, labeled say ℓ, and add two half-edges pending from this node, labeled $\ell' \neq \ell$ and $\ell'' \notin \{\ell, \ell'\}$. As a consequence, ϱ becomes even. Now, we pair the half-edges with identical labels, and connect them to form one edge. Parallel edges can be avoided, unless for some ℓ there are only two pending half-edges with label ℓ, and these are incident to the same edge. In this case the pair is connected by the gadget displayed in Figure 2, where $\ell = 0$. By labeling the edges of every gadget appropriately (as in the figure), we obtain a 3-homogeneous graph G^{hom}.

Claim. G^{hom} has at most 13 nodes more than G.

Proof. We added at most 1 node to correct the parities, and at most 3 gadgets to avoid parallel edges, each one with 4 nodes. Thus the total number of nodes added is at most 13. $\qquad \square$

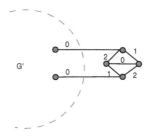

Fig. 2. The gadget for connecting half-edges

Remark. As for the simple trap extension, it is easy to check that one can construct an homogeneous extension for arbitrary degree δ, by using a specific gadgets for every δ.

2.2   Basic Properties

Consider a robot $A = (\Delta, \mathcal{S}, f, \hat{s})$. The transition function f defines a directed
labeled graph $G(A) = (\mathcal{S}, F)$ with node set $\mathcal{S}$ and arc set F, such that the arc
$(s, t) \in F$ iff $f_{st}(s) = t$, and the arc has label $f_\ell(s)$. Notice that the labeled graph
$G(A)$ together with the starting node $\hat{s}$ completely determine the robot A. We
assume in the rest of the paper that every state $s \in \mathcal{S}$ of A is reachable from $\hat{s}$;
unreachable states do not affect the behavior of A and can be ignored. Namely,
there is in $G(A)$ a directed path from $\hat{s}$ to every other node.

Each node of $G(A)$ has out-degree 1 because f is a function. It follows that
$G(A)$ consists of a simple, possibly empty path starting in $\hat{s}$ and ending in some
node s_1, followed by a simple cycle starting and ending in s_1. This is because we
assume that A has no unreachable states and $\mathcal{S}$ is finite. Thus, the arc labels of
the path define a *path word* W_0 over Δ, $|W_0| \geq 0$, and the arc labels of the cycle
define a *cycle word* W over Δ, $|W| \geq 1$. Clearly, $|W_0 W| = |\mathcal{S}|$. The *footprint* of
A is $fp(A) = W_0 W^*$. When A is placed on a node of a homogeneous graph G
in state $\hat{s}$, $fp(A)$ is the sequence of labels of edges traversed by A.

The next lemma says that once A reaches a node x of the graph in some state
s that belongs to the cycle of $G(A)$, the path that A follows in G is a closed path
that includes x; moreover, A returns to x in the same state s. A *configuration*
(x, s) denotes the fact that A is in node x in state s. Also, if $f_{st}(s) = s'$, $f_\ell(s) = i$,
and the label of the edge $\{x, x'\}$ is i then we write $(x, s) \to (x', s')$.

Lemma 1. *Consider a robot A with path and cycle words W_0, W traversing a
graph G. Let x be a node reached by A after at least $|W_0|$ steps, and assume A
is in state s when it is in x. Then A will eventually be back in (x, s).*

Proof. Assuming A is in state s when it is in x, consider the sequence of configurations starting with (x, s)

$$(x_0, s_0) \to (x_1, s_1) \to \cdots$$

where $(x, s) = (x_0, s_0)$. The sequence of configurations must contain two equal
configurations, say $(x_i, s_i) = (x_{i+k}, s_{i+k})$, for some $k > 1$, because both G and
A are finite. Assume k is as small as possible. If $i = 0$ we are done, so suppose
$i > 0$. We will prove that $(x_{i-1}, s_{i-1}) = (x_{i+k-1}, s_{i+k-1})$, which implies that
$(x_0, s_0) = (x_k, s_k)$, and the lemma follows.

Notice that A moves from x_{i-1} to x_i along the edge labeled $f_\ell(s_{i-1})$. Now,
when A eventually returns to the same configuration (x_{i+k}, s_{i+k}), the state
$s_{i+k-1} = s_{i-1}$ (all the states considered are in the cycle of $G(A)$ because the
state s belongs to the cycle of $G(A)$). Thus, $f_\ell(s_{i-1}) = f_\ell(s_{i+k-1})$. It follows
that the edge $\{x_{i+k-1}, x_{i+k}\}$ must be labeled $f_\ell(s_{i-1})$. Finally, $x_{i+k-1} = x_{i-1}$
since $x_{i+k} = x_i$ and G is edge-colored.   □

2.3   Reduced Robots

A robot A is *irreducible* if $G(A)$ satisfies two properties: (i) for any two consecutive (distinct) arcs $s \to s_1 \to s_2$, it holds $f_\ell(s) \neq f_\ell(s_1)$, and (ii) for two arcs

with the same end-node $s \to s_1$, $s_2 \to s_1$, it holds $f_\ell(s) \neq f_\ell(s_2)$. We show here how to obtain an irreducible robot A' from a robot A. The behavior of A and of A' on a graph will not be exactly the same, but will be related in the sense that the region of a graph traversed by A cannot be much larger than the region traversed by A'.

Let $\bar{G}(A)$ be the undirected graph corresponding to $G(A)$. Roughly speaking, we want the robot to be irreducible to construct a graph based on $\bar{G}(A)$ on which the robot will be moving. Since the constructed graph must be edge-colored, $\bar{G}(A)$ must be edge-colored. Then we can place A at the beginning of the path of $\bar{G}(A)$ and it will never try to go out of $\bar{G}(A)$. To obtain an irreducible robot A' from A we perform a series of reduction steps that modify its transition function and reachable states. When A and A' are placed on the same node of a graph, the path traversed by A' is contained in the path traversed by A; essentially A' skips some closed walks of A. These reductions are formally defined next.

A *reduction step* is the operation consisting of transforming a robot $A = (\Delta, \mathcal{S}, f, \hat{s})$ into another robot $A' = (\Delta, \mathcal{S}', f', \hat{s}')$, $\mathcal{S}' \subseteq \mathcal{S}$, where one of the above properties (i) or (ii) is enforced for two arcs. There are two types of reduction steps, corresponding to the two properties. The idea is to repeat type-i steps until no more are possible, and hence the robot satisfies property (i), and then if property (ii) is not satisfied, do a single type-ii step to enforce property (ii). Only type-i reductions change the path traversed by the robot.

Type-i Reduction. A type-i reduction step is *applicable* if $G(A)$ has two consecutive distinct arcs $s \to s_1 \to s_2$ with $f_\ell(s) = f_\ell(s_1)$. The basic idea is illus-

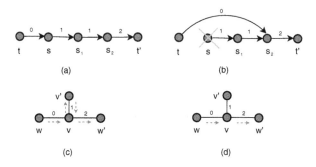

(a) (b) (c) (d)

Fig. 3. A type-i reduction

trated in Figure 3. In (a) there is a segment of $G(A)$ with the two consecutive arcs labeled 1, and in (b) there is the corresponding segment of $G(A')$ after the reduction. In this example s has only one in-neighbor, t, and hence s becomes unreachable. This is the basic idea behind the type-i reduction, but in the formal definition below we need to consider several special cases depending on the number of in-neighbors of s, and on where is the initial state $\hat{s}$.

The properties of a type-i reduction that we need are illustrated in Figure 3(c) and (d), where the path traversed by A and A' resp. is depicted in dotted arrows. If A is in node w of G in state t, it moves to node v in state s, and then it moves to v', change to state s_1, and move back to v, in state s_2 (since $f_\ell(s) = f_\ell(s_1) = i$, where $\{v, v'\}$ is colored i; in the figure $i = 1$). Thus, it is easy to check that a type-i reduction eliminates this v, v', v loop from the path traversed by the robot in the graph, and makes no other changes to the path; that is, if the path arrives to v from w and then proceeds to w' after traversing the v, v', v loop, after the type-i reduction the robot will go from w to v and then directly to w'. Therefore, before the reduction step, the robot explores a node at distance at most 1 from the nodes explored by the robot after the reduction.

Formally, a *type-i reduction* transforms A into A' by doing the following changes to f and by defining $\hat{s}'$ ($f'(\cdot) = f(\cdot)$ and $\hat{s}' = \hat{s}$ unless specified otherwise below). We consider four cases:

Case $s = s_2$: In this case the cycle is of length 2 with the same labels. Assume w.l.o.g. that s has no other in-neighbor besides s_1 (it is impossible that both s and s_1 have 2 in-neighbors). Let $f'(s_1) = (s_1, i)$, where $i = f_\ell(s_1)$. If $s = \hat{s}$ then $\hat{s}' = s_1$.

Otherwise, if $s \neq s_2$, it is possible that s has 0, 1, or 2 in-neighbors.

Case $s \neq s_2$, s has 0 in-neighbors: In this case $s = \hat{s}$. Let $\hat{s}' = s_2$.

Case $s \neq s_2$, s has 1 in-neighbor: Let t be the in-neighbor ($t \neq s_1$), with $f(t) = (s, i)$. Then let $f'(t) = (s_2, i)$. If $s = \hat{s}$ then let $\hat{s}' = s_2$.

Case $s \neq s_2$, s has 2 in-neighbors: Assume they are t_1, t_2, with $f(t_1) = (s, i)$, $f(t_2) = (s, j)$. Then $s \neq \hat{s}$. Let $f'(t_1) = (s_2, i)$ and $f'(t_2) = (s_2, j)$.

After doing these modifications, A' is obtained by removing any unreachable states. Notice that for each one of the previous four cases at least one unreachable state is removed, namely s. Thus, at most $K - 1$ type-i reductions are possible, starting from a K-state robot.

Lemma 2. *Let A' be the robot obtained from $A = (\Delta, \mathcal{S}, f, \hat{s})$ by applying a type-i reduction on arcs $s \to s_1 \to s_2$ with $f_\ell(s) = f_\ell(s_1)$. Then*

1. *The node s together with $s \to s_1$ does not appear in A'.*
2. *If A and A' start at the same node u of a graph in the same state s that belongs to their cycle, when A and A' are back in state s, they are placed in the same node v and A has traversed at most one edge more than A'.*

Proof. The first part of the lemma holds because state s becomes unreachable in A'. We now prove the second part of the lemma. We thus consider that A and A' are both started from a node x_0 in a state t_0 that belongs to their cycle.

Assume a type-i reduction is applied to $G(A)$ on the arcs $s \to s_1 \to s_2$ with $f_\ell(s) = f_\ell(s_1)$, to obtain A'. When s has one in-neighbor t, with $f(t) = (s, i)$, and $s \neq \hat{s}$, A' is equal to A except that $f'(t) = (s_2, i)$ (and hence s becomes unreachable).

Consider the sequence of configurations of $G(A)$ when starting in a node x_0,

$$(x_0, t_0) \to (x_1, t_1) \to \cdots$$

where $t_0 = \hat{s}$. Then the sequence of configurations of $G(A')$ is the same, except that each time A gets to state t, say in the i-th step

$$\cdots \to (x_i, t_i) \to (x_{i+1}, t_{i+1}) \to (x_{i+2}, t_{i+2}) \to \cdots$$

where $t_i = t$, and hence $t_{i+3} = s_2$ with $x_{i+1} = x_{i+3}$ (since $f_\ell(t_{i+1}) = f_\ell(t_{i+2})$), then the sequence of $G(A')$ is

$$\cdots \to (x_i, t_i) \to (x_{i+3}, t_{i+3}) \to \cdots$$

Therefore, the original path in the graph

$$x_0, x_1, \ldots, x_i, x_{i+1}, x_{i+2}, x_{i+3}, \ldots$$

becomes

$$x_0, x_1, \ldots, x_i, x_{i+3}, \ldots$$

and the loop $x_{i+1}, x_{i+2}, x_{i+3}$ ($x_{i+1} = x_{i+3}$) traversing the edge $\{x_{i+1}, x_{i+2}\}$ back and forth is eliminated from the path. $\qquad\square$

Type-ii Reduction. Once a type-i reduction step is not applicable in $G(A)$, a single type-ii reduction can be used. A type-ii reduction step is *applicable* if $G(A)$ has two states such that $f(s) = f(s_1)$, that is, $G(A)$ has two arcs with the same end-node $s \to t$, $s_1 \to t$, and $f_\ell(s) = f_\ell(s_1)$. See Figure 4 where $f_\ell(s) = f_\ell(s_1) = 1$; in part (a) there is $G(A)$, and in part (b) there is $G(A')$ after the reduction. A *type-ii reduction* transforms A into A' by doing the following

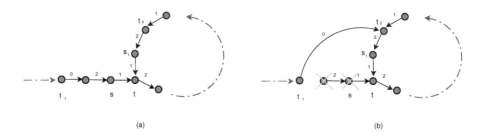

(a)   (b)

Fig. 4. A type-ii reduction

changes to f and by defining $\hat{s}'$ ($f'(\cdot) = f(\cdot)$ and $\hat{s}' = \hat{s}$ unless specified otherwise below). Exactly one of s, s_1 must be in the cycle of $G(A)$, let's say s_1. So there is a path from t to s_1. This path is of length at least 1, because otherwise $t = s_1$ and there is a loop from t to itself labeled $f_\ell(s)$, and a type-i reduction is applicable. Recall that $fp(A) = W_0 W^*$. Let W' be the longest common postfix of W_0 and W^*; $|W'| > 0$ by the type-ii assumption. Let t_2 be the node just before W' starts in the cycle of $G(A)$. In Figure 4, $W' = 21$. We consider two cases, in both cases A' is obtained from A by the following modifications, and removing any unreachable states:

Case $|W_0| > |W'|$: That is, W' is a strict postfix of W_0; let t_1 be the in-neighbor of the node just before W' starts in the simple path of $G(A)$. Thus $f_\ell(f_{st}(t_1)) = f_\ell(t_2)$ is the first letter of W'. Let $f'(t_1) = (t_2, f_\ell(t_1))$.

Case $|W_0| = |W'|$: Let $\hat{s}' = t_2$.

The following lemma is straightforward.

Lemma 3. *Let* $A' = (\Delta, \mathcal{S}', f', \hat{s}')$ *be the robot obtained from* $A = (\Delta, \mathcal{S}, f, \hat{s})$ *by applying a type-ii reduction on arcs* $s \to t$, $s_1 \to t$, *with* $f_\ell(s) = f_\ell(s_1)$, *and* s_1 *in the cycle of* $G(A)$. *Then*

1. *The node* s *together with* $s \to t$ *does not appear in* A'. *Moreover, a type-ii reduction is not applicable to* $G(A')$.
2. *If* A *and* A' *start at the same node of a graph, they both traverse the same path.*
3. *If a type-i reduction is not applicable to* $G(A)$ *then it is not applicable to* $G(A')$.

Using Lemma 2 and Lemma 3 it is easy to prove the following, summarizing the procedure to obtain an irreducible robot.

Lemma 4. *Let* $A' = (\Delta, \mathcal{S}', f', \hat{s}')$ *be the robot obtained from* $A = (\Delta, \mathcal{S}, f, \hat{s})$ *through the longest possible sequence of type-i reductions followed by a type-ii reduction (if applicable). Let* k *be the number of reduction steps in this sequence.*

1. A' *is irreducible.*
2. $|\mathcal{S}'| + k \le |\mathcal{S}|$.
3. *If* A *and* A' *start at the same node* u *of a graph in the same state* s *that belongs to their cycle, when* A *and* A' *are back in state* s, *they are placed in the same node* v *and* A *has traversed at most* k *edge more than* A'.

Proof. The first part of the lemma follows from Lemma 2(1) and from Lemma 3(1,3): if there are two arcs violating property (i), then a type-i reduction can be applied, and at least one state is removed in the process. Also, if there are two arcs violating property (ii) after all arcs satisfy property (i), then a type-ii reduction will eliminate the situation, without creating arcs that violate property (i).

The second part of the lemma follows because each type-i and type-ii reduction eliminates at least one state, as observed in Lemma 2(1) and Lemma 3(1).

The third part of the lemma follows from Lemma 2(2) and Lemma 3(2), by induction on k.   □

3   A Trap for a Single Robot

In this section, we focus on graph exploration by a single robot. We present a trap for a K-state robot of size $O(K)$. As explained in the Introduction, a similar result was presented in [12]. We consider a robot and an irreducible version of it. First we show how to construct a trap core for the irreducible robot, and then how to extend it to a trap for the original robot.

3.1   A Trap for an Irreducible Robot

Let $\widehat{A} = (\Delta, \mathcal{S}, f, \hat{s})$ be an irreducible robot with footprint $fp(\widehat{A}) = W_0 W^*$, $|W_0 W| = K$. Recall that its graph of state transitions $G(\widehat{A})$ consists of a directed path starting in the initial state $\hat{s}$, followed by a directed cycle. Thus, the corresponding undirected graph, $\bar{G}(\widehat{A})$, consists of a path P connected to a cycle C; let $\hat{x}$ be the initial node of P. If C is of length at least 3 then $\bar{G}(\widehat{A})$ is a simple edge-colored graph (no parallel edges and no self-loops), and it serves as a trap core (Definition 1) for $\widehat{A}$. If C is of length less than 3 we modify it a little to make it a simple edge-colored graph that is also a trap for $\widehat{A}$, denoted $\bar{G}_1(\widehat{A})$.

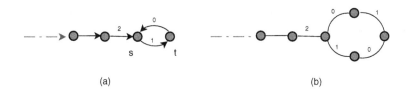

(a)   (b)

Fig. 5. Eliminating parallel edges

We construct the simple, edge-colored graph $\bar{G}_1(\widehat{A})$ from $\bar{G}(\widehat{A})$ as follows:

- Assume the directed cycle of $G(\widehat{A})$ is of length 2, with states s and t (Figure 5(a) illustrates this case with $W = 10$). Then the undirected cycle in $\bar{G}_1(\widehat{A})$ will have 4 edges, labeled WW, adding two new nodes as in Figure 5(b). The path is P, as in $\bar{G}(\widehat{A})$.
- Assume the directed cycle of $G(\widehat{A})$ is of length 1 with state s (Figure 6(a) illustrates this case with $W = 1$). Then the undirected cycle in $\bar{G}_1(\widehat{A})$ will have 4 edges, labeled $abab$, where a is equal to the single letter of W and b is different from a and from the last letter of W_0 (if any), as in Figure 6(b), where $abab = 1010$. The path is P, as in $\bar{G}(\widehat{A})$.

Notice that the only node of $\bar{G}_1(\widehat{A})$ of degree 3 is the node where the path and the cycle are joined. Thus, it is not homogeneous, and if we place a robot in one of its nodes, it could try to take an edge that does not exist in the graph. Clearly, this does not happen if we place $\widehat{A}$ at $\hat{x}$. Namely, starting at $\hat{x}$, $\bar{G}_1(\widehat{A})$ with any edge added is a trap core for $\widehat{A}$, with at most 3 nodes more than $\bar{G}(\widehat{A})$. We have the following straightforward lemma.

Lemma 5. *The graph $\bar{G}_1(\widehat{A})$ is simple and edge-colored, with at most $|\mathcal{S}| + 3$ nodes. Moreover, $\bar{G}_1(\widehat{A})$ is a trap core for $\widehat{A}$ when starting at $\hat{x}$.*

Fig. 6. Eliminating a self-loop

3.2 A Trap for the Original Robot

We present two different constructions of a trap for A. In both cases we use the graph $\bar{G}_1(\widehat{A})$ of Lemma 5, where $\widehat{A}$ is an irreducible robot obtained from A. The first method, described in Theorem 1, produces a smaller trap than the second, described in Theorem 2, but the second method will be useful in the following section.

Theorem 1. *For any robot $A = (\Delta, \mathcal{S}, f, s_0)$ there exist a trap of at most $|\mathcal{S}|+2$ nodes, and an homogeneous trap of at most $|\mathcal{S}| + 13$ nodes.*

Proof. Let $\widehat{A}$ be an irreducible robot obtained from A, and consider its undirected graph $\bar{G}(\widehat{A})$. By Lemma 5 the modified graph $\bar{G}_1(\widehat{A})$ is simple and edge-colored. Also, $\widehat{A}$ can be placed in the first node $\hat{x}$ of the path P of $\bar{G}_1(\widehat{A})$ in its initial state, and it never tries to take an edge not in the graph. Now, place A in $\hat{x}$ in its initial state. Each time A wants to take an edge with some label not in the graph, we add the edge (with the label) to the graph. By Lemma 2 the paths traversed by A (and not by $\widehat{A}$) are trees where A stays in states eliminated by the series of type-i reductions. Thus, the added edges form trees, and the nodes added correspond to the eliminated states, so we get back a graph with $|\mathcal{S}|$ nodes. Now, A never tries to take an edge not in the graph. For this graph, consider the simple extension of Definition 2, and the homogeneous extension as in Definition 3. The first is a trap $(G, \hat{x})$ for A with at most 2 additional nodes, while the second is a homogeneous trap $(G, \hat{x})$ for A with at most 13 additional nodes. □

Remark. Using extensions for arbitrary degree allows us to obtain a similar result as the one in [12]. In fact, the extension used in [12] outputs a graph which is neither simple nor homogeneous. It is however of smaller size: $|S| + 1$ nodes, independently from the considered degree.

The second way of constructing a trap uses the *K-tower method*. Assume an homogeneous graph H is given, together with one of its edges, say $\{v, v'\}$. Cut the edge to produce two pending half-edges e, e'. Add a "tower" of height $K + 1$ connected to e, e', and a gadget closing the tower as in Figure 7. The two internal nodes of the gadget at the top of the tower are denoted by v_1 and v_1'. Add labels to the tower and the gadget to make the whole graph edge-colored, and denote it G. Thus, G is homogeneous.

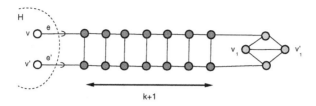

Fig. 7. The tower method

Theorem 2. *For any robot $A = (\Delta, \mathcal{S}, f, s_0)$ there exist an homogeneous trap (G, U) of at most $3|\mathcal{S}| + 22$ nodes.*

Proof. Let $\widehat{A}$ be an irreducible robot obtained from A, and consider its undirected graph $\bar{G}(\widehat{A})$. By Lemma 5 the modified graph $\bar{G}_1(\widehat{A})$ is simple and edge-colored. Also, $\widehat{A}$ can be placed in the first node $\hat{x}$ of the path P of $\bar{G}_1(\widehat{A})$ in its initial state, and it never tries to take an edge not in the graph. Consider the homogeneous extension H of $\bar{G}_1(\widehat{A})$, as in Definition 3, with at most 13 additional nodes. Pick any of the new edges added to H, say $\{v, v'\}$, and add a tower of height $K + 1$, $K = |\mathcal{S}|$, as described above, to obtain G. Now, place A in x_0 in its initial state. Notice that as G is homogeneous, A never tries to take an edge not in the graph. Finally, the edge $\{v_1, v_1'\}$ is not traversed by A. This is because $\widehat{A}$ does not traverse the edge $\{v, v'\}$, and hence it does not enter the tower. By Lemma 4, the trajectory of A is never at distance greater than K (where $K = |\mathcal{S}|$) from the trajectory of $\widehat{A}$. Thus, since the tower is of height $K + 1$, A never reaches the top of the tower. Therefore, A does not traverse the edge $\{v_1, v_1'\}$, and $(G, \hat{x})$ is a trap for A.

It remains to count the number of nodes of G. The graph $\bar{G}(\widehat{A})$ has at most K nodes, $\bar{G}_1(\widehat{A})$ has at most $K + 3$ nodes. Then, H has at most $K + 16$ nodes. The tower has $2K + 6$ nodes, so the total is at most $3K + 22$ nodes.   □

Corollary 1. *A robot that explores all graphs of size n requires at least $\Omega(\log n)$ memory bits.*

Using a different proof argument, [12] also proves a lower bound that depends on the diameter and the maximum degree of the graph, rather than just the number of nodes. It is, nevertheless, possible to use the trap core proof method to obtain a similar result. We recall the theorem from [12]:

Theorem 3. *A robot that explores all graphs of diameter D and maximum degree δ requires exactly $\Theta(D \log \delta)$ memory bits.*

4   A Trap for a Team of Non-cooperative Robots

In this section, we focus on graph exploration by a team of non-cooperative robots. (The independent robots may have different transition functions, hence they will follow different paths in the explored graph.) The main result of the section is the construction of a trap of size $O(qK)$ for any set of q non-cooperative K-state robots. This result is stated in Theorem 4. To prove this result, we first need an auxiliary lemma (Lemma 6) that shows how, given any automaton, any homogeneous graph can be transformed into a trap for this automaton. This result is used at every induction step of the proof of Theorem 4.

4.1   Trapping an Irreducible Robot

In this section we prove an auxiliary result for Theorem 4. Assume an irreducible K-state robot $\widehat{A} = (\Delta, \mathcal{S}, f, \hat{s})$ is placed at a node x_0 of a graph G in its initial state $\hat{s}$, and we want to create a trap core for $\widehat{A}$ by extending G at a given edge $\{v, v'\}$, and moreover, the extension should be of size $O(K)$. (if $\widehat{A}$ does not traverse the edge then there is nothing to be done.) See Figure 8. The extension

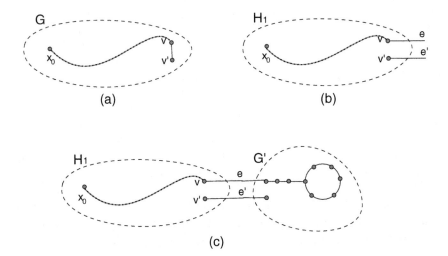

Fig. 8. Extending a graph at a single edge to trap a robot

is by cutting $\{v, v'\}$ to create two pending half-edges (Figure 8(b)); the node v is connected to a pending half-edge e and v' is connected to a pending half-edge e'. The resulting graph is H_1. A graph G' of $O(K)$ nodes is glued to e and e' (Figure 8(c)), such that $\widehat{A}$ does not traverse at least one of its edges. The resulting graph is called H. Actually, it turns out that the extension added to G is pretty simple: either adding a path connected to a cycle (based on $\bar{G}(\widehat{A})$ as illustrated in Figure 8(c)), or connecting e and e' by (an appropriately labeled) path.

Consider the footprint of $\widehat{A}$, $fp(\widehat{A}) = W_0 W^*$, $|W_0 W| = K$, where p_i is the i-th letter in $fp(\widehat{A})$. Consider the sequence of configurations of $\widehat{A}$

$$(x_0, s_0) \rightarrow (x_1, s_1) \rightarrow \cdots$$

where $s_0 = \hat{s}$. Let $p_{i+1} = f_\ell(s_i)$. Assume $\widehat{A}$ traverses $\{v, v'\}$ for the first time at step i, $i \geq 1$; *i.e.*, when going from (x_{i-1}, s_{i-1}) to (x_i, s_i); assume w.l.o.g. it traverses it at this time from v to v', *i.e.*, $x_{i-1} = v$ and $x_i = v'$. Thus, p_i is the label of $\{v, v'\}$. In other words, if we cut the edge to obtain the two pending half-edges e, e', then $\widehat{A}$ traverses e.

We consider two cases depending on when $\widehat{A}$ traverses e, during the simple path of $G(\widehat{A})$ or during the cycle of $G(\widehat{A})$.

Case 1. Assume $\widehat{A}$ traverses e at step $i \leq |W_0|$. Thus, p_i belongs to W_0. In this case we can use the undirected graph of $\widehat{A}$, $\bar{G}(\widehat{A})$, and construct the version with no parallel edges and self-loops, $\bar{G}^s(\widehat{A})$, as in Lemma 5, adding at most 3 new nodes. We glue the part of $\bar{G}^s(\widehat{A})$ that starts after p_i to e. Namely, we connect to e a path of length $|W_0| - i$ whose extremity is denoted by w. The edges of this path are labeled $p_{i+1}, \ldots, p_{|W_0|}$. At w, we add the ring of $\bar{G}^s(\widehat{A})$. The other half-edge e' is completed into an edge by adding to it one new node, and the graph obtained is H. Notice that we added at most $K + 4$ new nodes.

Case 2. If $\widehat{A}$ traverses e at step $i > |W_0|$, then it traverses e to get into some state s of the cycle in $G(\widehat{A})$; assume this is the j-th state of the cycle (recall that the cycle is assumed to start in the last state of the path of $G(\widehat{A})$). That is, after traversing e, $\widehat{A}$ would traverse edges labeled $p_{|W_0|+j}, p_{|W_0|+j+1}, \ldots$

Let x be the node of H_1 reached by $\widehat{A}$ after $|W_0|$ steps, let W^{-1} be the sequence W written in reverse order, and let $\widehat{A}^{-1}$ be the robot that traverses edges labeled $(W^{-1})^*$. Thus, when $\widehat{A}^{-1}$ starts at x and $\widehat{A}$ reaches x, $\widehat{A}^{-1}$ proceeds as $\widehat{A}$, but backwards. Let $\widehat{A}^*$ be the robot that traverses edges labeled W^*, *i.e.* the robot derived from $\widehat{A}$ by removing states and transitions that involved W_0.

Claim. Starting from x, $\widehat{A}^{-1}$ eventually traverses one of the half-edges pending at v or v'.

Proof. Assume for contradiction that $\widehat{A}^{-1}$ does not traverse any of the half-edges pending at v or v'. By Lemma 1, $\widehat{A}^{-1}$ returns to x in the same state, and hence its path in H_1 is a closed path. This path traversed backwards is exactly what $\widehat{A}^*$ traverses from x. So $\widehat{A}^*$ does not traverse any of the half-edges pending at v or v'. Thus, $\widehat{A}$ also does not traverse them, a contradiction. □

By Claim 4.1 we can consider the state reached by $\widehat{A}^{-1}$ after it traverses one of the pending half-edges; assume this is the k-th state of the cycle in $G(\widehat{A})$. We consider two sub-cases, depending on whether $\widehat{A}^{-1}$ traverses the same half-edge as $\widehat{A}$, or not.

Case 2.1. The robot $\widehat{A}^{-1}$ traverses the half-edge e pending at v (*i.e.*, the same as $\widehat{A}$). This implies that the k-th label in W is equal to the $(j-1)$-th label in W, which is the label of e. We consider the section of the cycle of $G(\widehat{A})$ from the j-th state to the k-th state. The end edges of this section have the label of e. We now consider the following word: $W' = W(j-1)W(j)W(j+1)\ldots W(k-1)W(k)W(k+1)\ldots W(j-1)W(j)W(j+1)\ldots W(k-1)W(k)W(k+1)\ldots W(j-1)W(j)W(j+1)\ldots W(k-1)W(k)$ (Note that $W(j-1) = W(k)$ and

$|W'| \geq 2 \times |W| + 2$). The two robots $\widehat{A}$ and $\widehat{A}^{-1}$ cannot follow the same path forever after crossing edge e: otherwise, it would mean that moving them both backwards, they would also follow the same path forever (which is impossible since the two robots took different paths at node x in the past). Moreover, the two robots must separate after at most $|W|$ steps, and since $|W'| \geq 2 \times |W| + 2$, they must separate after at least 1 step and at most $|W| - 1$ steps. Now, if the two robots separate from each other at some point after crossing edge e, let us consider the smallest l such that $W(j + l) \neq W(k - 1 - l)$, $i.e.$ the nearest place where the two robots separate from one another. Since $W(j - 1) = W(k)$, $l \geq 1$. By definition of l, we have $W(j + l - 1) = W(k - l)$. Since the considered robots are reduced, we also have $W(j + l - 1) \neq W(j + l)$. Still by definition of l, we get $W(j + l) \neq W(k - 1 - l)$. Finally, because we consider reduced robots and we have $W(j + l - 1) = W(k - l)$, we get $W(j + l - 1) \neq W(k - 1 - l)$. Overall, this means that $W(j + l - 1)$, $W(j + l)$, and $W(k - 1 - l)$ are pairwise disjoint. We are now ready to construct the following graph: from e, there is a chain that ends in $W(j + l - 1)$ at node w, and from this last node a circle W'' goes from $W(j + l)$ to $W(k - l - 1)$. Since $W(j + l) \neq W(k - 1 - l)$ (see above), $|W''| > 2$. We add at w a ring of length $|W''|$ labeled W'', starting and ending at w, so that once $\widehat{A}$ and $\widehat{A}^{-1}$ reach w, each one traverses this ring in the opposite direction, and gets back to w in the appropriate state to proceed along the path back to the half-edge e. The other half-edge e' is completed into an edge by adding to it one new node, and the graph obtained is H. Notice that we added at most $2K + 1$ new nodes.

Case 2.2. The robot $\widehat{A}^{-1}$ traverses the half-edge e' pending at v' ($i.e.$, not the same as $\widehat{A}$). Suppose when $\widehat{A}^{-1}$ goes through v' it is in state s. We consider again the section of the cycle of $G(\widehat{A})$ from the j-th state to the k-th state (if the section is of length 1, we extend it with W to make sure there is at least one internal node). We connect e and e' by a path with the labels of this section, to obtain H. Thus, when $\widehat{A}$ traverses the half-edge e, it follows the newly added path, and gets to v' in the appropriate state, namely s, to proceed along the same path of $\widehat{A}^{-1}$ but backwards, and return to x. Notice that we added at most $2K$ new nodes.

Lemma 6. *The graph H is simple and edge-colored. Also, H is a trap core for $\widehat{A}$ when starting in x_0, with at most $2K + 3$ nodes more than G.*

Proof. It follows directly from Lemma 5 that H is simple and edge-colored. The number of nodes of H is counted in the previous three cases.

The proof of Case 1 is as follows; the other cases are similar. Assume $\widehat{A}$ traverses e at step $i \leq |W_0|$. In this case $\widehat{A}$ does not traverse the edge e' of H. Observe that $\widehat{A}$ is trapped in the segment of $\bar{G}^s(\widehat{A})$ added to e. This follows because $\widehat{A}$ is in (x_{i-1}, s_{i-1}) before traversing e, and in (x_i, s_i) after traversing it. At this moment it is at the beginning of the segment of $\bar{G}^s(\widehat{A})$ added, so it will continue traversing this graph without trying to take an edge not defined, as in Lemma 5. $\qquad\square$

4.2   Trapping a Team of Robots

With the results of the previous subsection we are ready to prove the main result of this section.

Theorem 4. *For any set $\mathcal{A}$ of q non-cooperative K-state robots, there exist a 3-homogeneous graph G and two pairs of neighboring nodes $\{u, u'\}$ and $\{v, v'\}$ such that (1) the edge $\{u, u'\}$ is labeled 0, (2) starting at u or at u', any robot in $\mathcal{A}$ fails to traverse the edge $\{v, v'\}$, and (3) G has $O(qK)$ nodes.*

Proof. The proof is by induction on $q \geq 0$. The basic step is $q = 0$. The corresponding graph G is displayed on Figure 9.

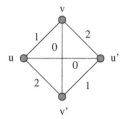

Fig. 9. Basic step of the induction

For the induction step, assume that Theorem 4 holds for q, and let us show that it holds for $q + 1$. Let $\mathcal{A}$ be a set of $q + 1$ non-cooperative K-state robots, and let $A \in \mathcal{A}$. By induction hypothesis, let G_q be an n-node 3-homogeneous graph (where n is $10qK + O(q)$) having two pairs of neighboring nodes $\{u, u'\}$ and $\{v, v'\}$ with the edge $\{u, u'\}$ labeled 0, such that, starting at u or at u', any robot in $\mathcal{A} \setminus \{A\}$ fails to traverse the edge $\{v, v'\}$. We construct a graph G_{q+1} that satisfies Theorem 4 for $\mathcal{A}$.

Let $\widehat{A}$ be an irreducible robot obtained from A as in Lemma 4. Consider its footprint $fp(\widehat{A}) = W_0 W^*$, $|W_0 W| \leq K$. We concentrate first our attention on $\widehat{A}$, and will come back later to the original robot A. Let us denote by p_i the i-th letter in $fp(\widehat{A})$. Recall that since $\widehat{A}$ is irreducible, its associated undirected graph $\bar{G}(\widehat{A})$ is edge-colored. Let us place $\widehat{A}$ at node u of G_q, and perform the construction of the previous subsection, Lemma 6, to obtain a graph H. Then, as in Theorem 2, construct an homogeneous graph H_1 from Definition 3, and add the tower at any of the newly added edges, say $\{v, v'\}$ of height $K + 1$ (see Figure 7), to obtain a graph H_2. The edge $\{v_1, v_1'\}$ in the gadget of the tower is not traversed by A when starting from u.

We repeat the same construction by considering the robot $\widehat{A}$ launched from u' in H_2. More precisely, we construct G_{q+1} from H_2 in the same way H_2 was constructed from G_q. In particular, there is a tower in G_{q+1}, and we define the nodes v_2 and v_2' of G_{q+1} as the two internal nodes of the gadget at the top of this tower. By construction G_{q+1} is 3-homogeneous. Also, any robot in $\mathcal{A}$ fails to traverse the edge $\{v_2, v_2'\}$ of G_{q+1} when starting from u or u'. This is because by induction hypothesis, starting from u or u', a robot in $\mathcal{A} \setminus \{A\}$ never traverses v, v' in G_q and so will never traverse any of the edges added to obtain G_{q+1}, and hence does not traverse the edge $\{v_2, v_2'\}$ of G_{q+1}. Starting from u, A fails to traverse the edge $\{v_1, v_1'\}$ of H_2. This edge being the one that is "opened" to construct G_{q+1} from H_2, A fails to reach any of the two nodes v_2 or v_2' in G_{q+1}. Finally, by construction of G_{q+1} from H_2, A fails to reach any of the two nodes v_2 or v_2' in G_{q+1} when starting from u', in the same way A fails to reach any of the two nodes w or w' in H_2 when starting from u.

To complete the proof, it just remains to compute the size of G_{q+1}. We claim $|G_{q+1}| \leq |G_q| + 10K + O(1)$. We give simple upper bounds on the size of the intermediate graphs. First, we have $|H| \leq |G_q| + 2K + 3$ by Lemma 6. Then, we have $|H_1|$ has 13 more nodes at the most, as in Definition 3, so $|H_1| \leq |G_q| + 2K + 16$. The tower has $2K + 6$ nodes (proof of Theorem 2), so $|H_2| \leq |G_q| + 4K + 22$. The same procedure for the starting node u' contributes to another $4K + 22$ additional nodes. The result follows. Therefore, $|G_{q+1}| \leq 8qK + O(q)$, which completes the proof of the theorem. $\qquad\square$

By simply rewriting Theorem 4, we derive a bound of the size of the smallest trap for a set of q non-cooperative K-state robots, improving the one by Rollik [16]:

Corollary 2. *For any set of q non-cooperative K-state robots, there exists a trap of size $O(qK)$.*

Corollary 3. *A team of q non cooperative robots that explores all graphs of size n requires at least $\Omega(\log \frac{n}{q})$ memory bits per robot.*

By simply plugging this latter bound in the construction by Rollik [16] for team of locally-cooperative robots, we get:

Corollary 4. *For any set of q locally-cooperative K-state robots, there exists a trap of size $\tilde{O}(K^{K^{\cdot^{\cdot^{\cdot^{K}}}}})$, with $q + 1$ levels of exponential.*

5   Bounds for Terminating Exploration

In this section, we consider the *terminating exploration* problem, in which a robot must traverse all edges of the graph, and eventually stop once this task has been achieved. A robot cannot solve this task in graphs with more nodes than its number of states, by Lemma 1. Thus, the robot is given pebbles that it can drop and take to/from any node in the graph. It is known that any finite robot with a finite source of pebbles cannot explore all graphs [16]. On the other hand, it is known that a robot with unbounded memory can explore all graphs, using only one pebble [8]. An important issue is to bound the size of the robot as a function of the size of the explored graphs.

A δ-*p-robot with a pebble* or simply *p-robot* when δ is understood, is an automaton $A = (\Delta, \mathcal{S}, f, s_0)$, with a finite set of states $\mathcal{S}$, $s_0 \in \mathcal{S}$, and

$$f : \mathcal{S} \times \{0, 1\} \rightarrow \mathcal{S} \times \Delta \times \{pick, drop\}.$$

Every state $s \in \mathcal{S}$ has a component $p(s) \in \{0, 1\}$ that indicates if A has the pebble, $p(s) = 1$, or not, $p(s) = 0$. Only if $p(s) = 1$ we allow f to be undefined; in such case we say s is a *stop* state. For the initial state s_0, $p(s_0) = 0$. Each node v of the graph is in some state $p(v) \in \{0, 1\}$ that indicates if the pebble is in v, $p(v) = 1$, or not, $p(v) = 0$. The initial state of the graph satisfies: $p(v) = 1$ for exactly one node v.

The movement of a δ-p-robot A on a δ-regular graph is represented by a sequence of *configurations*, each one consisting of the state of the robot and the state of the graph. For the initial configuration, A is placed on some node of the graph in state s_0, and the pebble is in exactly one node. In general, if A is in a node v in state s in some configuration, we compute $f(s, p(v)) = (s', i, b)$. In the next configuration A will be in the node v' such that the edge $\{v, v'\}$ is colored i, in state s'. Also in the next configuration: if $b = drop$ then $p(v) = 1$ and $p(s') = 0$, and if $b = pick$ then $p(v) = 0$ and $p(s') = 1$. It is assumed that b can be equal to $drop$ only if $p(s) = 1$ and b can be equal to $pick$ only if $p(v) = 1$.

A robot A performs terminating exploration of a graph if after starting in any node of the graph that has the pebble, it traverses all its edges and enters a stop state. A graph which A does not succeed terminating exploration is called a *trap* for A.

The next theorem shows that a p-robot that performs terminating exploration in all graphs of at most n nodes requires $\Omega(n^{1/3})$ states, or equivalently $\Omega(\log n)$ bits of memory.

Theorem 5. *For any K-state p-robot there exists a trap of size $O(K^3)$.*

Proof. Let $A = (\Delta, \mathcal{S}, f, s_0)$ be a K-state p-robot. We construct a trap of size $O(K^3)$ for A. For that purpose, we consider the restriction of A to states s such that $p(s) = 0$ and input 0 (on nodes with no pebble). This defines a robot (with no pebble, as in Section 2.1) except that some states may be unreachable from s_0. For every state s of this robot, we consider the robot A_s that has s as initial state, and includes only reachable states from s. Let $\mathcal{A} = \{A_s\}$ be the set of all these robots. Thus, $|\mathcal{A}| \leq K$.

Let G be a graph satisfying Theorem 4 for the set $\mathcal{A}$. Remove edges $\{u, u'\}$ and $\{v, v'\}$ from G. Consider two copies of the resulting graph, with the four nodes of degree 2 indexed by the index of the copy, 1 and 2. These nodes are re-connected as follows. Let c be the color of the deleted edge $\{v, v'\}$. Create two edges $\{v_1, v'_2\}$ and $\{v'_1, v_2\}$ with color c. The resulting graph is denoted by G_1 (see Figure 10).

Consider an infinite ternary tree modified as follows. Each node is replaced by a 6-cycle. Edges of the cycles are labeled alternatively 1 and 2. Then, edges of the infinite tree are replaced by two "parallel" edges labeled 0, as depicted on Figure 11. The resulting graph is denoted by T.

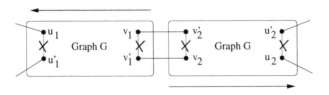

Fig. 10. The graph G_1

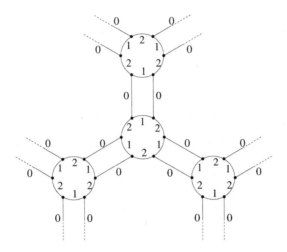

Fig. 11. The modified infinite tree T

The two graphs G_1 and T are composed by replacing every pair $\{\{x, y\},$ $\{x', y'\}\}$ of parallel edges in T by a copy of G_1. More precisely, x, y, x', y' are respectively connected to nodes u_1, u_2', u_1', u_2 in G_1. These new edges are labeled 0. The resulting graph is denoted by G_2. A "meta-edge" of G_2 is defined as a copy of G_1 replacing a parallel edge of T.

By definition of G and $\mathcal{A}$, the p-robot A is unable to traverse a meta-edge of G_2 without the help of the pebble[6]. We now modify G_2 to obtain a graph G_3 such that the p-robot A is unable to explore G_3, even with the pebble. G_3 contains $O(K)$ 6-cycles of T, and thus has at most $O(K^3)$ nodes. The transformation from G_2 to G_3 is technical and very similar to the transformation used in [12] and in [16]. Thus we only sketch the construction of G_3, skipping technical details. Since any p-robot cannot go from a 6-cycle to another 6-node cycle of G_2 without using the pebble, we define *key* steps as those for which the last time the p-robot leaves a 6-cycle with the pebble, go through a meta-edge, and enters another 6-cycle with the pebble. Because the number of states is finite, A will eventually be twice in the same state at these key steps, at two nodes w and w'. With the same technique as in [12], we identify the nodes w and w'. This leads to the

---

[6] Since the $\{u, u'\}$ edges are "open," the proof requires to consider the last time the p-robot is in a u node.

graph G_3 with the desired properties, that is G_3 has $O(K)$ 6-cycles, and thus $O(K)$ "parallel" edges. In each pair of "parallel" edges, there is a copy of G_1. Since G_1 has $O(K^2)$ nodes, then G_3 has $O(K^3)$ nodes. □

Corollary 5. *A robot that performs terminating exploration of all graphs of size* n *requires at least* $\Omega(\log n)$ *memory bits.*

Remark. This latter bound is tight, as proved in [13].

6   Conclusions

On the one hand, we have proved that terminating exploration (using one pebble) requires $\Omega(\log n)$ bits for the family of graphs with at most n nodes. On the other hand, we proved in [13] that there exists an terminating exploration algorithm using a robot with $O(D \log \Delta)$ bits of memory for the terminating exploration of all graphs of diameter at most D and degree at most Δ. The design of a tight bound for terminating exploration is still an open problem.

References

1. R. Aleliunas, R. Karp, R. Lipton, L. Lovasz and C. Rackoff Random walks, universal traversal sequences, and the time complexity of maze problems In Proc. of the 20th Annual Symposium on Foundations of Computer Science (FOCS), pages 218–223, 1979.
2. S. N. Bhatt, S. Even, D. S. Greenberg, R. Tayar. Traversing Directed Eulerian Mazes. *J. Graph Algorithms Appl.* **6**(2): 157–173, 2002.
3. M. Bender, A. Fernandez, D. Ron, A. Sahai and S. Vadhan. The power of a pebble: Exploring and mapping directed graphs. *Information and Computation* **176**: 1–21, 2002. (Prel. Version in STOC 1998.)
4. M. Bender and D. Slonim. The power of team exploration: Two robots can learn unlabeled directed graphs. In 35th Ann. Symp. on Foundations of Computer Science (FOCS), pages 75–85, 1994.
5. L. Budach. Automata and labyrinths. *Math. Nachrichten,* pages 195–282, 1978.
6. S. Cook and C. Rackoff. Space lower bounds for maze threadability on restricted machines. SIAM J. on Computing 9(3):636–652, 1980.
7. K. Diks, P. Fraigniaud, E. Kranakis, and A. Pelc. Tree Exploration with Little Memory. In 13th Annual ACM-SIAM Symp. on Discrete Algorithms (SODA), pages 588–597, 2002. (Full version to appear in J. of Algorithms.)
8. G. Dudek, M. Jenkins, E. Milios, and D. Wilkes. Robotic Exploration as Graph Construction. *IEEE Transaction on Robotics and Automation* **7**(6): 859–865, 1991.
9. S. Even, G. Itkis, S. Rajsbaum. On Mixed Connectivity Certificates. *Theor. Comput. Sci.* **203**(2): 253–269, 1998.
10. S. Even, A. Litman, P. Winkler. Computing with Snakes in Directed Networks of Automata. *J. Algorithms* **24**(1): 158–170, 1997.
11. S. Even and R.E. Tarjan. Computing an st -Numbering. *Theor. Comput. Sci.* **2**(3): 339–344, 1976.

12. P. Fraigniaud, D. Ilcinkas, G. Peer, A. Pelc, and D. Peleg. Graph Exploration by a Finite Automaton. In 29th International Symposium on Mathematical Foundations of Computer Science (MFCS), LNCS 3153, pages 451–462, 2004. (Full version to appear in Theoretical Computer Science.)

13. P. Fraigniaud, D. Ilcinkas, S. Rajsbaum, S. Tixeuil. Space Lower Bounds for Graph Exploration via Reduced Automata. In 12th International Colloquium on Structural Information and Communication Complexity (SIROCCO), Lecture Notes in Computer Science #3499, Springer, pp. 140–154, 2005.

14. M. Koucký, Universal Traversal Sequences with Backtracking, Proc. 16th IEEE Conference on Computational Complexity (2001), 21-26. (Also, to appear in J. Computer and System Sciences.)

15. O. Reingold. Undirected ST-Connectivity in Log-Space. To appear in 37th ACM Symp. on Theory of Computing (STOC), 2005.

16. H.-A. Rollik. Automaten in planaren graphen. *Acta Informatica* **13**: 287–298, 1980.

17. C.-E. Shannon. Presentation of a Maze-Solving Machine. In 8th Conf. of the Josiah Macy Jr. Found. (Cybernetics), pages 173–180, 1951.

Concurrent Zero-Knowledge with Timing, Revisited*

Oded Goldreich

Department of Computer Science and Applied Mathematics,
Weizmann Institute of Science, ISRAEL
oded.goldreich@weizmann.ac.il

Abstract. Following Dwork, Naor, and Sahai (*30th STOC*, 1998), we consider concurrent executions of protocols in a semi-synchronized network. Specifically, we assume that each party holds a local clock such that bounds on the relative rates of these clocks as well as on the message-delivery time are a-priori known, and consider protocols that employ time-driven operations (i.e., `time-out` in-coming messages and `delay` out-going messages).

We show that the constant-round zero-knowledge proof for $\mathcal{NP}$ of Goldreich and Kahan (*Jour. of Crypto.*, 1996) preserves its security when polynomially-many independent copies are executed *concurrently under the above timing model*.

We stress that our main result refers to zero-knowledge of *interactive proofs*, whereas the results of Dwork *et. al.* are either for zero-knowledge *arguments* or for a *weak notion* of zero-knowledge (called epsilon-knowledge) proofs.

Our analysis identifies two extreme schedulings of concurrent executions under the above timing model: the first is the case of *parallel execution* of polynomially-many copies, and the second is of concurrent execution of polynomially-many copies such that only a small (i.e., constant) number of copies are simultaneously active at any time (i.e., *bounded simultaneity*). Dealing with each of these extreme cases is of independent interest, and the general result (regarding concurrent executions under the timing model) is obtained by combining the two treatments.

1 Introduction

Zero-Knowledge proofs, introduced by Goldwasser, Micali and Rackoff [22, 23], are fascinating and extremely useful constructs. Their fascinating nature is due to their seemingly contradictory definition: they are both convincing and yet yield nothing beyond the validity of the assertion being proven. Their applicability in

---

* Preliminary version has appeared in the proceedings of the 34th *ACM Symposium on the Theory of Computing*, 2002. The current revision was prepared in memory of Shimon Even. I find it especially fitting that my wish to pay tribute to his memory has caused me to fulfill my duty (neglected for a couple of years) to produce a final version of the current work.

O. Goldreich et al. (Eds.): Shimon Even Festschrift, LNCS 3895, pp. 27–87, 2006.

the domain of cryptography is vast: they are typically used to force malicious parties to behave according to a predetermined protocol (which requires parties to provide proofs of the correctness of their secret-based actions without revealing these secrets). Such applications are based on the fact, proven by Goldreich, Micali and Wigderson [19], that any language in $\mathcal{NP}$ has a zero-knowledge proof system, provided that commitment schemes exist.[1] The related notion of a zero-knowledge *argument* was suggested (and implemented) by Brassard, Chaum and Crépeau [7], where the difference between proofs and arguments is that in the latter the soundness condition refers only to computationally-bounded cheating provers.

In this work we consider the preservation of zero-knowledge under restricted types of concurrent composition. Specifically, we consider multiple executions of a protocol under a naturally limited model of asynchronous computation (which covers synchronous computation as an important special case). We start by recalling the basic notion of zero-knowledge and providing a wider perspective on the question of its preservation under various forms of composition.

1.1   Zero-Knowledge Protocols

An interactive proof system for a language L is a (randomized) protocol for two parties, called verifier and prover, allowing the prover to convince the verifier to accept any common input in L, while guaranteeing that no prover strategy may fool the verifier to accept inputs not in L, except than with negligible probability. The first property is called completeness, and the second is called soundness. The prescribed verifier strategy is always required to be probabilistic polynomial-time. Furthermore, like in other application-oriented works, we focus on prescribed prover strategies that can be implemented in probabilistic polynomial-time given adequate auxiliary input (e.g., an NP-witness in case of NP-languages). Recall that the latter refers to the prover prescribed for the completeness condition, whereas (unlike in argument systems [7]) soundness must hold no matter how powerful the cheating prover is.

Zero-knowledge is a property of some prover-strategies. Loosely speaking, these strategies yield nothing to the verifier, beyond the fact that the input is in the prescribed language L. The fact that "nothing is gained by the interaction" is captured by stating that whatever the verifier can efficiently compute after interacting with the (zero-knowledge) prover on a specific common input, can be efficiently computed from the assertion itself, without interacting with anyone. Thus, the formulation of the zero-knowledge condition considers two ensembles of probability distributions, each ensemble associates a probability distribution to each input in L: The first ensemble represents the output distribution of the verifier after interacting with the specified prover strategy P, where the verifier is using an arbitrary efficient (i.e., probabilistic polynomial-time) strategy, not necessarily the one specified by the protocol. The second ensemble represents the output distribution of some probabilistic polynomial-time algorithm (which

---

[1] Or, equivalently [27, 24], that one-way functions exist.

does not interact with anyone). The basic paradigm of zero-knowledge asserts that for every ensemble of the first type there exist a "similar" ensemble of the second type. The specific variants differ by the interpretation given to the notion of 'similarity', and in this work (as in most of the literature) we focus on the most liberal interpretation. Under this (liberal) interpretation, similarity means computational indistinguishability (i.e., failure of any efficient procedure to tell the two ensembles apart). The ensembles $\{X_\alpha\}$ and $\{Y_\alpha\}$ are said to be computationally indistinguishable if, for every efficient procedure D (and every α), it holds that

$$|\Pr[D(\alpha, X_\alpha) = 1] - \Pr[D(\alpha, Y_\alpha) = 1]| \; < \; \mu(|\alpha|)$$

where μ is a *negligible function* (i.e., a function vanishing faster than the reciprocal of any positive polynomial). For a detailed treatment of zero-knowledge, the reader is referred to [16, Chap. 4].

1.2   Composition of Zero-Knowledge Protocols

A fundamental question regarding zero-knowledge proofs (and arguments) is whether the zero-knowledge condition is preserved under a variety of composition operations. Three types of composition operations were considered in the literature, and we briefly review these operations and what is known about the preservation of the zero-knowledge condition under each of them.

Sequential Composition. Here the protocol is invoked (polynomially) many times, where each invocation follows the termination of the previous one. At the very least, security (e.g., zero-knowledge) should be preserved under sequential composition, otherwise the applicability of the protocol is severely limited (because one cannot safely use it more than once).

Although the basic definition of zero-knowledge (as in the preliminary version of Goldwasser *et. al.* [22]) is not closed under sequential composition (cf. [18]), a minor augmentation of it (by auxiliary inputs) is closed under sequential composition (cf. [20]). Indeed, this augmentation was adopted in all subsequent works (as well as in the final version of Goldwasser *et. al.* [23]).

Parallel Composition. Here (polynomially) many instances of the protocol are invoked at the same time and proceed at the same pace. That is, we assume a synchronous model of communication, and consider (polynomially) many executions that are totally synchronized such that the ith round message in all instances is sent *exactly* at the same time. (One natural relaxation of this model is discussed below.)

Goldreich and Krawczyk [18] presented a simple protocol that is zero-knowledge (in a strong sense), but is not closed under parallel composition (even in a very weak sense).[2] At the time, their result was interpreted mainly in the context of *round-efficient error reduction*; that is, the construction of full-fledge zero-knowledge proofs (of negligible soundness error) by composing (in parallel) a basic zero-knowledge protocol of high (but bounded away from 1) soundness error. Since alternative ways of constructing constant-round zero-knowledge proofs (and arguments) were found (cf. [17, 15, 8]), interest in parallel composition (of zero-knowledge protocols) has died. In retrospect, as we argue in §1.4, this was a conceptual mistake.

We also consider a relaxed model of parallel composition. In this model (of "almost-parallel" composition), messages that are sent at the beginning of round i (according to the sender's local clock) are received before round $i + 1$ starts (according to the receiver's clock).

Concurrent Composition. This notion of concurrent composition general- izes both the notions of sequential composition and parallel composition. Here (polynomially) many instances of the protocol are invoked at arbitrary times and proceed at arbitrary pace. That is, we assume an asynchronous (rather than synchronous) model of communication.

In the 1990's, when extensive two-party (and multi-party) computations be- came a reality (rather than a vision), it became clear that it is (at least) desirable that cryptographic protocols maintain their security under concurrent compo- sition (cf. [12]). In the context of zero-knowledge, concurrent composition was first considered by Dwork, Naor, and Sahai [13]. Their actual suggestions refer to a model of naturally-limited asynchronicity (which certainly covers the case of parallel composition). Essentially, they assumed that each party holds a lo- cal clock such that the relative clock rates as well as the message-delivery time are bounded by a-priori known constants, and considered protocols that employ time-driven operations (i.e., `time-out` in-coming messages and `delay` out-going messages). This timing model is the main focus of the current paper (and we shortly discuss the pure asynchronous model in §1.4). The previously known main results *for the timing model* are (cf. [13]):

- Assuming the existence of one-way functions, every language in $\mathcal{NP}$ has a constant-round concurrent zero-knowledge *argument*.
- Assuming the existence of two-round perfectly-hiding commitment schemes (which in turn imply one-way functions), every language in $\mathcal{NP}$ has a con- stant-round concurrent epsilon-knowledge proof, where epsilon-knowledge means that for every noticeable function $\epsilon : \mathsf{N} \to (0, 1]$ a simulator working in time poly($n/\epsilon(n)$) can produce output that is ϵ-indistinguishable from the one of a real interaction. (For further discussion of epsilon-knowledge, see Section 1.6.)

---

[2] We comment that parallel composition is problematic also in the context of reducing the soundness error of arguments (cf. [3]), but our focus here is on the zero-knowledge aspect of protocols regardless if they are proofs, arguments or neither.

Thus, no *constant-round proofs* for $\mathcal{NP}$ were previously known to be concurrent zero-knowledge (under the timing model). We comment that proofs with *non-constant* number of rounds were known to be concurrent zero-knowledge (even in the pure asynchronous model; cf. §1.4).

1.3   Our Results

Our main result closes the gap mentioned above, by showing that a (known) constant-round zero-knowledge proof for $\mathcal{NP}$ is essentially *concurrent zero-knowledge under the timing model*. That is, we prove:

Theorem 1 *The* (five-round) *zero-knowledge proof system for $\mathcal{NP}$ of Goldreich and Kahan [17], augmented with suitable time-driven operations, is* concurrent zero-knowledge under the timing model.

Thus, the zero-knowledge property of the proof system (of [17]) is preserved under any concurrent composition that satisfies the timing model. In particular, *the zero-knowledge property is preserved under parallel composition*, a result which we consider of independent interest.

Recall that the proof system of [17] relies on the existence of two-round perfectly-hiding commitment schemes (which is implied by the existence of claw-free pairs of functions and implies the existence of one-way functions). Thus, we get:

Theorem 2 *Assuming the existence of two-round perfectly-hiding commitment schemes, there exists a* (constant-round) *proof system for $\mathcal{NP}$ that is* concurrent zero-knowledge under the timing model.

Using the same techniques, we can show that several other known (constant-round) zero-knowledge protocols remain secure under the concurrent timing-model. Examples include the (constant-round) zero-knowledge *arguments* of Feige and Shamir [15] and of Bellare, Jakobsson and Yung [4]. The latter protocol (referred to as the BJY-protocol) is of special interest because it is a four-round argument for $\mathcal{NP}$ that relies only on the existence of one-way functions. The above protocols are simpler (and use fewer rounds) than the argument systems previously shown (in [13]) to be concurrent zero-knowledge (under the timing-model), alas their security (under this model) is established by a more complex simulator. (See further details in Section 6.1.)

1.4   Discussion of Some Issues

We clarify some issues that underly our study. Some of these issues were mentioned explicitly above.

The Meaning of Composition. We stress that when we talk of composition of protocols (or proof systems) we mean that the honest users are supposed to follow the prescribed program (specified in the protocol description) that refers to a single execution. That is, the actions of honest parties in each execution are independent of the messages they received in other executions. The adversary, however, may coordinate the actions it takes in the various executions, and in particular its actions in one execution may depend also on messages it received in other executions.

Let us motivate the asymmetry between the independence of executions assumed of honest parties but not of the adversary. Coordinating actions in different executions is typically difficult but not impossible. Thus, it is desirable to use composition (as defined above) rather than to use protocols that include inter-execution coordination-actions, which require users to keep track of all executions that they perform. Actually, trying to coordinate honest executions is even more problematic, because one may need to coordinate executions of different honest parties (e.g., all employees of a big cooperation or an agency under attack), which in many cases is highly unrealistic. On the other hand, the adversary attacking the system may be willing to go into the extra trouble of coordinating its attack on the various executions of the protocol.

Important Zero-Knowledge Technicalities. We shortly discuss seemingly technical but actually fundamental variants on the basic definition of zero-knowledge. In particular, these variants play an important role in our work.

Auxiliary inputs and non-uniformity: As mentioned above, almost all work on zero-knowledge actually refer to zero-knowledge with respect to (non-uniform) auxiliary inputs. This work is no exception, but (as in most other work) the reference to auxiliary inputs is typically omitted. We comment that zero-knowledge with respect to auxiliary inputs "comes for free" whenever zero-knowledge is demonstrated (like in this work) via a black-box simulator (see below). The only thing to bear in mind is that allowing the adversary (non-uniform) auxiliary inputs means that the computational assumptions that are used need to be non-uniform ones. For example, when we talk of computational-binding (resp., computational-hiding) commitment schemes we mean that the binding (resp., hiding) property holds with respect to any family of polynomial-size circuits (rather than with respect to any probabilistic polynomial-time algorithm).

Black-box simulation: The definition of zero-knowledge (only) requires that the interaction of the prover with any cheating (probabilistic polynomial-time) verifier be simulateable by an ordinary probabilistic polynomial-time machine (which interacts with no one). A black-box simulator is one that can simulate the interaction of the prover with any such verifier when given oracle access to the strategy of that verifier. All previous zero-knowledge arguments (or proofs), with the exception of the recent (constant-round) zero-knowledge argument of Barak [1], are established using a black-box simulator, and our work is no exception (i.e., we also use a black-box simulator). Indeed, Barak demonstrated that (contrary to previous beliefs) non-black-box simulators may exist in cases where black-box ones do not exist [1]. However, black-box simulators, whenever they exist, are preferable to non-black-box ones, because the former offers greater security: Firstly, as mentioned above, black-box simulators imply zero-knowledge with respect to auxiliary inputs.[3] Secondly, black-box simulators imply polynomial bounds on the knowledge tightness, where *knowledge tightness* is the (inverse) ratio of the running-time of any cheating verifier and the running-time of the corresponding simulation [16, Sec. 4.4.4.2].[4]

Expected polynomial-time simulators: With the exception of the recent (constant-round) zero-knowledge argument of Barak [1], all previous *constant-round* arguments (or proofs) utilize an *expected* polynomial-time simulator (rather than a *strict* polynomial-time simulator). (Indeed our work inherits this "feature" of [17].) As recently shown by Barak and Lindell [2], this is no coincidence, because all the above (with the exception of [1]) utilize black-box simulators, whereas no *strict* polynomial-time black-box simulator can demonstrate the zero-knowledge property of a constant-round argument system for a language outside of $\mathcal{BPP}$.

Types of Concurrent Composition. We shortly discuss various types of asynchronous concurrent composition, starting with the pure asynchronous model and ending with the synchronous (or parallel) model.

Perspective: the pure asynchronous model. Regarding the pure asynchronous model, the current state of the art is as follows:

- Black-box simulator cannot demonstrated the concurrent zero-knowledge property of non-trivial proofs (or arguments) having significantly less than logarithmically many rounds (cf. Canetti *et. al.* [10]). By *non-trivial* proof systems we mean ones for languages outside $\mathcal{BPP}$, whereas by *significantly less than logarithmic* we mean any function $f : \mathbb{N} \to \mathbb{N}$ satisfying $f(n) = o(\log n / \log \log n)$.
- Under standard complexity assumptions, every language in $\mathcal{NP}$ has a concurrent zero-knowledge proof with almost-logarithmically many rounds, and

---

[3] In contrast, whether or not a non-black-box simulator implies zero-knowledge *with respect to auxiliary inputs*, depends on the specific simulator: In fact, in [1], Barak first presents (as a warm-up) a protocol with a non-black-box simulator that cannot handle auxiliary inputs, and later uses a more sophisticated construction to handle auxiliary inputs.

[4] That is, a protocol is said to have knowledge tightness $k : \mathbb{N} \to \mathbb{R}$ if for some polynomial p and every probabilistic polynomial-time verifier V^* the interaction of V^* with the prover can be simulated within time $k(n) \cdot T_{V^*}(n) + p(n)$, where T_{V^*} denotes the time complexity of V^*. In fact, the running-time of the simulator constructed by Barak [1] is polynomial in T_{V^*}, and so the knowledge tightness of his protocol is not bounded by any fixed polynomial.

this can be demonstrated using a black-box simulator (cf. [28], building upon [25], which in turn builds upon [29]).

- Recently, Barak [1] demonstrated that the "black-box simulation barrier" can be bypassed. With respect to concurrent zero-knowledge he only obtains partial results: constant-round zero-knowledge arguments (rather than proofs) for $\mathcal{NP}$ that maintain security as long as an a-priori bounded (polynomial) number of executions take place concurrently. (Barak's result also relies on standard complexity assumptions, and the length of the messages in his protocol grows linearly with this a-priori bound.)[5]

Thus, it is currently unknown whether constant-round arguments for $\mathcal{NP}$ may be concurrent zero-knowledge (in the pure asynchronous model).

On the timing model: The timing model consists of the *assumption* that talking about the actual timing of events is meaningful (at least in a weak sense) and of the *introduction of time-driven operations*. The *timing assumptions* amount to postulating that each party holds a local clock and knows a global bound, denoted $\rho \geq 1$, on the relative rates of the local clocks.[6] Furthermore, it is postulated that the parties know a (pessimistic) bound, denoted Δ, on the message-delivery time (which also includes the local computation and handling times). In our opinion, these timing assumptions are most reasonable, and are unlikely to restrict the scope of applications for which concurrent zero-knowledge is relevant. We are more concerned about the effect of the *time-driven operations* introduced in the timing model. Recall that these operations are the `time-out` of in-coming messages and the `delay` of out-going messages (and the protocol designer determines their duration). Typically (and in fact also in our work), the delay period is at least as long as the time-out period,[7] which in turn is at least

---

[5] We are quite sure that Barak's arguments remain zero-knowledge under concurrent executions that satisfy the *timing model*. But since these are arguments (rather than proofs) such a result will not improve upon the previously known result of [13] (which uses black-box simulations).

[6] Defining the rate of a clock is less straightforward than one may think. Firstly, clocks (or rather their reading) are typically discrete, and thus their relative rate is a ratio between pairs of reading (i.e., initial reading and final reading). Thus, rate must be computed with respect to sufficiently long time intervals. In particular, these intervals should be long enough such that the effect of a single change in the clock reading (i.e., a single "clock tick") can be neglected. Secondly, the clock rate may change with time, and so the aforementioned time intervals should not be too long. In the context of the current work, it is reasonable to measure the clock rate with respect to time intervals of length Δ. Thus, when we say that the relative rate of two clocks is ρ we mean that a time period of Δ units on one clock is measured as at least Δ/ρ (and at most $\rho\Delta$) units on the other clock.

[7] Following the conference presentation of this work, Barak and Micciancio raised the possibility of using a delay period that is smaller and yet linearly related to the time-out period. It seems plausible that, following their approach, security will deteriorate exponentially with the constant of the said proportion. We stress that so far their approach was not proved to work, but it does indicate that the common practice

Δ (i.e., the time-out period must be at least as long as the pessimistic bound on message-delivery time so not to disrupt the proper operation of the protocol). This means that such use of these time-driven operations yields slowing down the execution of the protocol (i.e., running it at the rate of the pessimistic message-delivery time rather than at the rate of the actual message-delivery time, which is typically much faster). Still, in absence of more appealing alternatives (i.e., a constant-round concurrent zero-knowledge protocol for the pure asynchronous model), the use of the timing model may be considered reasonable. (We comment that other alternatives to the timing-model include various set-up assumptions; cf. [9, 11].)

On parallel composition: Given our opinion about the timing model, it is not surprising that we consider the problem of parallel composition almost as important as the problem of concurrent composition in the timing model. Firstly, it is quite reasonable to assume that the parties' local clocks have approximately the same rate, and that clock drifting is corrected by occasional clock synchronization (which is transcendental to the model). Thus, it is reasonable to assume that the parties have approximately-good estimates of some global time. Furthermore, the global time may be partitioned into phases, each consisting of a constant (e.g., 5) number of rounds, so that each party wishing to execute the protocol just delays its invocation to the beginning of the next phase. Thus, concurrent execution of (constant-round) protocols in this setting amounts to a sequence of (time-disjoint) almost-parallel executions of the protocol. Consequently, proving that the protocol is preserves zero-knowledge under almost-parallel composition suffices for ensuring the preservation of zero-knowledge in the aforementioned concurrent setting. We stress that this setting assumes not only that the parties's clocks have practically the same rate, but also that the actual reading of their clocks at each time is essentially identical. (Note that this setting is covered by the notion of almost-parallel composition rather than parallel composition.)

1.5   Techniques

To discuss our techniques, let us fix a timing assumption (i.e., an a-priori bound ρ on the local clock rates and a bound Δ on the message-delivery time) and consider a c-round protocol that utilizes appropriately selected `time-out` and `delay` mechanisms (which depend on the above bounds; e.g., timing-out incoming messages after Δ time units). The reader may think of the bound on the relative rates of local clock as being close to 1 (or even just 1; i.e., equal rates), and of c as being a constant (in fact, we will use $c = 5$). Furthermore, suppose that *all prover's actions* in the protocol are time-driven (by the `time-out` and `delay` mechanisms, and that the corresponding time periods are $\Theta(\Delta)$).

   A key observation underlying our work is that a concurrent scheduling (of such protocol instances) under the timing model can be *decomposed* into a se-

---

(of using a delay period that is at least as long as the time-out period) may not be inherent to the model.

quence of parallel executions, called blocks, such that the number of simultaneously active blocks is bounded by $O(c)$. That is, each block consists of protocol
instances that are executed almost in parallel, and the number of blocks that are
(pairwise) active at the same time is $O(c)$, where two blocks are said to be active
at the same time if for some time t each block has a protocol instance that is
active at time t. The constant in the O-notation depends on the a-priori known
bound on the relative clock rates (as well as on the ratio between the time period
used in the time-driven operations). This decomposition applies whenever the
timing model is used (and is not restricted to the context of zero-knowledge),
and it may be useful towards the analysis of the concurrent execution of any set
of protocols under the timing model.

Let us clarify the above observation by providing a proof for a special (simple)
case. Our first simplifying assumption is that the clock rates are all equal. We
further assume that the prover utilizes equal delays between its messages, and
that these delays are four times the length of the time-out period, which is
defined as our basic time unit. Considering an arbitrary scheduling of protocol
instances, under the aforementioned timing model, we place a protocol instance
in the ith block if it is invoked during the ith time-interval (i.e., the time interval
$(i-1, i]$). Then, each block consists of an almost-parallel execution of instances
of the protocol (i.e., the $(j+1)$-st message in any instance of block i is supposed
to be sent at time $t+4j > i-1+4j$ and is timed-out at time $t+4j+1 < i+4j+1$,
where $t \in (i-1, i]$ is the invocation time of this instance). Clearly, the ith and j^{th}
blocks are simultaneously active (at some time) only if $|i-j| < 4c$, where c is the
number of rounds in the protocol. Thus, at most $8c+1$ blocks are simultaneously
active.

In view of the above, it is quite natural to conjecture that in order to analyze the concurrent composition of protocols under the timing model it suffices
to deal with two extreme schedulings: the parallel scheduling and the bounded-
simultaneity scheduling. Indeed, *this conjecture is essentially correct in the special cases considered in this work* (i.e., for certain zero-knowledge proofs).

Handling parallel composition. At first glance, one may be tempted to say that
the techniques used for proving that the Goldreich–Kahan (GK) protocol is
zero-knowledge (cf. [17] and Section 2.3) extend to showing that it remains zero-
knowledge under parallel composition. This would have been true if we were
handling *coordinated* parallel executions of the GK-protocol (where the prover
would abort all copies if the verifier decommits improperly in any of them).
However, this is not what we are handling here (i.e., parallel composition refers to
uncoordinated parallel execution of many copies of the protocol). Consequently,
a couple of new techniques are introduced in order to deal with the parallel
composition of the GK-protocol. We consider these simulation techniques to
be of independent interest, and note that they apply also for establishing the
preservation of zero-knowledge under almost-parallel composition.

Handling bounded-simultaneity concurrent composition. Experts in the area may
not find it surprising that the GK-protocol remains zero-knowledge under bounded-

simultaneity concurrent composition. In fact, previous works (e.g., [13]) suggest that the difficulty in simulating concurrent executions of the GK-protocol arises from the case in which a large number of instances is executed in a "nested" (and in particular simultaneous) manner.[8] Furthermore, the work of Richardson and Kilian [29] suggests that certain (related) protocols may be zero-knowledge under bounded-simultaneity concurrent composition. Still, to the best of our knowledge, such a technically-appealing result has not been proven before. We prove the result by using a rather straightforward approach, which nevertheless requires careful implementation. We stress that not every zero-knowledge protocol remains zero-knowledge under bounded-simultaneity concurrent composition (e.g., Goldreich and Krawczyk [18] presented a simple (constant-round) protocol that is zero-knowledge, but parallel execution of two instances of it is not zero-knowledge).

Handling the general case. Combining the techniques employed in handling the two extreme cases, we show that (augmented with suitable timing mechanisms) the GK-protocol is concurrent zero-knowledge under the timing model. This is shown by using the abovementioned decomposition, and applying the bounded-simultaneity simulator to the blocks while incorporating the parallel-composition simulator inside of it (i.e., to the individual blocks). Note that, by definition, the bounded-simultaneity simulator handles the special case in which each block contains a single copy, and does so by employing the single-copy simulator. Capitalizing on the high-level similarity of the parallel-composition simulator and the single-copy simulator, we just need to extend the bounded-simultaneity simulator by incorporating the former simulator in it. (Our presentation of the bounded-simultaneity simulator uses terminology that makes this extension quite easy.)

We stress that the combination of the treatments of parallel composition and bounded-simultaneity composition into a treatment of concurrent composition under the timing model *is not generic*, but rather refers to the specific structure of the GK-protocol (and its stand-alone simulator). Still we believe that our decomposition methodology may be useful in other settings.

1.6   Zero-Knowledge Versus Epsilon-Knowledge

Recall that epsilon-knowledge means that for every noticeable function (i.e., a reciprocal of some positive polynomial) $\epsilon : \mathbb{N} \to (0, 1]$ there exists a simulator working in time $\mathrm{poly}(n/\epsilon(n))$ that produces output that is ϵ-indistinguishable from the one of a real interaction, where n denotes the length of the input and the ensembles $\{X_\alpha\}$ and $\{Y_\alpha\}$ are said to be ϵ-indistinguishable if for every efficient procedure (e.g., a polynomial-time algorithm) D, it holds that

$$|\Pr[D(\alpha, X_\alpha) = 1] - \Pr[D(\alpha, Y_\alpha) = 1]| < \epsilon(|\alpha|) + \mu(|\alpha|)$$

---

[8] In fact, even if each level of nesting only multiplies the simulation time by a factor of 2, we get an exponential blow-up.

where μ is a negligible function. (Indeed, the standard notion of computational indistinguishability [21, 32] is a special case obtained by setting $\epsilon \equiv 0$.)

Indeed, as mentioned in [13], epsilon-knowledge does provide some level of security. However, this level of security is lower than the one offered by the standard notion of zero-knowledge, and more so when compared to simulators with bounded knowledge tightness (as discussed above; cf. [16, Sec. 4.4.4.2]).

Expected polynomial-time simulators versus epsilon-knowledge. The above discussion applies also to the comparison of epsilon-knowledge and zero-knowledge via expected polynomial-time simulators (rather than via strict polynomial-time simulators). Furthermore, simulation by an expected polynomial-time simulator implies an epsilon-knowledge simulator (running in strict time inversely proportional to the desired deviation).[9] The converse does not hold (e.g., consider a prover that, for $i = 1, 2...$, with probability 2^{-i} sends the result of a $\mathrm{BPTime}(2^{2i})$-complete computation).[10]

1.7   Relation to Shimon Even (A Personal Comment)

This work grew out of my sudden realization that the question of parallel composition of zero-knowledge protocols has not received the attention that it deserves. Specifically, when asked for a protocol that preserves zero-knowledge under parallel composition, one would have referred to the preservation of zero-knowledge under concurrent composition (possibly in the timing model). Thus, a potentially easier problem was reduced to a harder problem, which is not the 'right' way to go. Things were even worst because, as argued in §1.4, the preservation of zero-knowledge under parallel composition is a natural and important problem.

Readers that were fortunate to know Shimon well will immediately associate the attitudes underlying the previous paragraph with him. Indeed, the moment I reached the conclusion stated above, I got reminded of Shimon.

I then asked myself whether I already know of a simple protocol that preserve zero-knowledge under parallel composition, and my immediate conjecture was that this should be true of the GK-protocol. Once I proved this conjecture, which turned out to be harder to establish than I've originally thought, I asked myself whether this argument can be extended further (i.e., to concurrent composition under the timing model). Thus, I have established results similar to those known before, using a different approach that goes from a natural special case to the

---

[9] To obtain a deviation of at most ϵ, we may truncate the runs of the original simulator that exceed its expected running-time by a factor of $1/\epsilon$ (or so).

[10] We comment that even a stronger notion of ϵ-knowledge, by which the simulator's running-time is linear (rather than polynomial) in $1/\epsilon$ does not seem to imply zero-knowledge (via an expected polynomial-time simulator). Note that the naive attempt (of converting the former simulator into one that establishes zero-knowledge) fails: That is, selecting i with probability 2^{-i} and invoking the former simulator with $\epsilon = 2^{-i}$ does yield an expected polynomial-time simulator, but its output may not be computationally indistinguishable from the real interaction.

general case. This entire development reminds me again of Shimon, because he would have liked its course.

Finally, I wish to recall another connection to Shimon. In 1978, as an undergraduate, I attended his course *Graph Algorithms*. At some point, one student was annoyed at Shimon's "untraditional" way of analyzing algorithms and asked whether Shimon's demonstrations constituted a proof and if so what is a proof. Shimon answer was immediate, short and clear: *A proof is whatever convinces me.* A few years later, when first seeing the definition of interactive proofs, I was reminded of Shimon's answer. I think that interactive proofs are a perfect formalization of Shimon's intuition: interactive proofs are indeed convincing, and essentially any convincing demonstration is actually an interactive proof.

1.8   Organization

In Section 2, we recall some basic notions as well as review the *GK-protocol* (i.e., the five-round zero-knowledge proof system of Goldreich and Kahan [17]). In Section 3 we prove that the GK-protocol remains zero-knowledge under parallel composition. In Section 4 we prove that the GK-protocol remains zero-knowledge under bounded-simultaneity concurrent composition. The latter two sections can be read independently of one another, and are believed to be of independent interest.

In Section 5, we augment the GK-protocol with adequate `time-out` and `delay` mechanisms, and prove that the resulting protocol is concurrent zero-knowledge under the timing model. This is done by extending the simulator presented in Section 4, where the extension relies on the ideas underlying the simulator presented in Section 3. We conclude (cf. Section 6) by applying our techniques to the zero-knowledge argument system of Bellare, Jakobsson and Yung [4] and by presenting a class of protocols to which our techniques can be applied.

2   Background

Zero-knowledge is a property of some prover-strategies. Loosely speaking, it means that anything that is feasibly computable by (possibly improperly) interacting with the prover, can be feasibly computable without interacting with the prover. That is, the most basic definition of zero-knowledge (of a prover P w.r.t a language L) requires that, for every probabilistic polynomial-time verifier strategy V^*, there exists a probabilistic polynomial-time simulator M^* such that the following two probability ensembles are computationally indistinguishable:

1. $\{\langle P, V^* \rangle(x)\}_{x \in L} \stackrel{\text{def}}{=}$ the output of V^* when interacting with P on common input $x \in L$; and
2. $\{M^*(x)\}_{x \in L} \stackrel{\text{def}}{=}$ the output of M^* on input $x \in L$.

(The formulation can be easily extended to allow for auxiliary inputs to V^*; cf. Definition 3.) Recall that the ensembles $\{X_\alpha\}_{\alpha \in S}$ and $\{Y_\alpha\}_{\alpha \in S}$ are said to be computationally indistinguishable if, for every efficient procedure D, it holds that

$$|\Pr[D(\alpha, X_\alpha) = 1] - \Pr[D(\alpha, Y_\alpha) = 1]| < \mu(|\alpha|) \tag{1}$$

where μ is a negligible function. Recall that $\mu : \mathsf{N} \to [0,1]$ is called negligible if it vanishes faster than the reciprocal of any positive polynomial (i.e., for every positive polynomial p and all sufficiently large n, it holds that $\mu(n) < 1/p(n)$). We say that an event occurs with overwhelmingly high probability if it occurs with probability that is negligibly close to 1 (i.e., the event occurs with probability $1 - \mu$, where μ is a negligible function). Indeed, our entire treatment will refer to executions that are parameterized by some parameter, denoted n, which is polynomially related to the length of some relevant input.

2.1   Expected Polynomial-Time Simulation and Black-Box Simulation

As discussed in the introduction, we use two variants of the above definition (or definitional schema): On one hand, we allow the simulator to run in *expected* probabilistic polynomial-time (rather than *strict* probabilistic polynomial-time). On the other hand, we require the simulator to be implementable by a universal machine that gets oracle access to the (verifier) strategy V^*. See [16, Sec. 4.3.1.6] (resp., [16, Sec. 4.5.4.2] and [1]) for further discussion of the first (resp., second) issue.

Definition 3 (black-box zero-knowledge):

Next message function: *Let B be an interactive Turing machine, and x, z, r be strings representing a common-input, auxiliary-input, and random-input, respectively. Consider the function $B_{x,z,r}(\cdot)$ describing the messages sent by machine B such that $B_{x,z,r}(\overline{m})$ denotes the message sent by B on common-input x, auxiliary-input z, random-input r, and sequence of incoming messages $\overline{m}$. For simplicity, we assume that the output of B appears as its last message.*

Black-box simulator: *We say that an* expected *probabilistic polynomial-time oracle machine M is a* black-box *simulator for the prover P and the language L if for every polynomial-time interactive machine B, every probabilistic polynomial-time oracle machine D, every positive polynomial $p(\cdot)$, all sufficiently large $x \in L$, and every $z, r \in \{0,1\}^{p(|x|)}$:*

$$\left| \Pr\left[D^{B_{x,z,r}}(\langle P, B_r(z) \rangle(x)) = 1 \right] - \Pr\left[D^{B_{x,z,r}}(M^{B_{x,z,r}}(x)) = 1 \right] \right| < \frac{1}{p(|x|)}$$

where $B_{x,z,r}$ is the next-message function define above, and $B_r(z)$ denotes the interaction of machine B with auxiliary-input z and random-input r. That is, $\langle P, B_r(z) \rangle(x)$ denotes the output of B, having auxiliary-input z and random-input r, when interacting with P on common input x.

We say that P is black-box zero knowledge *if it has a black-box simulator.*

Note that an auxiliary-input for the verifier is explicitly incorporated in Definition 3, whereas an auxiliary input for the prover is only implicit in it. That is, P may be a probabilistic polynomial-time that is given an adequate additional information regarding the common input x as an auxiliary input (e.g., an NP-witness that $x \in L$, in case L is in $\mathcal{NP}$).

An important comment: Definition 3 suggests that it suffices to consider deterministic strategies for the adversary (verifier), because we quantify over all possible choices of the random-input r (just as we do for all possible choices of the auxiliary-input z). Thus, *throughout the rest of this work we only consider deterministic adversary strategies.*

A tedious comment: Definition 3 is equivalent to a form in which the distinguisher D is given (x, z) (or (x, z, r)) as an auxiliary input, which is more consistent with Eq. (1). In some sources, one consider distinguishers that get yet an additional auxiliary input that is not given to the verifier's strategy. It can be shown that Definition 3 is also equivalent to the latter form (e.g., by using adversaries that "typically" don't read their entire auxiliary-input, and yet enable the distinguisher (which runs for more time) to access this auxiliary-input).

2.2   Parallel and Concurrent Zero-Knowledge and the Timing Model

The definitions of parallel and concurrent zero-knowledge are derived from Definition 3 by considering appropriate adversaries (i.e., adversarial verifiers) that invoke multiple copies of the (basic) protocol. For simplicity, we will assume throughout this work that all copies are invoked on the same (common) input, but the our results extend easily to the case in which the adversary determines an arbitrary (common) input for each copy (on the fly). Each execution of such an individual copy is called a session. In case of parallel zero-knowledge, we consider adversaries that simultaneously invoke a polynomial number of sessions of the protocol, and interact with this multitude of sessions in a synchronized way (i.e., send their i^{th} message in all sessions at the same time). In case of concurrent zero-knowledge, we consider adversaries that invoke a polynomial number of sessions, and interleave their interaction with this multitude of sessions in an arbitrary way. Such adversaries may determine the scheduling of message delivery events at the various sessions in a dynamic manner (i.e., depending on the contents of all messages that have received so far); see Definition 4. In case of concurrent zero-knowledge under the timing model, the prescribed protocol may refer to time-driven operations (and the definition of the adversary may remain almost intact). Details follow.

Definition 4 (adversary for the study of unrestricted concurrent composition): *An* admissible adversary for concurrent composition *of a prover P for membership in L is a (deterministic) polynomial-time machine that, on input $x \in L$ and $z \in \{0, 1\}^{\text{poly}(|x|)}$, invokes polynomially many sessions of P, and interacts with* *them in an arbitrary order and manner. That is, based on (x, z) and all messages it has received so far, the adversary iteratively performs one of the following actions:*

1. *Invokes a new session of P on common input x.*
2. *Sends a message to one of the active sessions of P. It is assumed that this session responds immediately, and thus the response becomes part of the sequence of messages received by the adversary.*
3. *Halts with a final output.*

We stress that the active session of P selected in Case 2 is determined by the adversary. This means that the adversary has free control on the scheduling of messages received at the various sessions of P, and that it may schedule these messages adaptively (i.e., based on all information it has obtained so far).

Definition 4 may also be used in the study of concurrent composition under the *timing model*, but in such a case the adversary determines the exact timing of the events (in Cases 1 and 2) and not merely their relative order. That is, such an adversary annotates each action by a time value, where later actions are never assigned smaller time values than previous actions. An alternative and essentially equivalent formulation is presented next.

We recall that under the timing model, the prescribed prover strategy P may contain time-driven operations. Specifically, it is natural for P to halt (in a sesssion) when it detects that the verifier (it interact with) has violated the message-delivery bound. Thus, we may assume without loss of generality, that the adversary never violates the upper-bound on the message-delivery time; it may instead send an illegal message at the "latest possible adequate" time (to be discussed next). For simplicity, in this work, we consider only protocols in which the *prescribed verifier* does not delay its messages (but rather answers immediately).[11] We also assume, without loss of generality, that all local clocks are at most as fast as the real time, but they may be a factor $\rho > 1$ slower. In such a case, the aforementioned *latest possible adequate time* is the upper-bound on the message-delivery time as measured on a possibly slow local clock (where the adversary may determine the rate of the latter clock). With these conventions in mind, we re-define adversaries in the timing model as follows.

Definition 5 (adversary for the study of concurrent composition in the timing model): *Let Δ be an upper-bound on the message-delivery time and ρ be an upper-bound on the clock rate. An* admissible adversary in the timing model *(with parameters Δ and ρ) behave as in Definition 4, except that it responses to each message sent by each session of P within $2\rho\Delta$ units of time after P sent the said message. Formally, in each iteration, the adversary determines not only the next event (of Cases 1 or 2) but rather also at what time this event takes place,*

---

[11] Typically, the time-driven operations are employed by the *prescribed prover* in order to guarantee preservation of zero-knowledge in the timing-model. In this context, the verifier is not trusted anyhow, and thus there seems to be no benefit in having the prescribed verifier employ time-driven operations.

and if the event is the sending of a message to an active session of P in which an event took place at time t then the current (Case 2) *event is assigned time $t' \in [t, t + 2\rho\Delta]$.*

Note that the adversary determines the timing in which each session of P is invoked (i.e., Case 1) as well as the timing of each message delivery event (i.e., Case 2) for that session. The later timing is subject to avoiding detection (by P) of illegally slow message-delivery. Specifically, P may expects a response to its last message within 2Δ units of time (which accounts for the possible delay of its own message as well as the delay of the respose itself), but its own clock may be slowed down by a factor of ρ. Indeed, we assume that the adversary can determine P's clock rate, let alone know this rate.

An **adversary for the study of parallel composition** is obtained as a special case of Definition 5. Such an adversary invokes all (new) sessions at exactly the same time, and responses to all messages sent by these sessions exactly one unit of time after these messages were sent. (Indeed, in this case we assume that $\Delta \geq 1$.)

An important technicality. As discussed by Canetti *et. al.* [10], Definition 3 is too restrictive for serving as a basis for a definition of (unbounded) zero-knowledge composition, where the adversary B may invoke a (polynomial) number of sessions with P but this polynomial is not a-priori known. The problem is that the universal (black-box) simulator may invoke (the next message function associated with) B only for a fixed polynomial (expected) number of times, whereas B may describe a strategy that initiates a larger (polynomial) number of sessions with P. One solution is to consider for each polynomial a different universal simulator that can handle all adversaries that invoke at most a number of sessions (with P) that is bounded by that polynomial.[12] For simplicity, we adopt this solution here. We spell out the definition derived for the case of concurrent composition in the timing model.

Definition 6 (simulator for the study of concurrent composition in the timing model): *We say that P is* (black-box) **concurrent zero-knowledge** *for L in the timing model if for every polynomial p there exists an* expected *probabilistic polynomial-time oracle machine M such that for every p-time admissible (per Definition 5) adversary V^*, the following two probability ensembles are computationally indistinguishable:*

1. $\{\langle \||^P, V^*(z)\rangle(x)\}_{x \in L, z \in \{0,1\}^{p(x)}}$, where $\langle \||^P, V^*(z)\rangle(x)$ denotes the output of V^* after interacting with multiple sessions of P on common input x, where V^* uses the auxiliary input z.
2. $\{M^{V^*_{x,z}}(x)\}_{x \in L, z \in \{0,1\}^{p(x)}}$, where $V^*_{x,z}$ denotes the next message function associated with V^*.

As in Definition 3, the potential distinguishers are given oracle access to $V^*_{x,z}$.

Note that if V^* outputs the timing of the message delivery events (in its real interaction with the sessions of P) then a good simulator must do the same.

---

[12] An alternative solution is to provide the universal simulator with an auxiliary input that specifices (in unary) the running time of the verifier.

2.3   The Goldreich–Kahan (GK) Protocol

Loosely speaking, the Goldreich–Kahan (GK) proof system for Graph 3-Colorability (G3C) proceeds in four steps:

1. The verifier commits to a challenge (i.e., sequence of edges in the input graph).
2. The prover commits to a sequence of values (i.e., the colors of each vertex under several random relabelings of a fixed 3-coloring of the graph). This sequence is partitioned into subsequences, each corresponding to a different random relabeling of the coloring of the graph.
3. The verifier decommits (to the edge-sequence).
4. If the verifier has properly decommits then the prover decommits to a subset of the values as indicated by the decommitted challenge. Otherwise the prover sends nothing.

   Specifically, the challenge is a sequence of edges, each associated with an independently selected 3-coloring of the graph, and the prover responses to the ith edge by decommitting to the values in the ith committed coloring that correspond to the end-points of the ith edge.

A detailed description of the above protocol is provided in Construction 7 (below). We note that many of the specific details are not important to our analysis, and are provided merely for sake of clarity. We highlight a couple of points that are relevant to the analysis: Firstly, the prover's commitment is via a commitment scheme that is (perfectly-binding but only) computationally-hiding, and so commitments to different values are (only) computationally-indistinguishable (which considerably complicates the analysis; cf. [17]). Secondly, the verifier's commitment is via a commitment scheme that is (perfectly-hiding but only) computationally-binding, and so it is (only) infeasible for it to properly decommits in two different way (which slightly complicates the analysis).

Implementation Details: The Goldreich–Kahan protocol [17] utilizes two "dual" commitment scheme (see terminology in [16, Sec. 4.8.2]). The first commitment scheme, denoted C, is used by the prover and has a perfect-binding property. For simplicity, we assume that this scheme is non-interactive, and denote by $C(v)$ a random variable representing the output of C on input v (i.e., a commitment to value v).[13] The second commitment scheme, denoted $\mathcal{C}$, is used by the verifier and has a perfect-hiding property. Such a scheme must be interactive, and we assume that it consists of the receiver sending a random index, denoted α, and the committer responds by applying the randomized process $\mathcal{C}_\alpha$ to the value it wishes to commit to (i.e., $\mathcal{C}_\alpha(v) = \mathcal{C}(\alpha, v)$ represents a commitment to v relative to the receiver's message α). Consequently, Step 1 in the high-level description is implemented by Steps P0 and V1 below.

---

[13] Non-interactive perfectly-binding commitment schemes can be constructed using any *one-to-one* one-way function. In case one wishes to rely here only on the existence of one-way *functions*, one may need to use Naor's two-round perfectly-binding commitment scheme [27]. This calls for a minor modification of the description below.

Construction 7 (The GK zero-knowledge proof for G3C):

Common Input: *A simple (3-colorable) graph $G = (V, E)$.*
    Let $n \overset{\text{def}}{=} |V|$, $V = \{1, ..., n\}$, and $t \overset{\text{def}}{=} 2n \cdot |E|$.
Auxiliary Input to the Prover: *A 3-coloring of G, denoted ψ.*
Prover's preliminary step (P0): *The prover invokes the commit phase of the perfectly-hiding commitment scheme, which results in sending to the verifier a message α.*
Verifier's commitment to a challenge (V1): *The verifier uniformly and independently selects a sequence of t edges, $\bar{e} \overset{\text{def}}{=} ((u_1, v_1), ..., (u_t, v_t)) \in E^t$, and sends to the prover a random commitment to these edges. Namely, the verifier uniformly selects $s \in \{0, 1\}^{\text{poly}(n)}$, and sends $c \overset{\text{def}}{=} C_\alpha(\bar{e}, s)$ to the prover.*
    Motivating Remark: *At this point the verifier is effectively committed to a sequence of t edges. (This commitment is of perfect secrecy.)*
Prover's commitment step (P1): *The prover uniformly and independently selects a sequence of t random relabeling of the 3-coloring ψ, and sends the verifier commitments to the color of each vertex under each of these colorings. That is, the prover uniformly and independently selects t permutations, $\pi_1, ..., \pi_t$, over $\{1, 2, 3\}$, and sets $\phi_j(v) \overset{\text{def}}{=} \pi_j(\psi(v))$, for each $v \in V$ and $1 \leq j \leq t$. It uses the perfectly-binding commitment scheme to commit itself to the colors of each of the vertices according to each 3-coloring. Namely, the prover uniformly and independently selects $r_{1,1}, ..., r_{n,t} \in \{0, 1\}^n$, computes $c_{i,j} = C(\phi_j(i), r_{i,j})$, for each $i \in V$ and $1 \leq j \leq t$, and sends $c_{1,1}, ..., c_{n,t}$ to the verifier.*
Verifier's decommitment step (V2): *The verifier decommits the sequence $\bar{e} = ((u_1, v_1), ..., (u_t, v_t))$ to the prover. Namely, the verifier send $(s, \bar{e})$ to the prover.*
    Motivating Remark: *At this point the entire commitment of the verifier is revealed. The verifier now expects to receive, for each j, the colors assigned by the jth coloring to vertices u_j and v_j (i.e., the endpoints of the jth edge in $\bar{e}$).*
Prover's partial decommitment step (P2): *The prover checks that the message just received from the verifier is indeed a valid revealing of the commitment c made by the verifier at Step (V1) (i.e., it checks that $c = C_\alpha(\bar{e}, s)$ indeed holds and that $\bar{e} \in E^t$). Otherwise the prover halts immediately. Let us denote the sequence of t edges, just revealed, by $(u_1, v_1), ..., (u_t, v_t)$. The prover reveals (to the verifier), for each j, the jth coloring of vertices u_j and v_j, along with appropriate decommitment information. Namely, the prover sends to the verifier the sequence of four-tuples*

$$(r_{u_1,1}, \phi_1(u_1), r_{v_1,1}, \phi_1(v_1)), ..., (r_{u_t,t}, \phi_t(u_t), r_{v_t,t}, \phi_t(v_t))$$

Verifier's local decision step (V3): *The verifier checks whether, for each j, the values in the jth four-tuple constitute a correct revealing of the commitments $c_{u_j,j}$ and $c_{v_j,j}$, and whether the corresponding values are different. Namely, upon receiving $(r_1, \sigma_1, r_1', \tau_1)$ through $(r_t, \sigma_t, r_t', \tau_t)$, the verifier* *checks whether for each j, it holds that $c_{u_j,j} = C(\sigma_j, r_j)$, $c_{v_j,j} = C(\tau_j, r'_j)$, and $\sigma_j \neq \tau_j$ (and both are in $\{1, 2, 3\}$). If all conditions hold then the verifier accepts. Otherwise it rejects.*

Goldreich and Kahan proved that Construction 7 constitutes a (constant-round) zero-knowledge interactive proof for Graph 3-Colorability [17]. (We briefly review their simulator below.) Our first goal, undertaken in Section 3, is to show that the zero-knowledge property (of Construction 7) is preserved under parallel composition. We later extend the result to yield concurrent zero-knowledge under the timing-model.

High level description of the simulator used in [17]. The simulator (using oracle access to the verifier's strategy) proceeds in three main steps:

The Scan Step: The simulator emulates Steps (P0)–(V2), by using commitments to dummy values in Step (P1), and obtains the verifier's decommitment for Step (V2), which may be either proper or not. In case of improper decommitment the simulator *outputs the partial transcript just generated and* halts. Otherwise, it records the sequence $(u_1, v_1), ..., (u_t, v_t)$, just revealed, and proceeds as follows.

The Approximation Step: For technical reasons (discussed below), the simulator next approximates the probability that the first scan (or rather the emulation of Steps (P1)–(V2)) ended with a proper decommitment. (This is done by repeated trials, each as in the first scan, until some polynomial number of proper decommitments is found.)

The Generation Step: Using the (proper) decommitment information (i.e., the edge sequence $(u_1, v_1), ..., (u_t, v_t)$), obtained in the first scan, the simulator repeatedly tries to generate a full transcript by emulating Steps (P1)–(V2), using commitments to "pseudo-colorings" that do not "violate the coloring conditions imposed by the decommitted edges". That is, in each trial, the simulator sets $c_{i,j}$ to be a commitment to a dummy value if $i \notin \{u_j, v_j\}$, and sets $c_{u_j,j}$ and $c_{v_j,j}$ to be commitments to two different random values in $\{1, 2, 3\}$. The number of trials is inversely proportional to the probability estimated in the approximation step.

This completes the (high level) description of the simulator used in [17]. We conclude this section with a discussion of the purpose of the Approximation Step.

The purpose of the Approximation Step. The foregoing simulation procedure is a variant of the more natural (and in fact naive) procedure in which the Approximation Step is omitted and the Generation Step is repeated (indefinitely) untill a full transcript is generated. The problem with the naive variant is that the probability (denoted p) of proper verifier decommitment during the Scan Step is *not* identical to the probability (denoted p') of a proper verifier decommitment during the Generation Step. The difference is due to the fact that in the Scan Step we feed the verifier with commitments to dummy values, whereas in the

Generation Step we feed the verifier with commitments to "pseudo-colorings". Indeed, the hiding property of commitment schemes guarantees that $|p - p'|$ is negligible (in n), but this does not mean that p/p' is upper-bounded by a polynomial in n (e.g., $p = 2^{-n/3}$ and $p' = 2^{-n/2}$). Thus, the expected running-time of the naive simulation procedure (i.e., $(p/p') \cdot \mathrm{poly}(n)$) is *not* necessarily polynomial. This problem is resolved by the actual simulation procedure of [17] outlined above, whose running time is $p \cdot \frac{\mathrm{poly}(n)}{\widetilde{p}}$, where $\widetilde{p} = \Theta(p)$ is the approximation of p (obtained in the Approximation Step, and $\widetilde{p} = \Theta(p)$ holds with probability $1 - 2^{-\mathrm{poly}(n)}$).

An alternative approach. An alternative way of coping with the aforementioned problem is to use a different protocol that allows for the Scan Step to use the same distribution as in the Generation Step. This approach was recently pursued by Rosen [30], who suggested an alternative constant-round zero-knowledge proof for $\mathcal{NP}$ (by adapting the protocol of [29]). Rosen's protocol could be applied in the context of the current paper and yield a noticeable simplification of the proof of our main results (of Sections 3–5), but this will not allow to obtain the secondary results presented in Section 6 (which refer to protocols that do not satisfy the stronger property stated above). Furthermore, using Rosen's protocol avoids a natural problem that we would like to treat in the current paper, because this problem is likely to arise in future work (where, like in Section 6, it may not be avoided).

3   Simulator for the Parallel Case

Recall that the GK-protocol proceeds in four (abstract) steps:

1. The verifier commits to a challenge.
   (The actual implementation is by two rounds/messages.)
2. The prover commits to a sequence of values.
   (The challenge specifies a subset of the locations in the latter sequence.)
3. The verifier decommits to its challenge (either properly or not).
4. Pending on the verifier's proper decommitment, the prover decommits to the corresponding values.

The basic approach towards simulating this protocol (without being able to answer a random challenge) is to first run the first three steps with prover-commitments to arbitrary (dummy) values, obtaining the challenge, and then rewind to Step 2 and make a prover-commitment that passes this specific challenge (alas no other challenge). In case the verifier always decommits properly, this allows to easily simulate a full run of the protocol. In case the verifier always decommits improperly, things are even easier because in this case we only need to simulate Steps 1–3. The general case is when the verifier decommits with some probability. Intuitively, this can be handled by outputting the initial transcript of Steps 1–3 in case it contains an improper decommitment, and repeatedly trying to produce a full passing transcript (as in the first case) otherwise. Difficulties arise in case the probability of proper verifier decommitment is small but not negligible and furthermore when it depends (in a negligible way) on whether the prover commits to dummy or to "passing" values. Indeed, the focus of [17] is on resolving this problem (and their basic approach is to approximate the probability of proper decommitment in case of dummy values, and keep trying to produce a full passing transcript for at most a number of times that is inversely proportional to the latter probability).

The problem we face here is more difficult: several (say n) sessions of the protocol are executed in parallel and the verifier may properly decommit in some of them but not in others. Furthermore, the verifier decision regarding in which sessions it properly decommits may depend on the prover's messages in all sessions. That is, in the general case, each (parallel) execution of Steps 1-3 may yield a different configuration (out of 2^n possible ones) of proper/improper decommitment in the n sessions. Still, we need to simulate a transcript of all steps in sessions in which the verifier commits properly. Thus, the naive generalization of the case $n = 1$ (which consists of insisting on generating the same configuration as in the initial run) will not work.[14] Instead, referring to the n probabilities that correspond to proper decommitment in each of the n sessions, we add additional rewindings in which we try to obtain a proper decommit from all sessions that have at least as high a probability as the sessions that actually performed proper-decommitment in the initial simulated run. That is, letting p denote the minimum probability of proper-decommitment taken only over the sessions that have proper-decommitted in the initial run, we try to obtain the challenges of all sessions having proper-decommitment probability at least p. Once these challenges are obtained, we try to generate a parallel run in which only sessions having at least as high a probability (but not necessarily all of them!) properly decommit. Furthermore, in order not to skew the distribution (towards high proper-decommitment probabilities), we insist on having at least one session with a corresponding probability as low as some session in the initial run. That is, we try to generate a parallel run in which only sessions having proper-decommitment probability at least p perform proper-decommit, while insisting that at least one session having proper-decommitment probability approximately p performs proper-decommit.

One obvious problem with the above description is that we do not know the relevant proper-decommitment probabilities. Indeed, we may obtain good (multiplicative) approximation of them, but using these approximations in a straightforward manner will not do (because such approximations do not allow

---

[14] We refer to a procedure that obtains some challenges via an initial "dummy" execution of Steps 1-3, and next tries to produce an adequate simulation by repeatedly rewinding Steps 2-4 until one obtains again the same configuration. This may fail because all 2^n configurations may be equally likely, in which case the simulation is expected to make 2^n trials.

to rank the actual probabilities).[15] Instead, we cluster the n sessions according to the probability that each of them properly decommits, and try to obtain a proper decommit from all the sessions that are in the same (or heavier) cluster as the sessions that properly decommit in the initial simulated run. Once this is obtained, we try to generate a parallel run in which only sessions that belong to the above (or heavier) cluster (but not necessarily all of them) properly decommit. As one may expect, clusters are defined according to threshold probabilities, but picking these thresholds naively (e.g., as fixed quantities) is going to fail. Below, we will pick these thresholds at random from fixed intervals.

3.1   A High Level Description

Recall that our aim is to analyze the parallel execution of the GK-protocol. Specifically, we will consider n sessions of the protocol, being executed in parallel under the coordinated attack of an adversary (called a verifier) that plays the role of the verifier in all sessions. The parameter n is polynomially related to the length of the input to each of these sessions, and thus we deal with the general case of parallel composition (of the GK-protocol). When we say that some quantities are negligible or overwhelmingly high we refer to these quantities as a function of the parameter n.

The following basic notions are central to our analysis (of the parallel execution of the GK-protocol): An execution of a session (of the GK-protocol) is said to properly decommit if the verifier message in Step 3 is a valid decommitment to its (i.e., the verifier's) commitment in Step 1. In the first part of the simulation, we use prover's commitments to arbitrary values, which are referred to as commitment to dummy values. Later (in the simulation) we use commitments to values that will pass for a certain challenge (which is understood from the context). These are called commitment to passing values.[16] In addition, we also refer to the following more complex notions and notations:

– Let p_i denote the probability that the verifier properly decommits in the ith session (of the parallel run), when Step 2 is played with commitment to dummy values. Assuming that the adversary (verifier) is *deterministic* (see Section 2.1), we treat the Step 1 message as fixed.[17] Thus, the probability

---

[15] Consider, for example, the case that each of the sessions properly decommits with probability $(1/2) \pm \epsilon(n)$ for some negligible function ϵ or even for $\epsilon(n) = 1/t(n)$, where $t(n)$ is the running time of our approximation procedure.

[16] Recall that in the actual implementation (of the GK-protocol), challenges correspond to sequences of t edges (over the vertex-set $\{1, 2..., n\}$), and the prover commits to a sequence of $t \cdot n$ values in $\{1, 2, 3\}$ (i.e., a block of n values per each of the t edges). For a given edge sequence (i.e., a challenge), a passing sequence of values is one in which (for every i) the values assigned to the ith block are such that the endpoints of the ith edge (in the challenge) are assigned a (random) pair of distinct elements.

[17] In the actual implementation, we will fix a random value for the prover's initial choice of α, which in turn determines the transcript of Step 1.

space (underlying p_i) consists solely of the prover's actions (i.e., choice of commitment) in Step 2.

(When using other commitments (e.g., passing commitments) the probability of proper decommitment may be any p_i' such that $|p_i' - p_i|$ is negligible.)

- We will use a sequence of thresholds, denoted $t_1, ..., t_n$, that will be determined (probabilistically) on the fly such that with overwhelmingly high probability it holds that
  1. $t_j \in (2^{-(j+1)}, 2^{-j})$,
  2. no p_i lies in the interval $[t_j \pm (1/9n) \cdot 2^{-j}]$.

  Such t_j's exist and t_j can be found when given approximations of all p_i's up-to $(1/9n) \cdot 2^{-j}$ (or so). We also define $t_0 \stackrel{\text{def}}{=} 1$, and so $p_i \leq t_0$ for all i. We assume, without loss of generality, that for every i it holds that $p_i > 2^{-n}$, and so each p_i lies in one of the intervals $(t_j, t_{j-1}]$.

- For such t_j's, define $T_j = \{i : p_i > t_j\}$. (Indeed, $T_0 = \emptyset$, $T_{j-1} \subseteq T_j$ for all j, and $T_n = \{1, ..., n\}$.)

  Membership in T_j can be determined (probabilistically with negligible error probability) in time $\text{poly}(n) \cdot 2^j$, since t_j was selected to be sufficiently far-away from all the p_i's (i.e., $|t_j - p_i| = \Omega(2^{-j}/n)$).

- Referring to a specific run of the parallel execution, we denote by E_j the event that the verifier properly decommits to some session in $T_j \setminus T_{j-1}$ but to no session outside T_j. That is, we consider the set of sessions in which the verifier properly decommits in the specific run (of the parallel execution), and say that E_j holds if j is the minimum integer such that the said set contains an element of T_j. (Equivalently, j is the minimum integer such that the said set contains an element of $T_j \setminus T_{j-1}$.)

  Let $q_j = \Pr[E_j]$, when E_j refers to a random run with dummy values, where the probability is taken over the choice of prover's commitments to these dummy values. Note that $q_j \leq n \cdot t_{j-1}$ (because E_j mandates that the verifier properly decommits to some session in $T_j \setminus T_{j-1}$, which implies one of $|T_j \setminus T_{j-1}| \leq n$ events, each occuring with probability at most t_{j-1}). However, q_j may be much smaller than $t_j < t_{j-1}$, because the event E_j refers to n possibly dependent events (occurring in n sessions).

  Since $\{1, ..., n\} = T_n \supseteq T_{n-1} \cdots \supseteq \cdots T_1 \supseteq T_0 = \emptyset$, whenever the verifier properly decommits in some session, one of the events E_j (for $j \geq 1$) must hold. Otherwise (i.e., whenever the verifier decommits improperly in all sessions), we say that event E_0 holds.

We now turn to the simulator, which generalizes the one in [17]. All approximations referred to below are quite good w.v.h.p. (i.e., with $1 - 2^{-n}$ each approximation is within a factor of $(1 + (1/\text{poly}(n)))$ of the corresponding value). Loosely speaking, after fixing the verifier's coins (at random), the simulator proceeds as follows (while using the residual verifier strategy as a black-box):

Step S0: Obtain the verifier's commitments (of Step 1) in the n parallel sessions.

For more details on this and other steps, see Section 3.3.

Step S1: The purpose of this step is to generate an index $j \in \{0, 1, ..., n\}$ with distribution corresponding the probability that event E_j holds for a random parallel execution of the protocol, as well as to determine the sets T_j and T_{j-1} (as defined above, based on adequate thresholds t_j and t_{j-1}, which will be selected too). This has to be done in expected polynomial time. Recalling that event E_j occurs with probability $O(n/2^j)$, when we select a specific j, we may use $\text{poly}(n) \cdot 2^j$ steps.

We stress that we only determine the sets T_j and T_{j-1}, for the specific j that is selected, rather than determine all sets (i.e., $T_1, ..., T_n$). The sets T_j and T_{j-1} will allow us to determine (in subsequent steps) whether or not event E_j holds for other random parallel executions of the protocol.

The selection of j as well as the determination (or construction) of the sets T_j and T_{j-1} is achieved as follows:

1. First we simulate Steps 2–3 of the (parallel execution of the) protocol, while using (in Step 2) commitments to dummy values. Based on the verifier's decommitments in Step 3 (of the parallel execution), we determine the set $I \subseteq [n]$ of sessions in which the verifier has properly decommitted.

2. Next, we determine an appropriate sequence $t_1, ..., t_j$ of thresholds such that event E_j holds for the simulated run. Specifically, we determine the t_j's on the fly, starting with t_1, until we see that E_j holds. Thus, we stop without determining $t_{j+1}, ..., t_n$.

3. Finally, using t_{j-1} and t_j, we determine for each $i \in \{1, 2, ..., n\}$ whether or not $p_i > t_j$ (i.e., $i \in T_j$) and whether or not $p_i > t_{j-1}$ (i.e., $i \in T_{j-1}$).

Indeed, the above description (especially of the second sub-step) does not specify how the corresponding actions are performed (let alone within time $\text{poly}(n) \cdot 2^j$). We defer these crucial details to Section 3.2, where we show how to actually implement the current step within time $\text{poly}(n) \cdot 2^j$.

Step S2: For each session $i \in T_j$, we wish to obtain the challenge committed to in Step 1, while working within time $\text{poly}(n) \cdot 2^j$. This is done by rewinding and re-simulating Steps 2–3 for at most $\text{poly}(n) \cdot 2^j$ times, while again using (in Step 2) commitments to dummy values.

Step S3: For technical reasons[18], analogously to [17], we next obtain a good (i.e., constant factor) approximation of $q_j = \Pr[E_j]$. This approximation, denoted $\tilde{q}_j$, will be obtained within expected time $\text{poly}(n)/q_j$ by repeated rewinding and re-simulating Steps 2–3. (Specifically, we continue till we see some fixed polynomial number (say n^5) of runs in which event E_j holds.)

Step S4: We now try to generate a simulation of Steps 2–3 in which event E_j occurs. However, unlike in previous simulations, here we use (in Step 2) commitments to values that pass the challenges that we have obtained. This will allow us to simulate also Step 4, and complete the entire simulation. Specifically, we make at most $\text{poly}(n)/\tilde{q}_j$ trials to rewind and re-simulate Steps 2–3, while using (in Step 2 of each session in T_j) commitments to values

---

[18] We refer the reader to the end of Section 2.3 for a discussion of the purpose of the approximation step. Note that this step could have been eliminated if we had follows Rosen's alternative approach (also discussed at the end of Section 2.3).

that pass the corresponding challenge (which we obtained in Step S2). If the verifier answers (for Step 3) fit event E_j then we proceed to simulate Step 4 in the obvious manner. Otherwise, we rewind and try again (but never try more than $\text{poly}(n)/\tilde{q}_j$ times).

A more detailed description of the above steps is provided in Sections 3.2 and 3.3. A detailed analysis of the simulator is provided in Section 3.4, relying on the following observations:

1. Pending on the ability to properly implement Step S1, the (overall) expected running time of the simulation is some fixed polynomial, because each attempt (in Steps S2, S3, and S4) is repeated for a number of times that is inversely proportional to the probability of entering this repeated-attempts step. Specifically, each of these steps is repeated at most $(\text{poly}(n)/\tilde{q}_j) \approx (\text{poly}(n)/q_j)$ times (use $q_j = O(n \cdot 2^{-j})$ for Step S2), whereas j is selected with probability q_j.

2. The computational-binding property of $\mathcal{C}$ implies that we rarely get into trouble in Step S4; that is, only with negligible probability will it happen that in Step S4 the verifier properly decommits to a value different from the one to which it has properly decommitted in Step S2.

3. Since the probabilities of verifier's proper-decommitment (in Step 3) are almost unaffected by the prover's commitments (of Step 2) and since passing commitments look like commitments to truly valid values, the simulated interaction is computationally indistinguishable (cf. [21,32]) from the real one.

3.2   Setting the Thresholds and Implementing Step S1

One naive approach is to try to use fixed thresholds such as $t_j = 2^{-j}$. However, this may not allow to determine (for a given i), with high probability and within time $\text{poly}(n) \cdot 2^j$, whether or not p_i is smaller than t_j. (The reason being that p_i may be very close to 2^{-j}; e.g., $|p_i - 2^{-j}| = 2^{-2n}$.)

Instead, the t_j's will be selected in a more sophisticated way such that they are approximately as above (i.e., $t_j \approx 2^{-j}$) but also far enough (i.e., at distance at least $2^{-j}/9n$) from each p_i. This will allow us to determine, with high probability and within time $\text{poly}(n) \cdot 2^j$, whether or not p_i is smaller than t_j. The question is how to set the t_j's such that they are appropriately far from all p_i's. Since the p_i's are unknown probabilities (which we can only approximate), it seems infeasible to come-up with a deterministic setting of the t_j's. Indeed, we will settle for a probabilistic setting of the t_j's (provided that this setting is independent of other events).

Recall that Step S1 calls for the setting of $t_1, ..., t_j$ such that event E_j holds (for a random run), where whether or not event E_j holds depends on t_j and t_{j-1}. Furthermore, it is important that the setting of t_{j-1} in case event E_j holds be the same as the setting of t_{j-1} in case event E_{j-1} holds. Moreover, recalling that the setting of t_j must be performed in time $\text{poly}(n) \cdot 2^j$, we cannot afford to set all

t_k's whenever we set a specific t_j. Still, we provide below an adequate threshold-setting process. We start with the following key procedure, which selects $t_j \approx 2^{-j}$ such that with overwhelmingly high probability $|p_i - t_j| > 2^{-j}/9n$ for every i. We stress that the following procedure (has to run and indeed) runs in time $\mathrm{poly}(n) \cdot 2^j$, which requires a slightly non-straightforward implementation.[19]

Procedure $T(j, n)$, returns $t_j \in [(3/4) \pm (1/8)] \cdot 2^{-j} \subset (2^{-(j+1)}, 2^{-j})$: The procedure first approximates all p_i's sufficiently well, and then sets t_j in the desired interval such that t_j is sufficiently far from all the approximated values of the p_i's. A specific implementation follows.

1. For $i = 1, ..., n$, the procedure approximates p_i sufficiently well (in the following sense, which is motivated in Footnote 19). Specifically, with overwhelmingly high probability, the approximated value, denoted a_i, should satisfy:
   (a) If $p_i < 2^{-j-2}$ then $a_i < 2^{-j-1}$.
   (b) If $p_i > 2^{-j+1}$ then $a_i > 2^{-j}$.
   (c) If $2^{-j-2} \leq p_i \leq 2^{-j+1}$ then $|a_i - p_i| < (1/19n) \cdot 2^{-j}$.
   Each approximation is produced in time $\mathrm{poly}(n) \cdot 2^j$ as follows. First, we decide whether or not $p_i \geq 2^{-j-2}$. Actually, we distinguish with overwhelmingly high probability, between the case $p_i \geq 2^{-j-2}$ and (say) the case $p_i < 2^{-j-3}$, where in the intermediate range any decision is admissible. Likewise, we decide whether or not $p_i \leq 2^{-j+1}$ (i.e., distinguish between the case $p_i \leq 2^{-j+1}$ and the case $p_i > 2^{-j+2}$). These decisions can be made using $\mathrm{poly}(n) \cdot 2^j$ trials. In case we decided that $p_i \in [2^{-j-2}, 2^{-j+1}]$, we approximate p_i up-to an additive deviation of $(1/19n) \cdot 2^{-j}$, which can be implemented using $\mathrm{poly}(n) \cdot 2^j$ trials (because it calls for an approximation to within a factor of $1 \pm \Theta(1/n)$). Otherwise, we output the threshold value (i.e., $a_i = 2^{-j-2}$ if we decided that $p_i < 2^{-j-2}$ and $a_i = 2^{-j+1}$ if we decided that $p_i > 2^{-j+1}$).
   Note that if $p_i < 2^{-j-2}$ then both $a_i = 2^{-j-2}$ and $a_i = p_i \pm 2^{-j}/19n$ satisfy $a_i < 2^{-j-1}$. Similarly, if $p_i > 2^{-j+1}$ then both $a_i = 2^{-j+1}$ and $a_i = p_i \pm 2^{-j}/19n$ satisfy $a_i > 2^{-j}$. Finally, if $2^{-j-2} \leq p_i \leq 2^{-j+1}$ then we decided that $p_i \in [2^{-j-2}, 2^{-j+1}]$ and produced $a_i = p_i \pm 2^{-j}/19n$ as required.
2. Starting from a set of evenly spaced points in the desired interval (i.e., $\{(5/8), (5/8) + (1/4n), ..., (5/8) + (n/4n)\}$), we discard all points that are close to one of the a_i's obtained in Step 1. Specifically, the procedure determines

$$K \stackrel{\mathrm{def}}{=} \left\{ k \in \{0, 1, ..., n\} : (\forall i) \; a_i \notin \left(\frac{5}{8} + \frac{k}{4n} \pm \frac{1}{8n} \right) \cdot 2^{-j} \right\} \quad (2)$$

---

[19] The straightforward approach is to approximate each p_i up to an additive deviation of $\Theta(2^{-j}/n)$. The problem is that, in general, this requires $\Omega((2^{-j}/n)^{-2})$ samples. However, for $p_i \approx 2^{-j}$, such an additive approximation translates to a multiplicative approximation of $1 \pm \Theta(1/n)$, which can be obtained based on a sample of size $\mathrm{poly}(n)/p_i = \mathrm{poly}(n) \cdot 2^j$. We note that, for $p_i \notin [2^{-j-2}, 2^{-j+1}]$, a more crude approximation suffices, and can be obtained using a sample of size $\mathrm{poly}(n) \cdot 2^j$.

That is, a_i rules out the value k if $a_i \in (5n + 2k \pm 1) \cdot 2^{-j}/8n$. Note that K is not empty, because each a_i can rule out at most one element of K (whereas $|\{0, 1, ..., n\}| = n + 1$ and they are only n values of i).

Select an arbitrary (say at random or the first) $k \in K$. Output $t_j = ((5/8) + (k/4n)) \cdot 2^{-j}$.

By construction, $|t_j - a_i| \geq (1/8n) \cdot 2^{-j}$, for all i's. If $p_i \in [2^{-j-2}, 2^{-j+1}]$ then $|a_i - p_i| \leq (1/19n) \cdot 2^{-j}$ (with overwhelming probability), and it follows that p_i does not fall in the interval $t_j \pm (1/9n) \cdot 2^{-j}$ (because $|p_i - t_j| \geq |a_i - t_j| - |a_i - p_i| \geq ((1/8n) - (1/19n)) \cdot 2^{-j} > (1/9n) \cdot 2^{-j}$). Otherwise (i.e., if either $p_i < 2^{-j-2}$ or $p_i > 2^{-j+1}$), p_i does not fall in the interval $t_j \pm (1/9n) \cdot 2^{-j} \subset (2^{-j-2}, 2^{-j+1})$ (simply by the case hypothesis). We conclude that, with overwhelming probability, no p_i falls in the interval $t_j \pm (1/9n) \cdot 2^{-j}$.

Implementation of Step S1: Recall that the purpose of Step S1 is to generate an index $j \in \{0, 1, ..., n\}$ with distribution corresponding the probability that event E_j holds (for a random parallel run of the protocol), as well to determine the thresholds $t_1, ..., t_j$, and using these to determine for every $i = 1, ..., n$, whether or not $i \in T_j$ and whether or not $i \in T_{j-1}$. We thus start by generating a random run, and next determine all necessary objects with respect to it.

1. *Generating a* reference run: Simulate Steps 2–3 of the (parallel execution of the) protocol, while using (in Step 2) commitments to dummy values. Based on the verifier's decommitments in Step 3 (of the parallel execution), determine the set $I \subseteq [n]$ of sessions in which the verifier has properly decommitted.

2. *Determining the event E_j occuring in the reference run, as well as the sets T_j and T_{j-1}:*

   Case of empty I: Set $j = 0$ and $T_j = T_{j-1} = \emptyset$.

   Case of non-empty I: Set $t_0 = 1$ and $T_0 = \emptyset$. For $j = 1, ..., n$ do

   (a) $t_j \leftarrow T(j, n)$. (We stress that the value of t_j is set obliviously of I.)

   (b) Determine the set T_j by determining, for each i, whether or not $p_i > t_j$. We use approximations to each p_i (as computed in procedure $T(j, n)$), and rely on $|p_i - t_j| > (1/9n) \cdot 2^{-j}$. Recall that for each i, we obtain an approximation a_i such that $|a_i - p_i| < (1/9n) \cdot 2^{-j}$ if $2^{-j-2} \leq p_i \leq 2^{-j+1}$ and $a_i < 2^{-j-1} \leq t_j$ (resp., $a_i > 2^{-j} \geq t_j$) if $p_i < 2^{-j-2} < t_j$ (resp., if $p_i > 2^{-j+1} > t_j$). Thus, we may decide that $p_i > t_j$ if and only if $a_i > t_j$.

   (c) Decide whether or not event E_j holds for the reference run, by using T_{j-1} (of the previous iteration) and T_j (just computed). Recall that event E_j holds (for the reference run) if and only if both $I \subseteq T_j$ and $I \not\subseteq T_{j-1}$ hold.

(d) If event E_j holds then exit the loop with the current value of j as well as with the values of T_j and T_{j-1}. Otherwise, proceed to the next iteration.

Since we have assumed that $(\forall i)\ p_i > 2^{-n}$, some event E_j must hold.[20]

A key point in the analysis is that the values of the T_k's, as determined by Step S1 (i.e., $T_0, ..., T_j$), are independent of the value of j. Of course, which of the T_k's were determined does depend on the value of j. Thus, we may think of Step S1 as of an efficient implementation of the mental experiment in which all T_k's are determined, next j is determined accordingly (analogously to the above), and finally one outputs T_j and T_{j-1} for subsequent use.

3.3   A Detailed Description of the Simulator

For sake of clarity we present a detailed description of the simulator, before turning to its analysis. Recall that our aim is to simulate a parallel execution of n sessions of the GK-protocol. We start by selecting and fixing the verifier's coins at random. With respect to these fixed coins, we simulate the interaction of the *residual deterministic* verifier (with sessions of the predetermined prover) as follows:

Step S0: We simulate the parallel execution of Step 1 (i.e., Steps P0 and V1 of Construction 7) as follows. First, acting as the real prover in Step P0, we randomly generate messages $\alpha^1, ..., \alpha^n$ (one per each sessions). Invoking the verifier (as per Step V1), while feeding it with $\alpha^1, ..., \alpha^n$, we obtain its n commitments, $c^1, ..., c^n$, for the n sessions.

Step S1: As explained in Section 3.2, we determine (*for a random reference run*)[21] the index j for which E_j holds, as well as the sets T_j and T_{j-1}. Recall that this (and specifically procedure $T(\cdot, \cdot)$) involves $\mathrm{poly}(n) \cdot 2^j$ rewindings and re-simulations of Steps 2–3, while using commitments to dummy values. Each rewinding is performed as in Step S2 below.

In case $j = 0$, we may skip all subsequent steps, and just output the reference run produced in the current step.

Step S2: For each session $i \in T_j$, we wish to obtain the challenge (edge-sequence) committed to in Step 1, while working within time $\mathrm{poly}(n) \cdot 2^j$. This is done by rewinding and re-simulating Steps 2–3 (i.e., Steps P1 and V2 of Construction 7) for $\mathrm{poly}(n) \cdot 2^j$ times, while using commitments to dummy

---

[20] Removing this assumption enables the situation that no event E_j occurs. This may happen only if $p_i \le t_n < 2^{-n}$, for every $i \in I$. But the probability that the reference run corresponds to such a set I is at most $\sum_{i:p_i < 2^{-n}} p_i < n \cdot 2^{-n}$, and we may ignore this rare event. Alternatively, we may modify the verifier such that $p_i > 2^{-n}$ holds for all i, by making it properly decommit to all sessions with probability 2^{-n+1}, and note that the execution of the modified verifier is indistinguishable from the execution of the original verifier.

[21] Here and in the sequel, when referring to runs and steps of the protocol, we actually means steps in the parallel execution of the protocol.

values. (Actually, we may as well do the same for all i's (regardless whether $i \in T_j$ or not), but we are guaranteed to succeed only for i's in T_j. Furthermore, we may work on all i's at the same time.)

Specifically, each rewinding attempt proceeds as follows:

1. Generate n sequences of random (prover) commitments to a dummy value, say 0. That is, for every (session) $i = 1, ..., n$, select uniformly $r^i_{1,1}, ..., r^i_{n,t} \in \{0,1\}^n$, and compute $\bar{c}^i \overset{\text{def}}{=} (c^i_{1,1}, ..., c^i_{n,t})$, where $c^i_{k,\ell} = C(0, r^i_{k,\ell})$.

2. Feeding the verifier with (the n prover commitments) $\bar{c}^1, ..., \bar{c}^n$, obtain the verifier's n (Step 3) responses, denoted $(s^1, \bar{e}^1), ..., (s^n, \bar{e}^n)$.

3. For every properly decommitted session (i.e., i such that $c^i = \mathcal{C}_{\alpha_i}(s^i, \bar{e}^i)$), store the corresponding challenge (i.e., the edge sequence $\bar{e}^i$).

(Note that it is unlikely that we will obtain two conflicting proper decommitments to the same verifier commitment c^i.)[22]

Step S3: For technical reasons, analogously to [17], we next obtain a good approximation of $q_j = \Pr[E_j]$. This approximation, denoted $\tilde{q}_j$, will be obtained within expected time $\text{poly}(n)/\tilde{q}_j$ by repeated rewinding and re-simulating Steps 2–3 (i.e., Steps P1 and V2 of Construction 7). Specifically, we repeat the following steps until we obtain n^5 runs in which event E_j holds.

1. Perform Items 1 and 2 as in Step S2. Let I' denote the set of sessions in which the verifier has properly decommitted.

2. If I' fits event E_j (i.e., $I' \subseteq T_j$ and $I' \not\subseteq T_{j-1}$) then increment the "success counter" by one unit. (We proceed to the next iteration only if the "success counter" is still smaller than n^5.)

Suppose we have obtained n^5 successes while making τ trials. Then we set $\tilde{q}_j = n^5/\tau$.

Step S4: We now try to generate a simulation of Steps 2–3 of the protocol (i.e., Steps P1 and V2 of Construction 7) in which event E_j occurs. However, unlike in previous simulations, here we use (in Step 2) prover-commitments to values that pass the challenges that we have obtained. This will allow us to simulate also Step 4, and complete the entire simulation. Specifically, we make at most $\text{poly}(n)/\tilde{q}_j$ trials to rewind and re-simulate Steps 2–3, while using (in Step 2 of each session in T_j) commitments to values that pass the corresponding challenge (which we obtained in Step S2). Each attempt proceeds as follows:

1. Generate n sequences of random commitments to passing values (for sessions in T_j, and dummy values otherwise). Specifically, suppose that $i \in T_j$ (or more generally that we have obtained (in Step S2) a proper decommitment to c^i), and denote by $((u^i_1, v^i_1), ..., (u^i_t, v^i_t))$ the value of the decommitted challenge (edge sequence $\bar{e}^i$). Then, for every $\ell = 1, ..., t$, select uniformly $r^i_{1,\ell}, ..., r^i_{n,\ell} \in \{0,1\}^n$ and $a^i_\ell \neq b^i_\ell \in \{1,2,3\}$, and compute $c^i_{u^i_\ell,\ell} = C(a^i_\ell, r^i_{u^i_\ell,\ell})$, $c^i_{v^i_\ell,\ell} = C(b^i_\ell, r^i_{v^i_\ell,\ell})$, and $c^i_{k,\ell} = C(0, r^i_{k,\ell})$ for

---

[22] Unlike most probabilitistic statements in this section, the current statement refers to a probability space that contains also the possible (random) choice of $\alpha^1, ..., \alpha^n$.

$k \notin \{u_\ell^i, v_\ell^i\}$. Let $\bar{c}^i \stackrel{\text{def}}{=} (c_{1,1}^i, ..., c_{n,t}^i)$. For $i \notin T_j$ (or for i's for which we failed in Step S2), we produce $\bar{c}^i \stackrel{\text{def}}{=} (c_{1,1}^i, ..., c_{n,t}^i)$ as in (Item 1 of) Step S2.

2. Feeding the verifier with (the prover's commitments) $\bar{c}^1, ..., \bar{c}^n$, obtain the verifier's n (Step 3) responses, denoted $(s^1, \bar{e}^1), ..., (s^n, \bar{e}^n)$. Let $I' = \{i : \mathcal{C}_{\alpha_i}(s^i, \bar{e}^i) = c^i\}$ denote the set of sessions that have properly decommitted (in the current attempt). If I' does not fit event E_j (i.e., $I' \not\subseteq T_j$ or $I' \subseteq T_{j-1}$) then we abort this attempt. That is, we proceed only if I' fits event E_j.

3. For every properly decommitted session (i.e., $i \in I'$), we provide a proper decommitment (as in Step 4 of the protocol). This complete a full simulation of such a session, whereas improperly committed sessions are simulated by their transcript so far.

   Specifically, ignoring the rare case of conflicting proper decommitments, a proper decommitment to session $i \in I' \subseteq T_j$ must use the same challenge (edge sequence) as (found in Step S2 and) used in Item 1 (of the current attempt). Then, for every $i \in I'$ and $\ell = 1, ..., t$, we merely provide the 4-tuple $(r_{u_\ell^i, \ell}^i, a_\ell^i, r_{v_\ell^i, \ell}^i, b_\ell^i)$, where $((u_1^i, v_1^i), ..., (u_t^i, v_t^i))$ is the corresponding challenge. Indeed, this answer (like the prover's answer in Step 4 of the protocol) passes the verifier's check (since $a_\ell^i \neq b_\ell^i \in \{1, 2, 3\}$, $c_{u_\ell^i, \ell}^i = C(a_\ell^i, r_{u_\ell^i, \ell}^i)$, and $c_{v_\ell^i, \ell}^i = C(b_\ell^i, r_{v_\ell^i, \ell}^i)$).

In the rare case in which a conflicting proper decommitment is received, we proceed just as in case event E_j does not occur. If all $\text{poly}(n)/\tilde{q}_j$ trials fail then we output a special failure symbol.

For technical reasons, we modify the above simulation procedure by never allowing it to run more than 2^n steps. (This is easily done by introducing an appropriate step-count (which is implemented in linear or almost-linear time and so does not affect our running-time analysis).)

3.4   A Detailed Analysis of the Simulator

Lemma 8 (Simulator's running-time): *The simulator runs in expected polynomial-time.*

Proof: The key observation is that each repeated attempt to produce something is repeated for a number of times that is inversely proportional to the probability that we try this attempt at all. This reasoning is applied with respect to each of the main steps (i.e., Steps S1, S2, S3 and S4). Specifically:

- For Step S1: Recall that event E_j occurs in the reference run (generated at the onset of Step S1), with probability q_j. Letting $Q \stackrel{\text{def}}{=} T_j \setminus T_{j-1}$, we have $q_j \leq |Q| \cdot \max_{i \in Q}\{p_i\} \leq n \cdot t_{j-1} < n \cdot 2^{-(j-1)}$. Also, with probability at least $1 - 2^{-n}$, Step S1 correctly determines j. Pending on the latter (overwhelmingly high probability) event, the expected number of steps conducted in

Step S1 is

$$\sum_{j=0}^{n} q_j \cdot (\text{poly}(n) \cdot 2^j) < \sum_{j=0}^{n} (n \cdot 2^{-(j-1)}) \cdot (\text{poly}(n) \cdot 2^j) = \text{poly}(n) \quad (3)$$

Relaying on the fact that the simulator never runs for more than 2^n steps, we cover also the highly unlikely case (in which Step S1 determines a wrong j).

The same reasoning applies to Step S2. That is, again assuming that Step S1 correctly determines j, the expected number of steps made in Step S2 is as in Eq. (3).

- For Step S3: Assuming that $\tilde{q}_j = \Theta(q_j)$, the expected number of steps made in Step S3 is $\sum_{j=0}^{n} q_j \cdot (\text{poly}(n)/\tilde{q}_j) = \text{poly}(n)$. The above assumption holds with probability at least $1 - 2^{-n}$, and otherwise we rely on the fact that the simulator never runs for more than 2^n steps. The same reasoning applies to Step S4.

Thus, the overall expected running-time is polynomial (and this is proven without relying on any security properties of the commitment schemes). ∎

Lemma 9 (Simulator's output distribution): *Assume that the verifier's commitment scheme (i.e., $\mathcal{C}$) is computationally-binding and that the prover's commitment scheme (i.e., C) is computationally-hiding. Then the output of the simulator is computationally indistinguishable from the real parallel interaction.*

Recall that the assumption that $\mathcal{C}$ is perfectly-hiding and C is perfectly-binding is used in establishing (cf. [17, Sec. 4]) the soundness of the GK-protocol (as a proof system).

Proof: For sake of clarity of the analysis, one may consider an imaginary simulator that goes on to determine all t_j's (rather than determining only part of them as in Item 2 of Step S1). We may assume that all approximations made by the simulator are sufficiently good; that is, in Step S1 the simulator correctly determines j as well as T_j and T_{j-1}, and in Step S3 it obtains $\tilde{q}_j = \Theta(q_j)$. (Indeed, the assumption holds with probability at least $1 - 2^{-n}$.)

Next, we consider three unlikely events in the simulation:

1. In Step S2, the simulator fails to obtain a proper decommitment of some $i \in T_j$. This may happen only with exponentially vanishing probability, because we keep trying for $\text{poly}(n) \cdot 2^j$ times and each time a proper decommitment (for i) occurs with probability $p_i > t_j \geq 2^{-(j+1)}$.

2. In Step S4, the simulator fails to generate a simulation in which event E_j holds. We will show that this failure may happen only with negligible probability. Note that in order for this failure to occur, it must be that event E_j occurs in Step S1 but does not occur in the $\text{poly}(n)/\tilde{q}_j = O(\text{poly}(n)/q_j)$ trials conducted in Step S4, although event E_j may occur in each such trial with probability q'_j that is negligibly close to q_j. (Recall that q_j refers to the

probability that event E_j occurs for a "dummy" commitment, whereas q'_j refers to its probability for a "passing" commitment, and $|q_j - q'_j|$ is negligible because C is computationally-hiding (cf. [17, Clm. 3]).) Thus, the probability of this failure is upper-bounded by

$$\sum_{j=0}^{n} q_j \cdot (1 - q'_j)^{\text{poly}(n)/q_j} \tag{4}$$

Letting $\Delta_j \stackrel{\text{def}}{=} |q_j - q'_j|$, we consider two cases (cf. [17, Clm. 2]): in case $\Delta_j \leq q_j/2$, the corresponding term is exponentially vanishing (because $q'_j \geq q_j/2$ and $(1 - (q_j/2))^{2n/q_j} < \exp(-n)$), whereas in case $\Delta_j \geq q_j/2$ we simply bound the corresponding term by $q_j \leq 2\Delta_j$. Thus, in both cases, we obtain that each term in Eq. (4) is negligible (because it is upper-bounded by $\max(2\Delta_j, \exp(-n))$). Noting that Eq. (4) refers to the sum of $n+1$ such terms, the claim follows.

3. In Step S4, the simulator obtains a proper decommitment to some session such that the decommitted value is different from the one obtained for the same session in Step S2. (In such a case, the simulator may end-up outputting a failure symbol.) However, the hypothesis that C is computationally-binding implies that this bad event occurs only with negligible probability.

We conclude that, except with negligible probability, the simulator produces an output that looks syntactically fine. Needless to say, this is not enough: we need to prove that the simulator's output distribution, denoted S_n, is computationally indistinguishable from a random transcript of the real interaction, denoted R_n. The argument is analogous to the proof of [17, Clm. 4], but we present it a little differently based on an idea of Vadhan [31].

Intuitively, the computational indistinguishability of S_n and R_n should follow from the hypothesis that the commitment scheme C is computationally-hiding. The question is how exactly to transform a distinguisher of S_n and R_n into an algorithm that violates the hiding property of C. The presentation in [17, Clm. 4] takes the standard approach (which can be traced to [19]) of breaking the analysis into two cases that refer to some relevant events (such as our E_j's): in the first case one of these events occurs with significantly different probabilities in S_n and R_n, and in the second case each of these events occurs with essentially the same probability in S_n and R_n. The first case is easy to handle (i.e., in this case one can easily derive an algorithm that violates the hiding property of C), but the second case involves more work. Specifically, in the second case, one considers the conditional distributions of S_n and R_n subject to such an event (e.g., E_j) that occurs with noticeable probability, and uses the simulator to derive an algorithm that violates the hiding property of C. The latter derivation uses implicitly a hybrid simulator, which we shall discuss next. Vadhan [31, Sec. 2.2.3] suggests to *explicitly* introduce and analyze such a hybrid simulator.

The hybrid simulator is a mental experiment. It is given a 3-coloring of the input graph, and thus has no problem to emulate the real prover in a straightforward manner. However, the hybrid simulator acts as the real simulator, except that (in all steps) it uses commitments to (random relabelings of) the 3-coloring instead of commitments to dummy or to passing values (as used by the real simulator (in Steps S1-S3 and S4, respectively)). For the sake of clarity, we postulate that in case of conflicting verifier decommitments (as in the foregoing Item 3) the hybrid simulator also outputs a failure sybmol. We claim that the output of the hybrid simulator, denoted H_n, is indistinguishable from both S_n and R_n.

H_n versus R_n: Consider a modification of the hybrid simulator in which Step S2 and S4 are repeated indefinitely until they are successful (rather than being repeated $\text{poly}(n) \cdot 2^j$ and $\text{poly}(n)/\tilde{q}_j$ times, respectively). Then the output distribution of this *modified* hybrid simulator, denoted H'_n, is statistically close to R_n, where the statistical difference is due to conflicting verifier decommitments (as in Item 3). Specifically, event E_j occurs in Step S1 with exactly the same probability as in a real interaction, and (conditioned on not failing due to conflicting verifier decommitments) the conditional distribution in Step S4 is identical to the corresponding distribution in R_n. By the foregoing Items 1 and 2, the output of the hybrid simulator (i.e., H_n) is statistically close to the output of the modified hybrid simulator (i.e., H'_n), and thus H_n is statistically close to R_n.

H_n versus S_n: Recall that the hybrid simulator differs from the real simulator only in the prover commitments that it utilizes. Thus, intuitively, if H_n and S_n are computationally distinguishable then we can distinguish commitments with respect to the commitment scheme C. Indeed, combining the simulator with the said distinguisher, we obtain an algorithm that runs in expected polynomial-time and distinguishes commitments to 3-coloring of the graph from commitments to dummy and/or passing values. By truncating excessively long runs of the latter algorithm, we obtain a distinguisher that runs in strict probabilistic polynomial-time and maintains a non-negligible distinguishing gap. This distinguisher needs to get the said 3-coloring as auxiliary input, yielding a (non-uniform) family of polynomial-size distingiushing circuits, in violation of the computationally hidding property (as discussed in §1.4).

The lemma follows. ∎

Parenthetical Comment: Indeed, we wish to seize the opportunity and call the reader's attention to the elegant presentation technique suggested by Vadhan in [31, Sec. 2.2.3].

Combining Lemmas 8 and 9, we obtain

Theorem 10 *The* (constant-round) *GK-protocol is zero-knowledge under parallel composition.*

Recall that the GK-protocol is a proof system for $\mathcal{NP}$ (with exponentially vanishing soundness error) [17]. Thus, assuming the existence of claw-free pairs of functions, we have established the existence of *constant-round proof systems for $\mathcal{NP}$ that is zero-knowledge under parallel composition.*

Parenthetical Comment: Note that the foregoing simulator and its analysis hold also if we set $t = 1$ in Construction 7. But *under this setting of parameters*, Construction 7 only constitutes a weak type of interactive proof that rejects false assertions (only) with noticeable probability. Still, executing this protocol in parallel, for an adequate polynomial number of times, yields an alternative constant-round zero-knowledge proof system for $\mathcal{NP}$. Needless to say, the proof of the latter assertion is more complex than the analysis of the GK-protocol in the stand-alone setting (and in fact builds upon it).

3.5   An Extension

We relax the parallel execution condition to concurrent execution of polynomi-ally-many sessions (of the GK-protocol) that satisfy the following two conditions:

C1: No session enters Step 2 before all sessions complete Step 1.
C2: No session enters Step 4 before all sessions complete Step 3.

In other words, the concurrent execution proceeds in three phases:

Phase 1: All sessions perform Step 1 (in arbitrary order).
Phase 2: All sessions perform Steps 2 and 3 (in arbitrary order except for the obvious local timing condition (i.e., each session performs Step 3 after it has completed Step 2)).
Phase 3: All sessions perform Step 4 (in arbitrary order).

Our treatment of parallel executions extends to the above (concurrent) case. The reason being that the simulator treats Steps 2–3 as one unit, and so the fact that these steps may be interleaving among sessions is of no importance. Specifically, Step S0 of the extended simulator refers to Phase 1 (rather than to Step 1 of the protocol), its Steps S1–S3 refer to Phase 2 (rather than to Steps 2–3), and its Step S4 refers to Phases 2–3 (rather than to Steps 2–4).

4   Simulator for the Case of Bounded-Simultaneity

Recall that the GK-protocol proceeds in four (abstract) steps:

1. The verifier commits to a challenge (i.e., Steps (P0) and (V1) in the protocol).
2. The prover commits to a sequence of values (i.e., Step (P1) in the protocol).
3. (Step (V2):) The verifier decommits (either properly or not).
4. (Step (P2):) Pending on the verifier's proper decommitment, the prover de-commits to the corresponding values.

Here we consider (say n) concurrent executions in which up-to w sessions of the GK-protocol run simultaneously at any given time, where w may be any fixed constant.

4.1   Motivation

The case of $w = 1$ corresponds to sequential composition, and it is well-known that any zero-knowledge protocol maintains its security in this case. So let us turn (as a warm-up) to the case of $w = 2$. Trying to use the single-session simulator of [17] in this case, we encounter the following problem: when we try to deal with the simulation of one session (by using the single-session simulator), the verifier may invoke another session. A natural thing to do is to apply the single-session simulator also to the second session. The good news is that the verifier cannot initiate yet another session (before it terminates either the first or second session, because this would violate the bounded-simultaneity condition (for $w = 2$)). Instead, eventually (actually, in a few steps), one of two things will happen (first):

1. The verifier may execute Step 3 in the *second* session, in which case we make progress on treating the second session (towards completing a simulation of it, which would put us back in the one-session case).
2. Alternatively, the verifier may execute Step 3 in the *first* session, in which case we make progress on treating the first session. For example, if we were trying to get the decommitment value for the first session and we just got it, then we may abandon the treatment of the second session and proceed by rewinding the first session. (Note that in this case we lost all work done in the current simulation of the second session.) Similarly, if we were trying to simulate the full run of the first session then we just obtained one additional trial at a proper decommitment for Step 3 (which eventually will allow us to complete the simulation of the first session).

Thus, in each of these cases, we make progress. Intuitively, the cost of dealing with two simultaneous sessions is that we may have to invoke the single-session simulator (for the second session) per each operation of the single-session simulator (for the first session). As will be shown below, the above intuition remains valid also when we handle polynomially-many sessions such that at most two are running simultaneously. Furthermore, it extends also to the case that at most w sessions are running simultaneously, where w is any fixed constant. In that case, at most w sessions of the single-session simulator will be active at any point during the simulation. Specifically, each operation in the emulation of the i-th session will require invoking the single-session simulator (for simulating the $i + 1$st session). Thus, the time-complexity of the simulation will be exponential in w, where the base of the exponent is the time-complexity of the case where $w = 1$.

4.2   The Actual Simulation

We start with a high level description of the simulation, next provide detailed specification and implementation of the procedures used by the simulator, and finally analyze them. Throughout the rest of the description we fix a (deterministic) adversarial verifier (and use black-box access to it).

A High Level Description. In correspondence to the three main steps of the single-session simulator (cf. Section 2.3), we introduce three *recursive procedures*: Scan, Approx and Generate. Each of these procedures tries to handle a single session (just as done by the corresponding step of the single-session simulator), while making recursive calls when encountering a Step 2 message of some other session.[23] The recursive call will take place before executing this Step 2, and the execution of this Step 2 will be the first thing that the invoked procedure will do. The procedure *terminates* either upon completion of the task for which it was invoked (i.e., scanning or generating the transcript of the current session) or before doing so (e.g., reaching a problematic situation or completing the task for which a "ancesstor" recursive call was invoked). Note that encoutering Step 2 of some other session will cause any of these procedures to make a recursive call, whereas other steps of other sessions may be handled by these procedures themselves.

Here and throughout the description, when we say that a procedure encounters some step, we mean that this step is scheduled by the adversary (based on the simulation transcript). Formally, Steps 1 and 3 are determined by feeding the adversary with the current simulation transcript, and using its response (which is always either a Step 1 or a Step 3 action of some session). The corresponsing Steps 2 and 4 (which are prover actions) always follow immediately, but our description does not use this fact.

Before proceeding, let us recall the main steps of the *single-session simulator*, and slightly modify them to provide a more convenient basis for our generalization. In particular, in this modification, Step 1 (of the protocol) is simulated separately (rather than as part of the Scan Step), and the Generation Step is used also in case the Scan encountered an improper decommitment. The resulting simulation steps are as follows:

A straightforward simulation of Step 1: The simulator emulates Step 1 of the protocol by obtaining the verifier's commitment (of Step (V1), after emulating Step (P0) in a straightforward manner).

The Scan Step: The simulator emulates Steps 2–3 of the protocol by using commitments to dummy values in Step 2, and obtains the verifier's decommitment for Step 3, which may be either proper or improper. We call this proper/improper bit the type of the decommitment. The simulator records the type of the decommitment as well as the decommitment information in case of proper decommitment.

The Approximation Step: The simulator approximates the probability that a single scan (as performs in the Scan Step) ends with a decommitment of the recorded type. (This is done by repeated trials, each as in the Scan Step, until some polynomial number of decommitments of the recorded type is encountered.)

---

[23] This is no typo; we do mean Step 2, not Step 1. But indeed, being a prover step, Step 2 of a session is encountered immediately after the execution of the corresponding Step 1, which in turn is scheduled by the adversary.

The Generation Step: Using the decommitment information obtained in the Scan Step, the simulator repeatedly tries to generate a full transcript of the same type as encountered in the Scan Step. It does so by emulating Steps 2–4, using commitments to "pseudo-colorings" that do not "violate the coloring conditions imposed by the decommitted edges" (in case the Scan Step ended with a proper decommitment, and using commitments to dummy values otherwise). The number of trials is inversely proportional to the probability estimated in the approximation step, and if all fail then the simulator outputs a special failure symbol.

Analogously, the recursive procedures Scan, Approx and Generate, operate as follows, where the straightforward simulation of Step 1 (of each session) is performed "en route" (by one of these procedures, while handling a different session):

The Scan *procedure* is invoked to emulate Steps 2–3 of a certain session that is scheduled to perform Step 2 at the current point (i.e., just following the current "simulation transcript"), provided that the current "simulation record" contains no trace of a prior handling of Step 2 of this session. The procedure first emulates Step 2 of the said session by using commitments to dummy values, and the hope is that it will reach Step 3 of the current session and obtain the verifier's decommitment for this session, which may be either proper or improper. When this happens, the procedure returns the relevant information (i.e., the decommitment value in case of proper decommitment and a special symbol in case of improper decommitment). However, other things may happen (due to the other sessions, scheduled for action by the adversary):

- The procedure may encounter Step 1 of *some other session*, in which case it emulates it in a straightforward manner (which results in augmenting the simulation transcript). Next, the procedure continues handling the current session.
- The procedure may encounter Step 2 of *some other session*, in which case it invokes either Generate or Scan to handle this other session, depending on whether or not our current simulation record contains a trace of a prior handling of Step 2 of that session. We stress that the invoked procedure may return an answer that refers to a session that is not the one for which the procedure was invoked (i.e., the session to which the currently encountered Step 2 belongs). Following is a description of what the procedure does with the answer provided to it by the procedure it invokes, which indeed depends on which procedure was invoked.
    - When encountering a Step 2 of another session (denoted j) that was not handled before, we invoke Scan, and handle the answer (of Scan) according to whether or not it refers to session j. In the case that the answer relates to session $k \neq j$ (which includes the case that k equals the current session) we return the relevant information (as when we encounter Step 3 of the current session), otherwise (i.e.,

$k = j$) we record the information and continue (as when handling other steps of other sessions). In the latter case, we will next execute the following sub-case (which refers to the very same Step 2 (i.e., of session j)). We stress that, regardless of the answer of Scan, we do not extend the simulation transcript in the current sub-case (and thus, for $k = j$, an execution of the following sub-case referring to session j will necessarily follow the execution of the current sub-case).

- When encountering a Step 2 of another session that was already handled before, we repeatedly invoke Generate, until it either succeeds or an adequate number of trials was performed, and handle the answer (of Generate) as follows. If the answer provides an extension of the simulation transcript, we continue handling the current session using that transcript. Otherwise (e.g., the answer is a decommitment information of yet some other session) then we terminate returning this very answer.

(Indeed, Generate corresponds to a single trial of the Generation Step, and the repeated attempts are done by the procedure that invokes it.)

- The procedure may encounter Step 3 of *some other session*, which may happen when Step 2 of that session was handled by an invocation that preceded the current one in the recursion path. Again, the action depends on whether or not our current simulation record contains information regarding a prior handling of Step 3 of that session.
  - If no such prior handling exists (for this session) then the procedure returns the corresponding decommitment information (although it is not the session for which the current execution was invoked).
  - If such prior handling exists and the current emulation of Step 3 fits its type then the procedure augments the simulation transcript and continues handling the current session. If the type does not fit then the procedure returns a special failure symbol.

- The procedure may encounter Step 4 of *some other session*, which may happen when Step 2 of that session was handled by an invocation that preceded the current one in the recursion path. Furthermore, in that case the recorded information allows to emulate this step in a straightforward manner, and Scan continues handling the current session (after augmenting the simulation transcript).

Indeed, two key notions referred to above are the simulation transcript and the simulation record. The former is a prefix of a full transcript (of an execution) being generated by the simulator, and the latter provides auxiliary information regarding that (partial) transcript. In particular, the record contains information regarding sessions that appear in the transcript, where this information was obtained in previous invocations of various procedures on prefixes of this transcript. For example, a successful Scan returns information regarding the decommitment of a certain session.

The Approx procedure is invoked to approximate the probability that a certain invocation of Scan returns a certain value (i.e., the identity of the decommitting session and the type (i.e., proper or improper) of that decommitment).

This is done by repeated trials, where in each trial the procedure behaves similarly to Scan, until a sufficient number of trials return the value of interest.

The Generate *procedure* is invoked to emulate Steps 2–4 of a certain session that is scheduled to perform Step 2 at the current point, provided that the current "simulation record" contains information regarding a prior handling of Step 2 of this session (i.e., by Scan). The procedure behaves like Scan except that it emulates Step 2 using commitments to passing values (i.e., values that would pass w.r.t the corresponding proper decommitment, or arbitrary values in case the corresponding decommitment is improper). The hope is that the procedure will reach Step 4 of the current session, and that the verifier's behavior at the corresponding Step 3 fits the recorded information. When this happens, the procedure emulates these steps in a straightforward manner (relying on the fact that a proper decommitment yields a challenge that can be met by the "passing values" used in emulating Step 2). Once the emulation of Step 4 is completed, the procedure returns the corresponding simulation transcript. However, as in case of Scan, other things may happen:

- The procedure may encounter steps of other sessions. These are handled as in Scan.
- In addition, it may happen that Step 3 of the current session decommits differently than in the simulation record (i.e., differently with respect to the proper/improper bit). In this case, the procedure returns a special failure symbol.

As mentioned above, the three procedures maintain (and pass along) the state of the currently handled sessions as well as related auxiliary information. In particular, $\overline{h}$ will denote a partial transcript of the (concurrent) execution, and $\overline{a}$ will denote a corresponding list of currently active sessions together with auxiliary information regarding each of them (e.g., decommitment information obtained in previous related runs). For sake of clarity, although the the identity of the session that is responsible for the current procedure call (i.e., the session that encountered Step 2) is implicit in $\overline{h}$, we pass this identity explicitly. The (simulator's) main program merely consists of a special invocation of Generate with empty history (i.e., $\overline{h} = \overline{a} = \lambda$).

The Specification of the Procedures. Let us first elaborate on the structure of the auxiliary information $\overline{a}$, which consists of records, each corresponding to some encountered session of the protocol. The record corresponding to session i consists of three fields:

1. The verifier decommitment field (of session i) indicates whether the first encounter of Step 3 (i.e., the verifier's decommitment) of session i was proper or improper (i.e., the type of decommitment), and in the former case the field includes also the value of the decommitment. That is, if non-empty,

the field stores a pair (X, v), where $X \in \{\texttt{proper}, \texttt{improper}\}$ is a decommitment type and v is a decommitment value (which is meaningful only in case $X = \texttt{proper}$). This field (of the record of the ith session) is filled-up according to the answer returned by some invocation of $\texttt{Scan}(\overline{h}, \cdot, i)$.

2. The **decommitment probability field** (of session i) holds an approximation of the probability that an invocation (with parameters as the one that filled-up the first field) actually turns out returning same type. That is, suppose that the first field of record i (i.e., the record of the ith session) was filled-up according to the answer returned by $\texttt{Scan}(\overline{h}, \overline{a}, i)$, which resulted with a decommitment of type $X \in \{\texttt{proper}, \texttt{improper}\}$. Then the second field of record i should hold an approximation of the probability that $\texttt{Scan}(\overline{h}, \overline{a}, i)$ returns with an answer that encodes the same type of decommitment of session i. (Jumping ahead, we hint that $\texttt{Scan}(\overline{h}, \overline{a}, i)$ may return with a decommitment to some other session (or with failure), and so the sum of the two probabilities corresponding to the two types is not necessarily 1.)

3. The **prover decommitment field** (of session i) encodes the decommitment information corresponding to the prover's commitment in Step 2. This field (of the record of the ith session) is filled-up at the up-front of the execution of $\texttt{Generate}(\overline{h}, \overline{a}', i)$, which follows the invocation of $\texttt{Scan}(\overline{h}, \overline{a}, i)$, where $\overline{a}'$ is $\overline{a}$ augmented by the verifier decommitment information of session i and the prover's commitment is performed so to passed the latter.

As hinted above, the fields are filled-up in the order they appear above (i.e., the verifier decommitment field is filled-up first). In reading the following specifications, it may be instructive to consider the special case of a single session (in which case failure never occurs and $j = i$ always holds).

Specification of $\texttt{Scan}(\overline{h}, \overline{a}, i)$: This call produces a prefix of a "pseudorandom" execution transcript that extends the prefix $\overline{h}$, and *returns some related information*. The transcript is *pseudorandom* in the sense that it is computationally indistinguishable from a (prefix of a random) real continuation of $\overline{h}$ (by the adversary interacting with sessions of the prover).[24] The extended transcript is truncated (i.e., the extended prefix ends) at the first point where one of the following holds:

1. **Progress:** This is a case where the (extended) execution reaches Step 3 of some session j (possibly but not necessarily $j = i$) such that the first field of record j is empty. In this case, *the procedure should return the index j as well as the decommitment information* (provided in the current execution of Step 3 of session j). That is, the answer is a pair (j, y), where j is an index of a session and y is a decommitment information (which may be either proper or improper).

---

[24] The reader may wonder as to what will happen in case $\overline{h}$ itself is not consistent with any prefix of such a real interaction. The answer is that the extended execution will always be truncated before this fact becomes evident (i.e., we never perform Step 4 of a session unless Step 2 of that session was performed using commitments to passing values).

2. Failure: This is a case where the (extended) execution reaches Step 3 of some session $j \neq i$ such that the first field of record j encodes a decommitment type different than the one occuring in the current extension. That is, the first field of record j encodes decommitment type $X \in \{\texttt{proper}, \texttt{improper}\}$, whereas in the current execution Step 3 of session j has a decommitment type different from X (i.e., opposite to X). In this case, the procedure cannot continue (and should return a special $\texttt{failure}$ symbol).

(In contrast, in case the execution reaches Step 3 of some session $j \neq i$ such that the first field of record j encodes a decommitment type that equals the one occuring in the current execution, the procedure may continue handling session i.)

Furthermore, $\texttt{Scan}$ should make progress with overwhelmingly high probability (equivalently, should fail only with negligible probability).

Specification of $\texttt{Approx}(\overline{h}, \overline{a}, X, i)$*:* Always returns an approximation of the probability that $\texttt{Scan}(\overline{h}, \overline{a}, i)$ answers with a pair (i, y) such that y has type $X \in \{\texttt{proper}, \texttt{improper}\}$. The approximation is required to be correct to within a factor of 2, with probability at least $1 - 2^{-n}$.

Specification of $\texttt{Generate}(\overline{h}, \overline{a}, i)$*:* This call produces a prefix of a pseudorandom execution transcript that extends the prefix $\overline{h}$, and *returns either this extension or related information.* The notion of pseudorandom is the same as in case of $\texttt{Scan}$, and the extended transcript is truncated at the first point where one of the following holds:

1. Failure: Exactly as in the specification of $\texttt{Scan}$, except that here $j = i$ is possible too.
2. Progress: Here there are two sub-cases:
   (a) This is a case where the (extended) execution reaches Step 3 of some session j such that the first field of record j is empty. This sub-case is handled exactly as the Progress Case of $\texttt{Scan}$. (Unlike in $\texttt{Scan}$, here $j = i$ cannot not possibly hold.)
   (b) This is a case where the (extended) execution reaches Step 4 of session i. In this case, the procedure returns the currently extended transcript (including the execution of Step 4 of session i), along with a corresponding update to the auxiliary information $\overline{a}$.

Furthermore, $\texttt{Generate}$ should make progress with probability that is at most negligibly smaller than the probability approximated by the corresponding $\texttt{Approx}$-call. Thus, unlike in the presentation of the single-session simulator, here $\texttt{Generate}$ does not make progress almost always (not even in the case of a single session), but rather makes progress with probability that is close to the one approximated by the corresponding $\texttt{Approx}$-call. That is, $\texttt{Generate}$ is actually a generation-attempt, and the repetition of this attempt is made by the higher level invocation (rather than in the procedure itself).

The Implementation of the Procedures. We refer to the notion of a *passing commitment* as defined and used in Section 3. Recall that a passing commitment is a sequence of (prover's) commitments to values that pass for the corresponding challenge (encoded in the first field of the corresponding session): See Footnote 16.

We start with the description of Generate (although Generate$(\cdot, \cdot, i)$ is invoked after Scan$(\cdot, \cdot, i)$). We note that Generate$(\overline{h}, \overline{a}, i)$ is always invoked when the first field in the ith record in $\overline{a}$ is not empty (but rather encodes some decommitment, of arbitrary proper/improper type), and the third field is empty (and will be filled-up at the very beginning of the execution).

Procedure Generate$(\overline{h}, \overline{a}, i)$*:* Initializes $\overline{h}' = \overline{h}$ and $\overline{a}' = \overline{a}$, generates a passing commitment for (Step 2 of) session i, and augments $\overline{h}'$ and $\overline{a}'$ accordingly. Specifically:

1. The procedure generates a random sequence of values, denoted $\overline{v}$, that pass the challenge described in the first field of the ith record of $\overline{a}$. That is, $\overline{v}$ may be arbitrary if the said field encodes an improper decommitment; but in case of proper decommitment, $\overline{v}$ must pass with respect to the challenge value encoded in that field.
2. The procedure generates a random sequence of (prover's) commitments, denoted $\overline{c}$, to $\overline{v}$, augments $\overline{h}'$ by $\overline{c}$, and augments $\overline{a}'$ by placing the corresponding decommitment information in the third field of the ith record.

Next, the procedure proceeds in iterations according to the following cases that refer to the next step taken in the concurrent execution (as determined by the adversary).

Step 1 by some (new) session: Just augment $\overline{h}'$ accordingly (and proceed to the next iteration).

Step 2 by some session j (certainly $j \neq i$)**:** We consider two cases depending on whether or not $\overline{a}'$ contains the verifier's decommitment information for session j (i.e., whether or not the first field of the jth record is non-empty).

    1. In case $\overline{a}'$ does contain such information, we generate a corresponding passing commitment (i.e., a prover commitment to values that pass w.r.t challenge encoded in the first field of the jth record), augment $\overline{h}'$ and $\overline{a}'$ accordingly, and proceed to the next iteration. (Specifically, analogously to the up-front activity for (Step 2 of) the ith session, the third field in the jth record of $\overline{a}'$ is augmented by the decommitment information corresponding to this prover commitment, and $\overline{h}'$ is augmented by the commitment itself.)

    2. The case in which $\overline{a}'$ does not contain such information (i.e., the first field of the jth record is empty (and certainly $j \neq i$)), is the most involved part of the procedure. In this case, we proceed as follows:

       (a) We invoke Scan$(\overline{h}', \overline{a}', j)$, and consider its answer, which is either failure or a *progress pair* (k, y). In case of progress, we determine the type $X \in \{\texttt{proper}, \texttt{improper}\}$ of the decommitment information y (with respect to the corresponding Step 1 commitment in $\overline{h}'$).

(b) If the answer is either $\texttt{failure}$ or is a progress pair (k, y) with $k \neq j$ then we $\texttt{return}$ with the very same answer (i.e., either $\texttt{failure}$ or (k, y)).

(Here, in case of progress, $k \neq i$ must hold.)

(c) We reach this step only if the answer obtained from $\texttt{Scan}$ is a progress pair (k, y) with $k = j$. Letting X be the type of y, we let $\tilde{q} \leftarrow \texttt{Approx}(\overline{h}', \overline{a}', X, j)$, and update the jth record of $\overline{a}'$ placing (X, y) in the first field and $\tilde{q}$ in the second field. (Actually, it suffices to place (X, v) in the first field, where v is the decommitment value included in the decommitment information y.)

(We comment that in case $X = \texttt{improper}$, we could have skipped all subsequent sub-steps, and used instead the extended transcript generated by the above invocation of $\texttt{Scan}$, provided that $\texttt{Scan}$ were modified to return this information as well. However, avoiding this natural modification makes the extension in Section 5 more smooth.)

(d) Next, we repeatedly invoke $\texttt{Generate}(\overline{h}', \overline{a}', j)$ *until getting a progress, but not more than* $\text{poly}(n)/\tilde{q}$ *times.* (We will show that only with negligible probability can it happen that all calls return $\texttt{failure}$.) If all attempts have returned $\texttt{failure}$ then we $\texttt{return failure}$, otherwise we act according to the following sub-cases of the progress answer (of $\texttt{Generate}(\overline{h}', \overline{a}', j)$ as specified in §4.2), where the progress may be either a decommitment pair or an extended transcript:

  i. If the progress answer (of $\texttt{Generate}(\overline{h}', \overline{a}', j)$) provides a pair (k', y') (where certainly $k' \neq j$ as well as $k' \neq i$), then (analogously to sub-step 2b) we $\texttt{return}$ with the very same answer (k', y').

  ii. If the progress answer (of $\texttt{Generate}(\overline{h}', \overline{a}', j)$) provides an updated history $\overline{h}''$ (together with updated auxiliary information $\overline{a}''$) then update $\overline{h}'$ and $\overline{a}'$ (i.e., $\overline{h}' \leftarrow \overline{h}''$ and $\overline{a}' \leftarrow \overline{a}''$), and proceed to the next iteration. (Note that in this case $\overline{h}''$ ends with execution of Step 4 of session j.)

Note that in handling this case, we provide a full handling of session j, invoking all three procedures. Indeed, this handling is analogous to the single-session simulator.

Step 3 by session i: Just as the first sub-case in the next case (i.e., Step 3 by some session $j \neq i$ with a non-empty first field).

Step 3 by some session $j \neq i$: We consider two cases depending on whether or not $\overline{a}'$ contains the verifier's decommitment information for session j (i.e., the first field of the jth session is not empty).

  1. In case $\overline{a}'$ does contain such information, we consider sub-cases according to the relation of the contents of the the first field of the jth session, denoted $(X, \cdot)$, and the current answer of the verifier.

(a) If the decommitment type of the current Step 3 (of the jth session) fits X then we just augment $\overline{h}'$ accordingly (and proceed to the next iteration).

(b) Otherwise (i.e., the decommitment type of the current Step 3 does not fit X), `return failure`.

2. In case $\overline{a}'$ does not contain such information (i.e., the first field of the jth session is empty), obtain the relevant decommitment information from the adversary (it may be either an improper or proper decommitment), and `return` (as progress) with this information only. That is, `return` with (j, y), where y encodes the decommitment information just obtained from the adversary.

Step 4 by some session j (possibly $j = i$): We will show that, except with negligible probability, this step is reached only in case the corresponding (Step 2) prover commitment is passing and $\overline{a}'$ contains the corresponding decommitment (in the third field of the jth record). Using the latter prover's decommitment information, we emulate Step 4 in the straightforward manner (and augment $\overline{h}'$ accordingly). In case $j = i$, `return` with the current $\overline{h}'$ and $\overline{a}'$ (otherwise proceed to the next iteration).

Note that Step 2 of session i is handled up-front. In case of a single session i, the above procedure degenerates to the basic handling of Steps 2–4 of session i. In the fictitious invocation of `Generate` by the main program (i.e., with empty $\overline{h}$ and a fictitious i), only the handlings of Steps 2–4 for sessions $j \neq i$ are activated (whereas, in handling Step 2, sub-steps 2b and 2(d)i are never activated). We now turn to procedure `Scan`, which is similar to `Generate`, except for its handling of the steps of session i.

Procedure `Scan`$(\overline{h}, \overline{a}, i)$: Initializes $\overline{h}' = \overline{h}$ and $\overline{a}' = \overline{a}$, generates a dummy commitment for (Step 2 of) session i, and augments $\overline{h}'$ accordingly. (Specifically, the procedure generates a random sequence of commitments, $\overline{c}$, to dummy values, and augments $\overline{h}'$ by $\overline{c}$.) Next, the procedure proceeds in iterations according to the following cases that refer to the next step taken in the concurrent execution.

Step 1 by some (new) session: As in `Generate`.

Step 2 by some session j (certainly $j \neq i$): As in `Generate`.
  (We comment that unlike in sub-step 2b of `Generate`, here $k = i$ is possible.)

Step 3 by session i: Obtain the relevant decommitment information from the adversary (it may be either an improper or proper decommitment), and `return` (as progress) with this information. That is, `return` with (i, y), where y encodes the decommitment information just obtained from the adversary.

Step 3 by some session $j \neq i$: As in `Generate`.

Step 4 by some session $j \neq i$: As in `Generate`.

Note that we never reach Step 4 of session i (and that Step 2 of session i is handled up-front).

Procedure $\mathtt{Approx}(\overline{h}, \overline{a}, X, i)$: This procedure merely invokes $\mathtt{Scan}(\overline{h}, \overline{a}, i)$ until it obtains $m = \mathrm{poly}(n)$ invocations that return a pair that is a decommitment of type X for session i, and returns the fraction of m over the number of trials. Specifically, the procedure proceeds as follows:

Set $\mathtt{cnt}_{\mathrm{total}} = \mathtt{cnt}_{\mathrm{succ}} = 0$.
Until $\mathtt{cnt}_{\mathrm{succ}} = m$ do
    increment $\mathtt{cnt}_{\mathrm{total}}$ (unconditionally),
    $(j, y) \leftarrow \mathtt{Scan}(\overline{h}, \overline{a}, i)$,
    increment $\mathtt{cnt}_{\mathrm{succ}}$ if and only if $j = i$ and y is of type X.
Output: $m/\mathtt{cnt}_{\mathrm{total}}$.

Analysis of the Simulation. It is quite straightforward to show that the procedure $\mathtt{Approx}$ satisfies its specification. Ignoring the exponentially vanishing probability that any single approximation (by the procedure $\mathtt{Approx}$) is off by more than a factor of 2, we may bound the total expected running-time by using the recursive structure of the simulation. (We start with bounding the running-time, because we will have to use this bound in analyzing the output of the simulator.)

Running-time analysis. Towards the running-time analysis, it is useful to pass among the procedures also the corresponding path in the tree of recursive calls. For example, instead of saying that $\mathtt{Scan}(\overline{h}, \overline{a}, i)$ invokes $\mathtt{Generate}(\overline{h}', \overline{a}', j)$, we may say that $\mathtt{Scan}(\overline{h}, \overline{a}, i; \overline{p})$ invokes $\mathtt{Generate}(\overline{h}', \overline{a}', j; (\overline{p}, i))$, where $\overline{p}$ denotes the path of recursive calls leading to the calling invocation (i.e., $\mathtt{Scan}(\overline{h}, \overline{a}, i; \overline{p})$). Bounded-simultaneity implies that the depth of the recursive tree is a constant (i.e., equals the simultaneity bound w), because whenever a procedure is invoked with path $\overline{p}$ it must be the case that the sessions with indices in $\overline{p}$ are still active (i.e., the corresponding transcript does not contain their last message). The fact that the depth of the recursive tree is a constant is the key to the analysis of the running-time of the simulation.

Considering oracle calls to the adversary's strategy as atomic steps, the expected running-time of $\mathtt{Scan}(\overline{h}, \overline{a}, i; \overline{p})$ (resp., $\mathtt{Generate}(\overline{h}, \overline{a}, i; \overline{p})$) is dominated by the time spent by the recursive calls invoked by $\mathtt{Scan}(\overline{h}, \overline{a}, i; \overline{p})$ (resp., $\mathtt{Generate}(\overline{h}, \overline{a}, i; \overline{p})$). Such calls are made only when handling Step 2 of a session with no verifier decommitment information. Each of these handlings consists of first invoking $\mathtt{Scan}(\overline{h}', \overline{a}', j; (\overline{p}, i))$, where $\overline{h}'$ is the current extension of the transcript $\overline{h}$, and, pending on its not returning $\mathtt{failure}$, invoking $\mathtt{Approx}$ and $\mathtt{Generate}$ on $(\overline{h}', \cdot, j; (\overline{p}, i))$. (Specifically, the latter procedures are invoked only if $\mathtt{Scan}(\overline{h}', \overline{a}', j; (\overline{p}, i)) = (j, \cdot)$.) In particular, $\mathtt{Approx}(\overline{h}', \overline{a}', X, j; (\overline{p}, i))$ invokes $\mathtt{Scan}(\overline{h}', \overline{a}', j; (\overline{p}, i))$ for an expected number of times that is inversely proportional to the probability that $\mathtt{Scan}(\overline{h}', \overline{a}', j; (\overline{p}, i))$ answers with a type X decommitment to session j, and $\mathtt{Generate}(\overline{h}', \overline{a}', j; (\overline{p}, i))$ is invoked for the at most the same (absolute) number of times. That is, letting $\mathtt{Scan}'(\overline{h}', \overline{a}', j) \stackrel{\mathrm{def}}{=} (k, X)$ if $\mathtt{Scan}(\overline{h}', \overline{a}', j)$ answers with a type X decommitment to session k, we conclude that the expected number of recursive calls made (directly) by $\texttt{Scan}(\overline{h}, \overline{a}, i; \overline{p})$ (resp., $\texttt{Generate}(\overline{h}, \overline{a}, i; \overline{p})$) when handling a Step 2 message of Session j is

$$\sum_{X \in \{\texttt{proper}, \texttt{improper}\}} \Pr[\texttt{Scan}'(\overline{h}', \overline{a}', j) = (j, X)] \cdot \frac{\text{poly}(n)}{\Pr[\texttt{Scan}'(\overline{h}', \overline{a}', j) = (j, X)]} = \text{poly}(n)$$

(5)

The key point is that all these recursive calls (invoked by, say, $\texttt{Scan}(\overline{h}, \overline{a}, i; \overline{p})$) have the longer path $(\overline{p}, i)$. Furthermore, these calls refer to transcripts that are prefixes of one another (i.e., each recursive call refers either to the same transcript as the previous call or to an extension of it). Thus, each node in the (depth w) tree of recursive-calls has an expected polynomial number of children, and so the expected size of the tree is upper-bounded by $\text{poly}(n)^w$. It follows that, the simulation terminates in expected polynomial-time. That is:

Claim 11 *For any polynomial-time adversary and any constant w that bounds the number of simultaneously active sessions, the simulation terminates in expected polynomial-time.*

Output distribution analysis. We start the analysis (of the output distribution) by justifying the discarding of the (remote) possibility that during the (polynomial-time) simulation we ever get two conflicting proper decommitments to the same verifier commitment. (In fact, the above functional description suggests this assumption, although formally it is not needed in the functional description.) The justification is that the polynomial bound on the expected running-time implies that the computational-binding property of the verifier's commitment is violated during the simulation with negligible probability.

Next, we establish that the implementations of the various procedures satisfy the corresponding specification, by using backward induction on the depth of the recursive call. First, we establish that in sub-step 2d of the handling of a Step 2 message, it rarely happens that all invocations of $\texttt{Generate}$ return $\texttt{failure}$ (i.e., this bad event occurs with negligible probability). This is due to the specification of the procedures invoked at the current stage (assumed in the induction step or to the fact that no procedure is invoked in the base case of the induction). (Specifically, $\texttt{Generate}$ is invoked for a number of times that is inversely proportional to the probability it succeeds.) This holds for a single handling of a Step 2 message, and we infer the same for all handlings that take place in the recursion tree by using a union bound and relying on the polynomial bound on the expected number of handlings (implied by Claim 11). The analysis of the other sub-steps in the handling of a Step 2 message is straightforward (from the code and specification). The analysis of the handling of Step 3 messages is similar, and the analysis of other handlings is straightforward. Thus, we obtain:

Claim 12 *For any polynomial-time adversary and any constant w that bounds the number of simultaneously active sessions, the invocation of any procedure during the simulation behaves according to the corresponding specification.*

Recall that the specification allows for a negligible error probability and the output of Generate is required to be indistinguishable from a corresponding concurrent execution. Once Claim 12 is established, we look at the initial (fictitious) invocation of Generate, which cannot possibly return with failure, and conclude that the simulator's output is computational indistinguishable from a real interaction of the cheating verifier with sessions of the prover. Thus, we get

Theorem 13 *The* (constant-round) *GK-protocol is zero-knowledge under concurrent composition of bounded-simultaneity.*

5 Simulation Under the Timing Model

Recall that the timing assumptions refer to two constants, Δ and ρ, such that Δ is an upper bound on the message handling-and-delivery time, and $\rho \geq 1$ is a bound on the relative rates of the local clocks. Specifically (cf. Footnote 6), clock rates are measured with respect to time intervals of length Δ; that is, if during a real-time period of Δ units the reading of some local clock changed by Δ' units, then $\Delta/\rho \leq \Delta' \leq \rho\Delta$. For simplicity, we may assume without loss of generality that $\Delta/\rho \leq \Delta' \leq \Delta$ (i.e., that all clocks are at least as slow as the real time).[25]

5.1 The Time-Augmented GK-protocol

Recall that the GK-protocol proceeds in four abstract steps, but the actual implementation of the first step consists of the prover sending a preliminary message that is used as basis to the verifier's actual commitment. Thus, the GK-protocol is actually a 5-round protocol starting with a prover message. We augment this protocol with the following time-driven instructions, where all times are measured according to the prover's clock starting at the time of the invocation of the prover's program:

1. The prover time-outs Step 1 after $\Delta_1 \overset{\text{def}}{=} 2\Delta$ units of time (as measured on its clock).
   (By the timing assumption, this does not disrupt honest operation, because 2Δ real units of time suffice for the delivery of a message from the prover to the verifier and back.)
2. The prover delays its execution of Step 2 to time $\Delta_2 \overset{\text{def}}{=} \rho \cdot \Delta_1 + \Delta$. That is, it sends its message exactly when its clock shows that Δ_2 units of time have elapsed.
3. The prover time-outs Step 3 after $\Delta_3 \overset{\text{def}}{=} \Delta_2 + 2\Delta$ units of time.
   (Note that $\Delta_3 = (2\rho + 3) \cdot \Delta$.)
4. The prover delays its execution of Step 4 to time $\Delta_4 \overset{\text{def}}{=} \rho \cdot \Delta_3 + \Delta$.

---

[25] We comment that although our formulation looks different than the one in [13], it is in fact equivalent to it.

We comment that, compared to Dwork *et. al.* [13], we are making a slightly more extensive use of the time-out and delay mechanisms: Specifically, they only used the last two items and did so while setting $\Delta_3 = 4\Delta$ and $\Delta_4 = \rho\Delta_3$. On the other hand, our use of the time-out and delay mechanisms is less extensive than the one suggested by Section 1.5: We only guarantee that for two sessions that start at the same time, Step 2 (resp., Step 4) in one session starts after Step 1 (resp., Step 3) is completed in the other session, but we do not guarantee anything about the relative timing of Steps 2 and 3 (of different sessions). Relying on special properties of the GK-protocol (as analyzed in Section 3.5), we can afford doing so, whereas the description in Section 1.5 is generic and refers to any c-round protocol. (However, in the typical case where $\rho \approx 1$, the difference between the various time-augmentations of the GK-protocol is quite small.)

Parenthetical Comment: A more general treatment can be derived by introducing an auxiliary parameter, denoted $\delta > 0$, which (in the description above) we have set to equal Δ. In the general treatment, Step 2 uses delay $\Delta_2 \stackrel{\text{def}}{=} \rho \cdot \Delta_1 + \delta$, whereas Step 4 uses $\Delta_4 \stackrel{\text{def}}{=} \rho \cdot \Delta_3 + \delta$, where $\Delta_1 \stackrel{\text{def}}{=} 2\Delta$ and $\Delta_3 \stackrel{\text{def}}{=} \Delta_2 + 2\Delta$ (as above). Doing so, in the decomposition, one may partition time to intervals of length δ (rather than length Δ). For $\rho = 1$, the number of overlapping blocks in the forthcoming Claim 14 changes by a factor of $(3\Delta + \delta)/4\delta > 1/4$, whereas the execution time of the protocol changes by a factor of $(4\Delta + 2\delta)/6\Delta > 2/3$. Observe that we do not gain much by setting $\delta \neq \Delta$. Specifically, by setting $\delta \ll \Delta$ we may reduce the the execution time by not more than a factor of $2/3$, whereas the effect on the simulation time is devastating (because the latter depends exponentially on the number of overlapping blocks, which in turn grows by a factor of approximately $3\Delta/4\delta$ for $\delta \ll \Delta$). On the other hand, setting $\delta \gg \Delta$ does not make the simulation significantly faster, whereas it delays the execution time considerably (i.e., by a factor of approximately $\delta/3\Delta$ for $\delta \gg \Delta$). Thus, we chose to set $\delta = \Delta$.

5.2   The Simulation

As mentioned in the introduction, the simulation relies on a decomposition of any schedule that satisfies the timing model into sub-schedules such that each sub-schedule resembles parallel composition, whereas the relations among the sub-schedules resembles bounded-simultaneity concurrent composition. In fact, we can prove something stronger:

Claim 14 *Consider an arbitrary scheduling of concurrent sessions of the time-augmented GK-protocol that satisfy the timing assumption. Place a session in block i if it is invoked within the real-time interval $((i-1) \cdot \Delta, i \cdot \Delta]$. Then, for every i:*

1. *Each session in block i terminates Step 1 by real-time $i \cdot \Delta + \rho\Delta_1$, starts Step 2 after real-time $i \cdot \Delta + \rho\Delta_1$, terminates Step 3 by real-time $i \cdot \Delta + \rho\Delta_3$, and starts Step 4 after real-time $i \cdot \Delta + \rho\Delta_3$.*

2. *The number of blocks that have a session that overlaps with some session in block i is at most $16\rho^3$. That is, the number of $j \neq i$ such that there exists a time t, a session s in block i, and a session s' in block j such that s and s' are both active at time t is at most $16\rho^3$.*

The first item corresponds to Conditions C1 and C2 in Section 3.5, and the second item corresponds to bounded-simultaneity.[26]

Proof: The latest and slowest possible session in block i is invoked by real-time $i \cdot \Delta$, and takes $\rho\Delta$ units of real-time to measure Δ local-time units. It follows that such a session terminates Step 1 (resp., Step 3) by real-time $i \cdot \Delta + \rho \cdot \Delta_1$ (resp., $i \cdot \Delta + \rho \cdot \Delta_3$). On the other hand, the earliest and fastest possible session in block i is invoked after real-time $(i-1) \cdot \Delta$, and takes Δ units of real-time to measure Δ local-time units. It follows that such a session starts Step 2 (resp., Step 4) after real-time $(i-1) \cdot \Delta + \Delta_2 = i \cdot \Delta + \rho\Delta_1$ (resp., $(i-1) \cdot \Delta + \Delta_4 = i \cdot \Delta + \rho\Delta_3$). The first item follows.

For the second item, note that the earliest possible session in block i is invoked after real-time $(i-1) \cdot \Delta$, whereas the latest and slowest possible session in block i terminates by real-time $i \cdot \Delta + \rho\Delta_4 + \Delta = (i+1) \cdot \Delta + \rho \cdot (2\rho^2 + 3\rho + 1) \cdot \Delta$. Thus, all sessions of each block are active during a time interval of length $(2\rho^3 + 3\rho^2 + \rho + 2) \cdot \Delta$, and therefore these sessions may overlap sessions of at most $2 \cdot (2\rho^3 + 3\rho^2 + \rho + 2) \leq 16\rho^3$ other blocks. ∎

Given Claim 14, we extend the simulation strategy of Section 4 by showing how to handle blocks of "practically parallel" sessions rather than single sessions (which may be viewed as "singleton blocks"). For simplicity, the reader may think of the scheduling as being fixed such that the partition of sessions to blocks is fixed. However, the treatment actually holds also for a dynamic schedule where the membership of sessions in blocks is determined on-the-fly (i.e., upon their execution of Step 1). To motivate the final construction, we consider first the special case in which each block is a perfect parallel composition of some sessions.

Combining the Simulation Techniques – The Perfect Case. The key to the extension is to realize that all that changes is the types of verifier de-commitment events (corresponding to Step 3 messages). Recall that in case of a single session, there were two possible events (i.e., proper and improper de-commitment), and these were the two decommitment types we have considered. Here, for m parallel sessions (of some block), we may have 2^m possible events corresponding to whether each of the m sessions is proper or improper. However, the decommitment types we consider here are (not these 2^m events but rather) the $n+1$ events considered in Section 3: the events $E_0, E_1, ..., E_n$, where event E_j holds if all the properly decommitting sessions (in the current run)

---

[26] The second item is actually stronger than bounded-simultaneity, because it upper-bounds the total number of blocks that overlap with a given block (rather than upper-bounding the number of blocks that are (simultaneously) active at any given time).

have proper-decommitment probability above the threshold $t_j \approx 2^{-j}$ but not all these sessions have proper-decommitment probability above the threshold $t_{j-1} \approx 2^{-(j-1)}$. Indeed, E_0 is the event that all sessions have improperly decommitted in the current run. (It is important that the number of decommitment types is bounded by a polynomial; this will be reflected when trying to extend the analysis captured in Eq. (5).)

Given the new notion of decommitment types, the three procedures of Section 4 (Scan, Approx and Generate) are extended by using the corresponding operations in Section 3. We stress that, in case of progress, the extended Scan (as well as the first progress case in the extended Generate) returns the decommitment information, which includes the indication of whether each session has properly decommitted, but not the decommitment type. The latter will be determined as in Section 3 (which is far more complex than the trivial case handled in Section 4, where decommitment type equals the decommitment indicator bit). The decommitment type (rather than the sequence of decommitment indicators) is what matters in much of the rest of the activities of the modified procedures.

We focus on the most interesting modifications to the main procedures (Scan and Generate), and ignore straightforward extensions (which apply also to other steps):

1. *The handling of Step 2 messages by a block j with a non-empty first information field* is analogous to the treatment in the original procedure, and we merely wish to clarify what this means here. The point is that the first field of block j encodes a decommitment type E_k as well as decommitment information for all sessions that properly decommit with probability at least $t_k \approx 2^{-k}$. The prover commitment produced here is designed to pass with respect to these decommitment values. (The same applies to the initial actions in Generate.)

2. *The handling of Step 2 messages by a block j with an empty first information field* (i.e., the only case that invokes recursive calls). The following sub-steps correspond to the sub-steps in the original procedures (Scan and Generate):
   (a) We invoke Scan with a block index j (rather than with a session index), and consider its answer which is either failure or a progress pair (k, y), where k is a block index, and y is a list of decommitments corresponding to the various sessions of block k. We refer to the above invocation of Scan as to the initial one, and note that many additional invocations (with the same parameters) will take place in handling the current step. If (the initial invocation of) Scan returned with a progress pair (k, y) such that $k = j$, then we turn to the complex task of determining the decommitment type E_ℓ (which holds with respect to y) as well as the corresponding sets T_ℓ and $T_{\ell-1}$. (If $k \neq j$ then the following activity will not be conducted here, but rather be conducted by the instance that invoked Scan$(\cdot, \cdot, k)$.) The decommitment type E_ℓ as well as the corresponding sets T_ℓ and $T_{\ell-1}$ are determined analogously to the main part of Step S1 (of Section 3), which needs to be implemented in the current context. In particular, the implementation of Step S1 calls for the

approximation of the probabilities (denoted p_i's in Section 3) that each of the sessions properly decommits. This, in turn, amounts to multiple executions of Steps 2–3 of these sessions, which in our case should be handled by multiple invocation of $\mathtt{Scan}(\cdot, \cdot, j)$. Details follow.

Let $I \subseteq [n]$ denote the set of sessions in which the verifier has properly decommitted in y. (Recall we are in the case where the initial invocation of $\mathtt{Scan}(\overline{h}', \overline{a}', j)$ has returned the progress pair (j, y).) Our objective is to determine the corresponding event index ℓ as well as the sets T_ℓ and $T_{\ell-1}$. We consider the following cases (w.r.t I):

Case of empty I: Set $\ell = 0$ and $T_\ell = T_{\ell-1} = \emptyset$.

Case of non-empty I: Set $t_0 = 1$ and $T_0 = \emptyset$. We determine $\ell \geq 1$ (as well as T_ℓ), by iteratively considering $\ell = 1, ..., n$ (as in Section 3.2). That is, for $\ell = 1, ..., n$ do

  i. We obtain t_ℓ by invoking a procedure analogous to $T(\ell, n)$ (of Section 3.2).

  Specifically, we approximate each of the p_s's by $\mathrm{poly}(n) \cdot 2^\ell$ invocations of $\mathtt{Scan}(\overline{h}', \overline{a}', j)$. Recall that each call of $\mathtt{Scan}(\overline{h}', \overline{a}', j)$ specifies whether each session in Block j has properly decommitted, and approximations to the p_s's, denoted a_s's, are determined accordingly. We stress that p_s is the probability that $\mathtt{Scan}(\overline{h}', \overline{a}', j)$ returns a progress pair (j, y') such that Session s properly decommits in y' (e.g., p_s is upper-bounded by the probability that $\mathtt{Scan}(\overline{h}', \overline{a}', j)$ returns a progress pair $(j, \cdot)$). Once all a_s's are determined, we determine t_ℓ just as in the second step of $T(\ell, n)$.

  ii. Determine the set T_ℓ by determining, for each s, whether or not $p_s > t_\ell$. We use the above approximations to each p_s and rely on $|p_s - t_\ell| > (1/9n)2^{-\ell}$.

  iii. Decide if event E_ℓ holds for y by using $T_{\ell-1}$ (of the previous iteration) and T_ℓ (just computed). Recall that event E_ℓ holds for y if $I \subseteq T_\ell$ but $I \not\subseteq T_{\ell-1}$.

  iv. If event E_ℓ holds then exit the loop with the current value of ℓ as well as with the values of T_ℓ and $T_{\ell-1}$. Otherwise, proceed to the next iteration (i.e., the next value of ℓ).

In both cases (of I), we have determined the commitment type $X = E_\ell$ with respect to y (as obtained in the initial invocation of $\mathtt{Scan}$) as well as the corresponding sets T_ℓ and $T_{\ell-1}$.

(This corresponds to Step S1 of the simulator of Section 3.)

(b) Exactly as in the original sub-step 2b. (That is, if the initial answer is either a $\mathtt{failure}$ or is a progress pair (k, y) with $k \neq j$ then $\mathtt{return}$ with the very same answer.)

(c) Recall that we reach this sub-step only if the answer of the initial invocation of $\mathtt{Scan}$ is a progress pair (j, y), and that we have already determined the event E_ℓ that holds (for y). By $\mathrm{poly}(n) \cdot 2^\ell$ additional invocations of $\mathtt{Scan}$ (with the same parameters as above), we may obtain progress pairs of the form $(j, \cdot)$ several times. In each of these cases, the second component consists of a list of proper decommitment values. With overwhelmingly high probability, for each $s \in T_\ell$, we will obtain (from at least one of these lists) a proper decommitment for Session s (because $p_s > 2^{-\ell}$). Ignoring the question of what decommitment types hold in these lists,[27] we combine all these lists to a list v of all proper decommitment values (obtained in any of these lists). This list v together with T_ℓ and $T_{\ell-1}$ (as obtained in sub-step 2a) forms a new information string $z = (v, T_\ell, T_{\ell-1})$, which will be used below (i.e., recorded in $\overline{a}'$ for future use). (This corresponds to Step S2 of the simulator of Section 3.)

Next, analogously to the original sub-step 2c, we obtain an approximation to the probability that $\texttt{Scan}(\overline{h}', \overline{a}', j) = (j, y)$ such that E_ℓ holds in y. Specifically, we let $\tilde{q} \leftarrow \texttt{Approx}(\overline{h}', \overline{a}', (E_\ell, T_\ell, T_{\ell-1}), j)$, where procedure $\texttt{Approx}$ uses T_ℓ and $T_{\ell-1}$ in order to determine whether the event E_ℓ holds in each of invocations of $\texttt{Scan}(\overline{h}', \overline{a}', j)$. We update the jth record of $\overline{a}'$ by placing (E_ℓ, z) in the first field and $\tilde{q}$ in the second field. (This corresponds to Step S3 of the simulator of Section 3.)

(d) Finally, analogously to the original sub-step 2d, we invoke $\texttt{Generate}(\overline{h}', \overline{a}', j)$ up-to $\text{poly}(n)/\tilde{q}$ times and deal with the outcomes as in the original sub-step 2d. (This corresponds to Step S4 of the simulator of Section 3.)

3. *The handling of Step 3 messages by a block j* (possibly $j = i$) is analogous to the treatment in the original procedure, and we merely wish to spell out what this means: We consider two cases depending on whether or not $\overline{a}'$ contains the verifier's decommitment information for block j (i.e., the first field of the jth block is not empty).

(a) In case $\overline{a}'$ does contain such information, we consider sub-cases according to the relation of the contents of the the first field of the jth block, denoted (E_ℓ, z), and the current answer of the verifier. Specifically, we check whether the verifier's current answer is of type E_ℓ. We note that the type of the current verifier decommitment is determined using the sets T_ℓ and $T_{\ell-1}$ provided in z (i.e., $z = (v, T_\ell, T_{\ell-1})$, where v is a sequence of decommitment values not used here). The sub-cases (fit versus non-fit) are handled as in the original procedure.

(b) In case $\overline{a}'$ does not contain such information (i.e., the first field of the jth block is empty), we obtain the relevant decommitment information (i.e, a sequence of decommitments) from the adversary, and **return** (as progress) with this information only.

This completes the description of the modification to the main procedures for the current setting (of bounded-simultaneity of blocks of parallel sessions). We stress that here (unlike in Section 3) the events E_ℓ regarding the decommitment to block j are not the only things that may happen when we invoke $\texttt{Scan}$ with

---

[27] In particular, we do not care if the decommitment event happens to be of type E_ℓ or not. Furthermore, we may ignore y itself and not use it below (although we may also use y if we please).

block index j (which corresponds to Step S1 in Section 3). As in Section 4, the answer may be `failure` or progress with respect to a different block. Indeed, the latter may not occur in case there is only one block, in which case the above treatment reduces to the treatment in Section 3. It is also instructive to note that when each block consists of a single session, the above modified procedures degenerate to the original one (i.e., in Section 4).

To analyze the current setting (of bounded-simultaneity of blocks of parallel sessions), we plug the analysis of Section 3 into the analysis of Section 4. The only point of concern is that we have introduced additional recursive calls (i.e., in the handling of Step 2, specifically in the handling sub-step 2a). However, as shown in Section 3, the expected number of these calls is bounded above by a polynomial (i.e., it is $\sum_{\ell=0}^{n} \Pr[E_\ell] \cdot 2^\ell \text{poly}(n)$, whereas $\Pr[E_\ell] = O(n \cdot 2^{-\ell})$). Thus, again, the tree of recursive calls has expected $\text{poly}(n)$ branching and depth at most w. Consequently, again, the expected running-time is bounded by $\text{poly}(n)^w$.

Combining the Simulation Techniques – The Real Case. In the real case the execution decomposes into blocks of almost parallel sessions (rather than perfectly parallel ones) such that (again) bounded-simultaneity holds with respect to the blocks. In view of the extension in Section 3.5, the non-perfect parallelism *within* each block does not raise any problems (as far as a single block is concerned). What becomes problematic is the relation between the (non-perfectly parallel) blocks, and in particular our references to the ordering of steps taken by the different blocks. That is, our treatment of the perfect-parallelism case treats the parallel steps of each block as an atom. Consequently we have related to an ordering of these steps such that if one "block step" comes before another then all sessions in the the first block take the said step before any session of the other block takes the other step. However, in general, we cannot treat the parallel steps of each block as an atom, and the following problem arises: what if one session of block i takes Step A, next one session of block $j \neq i$ takes Step B, and then a different session of block i takes Step A. This problem seems particularly annoying if handling the relevant steps requires passing control between recursive calls. In general, the problem is resolved by treating differently the first (resp., last) session and other sessions of each block that reach a certain step. Loosely speaking, the first (or last) such session will be handled similarly to the atomic case (i.e., as in Section 5.2), whereas in some cases other sessions (of the block) will be handled differently (in a much simpler manner). In particular, recursive calls are made only by the first session, and control is returned only by either the first or last such sessions. For sake of clarity, we present below the modification to the procedure $\texttt{Generate}(\overline{h}, \overline{a}, i)$. Note that this procedure is invoked when the immediate extension of $\overline{h}$ calls for execution of Step 2 by the *first* session in block i (i.e., $\overline{h}$ contains no Step 2 by any session that belongs to block i).

Initialization (upon invocation) step: Initializes $\overline{h}' = \overline{h}$ and $\overline{a}' = \overline{a}$, generates a passing commitment for (Step 2 of) the current (i.e., first) session of block i, and augments $\overline{h}'$ and $\overline{a}'$ accordingly. Specifically, the commitment is generated so that it passes the challenge corresponding to the current session (as recorded in the first field of record i), and only the corresponding part of the third field of the ith record (in $\overline{a}'$) is updated.

In all the following cases, $\overline{h}'$ and $\overline{a}'$ denote the current history prefix and auxiliary information, respectively. (The following cases refer to the next message to be handled by the procedure, which handles such messages until it returns.)

Step 1 by some (new) session: Exactly as in the atomic case (i.e., augment $\overline{h}'$ and proceed to the next iteration).

Step 2 by the first session in block j (certainly $j \neq i$): Analogous to the atomic case (see Section 5.2). Specifically, the handling depends on whether or not $\overline{a}'$ contains the verifier's decommitment information for session j (i.e., whether or not the first field of the jth record is non-empty).

   1. In case $\overline{a}'$ does contain such information, we just generate a corresponding passing commitment (i.e., passing w.r.t the first field of the jth record), augment $\overline{h}'$ and $\overline{a}'$ accordingly, and proceed to the next iteration.

   2. In case $\overline{a}'$ does not contain such information (i.e., the first field of the jth record is empty), we try to obtain such information. This is done analogously to the atomic case (see Section 5.2). We stress that this activity will yield the necessary information for all sessions in the jth block, and not merely for the current (first) session in the block. Recall that the handling of this sub-case involves making recursive calls to the three procedures (with parameters $(\overline{h}', \overline{a}', j)$).

Step 2 by a non-first session in block j (here $j = i$ may hold): We consider two cases depending on whether or not $\overline{a}'$ contains the verifier's decommitment information for session j (i.e., whether or not the first field of the jth record is non-empty).

   1. In case $\overline{a}'$ does contain such information, we just generate a corresponding passing commitment, augment $\overline{h}'$ and $\overline{a}'$ accordingly, and proceed to the next iteration.
      (This is exactly as in the corresponding treatment of the first session of block j to reach Step 2.)

   2. In case $\overline{a}'$ does not contain such information (i.e., the first field of the jth record is empty), we generate a dummy commitment, augment $\overline{h}'$ accordingly, and proceed to the next iteration. (Recall that we count on the first session in the jth block to find out the necessary information (for all sessions in the block).)
      (This is very different from the treatment of the first session of block j to reach Step 2.)

Step 3 by a non-last session of block j (possibly $j = i$): Just augment $\overline{h}'$ accordingly (and proceed to the next iteration).
(This is very different from the treatment of the last session of block j to reach Step 3.)

Step 3 by the last session of block j (possibly $j = i$)**:** Analogous to the atomic case. We consider two cases depending on whether or not $\bar{a}'$ contains the verifier's decommitment information for block j (i.e., the first field of the jth block is not empty).

1. In case $\bar{a}'$ does contain such information, we consider sub-cases according to the relation of the contents of the the first field of the jth block, denoted (E_ℓ, z), and the Step 3 answer of the verifier (for all sessions in the jth block). Specifically, we should consider the answers to previous sessions in the jth block as recorded in $\bar{h}'$ and the answer to the last session in the block as just obtained. Recall that the type of the verifier decommitments (for the sessions in the jth block) is determined using the sets T_ℓ and $T_{\ell-1}$ provided in the first field of the jth block. The sub-cases (fit versus non-fit) are handled as in the original procedure. That is:

   (a) If the decommitment type of the Step 3 answers (of the jth block) fits E_ℓ then we just augment $\bar{h}'$ accordingly (and proceed to the next iteration).
   (b) Otherwise (i.e., the decommitment type of the current Step 3 does not fit E_ℓ), **return failure**.

   (As in the atomic setting this case must hold if $j = i$.)

2. In case $\bar{a}'$ does not contain such information (i.e., the first field of the jth block is empty), we obtain the relevant decommitment information as in the previous case, and **return** (as progress) with this information only. Specifically, the decommitment information for the previous sessions of the jth block is recorded in $\bar{h}'$, whereas the the decommitment information for the last session has just been obtained (from the adversary).

Step 4 by a session of block j (possibly $j = i$)**:** Using the prover's decommitment information (as recorded in the third field of the jth record), we emulate Step 4 in the straightforward manner (and augment $\bar{h}'$ accordingly). If this is the last session of block j and $j = i$, then **return** with the current $\bar{h}'$ and $\bar{a}'$ (otherwise proceed to the next iteration).

The modifications to procedure **Scan** are analogous. We stress that although the above description treats the schedule as if it is fixed, the treatment actually extends to a dynamic schedule where the membership of sessions in blocks is determined on-the-fly (i.e., upon their execution of Step 1). Also recall that by our assumption that the verifier never violates the time-out condition (cf. Sec. 2.2), the "last session in a block to reach a certain step" can be determined as well. The analysis of the perfect case can now be applied to the real case, and Theorem 1 follows. That is:

Theorem 15 *The Time-Augmented GK-protocol is concurrent zero-knowledge under the timing model.*

6 Other Applications of Our Techniques

As stated in Section 1.3, our techniques are applicable also to several well-known protocols that have a structure similar to the GK-protocol. Notable examples include the (constant-round) zero-knowledge *arguments* of [15] and [4] as well as the *perfect* (constant-round) zero-knowledge proof of [5]. In fact, our techniques are applicable also to protocols with less apparent similarity to the GK-protocol. One such example is provided by the protocols that result from the transformation of Bellare, Micali and Ostrovsky [6].

In Section 6.1, we show that our techniques can be applied to the four-round argument system of Bellare, Jakobsson and Yung [4]. In Section 6.2, we informally describe a general class of protocols to which our techniques are applicable.

6.1 Application to the BJY-protocol

We start by briefly recalling the BJY-protocol (due to Bellare, Jakobsson and Yung [4], which in turn builds upon the work of Feige and Shamir [15]). Their protocol uses an adequate three-round witness indistinguishable proof system (e.g., parallel repetition of the basic zero-knowledge proof of [19]). Specifically, we consider a three-round witness indistinguishable proof system (e.g., for $G3C$) of the form:

Step WI1: The prover commits to a sequence of values (e.g., the colors of each vertex under several 3-colorings of the graph). This commitment scheme is perfectly-binding (and non-interactive; see Footnote 13).

Step WI2: The verifier send a random challenge (e.g., a random sequence of edges).

Step WI3: The prover decommits to the corresponding values.

(The implementation details are as in Construction 7.) For technical reasons, it is actually preferable to use protocols for which demonstrating a "proof of knowledge" property is easier (e.g., parallel execution of Blum's basic protocol; cf. [16, Sec. 4.7.6.3] and [16, Chap. 4, Exer. 28]). Given the above, the (four-round) BJY-protocol (for any language $L \in \mathcal{NP}$) proceeds as follows:

1. The verifier sends many hard "puzzles", which are unrelated to the common input x. These puzzles are random images of a one-way function f, and their solutions are corresponding preimages. In fact, the verifier selects these puzzles by uniformly selecting preimages of f, and applying f to obtain the corresponding images. Thus, the verifier knows solutions to all puzzles he has sent.

   In the rest of the protocol, the prover will prove (in a witness indistinguishable manner) that either it knows a solution to one of (a random subset of) these puzzles or $x \in L$. The latter proof is by reduction to some instance of an NP-complete language.

2. The prover performs Step WI1 *in parallel* to asking to see a random subset of the solutions to the above puzzles. Specifically, the puzzles are paired, and the prover asks to see a solution to one (randomly *selected*) puzzle in each pair. Furthermore, in executing Step WI1, the prover refers to a statement derived from the reduction of the assertion $x \in L$ or some of the *non-selected* puzzles has a solution.

3. The verifier performs Step WI2 *in parallel* to sending the required solutions (to the selected puzzles).

4. The prover verifies the correctness of the solutions provided by the verifier, and in case all solutions are correct it performs Step WI3.

As shown in [4], the BJY-protocol is a four-round zero-knowledge argument system for L. The simulator is similar to the one presented for the GK-protocol. Specifically, it starts by executing Steps 1–3, while using dummy commitments (in Step 2). Such a partial execution is called proper if the adversary has revealed all solutions to the selected puzzles (and is called improper otherwise). In case the partial execution is improper, the simulator halts while outputting it. Otherwise, the simulator moves to generating a full execution transcript by repeatedly rewinding to Step 2 and trying to emulate Steps 2–4 using the fact that (unless it selects the same set of puzzles again (which is highly unlikely)) it already knows a solution to one of the puzzles not selected (by it) in the current execution (but rather selected in the initial execution of Steps 1–3). Using such a solution, which yields an NP-witness to the reduced instance, the simulator can emulate the WI proof. As in the simulation of the GK-protocol (cf. [17]), the number of repetitions must be bounded by the reciprocal of the probability of a proper (initial) execution (as approximated by an auxiliary intermediate step).[28]

Given the similarity of the two simulators (i.e., the one here and the one for the GK-protocol), it is evident that our treatment of concurrent composition of the GK-protocol applies also to the BJY-protocol. Thus, recalling that the BJY-protocol is only based on one-way functions, we obtain:

Theorem 16 *Assuming the existence of one-way function, there exists a* (four-round) *argument system for* $\mathcal{NP}$ *that is* concurrent zero-knowledge under the timing model.

6.2 Application to a General Class of Protocols

In this section, we informally describe a general class of protocols to which our techniques are applicable. These protocols proceed in four main abstract steps:

---

[28] Unfortunately, this technical issue is avoided by Bellare *et. al.* [4], but it arises here (i.e., in [4]) similarly to the way it arises in [17], and it can be resolved in exactly the same manner. (The issue is that the prover commitments in the initial scan are distributed differently (but computational-indistinguishably) than its commitments in the generation process.)

1. The verifier "commits" to some secret information. Indeed, this "commitment" may be (as in the case of the GK-protocol) the result of applying a commitment protocol to the said information, but need not be so (cf., e.g., the BJY-protocol).
2. Some initial sub-protocol takes place such that its execution can be easily simulated by a computationally-bounded party that is only given the public information (i.e., the common input and the transcript of Step 1).
   In the GK-protocol, this step consists of the prover's commitment to a sequence of 3-colorings and can be simulated by producing commitments to dummy values. In other cases (e.g., [6]), this step may be vacuous.
3. The verifier proves knowledge of the secret information it has committed to in Step 1.
   In the GK-protocol, this step amounts to performing the corresponding decommitment step.
4. Pending on the prover being convinced, some residual sub-protocol takes place. The two sub-protocols (of Steps 2 and 4) are such that they can be easily simulated by a computationally-bounded party that is given the verifier's secret (as well as the the public information).
   In the GK-protocol, these two steps can be simulated by first sending commitments to corresponding "pseudo-colorings" and next performing the corresponding decommitments.

The single-session simulation of the above abstract protocol is similar to the simulator used for the GK-protocol. Specifically, the simulator starts by performing Step 1, and then performs Steps 2–3 (by using the corresponding guarantee regarding Step 2). In case the transcript is unacceptable by the prover, the simulator halts outputting the truncated transcript. Otherwise, the simulator invokes the knowledge-extractor that is guaranteed for Step 3, and obtains the verifier's secret information.[29] Once the simulator has this secret information, it can simulate Steps 2–4 (by the corresponding guarantee). We warn that indeed the actual implementation of the simulation procedure is more complex than the above description (e.g., as in [17], in some cases an approximation sub-step needs to be added). Still, the interested reader may verify that the techniques applied in Sections 3–5 extend to the above (abstract) simulation scheme. We informally conclude that *every protocol of the above type is concurrent zero-knowledge under the timing model*.

Acknowledgments

We are grateful to Uri Feige and Alon Rosen for helpful discussions at the initial stages of this research, and to Salil Vadhan for such discussions at the final stages. We also wish to thank Rafi Ostrovsky for pointing out that that our techniques can be applied to the protocols in [5, 6], Boaz Barak and Daniele

---

[29] Actually, the simulator uses a knowledge-extractor that corresponds to Steps 2–3. Observe that if Step 3 is a proof-of-knowledge then so are Steps 2–3.

Micciancio for interesting discussions regarding the use of time-driven operations (see Footnote 7), and the anonymous referees for their extremely valuable comments. Our research was partially supported by the MINERVA Foundation, Germany.

References

1. B. Barak. How to Go Beyond the Black-Box Simulation Barrier. In *42nd FOCS*, pages 106–115, 2001.
2. B. Barak and Y. Lindell. Strict Polynomial-time in Simulation and Extraction. In *34th ACM Symposium on the Theory of Computing*, pages 484–493, 2002.
3. M. Bellare, R. Impagliazzo and M. Naor. Does Parallel Repetition Lower the Error in Computationally Sound Protocols? In *38th FOCS*, pages 374–383, 1997.
4. M. Bellare, M. Jakobsson and M. Yung. Round-Optimal Zero-Knowledge Arguments based on any One-Way Function. In *EuroCrypt'97*, Springer-Verlag LNCS Vol. 1233, pages 280–305.
5. M. Bellare, S. Micali, and R. Ostrovsky. Perfect Zero-Knowledge in Constant Rounds. In *22nd STOC*, pages 482–493, 1990.
6. M. Bellare, S. Micali, and R. Ostrovsky. The (True) Complexity of Statistical Zero Knowledge. In *22nd STOC*, pages 494–502, 1990.
7. G. Brassard, D. Chaum and C. Crépeau. Minimum Disclosure Proofs of Knowledge. *JCSS*, Vol. 37, No. 2, pages 156–189, 1988. Preliminary version by Brassard and Crépeau in *27th FOCS*, 1986.
8. G. Brassard, C. Crépeau and M. Yung. Constant-Round Perfect Zero-Knowledge Computationally Convincing Protocols. *Theoretical Computer Science*, Vol. 84, pages 23–52, 1991.
9. R. Canetti, O. Goldreich, S. Goldwasser, and S. Micali. Resettable Zero-Knowledge. In *32nd STOC*, pages 235–244, 2000.
10. R. Canetti, J. Kilian, E. Petrank and A. Rosen. Black-Box Concurrent Zero-Knowledge Requires (Almost) Logarithmically Many Rounds *SICOMP*, Vol. 32, No. 1, February 2002, pages 1–47. Preliminary version in *33rd STOC*, 2001.
11. I. Damgård. Efficient Concurrent Zero-Knowledge in the Auxiliary String Model. In *Eurocrypt'00*, pages 418–430, 2000.
12. D. Dolev, C. Dwork, and M. Naor. Non-Malleable Cryptography. *SICOMP*, Vol. 30, No. 2, April 2000, pages 391–437. Preliminary version in *23rd STOC*, 1991.
13. C. Dwork, M. Naor, and A. Sahai. Concurrent Zero-Knowledge. In *30th STOC*, pages 409–418, 1998.
14. C. Dwork, and A. Sahai. Concurrent Zero-Knowledge: Reducing the Need for Timing Constraints. In *Crypto98*, pages 442–457, Springer LNCS 1462.
15. U. Feige and A. Shamir. Zero-Knowledge Proofs of Knowledge in Two Rounds. In *Crypto'89*, Springer-Verlag LNCS Vol. 435, pages 526–544, 1990.
16. O. Goldreich. *Foundation of Cryptography – Basic Tools*. Cambridge University Press, 2001.
17. O. Goldreich and A. Kahan. How to Construct Constant-Round Zero-Knowledge Proof Systems for NP. *J. of Crypto.*, Vol. 9, No. 2, pages 167–189, 1996. Preliminary versions date to 1988.

18. O. Goldreich and H. Krawczyk. On the Composition of Zero-Knowledge Proof Systems. *SICOMP*, Vol. 25, No. 1, February 1996, pages 169–192. Preliminary version in *17th ICALP*, 1990.
19. O. Goldreich, S. Micali and A. Wigderson. Proofs that Yield Nothing but their Validity or All Languages in NP Have Zero-Knowledge Proof Systems. *JACM*, Vol. 38, No. 1, pages 691–729, 1991. Preliminary version in *27th FOCS*, 1986.
20. O. Goldreich and Y. Oren. Definitions and Properties of Zero-Knowledge Proof Systems. *J. of Crypto.*, Vol. 7, No. 1, pages 1–32, 1994.
21. S. Goldwasser and S. Micali. Probabilistic Encryption. *JCSS*, Vol. 28, No. 2, pages 270–299, 1984. Preliminary version in *14th STOC*, 1982.
22. S. Goldwasser, S. Micali and C. Rackoff. Knowledge Complexity of Interactive Proofs. In *17th STOC*, pages 291–304, 1985. This is a preliminary version of [23].
23. S. Goldwasser, S. Micali and C. Rackoff. The Knowledge Complexity of Interactive Proof Systems. *SICOMP*, Vol. 18, pages 186–208, 1989. Preliminary version in [22].
24. J. Håstad, R. Impagliazzo, L.A. Levin and M. Luby. A Pseudorandom Generator from any One-way Function. *SICOMP*, Vol. 28, No. 4, pages 1364–1396, 1999. Preliminary versions by Impagliazzo *et. al.* in *21st STOC* (1989) and Håstad in *22nd STOC* (1990).
25. J. Kilian and E. Petrank. Concurrent and resettable zero-knowledge in polylogarithmic rounds. In *33rd STOC*, pages 560–569, 2001.
26. J. Kilian, E. Petrank, and C. Rackoff. Lower Bounds for Zero-Knowledge on the Internet. In *39th FOCS*, pages 484–492, 1998.
27. M. Naor. Bit Commitment using Pseudorandom Generators. *J. of Crypto.*, Vol. 4, pages 151–158, 1991.
28. M. Prabhakaran, A. Rosen and A. Sahai. Concurrent Zero-Knowledge Proofs in Logarithmic Number of Rounds. In *43rd IEEE Symposium on Foundations of Computer Science*, pages 366–375, 2002.
29. R. Richardson and J. Kilian. On the Concurrent Composition of Zero-Knowledge Proofs. In *EuroCrypt99*, Springer LNCS 1592, pages 415–413.
30. A. Rosen. A Note on Constant-Round Zero-Knowledge Proofs for NP. In *1st TCC*, Springer LNCS 2951, pages 191–202, 2004.
31. S. Vadhan. Probabilistic Proof Systems – Part I. In *IAS/Park City Mathematics Series*, Vol. 10, pages 315–348, 2004.
32. A.C. Yao. Theory and Application of Trapdoor Functions. In *23rd FOCS*, pages 80–91, 1982.

Fair Bandwidth Allocation Without Per-Flow State

Richard M. Karp

International Computer Science Institute, Berkeley, USA
and University of California at Berkeley
karp@icsi.berkeley.edu

Abstract. A fundamental goal of Internet congestion control is to allocate limited bandwidth fairly to competing flows. Such flow control
involves an interplay between the behavior of routers and the behavior of end hosts. Routers must decide which packets to drop when their
output links become congested. End hosts must decide how to moderate
their packet transmissions in response to feedback in the form of acknowledgements of packet delivery (acks). Typically this is done according to
the TCP protocol, in which a host maintains a window (the number of
packets that have been sent but not yet acknowledged) that is increased
when an ack is received and decreased when a drop is detected.
Often the selection of packets to be dropped at a router depends on the
order of their arrivals at the router but not on the flows to which the
packets belong. An exception occurs when packets are stratified according to their quality of service guarantee; in this case packets at higher
strata are given priority, but within a stratum the packets from different
flows receive the same treatment. A number of methods have been proposed to ensure fairness by selectively dropping packets from flows that
are receiving more than their fair share of bandwidth. The most effective
known algorithms for detecting and selectively dropping high-rate flows
at a router are based on random hashing or random sampling of packets
and give only probabilistic guarantees. The known deterministic algorithms either require excessive storage, require packets to carry accurate
estimates of the rates of their flows, assume some special properties of
the stream of arriving packets, or fail to guarantee fairness. In a simplified theoretical setting we show that the detection and selective dropping
of high-rate flows can be accomplished deterministically without any of
these defects. This result belies the conventional wisdom that per-flow
state is required to guarantee fairness.
Given an arriving stream of packets, each labeled with the name of its
flow, our algorithm drops packets selectively upon arrival so as to guarantee that, in every consecutive subsequence of the stream of surviving
packets, no flow has significantly more than its fair share of the packets.
The main results of the paper are tight bounds on the worst-case storage
requirement of this algorithm. The bounds demonstrate that the storage and computation required to guarantee fairness are easily within the
capabilites of conventional routers.
It is important to acknowedge the limitations of this work. We have formulated the achievement of fairness at a router in terms of local informa-

O. Goldreich et al. (Eds.): Shimon Even Festschrift, LNCS 3895, pp. 88–110, 2006.
© Springer-Verlag Berlin Heidelberg 2006

tion on the stream of arriving packets at that router. The implications of such a locally optimal policy on the global stability of the Internet would require analyzing the Internet as a complex dynamical system involving interactions among routers and end hosts, of which some will be TCP-compliant and some will not. In work not reported here we have made an initial simulation study of this complex process, but such a study is outside the scope of the present paper.

1   Introduction

The allocation of bandwidth within the Internet involves a complex interaction among routers that must drop packets when congestion occurs on their output links, and end hosts that moderate the rate at which they transmit packets in response to acknowledgements of the delivery of their packets. In the TCP protocol, which is the canonical protocol expected of end hosts, the rate of each flow is determined by a dynamically changing window giving the number of packets that are allowed to be concurrently in transit. To establish a context for our work we present a highly simplified description of this protocol. Associated with a flow is a *round-trip time* (RTT) giving the nominal time from the transmission of a packet to the arrival of an acknowledgement (ack) of its delivery. In an initial *slow-start* phase a source increases its window by 1 every time it receives an ack, so that the window size doubles every RTT. After this initial phase the window is increased by 1 every RTT. However, whenever the source infers that a packet has been dropped, its window is halved. Different versions of TCP make this inference in different ways, for example by observing that a packet has not been acknowledged within the RTT, or by detecting a gap in the stream of acks. Unfortunately, many flows do not follow the TCP protocol, either by intent or because of misconfiguration at the end host.

Internet flow control is a complex system involving the interactions among large numbers of routers and end hosts. Understanding the dynamics of this system is beyond the scope of the present paper. Instead, we follow in the tradition of several previous algorithms [1,2,4,5,6,7,8,9] that aim to optimize the decisions at a router as to which packets to drop and which to forward, based strictly on information about the flows arriving at the router. The goal of these algorithms is to identify high-rate flows and selectively drop their packets, thus signaling that the rates of those flows need to be reduced.

Such algorithms are typically based on the following idealized model. Packets arriving at a router are to be transmitted along an output link that can accept data at a certain bit rate R. Each flow a transmits packets at a bit rate $r(a)$. let f be the largest real number such that $\sum_a \min(f, r(a)) \leq R$. Then, the *fairness criterion* stipulates that, for each flow a, a fraction $\min(1, f/r(a))$ of the packets for flow a should be transmitted, and the others dropped. This ensures that the bit rate of accepted packets does not exceed the bit rate of the output link.

This idealized model breaks down when the arrival process of packets is non-stationary, so that the bit rates of some flows are not well defined. Assuming that the rates are well defined, the fairness criterion can be satisfied by keeping, for each flow a of rate greater than f, a variable giving the amount by which the transmission rate of packets for flow a leads or lags the target rate f, and scheduling packet drops so as to keep the lead or lag as close to zero as possible; more simply, the criterion can be approximately satisfied by accepting each packet for each flow a independently with probability $\min(1, f/r(a))$. The amount of state needed to estimate the rates $r(a)$ can be excessive, and various randomized algorithms based on hashing or sampling techniques have been suggested for identifying the high-rate flows with high probability and estimating their rates without using excessive storage. These algorithms do not provide deterministic guarantees.

Another approach, called *core-stateless fair queueing* [12], requires that each router receive an estimate of the rates of each of its arriving flows and provide an estimate of the rate of each outgoing flow, derived from the flow's input rate and the fraction of its packets that are not dropped. If this rate information can be kept accurately then the fairness criterion can be met without requiring excessive state.

Finally, *fair queueing* is an algorithm that achieves fairness even when the rates are unknown or not well defined. Because of its large storage requirement it is not a practical general method, but it can serve as a yardstick to which other methods can be compared. In the case where all packet sizes are equal, fair queueing is particularly simple: for each flow, a queue of packets that have been received but not transmitted is maintained; the queues are polled in round-robin fashion, and, whenever a non-empty queue is polled, it transmits a packet. The large storage requirement of fair queueing has led to the folk belief that it is not possible to guarantee fairness without requiring per-flow state. To the best of our knowledge the present paper is the first to present theoretical results that belie this belief.

Section 2 gives our definition of fairness. Section 3 gives our algorithm for dropping packets so as to guaranteed fairness. The algorithm has two parameters: ϵ, a flow's fair share of the bandwidth of the output link, and r, the amount by which a flow is allowed to transiently exceed its fair share. These parameters, together with a packet arrival sequence, determine, for each flow, a dynamic variable called its *excess*, and the worst-case storage requirement of our fair drop policy corresponds to the maximum number of simultaneously positive excesses. Importantly, our policy differs radically from schemes such as fair queueing because it does not maintain any queues of packets.

The main body of the paper is devoted to characterizing the worst-case packet arrival sequences; i.e., the arrival sequences that maximize the number of simultaneously positive excesses. Section 4 gives some preliminary observations about the fair drop policy, and Section 5 then derives one of the main results of the paper, an exact determination of the maximum number of simultaneously positive excesses. This key quantity dictates the storage requirement of our fair drop policy. Sections 6 and 7 then extend this result by deriving tight bounds on the maximum number of simultaneously positive excesses in the time-bounded case, in which an upper bound T is placed on the number of packets accepted. The approach is to derive the exact worst-case storage requirement of a drop policy that is slightly more permissive than the policy of interest, and then bound the difference between the storage requirements of the two policies. The Appendix gives examples of some worst-case, or nearly worst-case, arrival sequences. In simplified form, the main results of the paper are that the maximum number of positive excesses is approximately $\frac{1}{\epsilon}(1 + \ln r)$ if no restriction is placed on the number of accepted packets, and the worst-case number of positive excesses is approximately $\frac{1}{\epsilon}(1 - (\frac{r-1}{r})^{T\epsilon - r} + \ln r)$ if the number of accepted packets is bounded above by T.

2   A Fair Drop Policy

Consider a simple router in which arriving packets enter a queue and are then transmitted in First-In First-Out order along an output link. Whenever the arrival of a packet would cause the queue to overflow the packet is dropped, irrespective of the flow it belongs to. Our fair drop policy would change the router by interposing a pre-filter that would examine each arriving packet and either drop it immediately if its acceptance would violate a certain fairness criterion, or else forward it to the queue. The implementation of the pre-filter does not require any temporary storage for pending packets, as each packet is dealt with immediately upon arrival. This property distinguishes our method from fair queueing. Our pre-filter bases its decision for each packet on a nonnegative integer variable called the *excess* associated with the flow to which the packet belongs, and stores no other data. Moreover, most of these excesses will be zero and need not be explicitly stored. The main result of the paper is an exact determination of the number of excesses that can be simultaneously positive. It follows that the amount of storage and processing needed to maintain our pre-filter is well within the capabilities of existing routers. Our pre-filter is a variant of the *token-bucket* schemes that have been suggested as mechanisms for traffic shaping and rate control [3,10,11,13,14], but the mathematical analysis of its storage requirement is new.

We next describe our fairness criterion and our algorithm for enforcing fairness. Consider a sequence of packets of equal size arriving at a router to be transmitted on a single output link. Each packet is labeled with the name of its flow. Let x_i denote the label of the ith packet. The sequence $\{x_i\}$ is called the *arrival sequence*. A *drop policy* is a rule for deciding, as each packet arrives, whether to drop the packet. For a given drop policy and arrival sequence $\{x_i\}$, let $\{y_t\}$ be the subsequence of $\{x_i\}$ consisting of the labels of the packets that are not dropped; $\{y_t\}$ is called the *output sequence* associated with the arrival sequence $\{x_i\}$, and we refer to the operation of placing a label in the output sequence as *hitting* the corresponding flow.

Informally, the output sequence is considered fair if no flow significantly exceeds it fair share in any period. The formal definition is in terms of a (typically small) positive constant ϵ defining the fair share of the output bandwidth available to any flow, and a positive constant r defining the maximum number of packets by which a flow may exceed its fair share in any period. Thus, the output sequence $\{y_t\}$ is *fair* if, for every consecutive subsequence of $\{y_t\}$ and every label a, the number of occurrences of a within the subsequence is at most $r + h\epsilon$, where h is the length of the subsequence; here ϵ represents the maximum fraction of total bandwidth that should be allocated to any flow in the long run, and r represents the maximum amount by which a flow may exceed that fraction over any period.

The *fair drop policy* is defined by the following rule: drop a packet if and only if accepting it would make the output sequence unfair. The main result of this paper is that this policy does not require an excessive amount of state to be maintained.

In practice the unequal sizes of packets should be taken into account and the fair share should be a dynamic variable that increases in periods of low congestion and decreases in periods of high congestion. In particular, no practical scheme should drop packets unless the output link is congested. To achieve this property one could use a variant of our policy that marks packets instead of dropping them. In the execution of the fair drop policy the marked packets would be treated as if they had been dropped, but they would be forwarded to the queue. Whenever the queue reached its capacity marked packets would be dropped in preference to unmarked ones. The low storage requirement of the fair drop policy opens up yet another alternative, in which the policy would be executed concurrently for several values of ϵ (for the same r).All packets would be forwarded to the queue, but each would be marked with the highest value of ϵ, if any, that caused it to be dropped. When the queue reaches capacity a packet with the highest mark would be dropped.

Although the present paper is restricted to the case of constant ϵ, similar upper bounds on the amount of state also hold when the fair share is a dynamic variable, provided that the variable never dips below a fixed value.

3   An Algorithm Implementing the Fair Drop Policy

We begin by reformulating our criterion for the fairness of the output sequence $\{y_t\}$. We assume that ϵ is the reciprocal of a positive integer d and that rd is a positive integer B. For any flow a and time step t let $\iota(a, t)$ be an indicator variable which is 1 if $y_t = a$, and 0 otherwise. Define $F(a, T)$ as $\max(0, \max_{T' \leq T} \sum_{T' \leq t \leq T}(d \cdot \iota(a, T) - 1))$. Then, as a direct consequence of the definition of fairness, the output sequence is fair if and only if, for all a and T, $F(a, T) \leq B$. $F(a, T)$ is called the *excess* of flow a at time T. A flow can receive a high excess either by having an intense burst of arrivals or by exceeding its fair share slightly over a long time period, but its excess is not permitted to exceed B.

The following lemma establishes a simple inductive algorithm for computing the excess for each flow at each time step:

Proposition 3.1. $F(a, T) = max(0, F(a, T - 1) + d \cdot \iota(a, T) - 1)$

Proof. $F(a,T) = \max(0, d\iota(a,T) - 1, \max_{T' \leq T-1} \sum_{T' \leq t \leq T}(d \cdot \iota(a,t) - 1)$
$= \max(0, d \cdot \iota(a,T) - 1, (d \cdot \iota(a,T)) - 1) + \max_{T' \leq T-1} \sum_{T' \leq t \leq T-1}(d\iota(a,t) - 1)$
$= \max(0, d \cdot \iota(a,T) - 1, d \cdot \iota(a,T) - 1 + F(a,T-1)$
$= max(0, F(a,T-1) + d \cdot \iota(a,T) - 1)$, where the last equality is valid because
$F(a,T-1) \geq 0$.

Thus we can implement the fair drop policy by maintaining the excess of each flow at each time step using the update rule
$F(a,T) = max(0, F(a,T-1) + d\iota(a,T) - 1)$, and dropping a packet if and only if accepting it would cause the excess of its flow to exceed B.

Even though our algorithm differs from fair queueing by not storing any packets, it might appear that the algorithm requires per-flow state in the form of a nonnegative integer excess for each flow. Hovever, one can radically decrease the storage requirement of the algorithm by maintaining a data structure that records only the currently positive excesses; if the excess of a flow a is not found in the data structure then its value is zero. The goal of the paper is to analyze the storage requirement of this modified implementation. One should note that, in contrast to fair queueing, the information being stored consists of small integers rather than packets.

Our main result is that the maximum possible number of flows with positive excesses is exactly $d + L(B - d + 1 - \lceil \frac{B+1}{d} \rceil)$ where $L(u)$ is a recursively defined nondecreasing function asymptotic to $d \ln(u/d)$. The maximum number of flows with positive excesses is thus roughly approximated by $d(1 + \ln \frac{B}{d})$. In terms of r and ϵ this number is $\frac{1}{\epsilon}(1 + \ln r)$.

We also derive lower and upper bounds on the number of flows with positive excesses when the length of the output sequence is fixed at a value T. Thus, the focus of this paper is on characterizing worst-case examples for the fair drop policy. The paper is written from the viewpoint of an adversary trying to demonstrate the weaknesses of the policy and, somewhat perversely, we refer to these worst-case examples as optimal.

Even though the number of positive excesses is small, it may be undesirable to decrement each positive excess explicitly at each step. This can be avoided by maintaining an approximate excess for each flow that becomes incorrect but gets restored to correctness every d steps. This is achieved by partitioning the universe of flow labels into d classes, and cyling through these classes so that at each step, the approximate excess of the arriving flow is increased by an appropriate amount, and then, instead of decreasing each positive excess by 1, all the positive approximate excesses in one class are decreased by d and then restored to zero if they became negative. Approximate excesses of zero are not recorded explicitly. The appropriate amount by which to increase the approximate excess of the arriving flow is $d + \max(0, s - e - 1)$ where e is the current approximate excess of the flow and s is the number of steps since the flow's class was last decremented. This amount is chosen so that the approximate excess of each flow becomes equal to the true excess just after the step in which it gets decreased by d. The term $\max(0, s - e - 1)$ compensates for those time steps at which the true excess would have been zero and therefore would not have been decremented. Some excessive storage cost is incurred because the delayed decrementation of its excess can cause a flow to linger in the data structure with an incorrect positive approximate excess for up to $d-1$ steps after its true excess has decreased to zero, but it is easy to see that the number of positive approximate excesses at any step does not exceed the number of positive true excesses $2d$ steps earlier by more than $2d$, so the device of using approximate excesses does not increase he worst-case storage bound by more than $2d$. .

The formal setting for our results can be described as follows. Consider a nondeterministic algorithm which, at each step, either hits a flow whose excess is already positive (an *incrementation step*), hits a new flow (an *initiation step*), or does not hit any flow (a *no-op*). A no-op can be thought of as adjoining to the output sequence a flow that is immune from the fairness requirement, or a flow whose rate is known *a priori* to be low. Alternatively, we can think of the link as accepting at most one packet per unit time, and a no-op as a time step at which no packet is available.

The progress of the algorithm can be described by a sequence of multisets $S_0, S_1, \ldots, S_n, \ldots$, where S_t denotes the multiset of positive excesses after the tth step. The evolution of the nondeterministic algorithm is as follows:

$S_0 = \phi$; for $t = 1, 2, \ldots$ do:

Set R equal to S_{t-1}. Nondeterministically, perform either an incrementation step, an initiation step or a no-op. Then set S_t equal to R.

Incrementation step Choose an element $x \in R$ such that $x \leq B - d + 1$, replace x by $x + d$, decrement each element by 1 and delete all occurrences of 0.

Initiation step Insert the element d into R, decrement every element of R by 1 and delete all occurrences of 0.

No-op decrement every element of R by 1 and delete all occurrences of 0.

For example, if $d = 5$ and $B = 9$ and the sequence of steps consists of three initiation steps followed by three incrementation steps, an initation step and a no-op, then the sequence of multisets would be:

$\phi, \{4\}, \{3, 4\}, \{2, 3, 4\}, \{1, 7, 3\}, \{6, 7\}, \{5, 6, 4\}, \{4, 5, 3\}$.

Throughout the paper we assume that $B \geq 2d - 1$. Let $N(B, d, t)$ denote the maximum possible cardinality of S_t and let $N(B, d) = \max_t N(B, d, t)$. Then $N(B, d, t)$ denotes the maximum possible number of positive excesses after t steps, and $N(B, d)$ denotes the maximum possible number of simultaneously positive excesses. In a context where B and d are fixed we abbreviate $N(B, d, t)$ to N_t and $N(B, d)$ to N. In Section   5 we determine $N(B, d)$ exactly, thus determining the worst-case storage requirement of the maximally permissive fair drop policy, and in Section   6 we derive a function $R(B, d, t)$ such that $N(B, d, t) \leq R(B, d, t) \leq N(B + d - 1, d, t)$, thus giving a fairly tight approximation to $N(B, d, t)$. For a restricted choice of B, d and t an even tighter bound is given in Section 7.

To aid the reader's understanding, the Appendix gives examples of some of the combinatorial constructions used in establishing these bounds.

4   Preliminary Observations

Before proving the main results let us make a few simple observations. Let $s(t)$ be the sum of all excesses, and let $n(t)$ be the number of positive excesses, just after y_t has been appended to the output sequence. Then $s(t) \leq s(t-1)+d-n(t)$. Thus, for any T, $s(T) - s(0) \leq dT - \sum_{t=1}^{T} n(t)$. Since $s(0) = 0$ and $s(T) \geq 0$ we obtain $\sum_{t=1}^{T} n(t) \leq dT$, which implies that the average number of positive excesses over the first T steps is at most d.

Next we deviate temporarily from our worst-case analysis to consider a stationary probabilistic model in which the labels in the arrival sequence are iid random variables. Let $p(a)$ be the probability of label a. We shall prove that, at any time t, the expected number of positive excesses is at most d. This is an immediate consequence of the following lemma.

Proposition 4.1. *Let $e(a,t)$ be the excess of label a at time t. For all a and t, $P(e(a,t)) > 0) \leq dp(a)$.*

Proof. We may restrict attention to the case where $dp(a) < 1$. Let $q(a) = 1-p(a)$. For a fixed a the sequence $\{e(a,t)\}$ is a Markov chain with initial state 0, state set $\{0, 1, \cdots , B\}$ and the following nonzero transition probabilities:

1. $p(0,0) = q(a)$ and $p(0, d-1) = p(a)$.
2. For $1 \leq i \leq B - d + 1$, $p(i, i-1) = q(a)$ and $p(i, i+d-1) = p(a)$.
3. For $B - d \leq i \leq B$, $p(i, i-1) = q(a)$ and $p(i, i) = p(a)$.

Let us compare this Markov chain with the following Markov chain in which the state set is the nonnegative integers, the initial state is 0 and the nonzero transition probabilities are as follows:

1. $p(0,0) = q(a)$ and $p(0, d-1) = p(a)$.
2. For $i \geq 1$, $p(i, i-1) = q(a)$ and $p(i, i+d-1) = q(a)$.

We shall analyze this positive-recurrent infinite-state chain. Let $t(i,j)$ be the expected time to reach state j from state i. We make the following simple observations:

For all $i > 0$, $t(i, i-1) = t(1,0)$;

For all (i,j) with $j < i$, $t(i,j) = (i-j)t(1,0)$.

$t(1,0) = 1 + p(a)t(d,0) = 1 + p(a)(dt(1,0)) = \frac{1}{1-dp(a)}$;

therefore, for all $i > 0$, $t(i,0) = \frac{i}{1-dp(a)}$. $t(0,0) = 1 + p(a)t(d-1,0) = 1 + \frac{(d-1)p(a)}{1-dp(a)} = \frac{1-p(a)}{1-dp(a)}$.

The stationary probability of state 0 is $\frac{1}{t(0,0)} = \frac{1-dp(a)}{1-p(a)}$, which is greater than $1 - dp(a)$

Next we compare the two Markov chains. Let M be the transition probability matrix of the finite-state chain and M_∞ the transition probability matrix of the infinite-state chain. By induction on t, the following holds for all t and all i and j such that $i \le j$: $M_{i,0}^t \ge M_\infty^{t_j,0}$. Since M is started in state 0 it follows that the probability that M is in state 0 at time t is greater than or equal to the stationary probability of state 0 in the infinite-state chain, and hence greater than or equal to $1 - dp(a)$. Thus the probability that $e(a,t) > 0$ is at most $dp(a)$.

In this section we have shown that the number of positive excesses, averaged over the first T arrivals of accepted packets, is at most d, and that, under a simple probabilistic model, the expected number of positive excesses at any step is at most d. The worst-case analysis to follow provides stronger guarantees that the storage requirement of our policy will be small.

5   Maximum Number of Positive Excesses

The results of the previous section suggest that the number of positive excesses will seldom be much greater than d. However, it is possible to construct an adversarial sequence of packet arrivals for which the number of positive excesses becomes much larger than d. In this section we derive an exact formula for N, the maximum number of simultaneously positive excesses. The formula involves a function $L(u)$, where u ranges over the positive integers. The function is defined recursively as follows: $L(0) = L(1) = 0$
$L(u) = 1 + L(u - \lceil \frac{u+1}{d} \rceil)$, $u \ge 1$.
   The function $L(u)$ is asymptotic to $d \ln(u/d)$.

Theorem 5.1.
   $N = d + L(B - d + 1 - \lfloor (\frac{B+1}{d}) \rfloor)$.

Before proving the theorem we give a schedule that achieves the bound given in the theorem. Instead of describing it as a sequence $\{S_t\}$ of multisets, we describe it in terms of its output sequence. In this description, hitting a flow means incrementing its excess by d. Each step consists of hitting a flow, decrementing the excesses of all flows, and discarding excesses that are less than 1. A flow may be hit only if its excess is less than or equal to $B - d + 1$.
   The schedule consists of three phases.
   An Output Sequence with the Maximum Number of Positive Excesses
   Phase 1 Repeat $B - d + 1$ times: successively hit flows $1, 2, \ldots, d - 1$. At this point, the excesses are $B, B - 1, \ldots, B - d + 2$.
   Phase 2 Repeat until flow d has the smallest excess among flows $1, 2, \ldots, d$ and the difference between the largest and smallest positive excesses is at most $d - 1$: Hit the flow with the smallest excess among flows $1, 2, \ldots, d$.
   Phase 3 Set T equal to the excess of flow d.
   For $t = T - 1$ down to 1 do:
   If every flow with positive excess has an excess greater than or equal to t then hit a new flow, else hit a flow with the smallest positive excess.

The proof that this output sequence maximizes the number of simultaneously positive excesses will be given as part of the proof of Theorem 5.1.

As an example, in the case $d = 3$, $B = 5$ the sequence of multisets is
$\phi, \{2\}, \{1, 2\}, \{3, 1\}, \{2, 3\},$
$\{4, 2\}, \{3, 4\}, \{5, 3\}, \{4, 5\}, \{3, 4, 2\}, \{2, 3, 1, 2\},$
the output sequence is $1, 2, 1, 2, 1, 2, 1, 2, 3, 4$ and $N = 4$.

Throughout the paper we assume that once a flow is initiated its excess remains positive at all steps. This is not a significant restriction, since we can transform any schedule into one satisfying this property by repeatedly applying the following transformation, which does not affect the score: if the excess of a flow drops to zero for the last time at step t, substitute no-ops for all hits on that flow preceding step t.

We approach the proof of Theorem 5.1 through a series of lemmas.

Lemma 5.2. *If all the flows in a set S have positive excesses at some step T and collectively have had exactly u packet arrivals up to and including that step, then the cardinality of S is at most $L(u)$.*

Proof. The proof is by induction on u, with base case $u = 1$. Consider x, the first flow in S to experience an arrival. From the time of this arrival up to time T at least u steps must occur. At each step the excess of x will be increased by $d - 1$ if it has an arrival or decreased by 1 if it does not. Therefore, in order to have a positive excess at time T, x must experience at least $\lceil \frac{u+1}{d} \rceil$ arrivals, leaving at most $u - \lceil \frac{u+1}{d} \rceil$ arrivals for the flows in $S - x$. Since $L(\cdot)$ is a monotone nondecreasing function it follows by induction hypothesis that the cardinality of $S - x$ is at most $L(u - \lceil \frac{u+1}{d} \rceil)$. Therefore the cardinality of S is at most $1 + L(u - \lceil \frac{u+1}{d} \rceil)$, which is equal to $L(u)$.

Corollary 5.3. *If all the flows in a set S have positive excesses at some step T and collectively have had at most u packet arrivals up to and including that step, then the cardinality of S is at most $L(u)$.*

The corollary follows from the monotonicity of the function $L(\cdot)$.

Lemma 5.4. *At any step, the i-th largest excess is at most $B - i + 1$.*

Proof. Arrange the flows in decreasing order of the times of their most recent arrivals. Each of these flows had an excess of at most B at the time of that arrival, and its excess was decreased by 1 at each subsequent step. Thus the excess of the i-th flow in the ordering at time T is at most $B - i + 1$ It follows that at most $i - 1$ flows can have excesses greater than or equal to $B - i + 1$.

Proof of Theorem 5.1
Consider t^*, the last step after which fewer than d flows have positive excesses. Let Q be the set consisting of the d flows with positive excesses after step $t^* + 1$.

Number the flows in Q in decreasing order of their excesses just before t^*, with the number d assigned to the flow initiated for the first time at step $t^* + 1$. After any step t, let $e_i(t)$ be the excess of the ith flow in this fixed numbering. Then, for $i = 1, \ldots, d - 1$, $e_i(t^*) \leq B - i + 1$, and $e_d(t^*) = 0$. At any step after t^* the sum of the excesses of the flows in Q is unchanged if the step hits a flow in Q, and decreases by d if the step hits a flow not in Q. Let T be the last step of the schedule. Then the number of steps executed after step t^* by flows not in Q is $\frac{\sum_{i=1}^{d}(e_i(t^*)-e_i(T))}{d}$ We shall derive an upper bound u^* on this quantity. It will then follow from Corollary 5.3 that the number of flows whose first initiation occurs after step t^* is at most $L(u^*)$, and hence that the total number of positive excesses at the end of the schedule is at most $d + L(u^*)$. We observe that, at every step, the excess of flow i is either decreased by 1 or increased by $d - 1$; in both cases it is decremented by 1 modulo d. We therefore have the following invariant: for $i = 1, \ldots, d$, $j = 1, \ldots, d$ and $t = t^*, \ldots, T$, $e_i(t) - e_j(t) = e_i(t^*) - e_j(t^*) \bmod d$.

Thus it follows that u^* is bounded above by the optimal value of the following integer maximization problem:

Maximize $\frac{\sum_{i=1}^{d}(e_i(t^*)-e_i(T))}{d}$ subject to

$e_i(t^*) \leq B - i + 1$, $i = 1, \ldots, d - 1$
$e_d(t^*) = 0$
$e_i(T) \geq 1$, $i = 1, \ldots, d$
$e_i(T) - e^j(T) = e_i(t^*) - e_j(t^*)$, $1 \leq i < j \leq d$

Clearly, in a maximizing solution of this problem, every variable $e_i(T^*)$ lies between 1 and d. For any integer y, let $M(y)$ be the least positive integer congruent to y modulo d. Then in a maximizing solution $e_i(T) = M(e_i(t^*)+e_d(T))$.

Moreover, for a fixed choice of $e_d(T)$, decrementing any variable $e_i(t^*)$ cannot decrease $e_i(T)$ by more than one, and thus cannot increase the objective function value. It follows that there is a maximizing solution in which each variable $e_i(t^*)$, for $i = 1, \ldots, d - 1$, is equal to its maximum possible value, $B - i + 1$. Fixing these values, the above maximization problem reduces to:

minimize $\sum_{i=1}^{d} e_i(T)$
subject to $e_i(T) = M(B - i + 1 + e_d(T))$, $i = 1, \ldots, d - 1$.

Given this explicit formula it is easy to see that, in the maximizing solution, $\sum_{i=1}^{d} e_i(T) = \frac{d(d+1)}{2} + e_d(T) - M(B+1+e_d(T)))$. This expression is minimized when $e_d(T) = 1$, in which case it is equal to $\frac{d(d+1)}{2} + 1 - M(B + 2)$.

The optimal value for the maximization problem (*) is therefore
$\frac{\sum_{i=0}^{d-1}(B-i+1)- \frac{d(d+1)}{2} -1+M(B+2)}{d}$

This expression simplifies to $B - d + 1 - \lfloor \frac{B+1}{d} \rfloor$.

This establishes the upper bound of $d + L(B - d + 1 - \lfloor \frac{B+1}{d} \rfloor)$ on N.

To prove that the output sequence given above achieves the upper bound, we note the following:

After phase 1 the $d - 1$ excesses are $B, B - 1, B - 2, \ldots, B - d + 2$. After phase 3 flow d has an excess of 1 and, for $i = 1, 2, \ldots, d - 1$, flow i has an excess of $M(B - i + 2)$ It follows that the number of steps at which flows other than $1, 2, \ldots, d$ are hit is u^*. Since those additional flows are hit in consecutive steps, the number of additional flows with positive excesses after the last step is $L(u^*)$ and the total number of positive flows after the last step is $d + L(u^*)$, which is equal to the upper bound.

6   Time-Bounded Schedules

We now investigate N_T, the maximum number of positive excesses at time T, As before, B and d are fixed. We assume throughout this section that $B \geq 2d - 1$. Define the *score* of a T-step schedule as the number of positive excesses after step T. Instead of specifying a schedule as a sequence $\{S_t\}$ of multisets, we specify it by its output sequence; i.e., the sequence in which flows are hit.

The earlier a flow is first hit, the larger is the total number of hits the flow must experience. This suggests that it is desirable to defer the initiation of a new flow whenever possible, leading to the conjecture that there is an optimal T-step schedule in which a new flow is initiated only when there does not exist a flow with a positive excess that requires at least one more hit. However, this conjecture is false. The parameter choice $B = 9, d = 7, T = 11$ is a counterexample. The conjectured policy would yield the output sequence $1, 2, 1, 2, 3, 4, 3, 4, 5, 6, 7$ for a score of 7, but the output sequence $1, 2, 3, 1, 2, 3, 4, 5, 6, 7, 8$ achieves a score of 8.

We will give a simple construction that produces a T-step schedule with threshold $B + d - 1$ and increment d whose score is at least as great as that of any T-step schedule with increment d and threshold B. Thus the result is not quite optimal because we allow a slight relaxation of the threshold. It remains an open problem to characterize the optimal T-step schedules.

Consider an arbitrary T-step schedule. Number the steps in order of increasing time, so that the first step is numbered 1 and the last step is numbered T. Flow a is said to be hit at step t if $y(t) = a$. Let $e(a, t)$ be the excess of flow a just after step t. Define the function $f(x, t) = \lceil (T - t + 1 - x)/d \rceil$. Define $hit(a, t)$ as $f(e(a, t), t)$. Then $hit(a, t)$ is the number of times flow a must be hit in steps $t + 1, t + 2, \cdots, T$, given that its excess after step t is $e(a, t)$ and its excess after step 1 is required to be positive. Let $start(t) = \max(0, f(d - 1, t))$ and $low(t) = \max(0, f(B, t))$. Then $start(t) + 1$ is the total number of times that a flow must be hit if its first hit occurs at step t, and $low(t)$ is a lower bound on the number of times that a flow is hit during the last $T - t$ steps if its excess is positive at step t and all subsequent steps.

For any T-step schedule (with increment d and threshold B) the function $hit(a, t)$ has the following easily checked properties, which hold for all a and t:

1. If flow a is hit for the first time at step t then $hit(a, t) = start(t)$;
2. For all a and all t, $hit(a, t) \geq low(t)$;
3. $hit(a, t) = hit(a, t - 1)$ if $y(t) \neq a$ and $hit(a, t) = hit(a, t - 1) - 1$ if $y(t) = a$.
4. $hit(a, T) = 0$

The requirement that $hit(a,T) = 0$ for all a is implied by the requirement that all excesses must be positive at the last step.

We now relax our scheduling problem by considering all T-step schedules for which the function $hit(a,t)$ satisfies the above four properties. We call these schedules *relaxed T-step schedules*. To distinguish the schedules previously defined from relaxed schedules we sometimes refer to the original schedules as *strict schedules* Every strict T-step schedule also qualifies as a relaxed T-step schedule (more precisely, it has the same output sequence as some relaxed T-step schedule).

Lemma 6.1. *In every relaxed schedule the following holds for all flows a and times t: $e(a,t) \leq B + d - 1$.*

Proof. $low(t) \leq hit(a,t)$ Therefore, $f(B,t) \leq f(e(a,t),t)$. Using the definition of f we obtain: $\frac{T-t+1-B}{d} \leq \lceil \frac{T-t+1B}{d} \rceil \leq \lceil \frac{T-t+1-e(a,t)}{d} \rceil \leq 1 \leq \frac{T-t+1-e(a,t)+d-1}{d}$. Hence $T - t + 1 - B \leq T - t + 1 - e(a,t) + d - 1$, giving $e(a,t) \leq B + d - 1$.

We will exhibit a family $\mathcal{F}$ of optimal relaxed T-step schedules which follow the natural rule that a new flow is initiated only when it is not possible to hit any existing flow (recall that this natural rule is *not* optimal under the original definition of a schedule). Moreover, the optimal relaxed schedules in $\mathcal{F}$ have an important property that is not shared by relaxed schedules in general: once a flow has been initiated, its excess remains positive at all subsequent steps. Thus, for these special relaxed schedules, the score coincides with the number of positive excesses after the last step, and the only deviation from the original definition of a strict schedule is that the excess of a flow may exceed the threshold B by as much as $d - 1$. Because of this property we can derive near-optimal strict schedules with threshold B by constructing optimal relaxed schedules for threshold $B - d + 1$.

Call a flow a *eligible* at time t if $t > 1$ and $hit(a, t - 1) > low(t)$. A schedule lies in $\mathcal{F}$ if:

1. At any step at which an eligible flow exists, an eligible flow with the largest possible hit-value is hit.
2. If there exists no eligible flow at time t then a new flow is initiated, and this flow is hit in consecutive steps as long as it remains eligible. This series of steps is called the *initial run* of the flow.

In the next two sections we prove that every scedule in $\mathcal{F}$ maintains positive excesses at all times. Then, using long but conceptually simple interchange arguments, we shall show that there is an optimal relaxed schedule that never starts a new flow when an eligible flow exists, or devotes a step to a no-op. It will then follow that that the schedules in $\mathcal{F}$ are optimal among relaxed schedules.

6.1 $\mathcal{F}$ Maintains Positive Excesses

Theorem 6.2. *Eery relaxed schedule in $\mathcal{F}$ has the property that, once a flow has been initiated, its excess remains positive at all subsequent steps.*

In preparation for the proof of the theorem, define the jth *epoch* as the set of (consecutive) time slots t such that $low(t) = j$. The following properties follow easily from the definitions. The $low(T)$th epoch occurs first, followed successively by the remaining epochs in decreasing order of their $low(\cdot)$ values. Epoch 0 contains B time slots, epochs $1, 2, \cdots, low(T) - 1$ contain exactly d time slots each, and epoch $low(T)$ contains at most d time slots. Define $dif(t)$ as $start(t) - low(t)$. Then over the domain $\{B - d + 1, \ldots, T\}$ the range of $dif(t)$ consists of two consecutive integers, and $dif(t)$ is periodic with period d and nonincreasing within each epoch. Define $gap(a, t)$, the *gap* of flow a at time t as $hit(a, t) - low(t)$.

Note that, if $hit(a, t) < start(t)$ then $e(a, t) \geq d$. Thus, $e(a, t)$ can be non-positive only if $hit(a, t) \geq start(t)$ which is equivalent to $gap(a, t) \geq dif(t)$. We shall show that this inequality can hold only during the initial run of a flow. The theorem will follow, since it is clear that the excess of a flow during its initial run is positive.

To show that $gap(a, t) \geq dif(t)$ only during the initial run of a flow we consider how the gap of a flow varies from step to step. When the flow is first initiated at some step t, its gap becomes $dif(t)$. Therafter, its gap decreases by 1 whenever it is hit, increases by 1 after the completion of each epoch, and never becomes negative. The initial run of a flow continues until its gap becomes zero, and during the initial run the flow is hit at every step.

We first consider time steps t that are outside epoch 0 (i.e., we assume that $t \leq T - B + 1$). Let g be the maximum, over all $t \leq T - B + 1$, of $dif(t)$. Then, for all $t \leq T - B + 1$, $dif(t)$ is either g or $g - 1$. Thus, $e(a, t) = 0$ only if $gap(a, t) \geq g - 1$.

Call step t a *renewal step* if each flow has a gap of zero.

Lemma 6.3. *At every renewal step outside epoch 0, the number of active flows is less than d.*

Proof. Define the *state* of a schedule in $\mathcal{F}$ just after any step as (X, Y), where X is the sum of the gaps of all flows and Y is the number of flows that have been initiated. We shall prove by induction that, upon the completion of any epoch except epoch 0, $Y \leq d$ and $X = 0$ if $Y = d$. This holds for epoch $low(T)$, which is the base case of the induction. For the induction step, let (X, Y) be the state after the completion of epoch $j + 1$, where $j \geq 1$, and let (X', Y') be the state after the completion of epoch j. We may assume that $X \geq 1$, since at least one flow will have been initiated before the completion of epoch $j + 1$. If $X + Y \geq d$ then $X' = X + Y - d$ and $Y' = Y$. In this case, $Y' \leq d$ and it is not possible that $(X', Y') = (0, d)$, since that would imply that $(X, Y) = (0, d)$, contradicting the induction hypothesis. If $X + Y < d$ then the state will be $(0, Y)$ after the first $X + Y$ steps of epoch j, and the remaining $d - X - Y$ steps will be given over to the initial runs of new flows. From the inequality $B \geq 2d - 1$ it follows that, for every $t \geq d$, $start(t) > low(t)$, so every initial run contains at least two steps. It follows that either $X + Y = d - 1$, in which case $X' = 1$ and $Y' = Y + 1$, or $X + Y < d - 1$, in which case the number of flows initiated during the last $d - X - Y$ steps of epoch j is strictly less than $d - X - Y$, implying that $Y' < Y + (d - X - Y) = d - X$, so $Y' < d$. In both cases, the induction step succeeds.

Having verified the inductive claim we may now assume that, at the end of epoch $j + 1$, $Y \leq d$ and $X > 0$ if $Y = d$. If $X + Y > d$ then there is no renewal step in epoch j. If $X + Y = d$ then $X > 0$ and the only renewal step is at the end of epoch j, when the state is $(0, Y)$ with $Y < d$. If $X + Y = d - 1$ then the only renewal step in epoch j occurs at the penultimate step of the epoch, and the state is $(0, d - 1)$. If $X + Y < d - 1$ then $Y' < d$ and the state $(0, d)$ cannot occur during epoch j.

We proceed to the proof of Theorem 6.2.

Proof. Consider a relaxed schedule in $\mathcal{F}$. We shall prove that, for every flow a, $gap(a, t) \geq 1$ at all steps that are not within the initial run of flow a. From this it follows, as explained above, that, once a flow has been initiated, its excess never dips below 1. We first consider the time steps outside epoch 0. By Lemma 4, the number of active flows at a renewal step is at most $d - 1$. The initiation of a new flow occurs immediately after a renewal step. Its initial gap is at most g, and it drops by 1 at each step of the initial run except for the first step of an epoch, when the gap does not change. The initial run ends when the gap becomes zero. It follows that at most $\lfloor \frac{g}{d-1} \rfloor$ new epochs can start during the initial run of a flow, since d steps intervene between the starts of successive epochs. Thus, after the initial run of the new flow, each of the other flows has a gap of at most $\lfloor \frac{g}{d-1} \rfloor$, since these gaps are incremented only at the start of a new epoch. There then follows a sequence of steps in which the d or fewer active flows are hit, during which the flow with the largest gap is hit at each step. This continues until all gaps are reduced to 0, producing a renewal step. During this sequence no gap will exceed $\lfloor \frac{g}{d-1} \rfloor + 1$ and, provided this value is less than $g - 1$ (a lower bound on $dif(t)$ outside epoch 0) $gap(a, t)$ will remain less than $dif(t)$ except possibly during the initial run of a flow, implying that the excess of a flow never dips below 1 after the flow has been initiated. The contrary inequality $\lfloor \frac{g}{d-1} \rfloor + 1 \geq g - 1$ holds only if $g \leq 2$ or $d \leq 4$. These cases can be checked by a separate case analysis, showing that $hit(a, t) \geq start(t)$ only during the initial run of a flow, as required for the proof. The case where t lies in epoch 0 is simpler, because $gap(a, t)$ cannot increase during epoch 0, and yields to a similar calculation.

6.2 $\mathcal{F}$ Is Optimal

Next we show that every relaxed schedule in $\mathcal{F}$ is optimal among relaxed schedules. To do so, we shall show that there is an optimal relaxed schedule that never starts a new flow when an eligible flow exists, or devotes a step to a no-op. This is done by showing how to transform any relaxed schedule into one satisfying these requirements, without reducing its score.

Define the *potential* of flow a at time t as $hit(a, t-1) - low(t)$. Define the *state* of a relaxed schedule before step t as an ordered pair $(X(t), Y(t))$, where $Y(t)$ is the number of flows that have been initiated at previous steps and $X(t)$ is the sum of the potentials of those flows at step t. Note that this definition of state differs from the definition used in the proof of Theorem 6.2

The state evolves according to the following rules:

1. $(X(0), Y(0)) = (0, 0)$.
2. There is an eligible flow at step t if and only if $X(t) > 0$;
3. If t is not the last time step of an epoch then
   (a) If an eligible flow is hit at time t then $(X(t+1), Y(t+1)) = (X(t)-1, Y(t))$.
   (b) If there is a no-op at time t then $(X(t+1), Y(t+1)) = (X(t), Y(t))$.
   (c) If a flow is initiated at time t then $(X(t+1), Y(t+1)) = (X(t) + dif(t), Y(t) + 1)$
4. If t is the last time step of an epoch then
   (a) If an eligible flow is hit at time t then $(X(t+1), Y(t+1)) = (X(t) + Y(t) - 1, Y(t))$.
   (b) If there is a no-op at time t then $(X(t+1), Y(t+1)) = (X(t)+Y(t), Y(t))$.
   (c) If a flow is initiated at time t then $(X(t+1), Y(t+1)) = (X(t)+Y(t) + dif(t), Y(t) + 1)$

Motivated by these properties we define, a *trace* of duration T as a sequence of states $\{(X(t), Y(t))\}$, for $t = 0, 1, \ldots, T$, satisfying the following properties:

1. $(x(0), y(0)) = (0, 0)$;
2. If t is not the last step of an epoch and $X(t) = 0$ then $(X(t+1), Y(t+1)) \in \{(0, Y(t)), (dif(t), Y(t) + 1)\}$;
3. If t is not the last step of an epoch and $X(t) > 0$ then $(X(t+1), Y(t+1)) \in \{(X(t) - 1, Y(t)), (X(t), Y(t)), (X(t) + dif(t), Y(t) + 1)\}$;
4. If t is the last step of an epoch and $X(t) = 0$ then $(X(t+1), Y(t+1)) \in \{(Y(t), Y(t))), (dif(t) + Y(t), Y(t) + 1)\}$;
5. If t is the last step of an epoch and $X(t) > 0$ then $(X(t+1), Y(t+1)) \in \{(X(t)+Y(t)-1, Y(t)), (X(t)+Y(t), Y(t)), (X(t)+dif(t)+Y(t), Y(t)+1),\}$;
6. $X(T) = 0$.

We define the *score* of a trace as $Y(T)$.

Lemma 6.4. *The state sequence of any relaxed schedule is a trace with the same score as the relaxed schedule, and for every trace there exists a relaxed schedule with the same score as the trace.*

Call two relaxed schedules *trace-equivalent* if they share the same trace. All relaxed schedules in $\mathcal{F}$ are trace-equivalent, and their trace-equivalence class includes additional schedules, some of which do not have the property that, once a flow has been initiated, its excess remains positive.

In view of the correspondence between relaxed schedules and traces, the problem of finding the maximum score of a relaxed schedule is reduced to the problem of finding the maximum score of a trace. We will prove that the traces of the relaxed schedules in $\mathcal{F}$ have the maximum score among T-step traces. The proof will be inductive, where the induction step involves consideration of traces of length less than T, and with initial states other than $(0,0)$.For this purpose we define a U-step *generalized trace* as a sequence of states $(X(t), y(T))$, $t = 0, 1, \ldots, U$ having all the properties required for a trace except that the initial state $(X(0), Y(0))$ is an arbitrary ordered pair of nonnegative integers. A generalized trace is *optimal* if it has the maximum score among generalized traces with the same number of steps and initial state. For any initial state, the *canonical generalized trace* is the generalized trace determined by following the set of rules defining $\mathcal{F}$, starting at the given initial state. For certain initial states this leads to a final state with $X(U) > 0$. in which case no canonical generalized trace exists. We will find that, in such cases, no generalized trace exists , as a state $(X(U), Y(0))$ with $X(U) = 0$ is unattainable from the initial state.

For any initial state $(X(0), Y(0))$ define $score((X(0), Y(0)), U)$ as the score of an optimal U-step generalized trace U with initial state $(X(0), Y(0))$, or 0 if no such optimal generalized trace exists.

We call the following result the Monotonicity Lemma.

Lemma 6.5. *For any initial state* $(X(0), Y(0))$, $score(X(0), Y(0)), U) \geq score(X(0) + 1, Y(0)), U)$.

Proof. Consider an optimal generalized trace τ of duration U with initial state $(X(0) + 1, Y(0))$. The generalized trace τ must include a step t at which $X(t) < X(t-1)$. We construct a trace τ' with initial state $X(0), Y(0)$ with the same score as τ. The two traces are identical except at step t, where τ' has a no-op and τ hits an eligible flow.

A generalized trace is called *initiation-avoiding* if it never initiates a new flow at a step where an existing flow is eligible.

Lemma 6.6. *For every initial state* $(X(0, Y(0))$ *such that an optimal U-step generalized trace exists, there is an optimal U-step generalized trace that is initiation-avoiding.*

Proof. The proof is by induction on U. For the base case $U = 1$ the only situation of interest is when $X(0) = 1$. In this case, the only action that leads to a generalized trace is to hit the eligible flow. Thus, in the base case $U = 1$, whenever an optimal generalized trace exists there is an optimal initiation-avoiding generalized trace.

For the induction step it suffices to show that, for every initial state $(X(0), Y(0))$ where $X(0) > 0$ such that an optimal U-step generalized trace exists, there exists an optimal U-step generalized trace whose first action is to hit an eligible flow. For contradiction, suppose that this is not true for initial state $(X(0), Y(0))$. The state after step 1 resulting from hitting an eligible flow is $(X(0) - 1, Y(0))$ and the state after step 1 resulting from starting a new flow is $(X(0) + dif(0), Y(0) + 1)$. We derive a contradiction by showing that the score achievable from $(X(0) - 1, Y(0))$ by an optimal $(U - 1)$-step generalized trace is at least as great as the score achievable from $(X(0) + dif(0), Y(0) + 1)$ by an optimal $(U - 1)$-step generalized trace. By induction hypothesis we may assume that there is an optimal $(U - 1)$-step generalized trace τ with initial state $(X(0) + dif(0), Y(0) + 1)$ that is initiation-avoiding. We shall construct a $(U - 1)$-step generalized trace σ starting at $(X(0) - 1, Y(0))$ that achieves at least as high a score as τ. As long as σ can emulate the actions of τ it does so. This will continue until some step t at which σ is in state $(0, Y(t))$ and τ is in state $(dif(0) + \Delta, Y(t) + 1)$, where Δ is the number of epochs that intersect the interval $[0, t]$. At this point τ hits an eligible flow and enters the state $(dif(0) + \Delta - 1, Y(t) + 1)$ and σ starts a new flow and enters the state $(dif(t), Y(t) + 1)$. But $|Delta \geq 1|$ and, using the fact that the function dif is periodic with period d outside epoch 0 and nonincreasing within each epoch, it follows that $dif(0) \geq dif(t) - 1$ and, if $\Delta = 1$ $dif(0) \geq dif(t)$. Thus, $dif(0) + \Delta - 1 \geq dif(t)$, so, by the monotonicity lemma, the state of σ is at least as favorable as the state of τ, and it follows that σ can be completed so as to achieve at least as high a score as τ. This contradiction completes the induction step.

Our next goal is to show that, from initial state $(0, 0)$, there is an optimal initiation-avoiding trace that contains no no-ops.

Lemma 6.7. *In an initiation-avoiding trace starting at state $(0, 0)$, it is not possible to reach a state $(0, Y)$ with $y \geq d$ during epochs $low(T), low(T - 1), \ldots, 1$.*

The inductive proof is similar to the proof of Lemma 6.3.

Theorem 6.8. *If there exists an optimal trace, then there exists an optimal initiation-avoiding trace without no-ops.*

Proof. Lemma 6.6 states that, if there exists an optimal trace, then there exists an optimal initiation-avoiding trace. Let τ be an optimal initiation-avoiding trace containing at least one no-op. We shall show that there is an optimal initiation-avoiding trace in which the last no-op in τ is replaced by an optimal initiation-avoiding trace τ' which coincides with τ up to the step t where the last no-op in τ occurs, and does not have a no-op at step t. Repetition of this transformation eventually leads to an optimal initiation-avoiding trace free of no-ops.

Let the last no-op in τ be a repetition of the state X, Y at times $t - 1$ and t. We carry out the following case analysis.

$X > 0$ Since $X > 0$ there is an eligible flow at step t and, by hitting such a flow, state $(X - 1, Y)$ is reached at step t. By the monotonicity lemma, this state has at least as high a $(T - t)$-step score as state (X, Y). and so optimality is not lost by moving to state $(X - 1, Y)$ at step t. Thus τ can be replaced by an optimal initiation-avoiding trace in which the states at steps $1, 2, \ldots, t-1$ are unchanged and there is no no-op at step t.

$X = 0$ **and t is not the last step of its epoch** Since the last no-op in τ occurs at step t, τ starts a new flow at step $t + 1$, arriving at the state $(X + dif(t + 1), Y + 1)$. Instead it is possible to start a new flow at step t, arriving at the state $X + dif(t), Y + 1$, and then hitting that flow, arriving at state $(X + dif(t) - 1, Y + 1)$ which, by the monotonicity lemma, has at least as high a score as $(X + dif(t+1), Y + 1)$, since $dif(t) - dif(t+1) \le 1$. Thus the no-op at step t can be removed without loss of optimality, and the trace remains initiation-avoiding.

$X = 0$ **and t is the last step of its epoch** In this case, instead of ending the epoch in the state $(0, Y)$ it is possible to replace the no-op at step t by the initiation of a flow, producing the state $(dif(t), Y + 1)$. We now invoke Lemma 6.7, which tells us that $Y \le d - 1$. If the epoch ends in state $(0, Y)$ then, since the trace is initiation-avoiding and has no further no-ops, the next epoch starts with Y hits on eligible flows followed by an initiation, leading to the state $dif(t+Y+1, Y+1)$ after step $t+Y+1$. On the other hand, if the state after step t is $(dif(t), Y + 1)$ then the next epoch starts with $dif(t) + Y + 1$ hits on eligible flows. After the first $Y + 1$ steps of the epoch the state will be $(dif(t), Y+1)$. By the monotonicity of dif within an epoch and the fact that it is periodic with period d, it follows that $dif(t) \le dif(t+Y+1)$, so, by the monotonicity lemma, the state $(dif(t), Y + 1)$ has at least as high a score as the state $(dift+Y+1, Y+1)$, and thus it is possible to eliminate the no-op at step t without losing optimality or the property of being initiation-avoiding.

Since the canonical T-step trace is completely characterized by the properties of being initiation-avoiding and free of no-ops, it follows that it is optimal among T-step traces. Because of the equivalence between the score of a relaxed schedule and the score of its trace, it follows that the relaxed schedules in $\mathcal{F}$ are optimal. Moreover, they have positive excesses at every step, so their only defect is that an excess may be as high as $B + d - 1$ whereas, in a strict schedule, the excesses are bounded above by B.

7   Bounds on the Optimal Score

Since every strict schedule can be viewed as a relaxed schedule with the same parameters, $N(B, d, T) \le R(B, d, T)$. Since every relaxed schedule in $\mathcal{F}$ with parameters B and d can also be viewed as a strict schedule with parameters $B + d - 1$ and d, $N(B, d, T) \ge R(B + d - 1, d, T)$.

The following theorem shows that tighter lower bounds on $N(B, d, T)$ can be obtained in a special case.

Theorem 7.1. *If there are integers $s \ge 2$ and $k \ge d$ such that $B = ds - 1$ and $T = (k + d)s - 1$ then $R(B, d, T) \le (d+1)(1 - (\frac{s-1}{s})^k)\frac{s-1}{s} + dH_s$, $N(B, d, T) \ge (d - s + 1)(1 - (\frac{s-1}{s})^k\frac{s-1}{s} + (d-2)H_s$. and thus $R(B, d, T) - N(B, d, T) \le s + 2H_s$.*

The motivation for the restriction on B is as follows. Recall that, within a relaxed schedule in $\mathcal{F}$, if a flow is initiated at step t and its initial run takes place entirely within a single epoch, then it is hit $start(t) - low(t) + 1$ times in its initial run. The stated restriction implies that $(d-1)(start(t) - low(t) + 1) \leq B$, ensuring that the excess of the flow upon the completion of such an initial run does not exceed B. This suggests a way of modifying the relaxed schedules in $\mathcal{F}$ so as to avoid reaching an excess greater than B, simply by refraining from executing any initial run that cannot be completed within a single epoch. The resulting schedule will be strict and will have a score close to $R(B, d, T)$, the score achieved by the relaxed schedules in $\mathcal{F}$. This idea is the basis for the following proof.

Proof. We describe a strict T-step schedule ρ with threshold B and increment d that achieves a score of at least $R(B, d, T) - (s + 2H_s)$. The schedule satisfies the following inductive assertion: for $j = low(T), low(T) - 1, \ldots 1$, every flow a that is eligible just before the first step t of epoch j satisfies $h(a, t+1) = low(t) + 1$; i.e., each eligible flow has a potential of 1. Let $a(j)$ be the number of such eligible flows. The last $a(j)$ steps of epoch j are used to hit each of these eligible flows once. During these steps the flows are hit in increasing order of their excesses. The first $d - a(j)$ steps of epoch j are used for the initial runs of new flows. However, no initial run is started unless it can be completed within the same epoch. No-ops are assigned to those time steps that are used neither for initial runs nor for hitting eligible flows. Epoch 0 proceeds as follows: at any step at which an eligible flow exists, an eligible flow with the smallest possible excess is hit; otherwise, a new flow is initiated and is hit at consecutive steps as long as it remains eligible.

Under the stated restriction on B, all excesses occurring in the course of schedule ρ lie between 1 and B. Thus ρ is a strict schedule. The main difference in performance between ρ and the optimal relaxed schedules in $\mathcal{F}$ is that ρ never undertakes an initial run unless that run can be completed in the same epoch, whereas the schedules in $\mathcal{F}$ allow initial runs that cross epoch boundaries.

We now estimate $R(B, d, T)$ and compare it with the count of the strict schedule ρ. Under the stated restriction ($B = sd - 1$ and $T = kd + B$) the successive epochs are indexed $k, k-1, \cdots, 0$. Epochs $k, k-1, \cdots, 1$ each consist of d time steps, and at every time step t within these epochs, $start(t) - low(t) = s - 1$, so the length of an initial run is s if the run is entirely within an epoch, and $s + 1$ if the run crosses an epoch boundary. Epoch 0 has B time steps, and consists of s successive subintervals, $I_s, I_{s-1}, \cdots, I_1$. Each subinterval except the final one, I_1, is of length d, and I_1 is of length $d - 1$. For any i, the initial runs initiated in subinterval I_i are of length i.

The strict schedule ρ and the relaxed schedules in $\mathcal{F}$ achieving the count $R(B, d, T)$ behave similarly throughout epoch 0. In each case, let x denote the number of flows whose initial runs are completed during epochs $k, k-1, \ldots, 1$. Then $x \leq d$. Each of these flows gets hit once in each epoch after the completion of its initial run and has a gap of 1 at the beginning of epoch 0. Epoch 0 begins by hitting each of these x flows once, and then continues with a series of initial runs. The final score lies between $x\frac{s-1}{s} + (d-2)H_s$ and $x\frac{s-1}{s} + dH_s$ where $H_s = 1 + 1/2 + 1/3 + \ldots 1/s$. Thus the problem of determining the scores achieved by these schedules revolves around finding x, the number of flows that complete their initial runs in the first k epochs.

To estimate $R(B, d, T)$ we consider a relaxed schedule that is initiation-avoiding and free of no-ops; such schedules are optimal. To get an upper bound we pretend that an initial run in epochs $k, k-1, \ldots, 1$ is of length s whether or not it crosses the boundary between two epochs. This inaccuracy can only increase the score. Under this inaccurate assumption, let a_j be the number of steps devoted to initial runs during epochs $k, k-1, \ldots, j$. Then $\lfloor \frac{a_j}{s} \rfloor$ is an upper bound on the number of initial runs actually completed during those epochs. Each of the $\lfloor \frac{a_j}{s} \rfloor$ flows whose initial run is completed prior to epoch $j-1$ is hit once in epoch $j-1$, and the remaining $d - \lfloor \frac{a_j}{s} \rfloor$ time steps in epoch $j-1$ are devoted to initial runs. Thus we obtain the recurrence $a_{j-1} = a_j + d - \lfloor \frac{a_j}{s} \rfloor$ with the initial condition $a_k = d$. Then $a_{j-1} \le \frac{s-1}{s} a_j + d + 1$, and it follows that $a_j \le (d+1)s(1 - (\frac{s-1}{s})^{k-j+1})$. Therefore the number of initial runs completed by the end of epoch 1 is at most $(d+1)(1 - (\frac{s-1}{s})^k)$.

Now consider the first k epochs of schedule ρ. In this schedule each initial run is of length s, but no initial run is allowed to cross an epoch boundary, so up to $s-1$ time steps in each epoch may be wasted on no-ops. Let b_j denote the number of steps devoted to initial runs during epochs $k, k-1, \ldots, j$. Then $b_{j-1} \ge b_j + d - s + 1 - \lfloor \frac{b_j}{s} \rfloor$ with the initial condition $b_{k+1} = 0$. Then $b_{j-1} \ge \frac{s-1}{s} b_j + d - s + 1$, and it follows that $b_j \ge (d - s + 1)s(1 - (\frac{s-1}{s})^{k-j+1})$. Therefore the number of initial runs completed by the end of epoch 1 is at least $(d - s + 1)(1 - (\frac{s-1}{s})^k)$.

Combining the bounds for an optimal relaxed schedule and for ρ we conclude that $R(B, d, T) - N(B, d, T) \le s + 2H_s$, completing the proof.

Expressing the bound of Theorem 7.1 in terms of B, d and T we find that $R(B, d, T)$ is approximately $d(1 - (\frac{B-d}{B})^{\frac{T-B}{d}})\frac{B-d}{B} + H_{B/d})$. and $R(B, d, T) - N(B, d, T) \le \frac{B+1}{d} + 2H_{\frac{B+1}{d}}$.

The key property that makes Theorem 7.1 work is that an initial run of length $start(t) - low(t) + 1$ does not create an excess greater than B. This property holds whenever, for some positive integer $s \ge 2$, $s(d-1) \le B \le sd - 1$ and $T \ge 2B$. Theorem 7.1 can be extended to show that, in all such cases, $R(B, d, T) - N(B, d, T) \le s + O(H_s)$. The proof is omitted.

8   Conclusion

This paper analyzes a scheme for selectively dropping arriving packets to ensure that in each interval within the sequence of accepted packets, no flow may exceed its fair share of packets by more than r. Here the fair share of a flow within an interval is defined as ϵ times the length of the interval, and the parameter r specifies the amount by which the fair share can be exceeded transiently. The goal of the paper is to determine the maximum number of flows for which state must be kept simultaneously in order to implement this scheme; this is equivalent to the number of simultaneously positive excesses, where the excess of a flow represents the amount by which it currently exceeds its fair share. An extensive combinatorial analysis is carried out to determine the worst-case output sequences and their storage requirement, leading to the final conclusion that the maximum number of simultaneously positive excesses is approximately $\frac{1}{\epsilon}(1 + \ln r)$ if no restriction is placed on the number of accepted packets, and is approixoximately $\frac{1}{\epsilon}(1 - (\frac{r-1}{r})^{T\epsilon - r} + \ln r)$ if the number of accepted packets is bounded above by T. These results belie the folk belief that guaranteeing fairness requires per-flow state, and demonstrate that fairness can be guaranteed by a policy that can easily be implemented within the storage and comptational capabilities of a conventional router.

Appendix

For the case $B = 39$, $d = 10$, $T = 69$ we present the output sequences of an optimal relaxed schedule in $\mathcal{F}$, the strict schedule ρ and a strict schedule based on the following heuristic: if there is an excess less than or equal to $B - d + 1$ hit the flow of lowest excess, else start a new flow. Letters of the alphabet are used as flow labels, and $-$ denotes a no-op. Each output sequence is given as a series of rows, where each row corresponds to an epoch or a phase within epoch 0.The schedules establish that $R(39, 10, 69) = 25$ and $23 \le N(39, 10, 69) \le 25$.

A Relaxed Schedule in $\mathcal{F}$

A A A A B B B B C C
C C C A B D D D D E
E E E E A B C D F F
F F F G G G G H H H
H I I I J J J K K K
L L M M N N O O P P
Q R S T U V W X Y

The Strict Schedule ρ

A A A A B B B B - -
C C C C D D D D A B
E E E E - - A C B D
F F F F - A C B D E
G G G H H H I I I J
J J K K L L M M N N
O P Q R S T U V W

A Heuristic Strict Schedule

A A A A B B B B C C
C C A D D D D B C E
E E E A E B F F F F
C E A D B G G G G F
C E H H H I I I J J
J K K L L M M N N O
O P Q R S T U V W

Acknowledgement The author wishes to thank Jacob Scott for valuable discussions.

References

1. C. Estan and G. Varghese. New directions in traffic measurement and accounting: Focusing on the elephants, ignoring the mice. In *ACM Transactions on Computer Systems* Vol. 21(3):pp. 270-313 (2003).
2. W.-C. Feng, D. D. Kandlur, D Saha, K. Shin. Blue: A New Class of Active Queue Management Algorithms. Technical Report CSE-TR-387-99, University of Michigan, April 1999.
3. P. Ferguson and G. Huston. Quality of Service: Delivering QoS on the Internet and in Corporate Networks. John Wiley and Sons, 1998.
4. S. Floyd and K. Fall. Promoting the Use of End-to-End Congestion Control in the Internet. *IEEE/ACM Transactions on Networking*, Vol. 7(4) pp458-573, August 1999.
5. S. Floyd and V. Jacobson. Random Early Detection Gateways for Congestion Avoidance. *IEEE/ACM Transactions on Networking*, Vol. 1(4):pp. 397-412, August 1993.
6. R. Mahajan and S. Floyd. Controlling High-Bandwidth Flows at the Congested Router. Technical Report TR-01-001, International Computer Sciences Institute.
7. R. Mahajan, S. Floyd, and D. Wetherall. Controlling High-Bandwidth Flows at the Congested Router. In *ACM ICNP*, November 2001.
8. R. Pan, L. Breslau, B. Prabhakar, and S. Shenker. Approximate Fairness through Differential Dropping. In *ACM Computer Communication Review* , July 2003.
9. R. Pan, B. Prabhakar, and K. Psounis. CHOKe, A Stateless Active Queue Management Scheme for Approximate Fair Bandwidth Allocation. In *IEEE INFOCOM*, March 2000.
10. C. Partridge. Gigabit Networking. Addison-Wesley, 1994.
11. S. Shenker, C. Partridge, and R. Guerin. Specification of Guaranteed Quality of Service. IETF RFC 2212, Sept., 1997.
12. I. Stoica, S. Shenker, and H. Zhang. Core-Stateless Fair Queuing:Achieving Approximately Fair Bandwidth Allocations in High Speed Networks. In *ACM SIG-COMM*, September 1998
13. A. Tanenbaum. Computer networks, 3d Edition. Prentice-Hall, 1996.
14. J. Wrochalski. Specification of Controlled-Load Network Element Service. IETF RFC 2211, September, 1997.

Optimal Flow Distribution Among Multiple Channels with Unknown Capacities*

Richard Karp[1], Till Nierhoff[2]**, and Till Tantau[3]**

[1] International Computer Science Institute, Berkeley, USA
`karp@icsi.berkeley.edu`
[2] Institut für Informatik, Humboldt-Universität zu Berlin, Germany
`nierhoff@informatik.hu-berlin.de`
[3] Fakultät für Elektrotechnik und Informatik, TU Berlin, Germany
`tantau@cs.tu-berlin.de`

Abstract. Consider a simple network flow problem in which a flow of value D must be split among n channels directed from a source to a sink. The initially unknown channel capacities can be probed by attempting to send a flow of at most D units through the network. If the flow is not feasible, we are told on which channels the capacity was exceeded (binary feedback) and possibly also how many units of flow were successfully sent on these channels (throughput feedback). For throughput feedback we present optimal protocols for minimizing the number of rounds needed to find a feasible flow and for minimizing the total amount of wasted flow. For binary feedback we present an asymptotically optimal protocol.

1 Introduction

Suppose you are daily producing 1000 copies of a newspaper, and you do not know how many will be bought at each of the newspaper stands in a new city. On a given day you place q_i copies at the ith newspaper stand and at the end of the day you get reports on how many have been sold at each stand. Based on that, you update the q_i for the next day. How many days does it take before you figure out the optimal way of distributing your newspapers? How do you minimize the number of unsold newspapers? Can you minimize both the number of days and the total number of unsold newspapers at the same time?

In the present paper we study different variants of these questions and propose optimal protocols for solving them. An Internet service provider might use such protocols to distribute bandwidth among channels with unknown capacities [1]. On a smaller scale, a user of a peer-to-peer network who tries to service parallel download requests from different peers also faces the problem of distributing her fixed, often small, bandwidth among the peers.

---

* An extended abstract of this paper was presented at the GRACO 2005 conference [6].
** Supported through DAAD (German academic exchange service) postdoc fellowships.

O. Goldreich et al. (Eds.): Shimon Even Festschrift, LNCS 3895, pp. 111–128, 2006.

1.1 Problem Statement

Consider the network flow problem in which there are n channels directed from a source to a sink, and we wish to determine a feasible flow of value D from the source to the sink. Each channel i has a capacity c_i. Initially, these capacities are unknown, but we know that they are nonnegative, integral, and sum up to some $C \geq D$. We consider only the static case in the present paper, that is, we assume that capacities do not change over time. To determine a feasible flow we proceed in rounds $t = 1, 2, 3, \ldots, T$. In round t we choose a *query vector* $q(t) = \big(q_1(t), q_2(t), \ldots, q_n(t)\big)$ and, simultaneously for all i, attempt to send $q_i(t)$ units of flow through channel i. The queries are nonnegative, rational, and sum up to at most D. We then receive feedback about the success of our attempts in the form of a *feedback vector* $f(t) = \big(f_1(t), \ldots, f_n(t)\big)$. In the case of *binary feedback* we learn, for each channel i, whether all of our flow reached the sink; that is, $f_i(t) = success$ when $q_i(t) \leq c_i$ and $f_i(t) = failure$ otherwise. In the case of *throughput feedback* we learn how much flow was delivered through each channel; that is, $f_i(t) = \min\{q_i(t), c_i\}$.

We study the efficient choice of the successive query vectors. We may be interested in minimizing the number of rounds required to determine a feasible flow, or in finding a feasible flow with minimum total waste. In the latter case, the *throughput* $P(t) = \sum_{i=1}^n \min\{q_i(t), c_i\}$ of round t is the total amount of flow that reaches the sink, the *waste* $W(t)$ in round t is $D - P(t)$, and the *total waste* is the sum of the waste over all rounds.

Our aim is to find optimal protocols, for each type of feedback, that output a feasible solution in a minimal number of rounds or cause a minimal amount of total waste. We introduce functions that describe how well such optimal protocols perform: Let ROUNDS-BF(n, D, C) denote the minimal number of rounds for binary feedback needed by any protocol to find a feasible solution for n channels, a demand D, and a network capacity C. Similarly, we define ROUNDS-TF(n, D, C) for throughput feedback. The functions WASTE-BF(n, D, C) and WASTE-TF(n, D, C) tell us how much total waste any protocol must cause before it finds a solution. Finally, we also introduce four sibling functions that omit the third parameter C as in ROUNDS-BF(n, D). For these functions the total capacity C is not given to the protocol, but is known to be at least D. Any protocol for this model can also be used in an alternative model where it is not guaranteed that $C \geq D$ and the task is to either find a feasible solution or determine that $C < D$. In this case one additional round may be required to check whether the final solution produced by the protocol on the assumption that $C \geq D$ is in fact feasible.

The problems and protocols presented in the present paper are "in the middle" between the protocols for optimally probing a single channel, studied for example in [2, 4, 5], and the protocols for routing and congestion control in large networks like the Internet, studied for example in [3], see [7] for a survey.

1.2   Our Contribution

The only previous result related to our complexity measures is given in [1], where it is shown that ROUNDS-BF$(n, D, D) \leq \log_2 D + \frac{\log_2 n}{2}$. Our main results can be summarized by the inequalities and the equality shown in Table 1. In all cases, the lower bounds are established through adversary strategies. The upper bounds are established by analyzing the performance of efficient protocols. In the equations, and in the following, $o(1)$ always refers to the parameter n.

Table 1. Summary of lower and upper bounds established in the present paper.

WASTE-TF(n, D)	$\geq \left(1 - o(1)\right) D \frac{\ln n}{\ln \ln n}$
WASTE-TF(n, D)	$\leq \left(1 + o(1)\right) D \frac{\ln n}{\ln \ln n}$
ROUNDS-TF(n, D)	$\geq \left(1 - o(1)\right) \frac{\ln n}{\ln \ln n}$
ROUNDS-TF(n, D)	$\leq \left(1 + o(1)\right) \frac{\ln n}{\ln \ln n}$
WASTE-BF(n, D, C)	$\geq \left(1 - o(1)\right) D \frac{\ln n}{\ln \ln n}$
WASTE-BF(n, D, C)	$\leq \left(2 + o(1)\right) D \frac{\ln n}{\ln \ln n}$
ROUNDS-BF(n, D, D)	$\geq \left(1 - o(1)\right) \left(\log_2 D - \log_2 n\right)$
ROUNDS-BF(n, D, D)	$\leq \log_2 D + \left(1 + o(1)\right) \frac{\ln n}{\ln \ln n}$
ROUNDS-BF(n, D, C)	$\geq \left(1 - o(1)\right) \left(\log_2 \frac{C}{C - D + 1} - \log_2 n\right)$
ROUNDS-BF(n, D, C)	$\leq \log_2 \frac{C}{C - D + 1} + \left(1 + o(1)\right) \log_2 n$
ROUNDS-BF$(2, D, D)$	$= \lceil \log_3 D \rceil$

1.3   Organization of This Paper

In Section 2 we introduce notions that are common to the analysis of all variants of the problem. Section 3 presents a key protocol that is useful both in the case of binary feedback and in the case of throughput feedback. Section 4 treats throughput feedback, and we establish matching upper and lower bounds, simultaneously for the minimal number of rounds and the minimal waste. In Section 5 we study binary feedback. There, we treat waste and rounds separately and pay special attention to the case of two channels.

2   Basic Protocol Analysis Tools

In this section we introduce basic ideas and terminology that will be used in all of our analyses.

2.1   Maintaining the Pinning Box

For any protocol, at any point during a run of the protocol we will have gathered certain information about the (unknown) capacities of the channels. For each channel i, from the answers to the previous t queries we will have deduced an *upper bound* $h_i(t)$ and a *lower bound* $l_i(t)$ for the channel capacity c_i. Thus $c_i \in [l_i(t), h_i(t)]$, called the *pinning interval*. The cross product of the pinning intervals will be called the *pinning box*. If, at any point, $l_i(t) = h_i(t)$, we obviously know c_i. At the beginning of a run, we know the trivial bounds $l_i(0) = 0$ and $h_i(0) = \infty$. A better upper bound is given by $h_i(0) = C$, but we may not know C. The sum of the $l_i(t)$ at time t will be denoted $L(t)$. Similarly, the sum of the $h_i(t)$ will be denoted $H(t)$.

We query a vector $q(t) = (q_1(t), \ldots, q_n(t))$ at time step t. The feedback $f(t) = (f_1(t), \ldots, f_n(t))$ may allow us to improve some or perhaps even all of our pinning intervals. For binary feedback, we can perform the following updating: if $f_i(t) = \textit{failure}$, set $h_i(t) = \min\{h_i(t-1), \lceil q_i(t) \rceil - 1\}$; if $f_i(t) = \textit{success}$, set $l_i(t) = \max\{l_i(t-1), \lceil q_i(t) \rceil\}$. Note that transmitting more than the upper bound makes little sense, but transmitting less than the lower bound can be useful: the demand "saved" by not transmitting it on a certain channel might be used to probe the capacity of other channels more quickly. For throughput feedback, we can always set $l_i(t) = \max\{l_i(t-1), f_i(t)\}$; and if $f_i(t) < q_i(t)$, we even know $l_i(t) = h_i(t) = f_i(t) = c_i$.

2.2   Effects of Increasing the Capacity

In certain situations an increase in the total capacity C affects the number of rounds or the waste needed to find a solution. Intuitively, a bigger capacity C should make it easier to find a solution—or at least not harder. However, a protocol that works fine for a capacity of, say, $C = D$ might try to exploit this fact to its advantage. For example for $n = 2$ and $C = D$, if we know $l_1 = \frac{3}{4}D$, then we can conclude that the capacity on the second channel can be at most $\frac{1}{4}D$, *but we cannot conclude this if we only know* $C \geq D$. Nevertheless, the following theorem shows that our first intuition is correct in the case of binary feedback.

Theorem 2.1. *Let $D \leq C \leq C'$. Then*

$$\text{ROUNDS-BF}(n, D, C) \geq \text{ROUNDS-BF}(n, D, C'),$$
$$\text{WASTE-BF}(n, D, C) \geq \text{WASTE-BF}(n, D, C').$$

The first inequality is proper in many cases. For example, later on we will see that $\text{ROUNDS-BF}(n, D, D) = \Theta(\log D)$ while $\text{ROUNDS-BF}(n, D, 2D) = \Theta(1)$ for fixed n.

Proof (of Theorem 2.1). Let P be a protocol that minimizes the number of rounds for n channels, a demand D, and a guaranteed capacity of C. We give a protocol P' that will need at most as many rounds as P and will work for any capacity $C' \geq C$. It does not even need to know C'.

Protocol P'.

1   **in** *round* $t \leftarrow 1, 2, 3, \ldots$ **do**

2   **let** $q(t)$ *be the query vector that protocol P would pose in round t*
          if it had seen the same results to our previous queries as we
          have seen

3   **query** $q(t)$ *and* **compute** *the new $l_i(t)$ and $h_i(t)$*

4   **let** $B := \{(c_1, \ldots, c_n) \mid l_i(t) \leq c_i \leq h_i(t), \sum_{i=1}^{n} c_i = C\}$

5   **let** $(m_1, \ldots, m_n) := \left(\min_{(c_1, \ldots, c_n) \in B} c_1, \ldots, \min_{(c_1, \ldots, c_n) \in B} c_n\right)$

6   **if** $\sum_{i=1}^{n} m_i \geq D$ **then**

7   **output** *some* $(d_1, \ldots, d_n)$ *with* $\sum_{i=1}^{n} d_i = D$ *and* $l_i(t) \leq d_i \leq m_i$

8   **stop**

In essence, for an unknown capacity vector $c' = (c'_1, \ldots, c'_n)$ summing up to C', Protocol P' runs Protocol P, "pretending" that the capacity is C. It interrupts the simulation once it has found a vector $m = (m_1, \ldots, m_n)$ that sums up to at least D and that, in a certain sense, lies "beneath" all vectors summing up to C inside the pinning box.

Our first claim is that the output of the protocol is, indeed, a solution. There exists a vector $c \in B$ that is componentwise below the real capacity vector c'. It can be obtained, for example, by successively dropping the components of c' to their established lower bounds until we can drop some components exactly as much as is needed to make the resulting vector c sum up to C. Then $c \in B$. Since $c \in B$, the vector m will be componentwise below c. Since the output is componentwise below m in turn, we conclude that the output is componentwise below the capacity vector c' and is hence a solution.

Our second claim is that Protocol P' runs for at most ROUNDS-BF(n, D, C) rounds. Consider the situation the protocol faces at round ROUNDS-BF(n, D, C). The crucial observation at this point is that all elements of the set B produce the exact same answers to all the queries that Protocol P' (and hence also P) has posed until now. Since Protocol P always finishes within ROUNDS-BF(n, D, C) rounds, it must be able to output a solution that is componentwise less than or equal to every element of B. But such a solution must necessarily be componentwise less than or equal to m, which must thus sum up to at least D. Thus $\sum_{i=1}^{n} m_i \geq D$.

For the claim WASTE-BF$(n, D, C) \geq$ WASTE-BF(n, D, C') just note that Protocol P' also wastes at most as much as Protocol P does.   □

Corollary 2.2. *For all n and D we have*

$$\text{ROUNDS-BF}(n, D, D) = \text{ROUNDS-BF}(n, D)$$
$$\text{WASTE-BF}(n, D, D) = \text{WASTE-BF}(n, D).$$

Proof. Having more information can never hurt us, which means

$$\text{ROUNDS-BF}(n, D, D) \leq \text{ROUNDS-BF}(n, D),$$
$$\text{WASTE-BF}(n, D, D) \leq \text{WASTE-BF}(n, D).$$

On the other hand, we just saw that an optimal protocol for $C = D$ can be made to work for all $C \geq D$. $\qquad\square$

For throughput feedback, the situation is simpler.

Theorem 2.3. *For all $C \geq D$ we have*

$$\text{ROUNDS-TF}(n, D, C) = \text{ROUNDS-TF}(n, D),$$
$$\text{WASTE-TF}(n, D, C) = \text{WASTE-TF}(n, D).$$

Proof. As in the proof of Theorem 2.1, we take a protocol P that minimizes the number of rounds or the waste when C and D are given and construct a protocol P' that simulates this protocol given D but not C and stops when the sum of the lower bounds is at least D. Observe that until this happens, there always exists a capacity vector with sum less than D that produces the exact same feedback as the one we have seen. Furthermore, for this capacity vector, at least one channel will not yet have been bounded from above. This is true because, in the case of throughput feedback, once a channel is bounded from above, its capacity is determined and its lower bound is equal to that capacity. Thus, if all channels were bounded from above, the sum of the lower bounds would be C, which is greater than or equal to D. Thus, as long as the sum of the lower bounds is less than or equal to D, we can find a vector summing up to any value $C > D$ that produces the exact same feedback as we have seen. $\qquad\square$

3   A Key Protocol

In this section we present a key protocol that is useful both in the case of binary feedback and in the case of throughput feedback. It is well defined in both cases but plays different roles. In the case of throughput feedback it is a complete protocol, and its analysis yields tight upper bounds on ROUNDS-TF(n, D) and WASTE-TF(n, D). In the case of binary feedback it is not a complete protocol, but serves as the first stage of a complete protocol that gives our best upper bound on WASTE-BF(n, D, C).

The key protocol seeks to bound from above as many channels as possible as quickly as possible. A channel is said to be *bounded* if some query on that channel has exceeded its capacity. For throughput feedback the capacity of a channel is determined as soon as it becomes bounded.

Protocol 3.1 (Key Protocol). *In each round send 0 on each bounded channel and divide a flow of D equally among the unbounded channels. Stop as soon as $L(t)$ is at least D or all channels have become bounded.*

Note that in the case of binary feedback (and only in that case) Protocol 3.1 is not a complete protocol, that is, when it stops we may not yet have found a feasible flow. The reason is that when the protocol stops because all channels have become bounded but $L(t) < D$, there may not be enough information for the protocol to produce a feasible flow of value D.

Theorem 3.2. *Let Protocol 3.1 stop after T_0 rounds. Then, both for binary and for throughput feedback, we have*

$$T_0 \leq \left(1 + o(1)\right) \frac{\ln n}{\ln \ln n}.$$

If the protocol stops because it has bounded all channels from above, then $H(T_0) \leq D\left(1 + o(1)\right) \frac{\ln n}{\ln \ln n}$ for binary feedback and $H(T_0) = C$ for throughput feedback.

Proof. The key idea in proving the upper bound on the number T_0 of rounds until the protocol stops is to establish a tradeoff between the time needed to bound all channels from above and the growth rate of $L(t)$.

Consider a run of the protocol. We focus on rounds $t = 1, \ldots, T_0 - 1$. Let $g(0) = n$ and let $g(t)$ denote the number of unbounded channels after round t. Let $\alpha(t) := g(t)/g(t-1)$ and let $t_{\max} \in \{1, \ldots, T_0 - 1\}$ be an index such that $\alpha(t_{\max}) \geq \alpha(t)$ for all t. In other words, in round $t_{\max}$ we see a maximal ratio of still-unbounded channels to previously-unbounded channels. Our two intermediate goals are to obtain an upper bound on the arithmetic mean and a lower bound on the geometric mean of the $\alpha(t)$ for $t \neq t_{\max}$. Applying the inequality relating the arithmetic and geometric means will then yield the claim.

For our first goal, the upper bound on $\sum_{t \neq t_{\max}} \alpha(t)$, we start with some simple observations concerning the lower bounds on the capacities of the $g(t)$ channels that have not yet been bounded from above. After the first round, the capacities of the $g(1)$ unbounded channels will each be bounded from below by $D/g(0)$, which means $L(1) \geq \alpha(1)D$ or, equivalently,

$$\frac{L(1)}{D} \geq \alpha(1).$$

Similarly, after round $t \geq 2$ the lower bounds on the capacities of $g(t)$ channels increase from $D/g(t-2)$ to $D/g(t-1)$. Thus, $L(t) - L(t-1) \geq g(t) \cdot \left(D/g(t-1) - D/g(t-2)\right)$ or, equivalently,

$$\frac{L(t) - L(t-1)}{D} \geq \frac{g(t)}{g(t-1)} - \frac{g(t)}{g(t-2)} = \alpha(t)\left(1 - \alpha(t-1)\right).$$

Telescoping this inequality and then "reindexing around $t_{\max}$" yields

$$\frac{L(T_0 - 1)}{D}$$

$$\geq \alpha(1) + \sum_{t=2}^{T_0-1} \alpha(t)\left(1 - \alpha(t-1)\right)$$

$$= \alpha(t_{\max}) + \sum_{t=1}^{t_{\max}-1} \alpha(t)\left(1 - \alpha(t+1)\right) + \sum_{t=t_{\max}+1}^{T_0-1} \alpha(t)\left(1 - \alpha(t-1)\right)$$

$$\geq \alpha(t_{\max}) + \left(1 - \alpha(t_{\max})\right) \sum_{t \neq t_{\max}} \alpha(t). \tag{1}$$

Since all terms in (1) are nonnegative, we conclude $L(T_0 - 1)/D \geq \alpha(t_{\max})$. We know that the protocol does not stop after query $T_0 - 1$, so $L(T_0 - 1) < D$ or, equivalently, $1 > L(T_0 - 1)/D$. We conclude that $0 < \alpha(t_{\max}) < 1$ and using (1) we get $1 - \alpha(t_{\max}) \geq 1 - \frac{L(T_0-1)}{D} > (1 - \alpha(t_{\max})) \sum_{t \neq t_{\max}} \alpha(t)$. Then, by dividing by the positive term $1 - \alpha(t_{\max})$, we obtain an upper bound on the arithmetic mean of the $\alpha(t)$:

$$\sum_{t \neq t_{\max}} \alpha(t) < 1. \tag{2}$$

For our second goal, a lower bound on $\prod_{t \neq t_{\max}} \alpha(t)$, we once more start with the observation that the protocol does not stop after query $T_0 - 1$. This implies $g(T_0 - 1) \geq 1$ since there is still an unbounded channel after round $T_0 - 1$. We can write $g(T_0 - 1)$ as $n \prod_{t=1}^{T_0-1} \alpha(t)$, which yields

$$\frac{1}{n} \leq \frac{g(T_0 - 1)}{n} = \prod_{t=1}^{T_0-1} \alpha(t) < \prod_{t \neq t_{\max}} \alpha(t). \tag{3}$$

For our final sequence of inequalities, we first use the lower bound (3) on the geometric mean, then the inequality $\prod x_i^{1/m} \leq \frac{1}{m} \sum x_i$ that relates the arithmetic and geometric means, and finally the upper bound (2) on the arithmetic mean:

$$\sqrt[T_0-2]{\frac{1}{n}} < \sqrt[T_0-2]{\prod_{t \neq t_{\max}} \alpha(t)} \leq \frac{1}{T_0 - 2} \sum_{t \neq t_{\max}} \alpha(t) < \frac{1}{T_0 - 2}.$$

This inequality is equivalent to $(T_0 - 2)^{T_0-2} < n$, which in turn implies $T_0 \leq (1 + o(1)) \frac{\ln n}{\ln \ln n}$.

It remains to show the upper bounds on $H(T_0)$ when the protocol stops because it has bounded all channels. For throughput feedback, once every channel has been bounded, we *know* the capacities of all channels and $H(T_0) = C$. For binary feedback, the key observation is that each channel is bounded from above exactly once. Consider the channels that are bounded in round t. The number of such channels is $g(t - 1) - g(t)$. The protocol issues a query of $D/g(t-1)$ and gets that value as an upper bound on their capacity. Thus, the upper bounds established in round t sum up to at most $(g(t - 1) - g(t))D/g(t - 1) = D(1 - g(t)/g(t-1)) \leq D$. So in each round we get a contribution of at most D to $H(T_0)$ and we have already shown that the number of rounds is at most $(1+o(1))\frac{\ln n}{\ln \ln n}$. This yields the claim. □

4   Throughput Feedback

In this section we establish matching upper and lower bounds on the number of rounds and the waste needed to find a solution when we get throughput feedback.

The results of this section can be summed up by the following inequalities from Table 1:

$$\left(1 - o(1)\right) D \frac{\ln n}{\ln \ln n} \leq \text{WASTE-TF}(n, D) \leq \left(1 + o(1)\right) D \frac{\ln n}{\ln \ln n},$$
$$\left(1 - o(1)\right) \frac{\ln n}{\ln \ln n} \leq \text{ROUNDS-TF}(n, D) \leq \left(1 + o(1)\right) \frac{\ln n}{\ln \ln n}.$$

4.1   Upper Bounds on Rounds and Waste for Throughput Feedback

Theorem 4.1. *For all n and D we have*

$$\text{WASTE-TF}(n, D) \leq (1 + o(1)) D \frac{\ln n}{\ln \ln n},$$
$$\text{ROUNDS-TF}(n, D) \leq (1 + o(1)) \frac{\ln n}{\ln \ln n}.$$

Proof. Theorem 3.2 states that if we run Protocol 3.1 for $T_0 \leq (1 + o(1)) \frac{\ln n}{\ln \ln n}$ rounds, then either $L(T_0) \geq D$ or $H(T_0) = C$. In either case we can directly specify a feasible flow. The upper bound on WASTE-TF(n, D) now follows from the fact that the waste in each round is at most D.   □

4.2   Lower Bounds on Rounds and Waste for Throughput Feedback

Theorem 4.2. WASTE-TF$(n, D) \geq \left(1 - o(1)\right) D \frac{\ln n}{\ln \ln n}$.

Proof. Consider an optimal protocol. We describe an adversary strategy that causes it to waste at least the amount stated in the theorem. The rough idea is as follows: The adversary maintains a list of channels on which it has *committed* itself. This list gets larger in each round, but the adversary tries to make it grow as slowly as possible. A commitment occurs for channel i when the adversary answers, for the first time in some round t, that a query $q_i(t)$ is larger than the capacity c_i. Once the adversary has committed itself on channel i, it will answer all subsequent queries $q_i(t)$ with the feedback $\min\{q_i(t), c_i\}$. However, if the adversary has not yet committed itself on a channel, it is still free to choose the capacity of the channel in any way it likes later on, as long as the capacity is at least as large as the largest query asked on the channel.

In order to commit itself as slowly as possible, in each round t the adversary analyzes the queries. For the channels on which it has already committed itself it simply answers $\min\{q_i(t), c_i\}$, where c_i is its previous commitment. For the channels on which it has not yet committed itself, it sorts the queries according to their value and then commits itself on the $r(t)$ *largest* ones to a certain value $z(t)$. Committing itself on the largest channels ensures that the only good strategy against the adversary is to evenly distribute the demand on the channels on which the adversary has not yet committed itself. Any strategy that poses small queries for certain uncommitted channels and large queries on other uncommitted channels will cause the adversary to answer, essentially, "the small queries succeed and cause a lot of waste since they are too small, and the large ones fail and cause a lot of waste since they are too high."

The two key parameters of the adversary's strategy are the number $r(t)$ of channels on which it commits itself in round t and the value $z(t)$ to which it newly commits itself. These parameters are chosen as follows:

$$r(t) := \lceil \alpha^{t-1}n \rceil - \lceil \alpha^t n \rceil \quad \text{for } t \in \{1, \ldots, T\},$$

$$z(t) := \frac{D}{(1-\alpha)\alpha^{t-2}n} \quad \text{for } t \in \{2, \ldots, T\},$$

where $\alpha := 1/\ln n$ and $T := \left\lceil \frac{\ln n}{\ln \ln n} \right\rceil$. Special cases occur in the first round and round $T + 1$. In the first round $z(1) := 0$. In round $T + 1$, we set $r(T + 1) := 1$ and the adversary chooses $z(T + 1)$ in any way that ensures that the capacities add up to $C = D$. Note that the choice of α and T ensures $\alpha^{t-1}n > 1$ for $t \leq T$ and $\alpha^T n \leq 1$.

We must now show the following:

1. During the first T rounds the answers of the adversary are consistent. This means that there exists a capacity vector, summing up to $C = D$, for which exactly the same answers to the queries would be given as the ones that the adversary gave.
2. During the first T rounds, we have $L(t) < D$. In particular, no way has yet been found to distribute the whole flow D.
3. The total waste is $\left(1 - o(1)\right)D\frac{\ln n}{\ln \ln n}$.

The first claim follows directly from the commitment rule. To prove the second claim, consider the total capacity of the channels on which the adversary newly commits itself in round t. This capacity is given by the product $r(t)z(t)$, which can be bounded as follows for $t \leq T$:

$$r(t)z(t) \leq \frac{\lceil \alpha^{t-1}n \rceil - \lceil \alpha^t n \rceil}{(1-\alpha)\alpha^{t-2}n}D \leq \frac{1 + \alpha^{t-1}n - \alpha^t n}{(1-\alpha)\alpha^{t-1}n}\alpha D$$

$$= \underbrace{\frac{1 + (1-\alpha)\alpha^{t-1}n}{(1-\alpha)\alpha^{t-1}n}}_{\leq 3}\alpha D \leq 3\alpha D. \tag{4}$$

To see that the fraction is, indeed, less than or equal to 3 for large n, recall that $\alpha^{t-1}n > 1$. This shows that the fraction is bounded from above by $1 + \frac{1}{1-\alpha}$, and since $\alpha \leq 1/2$ for $n \geq 4$, it follows that $1 + \frac{1}{1-\alpha} \leq 3$ for $n \geq 4$.

Consider the lower bounds established in round T. The number of uncommitted channels after round T is $\lceil \alpha^T n \rceil$, which is equal to 1 since $0 < \alpha^T n \leq 1$. On the one uncommitted channel, the lower bound will be at most $z(T)$ since we have never had a query above this on this channel. This allows us to bound $L(T)$ as follows, which proves the second claim:

$$L(T) < z(T) + \sum_{t=1}^{T} r(t)z(t) \leq \frac{\alpha D}{1 - \alpha} + 3\alpha T D = o(D).$$

For the third claim, we must establish a lower bound on the waste. For this, first consider the throughput $P(t)$ in round t. It is at most

$$P(t) \leq g(t)z(t) + r(t)z(t) + \cdots + r(1)z(1), \tag{5}$$

where $g(t) = \lceil \alpha^t n \rceil$ is the number of uncommitted channels in round t. Here $g(t)z(t)$ is the sum of the lower bounds for the uncommitted channels and each term $r(i)z(i)$ is the sum of the lower bounds for the channels that became committed in round i. We know already, from (4), that we can upper bound each $r(i)z(i)$ by $3\alpha D$ (and we also know $z(1) = 0$). A bound on $g(t)z(t)$ can be obtained as follows:

$$g(t)z(t) = \lceil \alpha^t n \rceil \frac{D}{(1-\alpha)\alpha^{t-2}n} \leq \frac{1+\alpha^t n}{(1-\alpha)\alpha^{t-2}n} D = \underbrace{\frac{1+\alpha\alpha^{t-1}n}{(1-\alpha)\alpha^{t-1}n}}_{\leq 3} \alpha D \leq 3\alpha D.$$

To see that the bound of 3 on the large fraction is, indeed, correct, compare the fraction to the corresponding fraction in (4) and note that $\alpha \leq 1 - \alpha$.

The final step is to bound the total waste. Applying the bounds on $r(i)z(i)$ and $g(t)z(t)$ to inequality (5) yields $P(t) \leq 3t\alpha D$. This implies that the total waste is at least

$$\sum_{t=1}^{T-1} (D - P(t)) \geq (T-1)D - 3\binom{T}{2}\alpha D = \left(1 - O(T\alpha)\right)TD. \tag{6}$$

Using $\alpha T = \frac{1}{\ln \ln n} + o(1)$, the theorem follows. $\square$

Once more, by observing that in any round we can waste at most D, we get a corollary.

Corollary 4.3. ROUNDS-TF$(n, D) \geq \left(1 - o(1)\right)\frac{\ln n}{\ln \ln n}$.

5   Binary Feedback

The waste function WASTE-BF(n, D, C) for binary feedback behaves similarly to the waste function for throughput feedback and almost the same bounds are established below. Opposed to this, ROUNDS-BF(n, D, C) behaves somewhat differently from the corresponding function for throughput feedback. We show that for binary feedback the number of rounds does not only depend on n and D, but also on C. Our upper and lower bounds for ROUNDS-BF(n, D, C) do not quite match for general n, but we solve the special case of two channels completely: ROUNDS-BF$(2, D, D) = \lceil \log_3 D \rceil$.

5.1   Minimizing Waste for Binary Feedback

Recall that the waste caused by an optimal protocol for throughput feedback is $\Theta(D\frac{\ln n}{\ln \ln n})$. Our analysis gave even more precise bounds: The ratio between

ROUNDS-TF(n, D, C) and $D\frac{\ln n}{\ln \ln n}$ approaches 1 as n tends to infinity. For binary feedback we establish the same Θ-bound as for throughput feedback, but the exact upper and lower bounds do not match as tightly:

$$\left(1 - o(1)\right) D \frac{\ln n}{\ln \ln n} \leq \text{WASTE-BF}(n, D, C) \leq \left(2 + o(1)\right) D \frac{\ln n}{\ln \ln n}.$$

The lower bound follows from Theorem 4.2. To prove the upper bound, we consider a protocol that begins by executing Protocol 3.1, which terminates when $L(t) \geq D$ or all channels have become bounded. In the former case, one can immediately specify a feasible flow of value D, but in the latter case, because binary feedback is weaker than throughput feedback, Protocol 3.1 may leave us with some pinning box. We use a second protocol to reduce the sum of all the pinning interval gaps to zero.

Intriguingly, this second protocol is the *proportional allocation protocol*, which was originally introduced in [1] in the context of *round* minimization. Our analysis shows that this protocol reduces the sum of the pinning intervals to zero without wasting more than this sum. The idea of the proportional allocation protocol is, at each step, to choose queries summing to D that split all the pinning intervals in the same proportions: for each round t and each channel i, $q_i(t) - l_i(t - 1) = \rho(t) \cdot \left(h_i(t - 1) - l_i(t - 1)\right)$, where $\rho(t)$ is chosen to make the sum of the queries equal to D.

Protocol 5.1 (Proportional Allocation, [1]). *Assume that some protocol has been used during rounds $t = 1, 2, 3, \ldots, T_0$ to establish (possibly nonoptimal) pinning intervals $[l_i(T_0), h_i(T_0)]$ for some or for all channels i. If no bounds have yet been established for some channel i, let $l_i(T_0) = 0$ and $h_i(T_0) = \infty$.*

1   **foreach** $i \in \{1, \ldots, n\}$
2       $h_i(T_0) \leftarrow \min\{h_i(T_0), D\}$
3   **in** round $t \leftarrow T_0 + 1, T_0 + 2, T_0 + 3, \ldots$ **do**
4       $\rho(t) \leftarrow \frac{D - L(t-1)}{H(t-1) - L(t-1)}$
5   **foreach** $i \in \{1, \ldots, n\}$ **do**
6           $q_i(t) \leftarrow l_i(t - 1) + \rho(t) \cdot \left(h_i(t - 1) - l_i(t - 1)\right)$
7   **query** $\left(q_1(t), \ldots, q_n(t)\right)$
8   **if** $H(t) = C$ or $L(t) = D$ **then output** last query vector; **stop**

Theorem 5.2. *Let Protocol 5.1 be started in round $T_0 + 1$ with a certain pinning box already established. Then it will find a solution wasting no more than $H(T_0) - L(T_0)$.*

Proof. Let $\Delta(t) := H(t) - L(t)$ denote the sum of the pinning interval gaps. Our aim is to show the following claim: *If the protocol wastes $W(t)$ in round t, we have $\Delta(t + 1) \leq \Delta(t) - W(t)$.* In other words, we reduce the pinning interval gap by at least the amount we waste. If this claim holds, we clearly cannot waste more than $\Delta(T_0)$ before the gap drops to $C - D \geq 0$.

In each round t we distinguish two cases, depending on whether $\rho(t) \leq 1/2$ or $\rho(t) > 1/2$ for this round. For the case $\rho(t) \leq 1/2$, consider the values $w_i(t) := \max\{0, q_i(t) - c_i\}$. We claim $W(t) = \sum_{i=1}^{n} w_i(t)$. This can be seen as follows:

$$W(t) = D - \sum_{i=1}^{n} \min\{c_i, q_i(t)\}$$

$$= D - \sum_{i=1}^{n} (q_i(t) - \max\{0, q_i(t) - c_i\}) = \sum_{i=1}^{n} w_i(t).$$

We used the fact that Protocol 5.1 always distributes the complete demand D, that is, $\sum_{i=1}^{n} q_i(t) = D$. Consider a channel i with $w_i(t) > 0$ and thus $q_i(t) > c_i$. Since the query was a failure, the upper bound $h_i(t)$ will be decreased to $q_i(t)$. Thus the pinning interval changes from $[l_i(t-1), h_i(t-1)]$ to $[l_i(t-1), q_i(t)]$ and its size changes from $h_i(t-1) - l_i(t-1)$ to $q_i(t) - l_i(t-1) = \rho(t) \cdot (h_i(t-1) - l_i(t-1))$. Thus, channel i causes the gap Δ to shrink by at least $(1 - \rho(t))(h_i(t-1) - l_i(t-1))$ while it causes a waste of at most $w_i(t) = q_i(t) - c_i \leq q_i(t) - l_i(t-1) = \rho(t) \cdot (h_i(t-1) - l_i(t-1))$. Since this argument is true for all channels, we conclude that $W(t)$ is at most the decrease of the pinning interval gaps.

For the case $\rho(t) > 1/2$, we argue similarly, but consider the values $w_i'(t) := \max\{0, c_i - q_i(t)\}$. The sum of these values is an upper bound on the waste:

$$W(t) = D - \sum_{i=1}^{n} \min\{c_i, q_i(t)\}$$

$$= D - \sum_{i=1}^{n} (c_i - \max\{0, c_i - q_i(t)\}) = D - C + \sum_{i=1}^{n} w_i'(t) \leq \sum_{i=1}^{n} w_i'(t).$$

It remains to argue that on each channel we decrease the size of the pinning interval by at least $w_i'(t)$. Suppose $w_i'(t) > 0$. Then $c_i > q_i(t)$ and the pinning interval changes from $[l_i(t-1), h_i(t-1)]$ to $[q_i(t), h_i(t-1)]$. Its size changes from $h_i(t-1) - l_i(t-1)$ to $h_i(t-1) - q_i(t) = (1 - \rho(t))(h_i(t-1) - l_i(t-1))$ and thus its size is reduced by at least $\rho(t) \cdot (h_i(t-1) - l_i(t-1))$. Since $w_i'(t) = c_i - q_i(t) \leq h_i(t-1) - q_i(t) = (1 - \rho(t))(h_i(t-1) - l_i(t-1))$ and since $\rho(t) > 1/2$, we conclude that the reduction of the interval size is larger than the waste. □

Note that, if we omitted Protocol 3.1 and only used Protocol 5.1 starting with $T_0 = 0$, then we would have $H(0) = nD$, $L(0) = 0$, and the theorem would merely guarantee an upper bound on waste of nD. By using Protocol 3.1 followed by Protocol 5.1 we obtain a much better bound.

Theorem 5.3. WASTE-BF$(n, D, C) \leq (2 + o(1)) D \frac{\ln n}{\ln \ln n}$.

Proof. We run Protocol 3.1 for T_0 rounds until it stops, and then start using Protocol 5.1 from round $T_0 + 1$ on. Theorem 3.2 tells us that Protocol 3.1 terminates within at most $(1 + o(1)) \frac{\ln n}{\ln \ln n}$ rounds. During each round at most D can be wasted, which yields to a total waste of at most $(1 + o(1)) D \frac{\ln n}{\ln \ln n}$ during the execution of Protocol 3.1. Theorem 3.2 also tells us that after the execution of Protocol 3.1 we are left with upper bounds $H(T_0) \leq (1 + o(1)) D \frac{\ln n}{\ln \ln n}$. Thus, when Protocol 5.1 starts in round $T_0 + 1$, the size of the pinning box is at most

$H(T_0) - L(T_0) \le H(T_0) \le (1 + o(1))D\frac{\ln n}{\ln \ln n}$. By Theorem 5.2, during the execution of Protocol 5.1 we waste at most the size of the pinning box, which yields the claim.   □

5.2 Minimizing Rounds for Binary Feedback

For waste it makes little difference what kind of feedback is used: using binary feedback at most doubles the waste. The situation is quite different for rounds. For throughput feedback the optimal number of rounds $\frac{\ln n}{\ln \ln n}$ does not depend on either D or C. For binary feedback the optimal number of rounds depends both on D and on C. In particular, an increased total capacity C allows us to find solutions more quickly. In detail, we show the following:

$$\text{ROUNDS-BF}(n, D, C) \ge (1 - o(1))\left(\log_2 \frac{C}{C-D+1} - \log_2 n\right),$$

$$\text{ROUNDS-BF}(n, D, C) \le \log_2 \frac{C}{C-D+1} + (1 + o(1))\log_2 n.$$

It is proven in [1] that Protocol 5.1 started with a certain pinning box already established in round T_0 will find a solution within $\log_4 (H(T_0) - D)(D - L(T_0))$ additional rounds. Chandrayana et al. infer from this, using the initial bounds $h_i(0) = D$ and $l_i(0) = 0$, that Protocol 5.1 will find a solution within $\log_2 D + \frac{\log_2 n}{2}$ rounds. Using the same protocol composition trick as for Theorem 5.3, this bound can be improved as follows:

Theorem 5.4. $\text{ROUNDS-BF}(n, D, D) \le \log_2 D + (1 + o(1))\frac{\ln n}{\ln \ln n}$.

Proof. We run Protocol 3.1 followed by Protocol 5.1. By Theorem 3.2, when the proportional allocation protocol starts in round T_0, we have $H(T_0) \le (1 + o(1))D\frac{\ln n}{\ln \ln n}$, which implies $(H(T_0) - D)(D - L(T_0)) \le (1+o(1))D^2\frac{\ln n}{\ln \ln n}$. Thus the number of rounds for the second protocol will be at most

$$\log_4\left((1 + o(1))D^2\frac{\ln n}{\ln \ln n}\right) = \log_2 D + o(\ln \ln n).$$

Since Protocol 3.1 requires only $(1 + o(1))\frac{\ln n}{\ln \ln n}$ rounds, we get the claim.   □

Chandrayana et al. prove a lower bound that matches this upper bound in the sense that they show $\text{ROUNDS-BF}(n, D, D) = \Theta(\log_2 D)$ for fixed n. However, the protocol is far from optimal if C is larger than D and if we know this. For example, $\text{ROUNDS-BF}(2, D, 2D) = 2$ for all D since $(D, 0)$ or $(0, D)$ is always a solution.

We present a modification of Protocol 5.1 that finds a solution in a number of rounds that depends on n and ϵ but not on D, if we know $C \ge (1 + \epsilon)D$.

Protocol 5.5 (Scaled Proportional Allocation).
Let $\Delta := \lfloor (C - D)/n \rfloor + 1$ and let $D' := \lceil D/\Delta \rceil$. Scaled proportional allocation simulates Protocol 5.1 from round $T_0 + 1$ onward, but

1. *we try to distribute a demand of D' instead of D and*
2. *whenever proportional allocation wishes to query $(q'_1(t), \ldots, q'_n(t))$ we query the vector $(\Delta q'_1(t), \ldots, \Delta q'_n(t))$ instead.*

Theorem 5.6. ROUNDS-BF$(n, D, C) \leq \log_2 \frac{C}{C-D+1} + (1 + o(1)) \log_2 n$.

Proof. We first run Protocol 3.1 (without any scaling) for T_0 rounds to establish an initial pinning box. Now consider the actual capacities $(c_1, \ldots, c_n)$ and let $c'_i := \lfloor c_i/\Delta \rfloor$. A run of Protocol 5.5 will produce the same queries as running the original Protocol 5.1 for the capacity vector $(c'_1, \ldots, c'_n)$ and for the demand D': we have $\Delta q'_i(t) \leq c_i$ if and only if $\Delta q'_i(t) \leq \Delta \lfloor c_i/\Delta \rfloor$, which is in turn equivalent to $q'_i(t) \leq \lfloor c_i/\Delta \rfloor$. Thus a query $\Delta q'_i(t)$ will be answered with *success* in Protocol 5.5 if and only if the query $q'_i(t)$ is answered the same way in Protocol 5.1 for the scaled capacities and the scaled demand.

There exists a solution with respect to the capacities $(c'_1, \ldots, c'_n)$ and the demand D' since

$$\sum_{i=1}^{n} c'_i = \sum_{i=1}^{n} \left\lfloor \frac{c_i}{\Delta} \right\rfloor \geq \sum_{i=1}^{n} \frac{c_i - \Delta + 1}{\Delta} = \frac{C - n\Delta + n}{\Delta} = \frac{D}{\Delta}.$$

The left-hand side is an integer and we even have $\sum_{i=1}^{n} c'_i \geq \lceil D/\Delta \rceil = D'$.

So far, we have shown that Protocol 3.1 followed by 5.5 will need as much time to find a solution as Protocol 3.1 followed by 5.1 will need for the scaled capacity vector and the scaled demand. Thus we can apply Theorem 5.4 with D replaced by D'. The interesting term $\log_2 D'$ can be bounded as follows:

$$\log_2 D' = \log_2 \left\lceil \frac{D}{\lfloor (C-D)/n \rfloor + 1} \right\rceil = \log_2 \left\lceil \frac{nD}{n \lfloor (C-D)/n + 1 \rfloor} \right\rceil$$
$$\leq \log_2 \left\lceil \frac{nD}{C-D+1} \right\rceil \leq \log_2 \frac{C}{C-D+1} + \log_2 n.$$

This proves the theorem.   □

Theorem 5.7. ROUNDS-BF$(n, D, C) \geq (1 - o(1))\left(\log_2 \frac{C}{C-D+1} - \log_2 n\right)$.

Proof. We present an adversary strategy against an optimal protocol for given numbers n, D, and C. The adversary keeps track of a set X of capacity vectors summing up to C that are consistent with all the answers the adversary has provided until now. Initially, X contains all vectors of nonnegative integers summing up to C and thus has size $\binom{C+n-1}{n-1}$. When the protocol produces a query vector, the 2^n possible answer vectors partition X into 2^n sets whose elements are consistent with one answer vector. At least one of these sets has size at least $|X|/2^n$ and the adversary returns the answer vector corresponding to this set.

We claim that the protocol cannot produce its final output before $|X|$ has dropped to $\binom{C-D+n-1}{n-1}$. To see this, note that every vector summing up to D is componentwise below at most $\binom{C-D+n-1}{n-1}$ many vectors in X. We conclude that the number T of rounds needed by the optimal protocol to produce its solution must satisfy $\binom{C+n-1}{n-1} / (2^n)^T \leq \binom{C-D+n-1}{n-1}$ and thus

$$
\begin{aligned}
T &\geq \frac{1}{n}\log_2 \frac{\binom{C+n-1}{n-1}}{\binom{C-D+n-1}{n-1}} \\
&\geq \frac{n-1}{n}\log_2 \frac{C+n-1}{C-D+n-1} \geq \frac{n-1}{n}\log_2 \frac{C}{n(C-D+1)}.
\end{aligned}
$$

The last term equals $(1 - o(1))\left(\log_2 \frac{C}{C-D+1} - \log_2 n\right)$, which proves the claim.

$\square$

5.3 Minimizing Rounds for Two Channels with Binary Feedback

The upper and lower bounds proved in the previous section do not quite match: the upper bound contains a positive $\log_2 n$ term, the lower bound a negative $\log_2 n$ term. A first step toward closing this gap is the following optimal protocol for $n = 2$. The performance of this protocol was surprising to us since even a binary search needs $\log_2 D$ rounds to find a solution:

Protocol 5.8 (Two Channel Protocol).

```
1    l ← 0, h ← D
2    in round t ← 1, 2, 3, . . . do
3         q₁(t) ← (h − l)/3 + l
4         q₂(t) ← (h − l)/3 + D − h
5         query (q₁(t), q₂(t)) receiving (f₁(t), f₂(t))
6         if H(t) = D or L(t) = D then output last query vector; stop
7         if f₁(t) = failure then h ← q₁(t) else l ← q₁(t)
8         if f₂(t) = failure then l ← max{l, D − q₂(t)}
9                              else h ← min{h, D − q₂(t)}
```

Theorem 5.9. ROUNDS-BF$(2, D) = \lceil \log_3 D \rceil$.

Proof. By Corollary 2.2 it suffices to show ROUNDS-BF$(2, D, D) = \log_3 D$. We begin with an adversary strategy that ensures that any fixed protocol cannot find the solution in less than $\log_3 D$ rounds. This will show the inequality ROUNDS-BF$(2, D, D) \geq \log_3 D$.

The aim of the adversary is to keep the protocol in the dark about the capacity c_1. The adversary keeps track of a pinning interval $[l, h]$ that gets smaller in each round. In any round t, the answers of the adversary up to then will be consistent with every capacity vector (c_1, c_2) with $c_1 \in [l, h]$ and $c_2 = D - c_1$. Initially, $l = 0$ and $h = D$, which clearly fulfills the requirements. In round t, consider a query vector $q(t) = (q_1(t), q_2(t))$ and consider where the two numbers $q_1(t)$ and $D - q_2(t)$ lie in the interval $[l, h]$. They can split the interval into at most three intervals and at least one of them must have size at least $(h - l)/3$. The adversary answers such that any value within this largest interval is permissible. For simplicity, assume $l \leq q_1(t) \leq D - q_2(t) \leq h$—other cases are similar. Then, in detail, if the largest interval is $[l, q_1(t)]$, the adversary answers (*failure*, *success*). If the largest interval is $[q_1(t), D - q_2(t)]$, the adversary answers (*success*, *success*). If the largest interval is $[D - q_2(t), h]$, the adversary answers (*success*, *failure*). In each round, the size of the interval is reduced by a factor of at most 3. Thus, the adversary does not have to settle on a capacity distribution before $\log_3 D$ rounds have passed.

To prove the inequality ROUNDS-BF$(2, D, D) \leq \log_3 D$, consider Protocol 5.8. It implements a strategy against the just-given adversary by keeping track of the interval $[l, h]$ for which it knows $c_1 \in [l, h]$ and thus $c_2 \in [D - h, D - l]$. In each round it poses two queries $\big(q_1(t), q_2(t)\big)$ such that $q_1(t)$ and $D - q_2(t)$ cut the interval into three equal parts. For every possible answer vector $\big(f_1(t), f_2(t)\big)$ the interval size will be reduced by a factor of 3, which proves the claim.   □

6   Conclusion and Open Problems

We proposed a framework for studying different ways of distributing a flow in a simple network with unknown capacities. We studied two kinds of feedback, namely binary and throughput feedback. For the latter type of feedback we presented a protocol that is optimal both with respect to the number of rounds and the waste produced. For binary feedback there is still a gap between the upper and lower bounds when the number n of channels is also taken into account. The main open problem of this paper is closing this gap. For the special case of two channels we establish an optimal protocol and note that it outperforms binary search.

Experimental work done by Chandrayana et al. [1] has shown that the (unscaled) proportional allocation protocol performs well on real data with respect to the number of rounds needed. Our theoretical work backs these findings, but we showed that scaling can significantly improve the performance of the protocol if the available capacity is larger than the demand. This opens the intriguing possibility that the scaled proportional allocation protocol might be an optimal protocol for minimizing both rounds and waste.

We did not address the computational complexity of protocols within our models. However, reviewing the protocols that we used for upper bounds, we see that they are both easy to implement and have low computational complexity.

For further research we suggest studying dynamic versions of the problem, in which the capacities may change from round to round, either under deterministic constraints or under a stochastic fluctuation model. For the case $n = 1$, some results in the framework of competitive analysis of on-line algorithms are given in [5]. For general n no theoretical results are available, but Chandrayana et al. [1] have proposed n-channel protocols for the dynamic case that perform well in practice.

We also suggest studying optimal protocols for a different throughput model: let the throughput on channel i be $q_i(t)$ if $q_i(t) \leq c_i$, and 0 otherwise. This model is suggested by Internet congestion control protocols in which, whenever a packet is dropped in a round, it cannot be guaranteed that any packets are delivered. For the case $n = 1$, this model is studied in [5].

References

1. K. Chandrayana, Y. Zhang, M. Roughan, S. Sen, and R. Karp. Search game in inter-domain traffic engineering. Manuscript, 2004.
2. D. Chiu and R. Jain. Analysis of the increase/decrease algorithms for congestion avoidance in computer networks. *J. of Comput. Networks and ISDN*, 17(1):1–14, 1989.
3. N. Garg and N. E. Young. On-line end-to-end congestion control. In *Proc. of FOCS '02*, pages 303–312, 2002.
4. V. Jacobson. Congestion avoidance and control. In *Proc. of ACM SIGCOMM '88*, pages 314–329, 1988.
5. R. Karp, E. Koutsoupias, C. Papadimitriou, and S. Shenker. Combinatorial optimization in congestion control. In *Proc. of FOCS '00*, pages 66–74, 2000.
6. R. Karp, T. Nierhoff, and T. Tantau. Optimal Flow Distribution Among Multiple Channels with Unknown Capacities. In *Proc. of GRACO '05*, volume 19 of Electronic Notes in Discrete Mathematics, pages 225–231. Elsevier, 2005.
7. S. H. Low, F. Paganini, and J. C. Doyle. Internet congestion control: An analytical perspective. *IEEE Control Systems Magazine*, pages 28–43, Feb. 2002.

Parceling the Butterfly and the Batcher Sorting Network

Ami Litman

Computer Science Department, Technion,
Haifa 32000, ISRAEL
litman@cs.technion.ac.il

Abstract. This paper proposes a new metric that aims to express the cost of manufacturing large-scale, communication-intensive digital systems. These systems are modeled by networks with internal and external edges, where the latter are input/output edges connecting the system with the external world. A *k–parceling* of such a network is a partition of the network into components each having at most k non-internal edges. (Such a partition is of interest when the number of the external edges is much larger than k.) The *k–parceling number* of a network is the minimal number of components in a k–parceling.

We argue that the parceling number of a large-scale, communication-intensive network expresses the cost of such a system better than the contemporary prevalent metrics and therefore it can guide the designers of such systems better than these metrics.

The paper studies the parceling of two important networks, the Butterfly and the Batcher Bitonic sorting network. It establishes explicit (rather than asymptotic) lower and upper bounds on the parceling number of both networks.

1 Introduction

In the theoretical study of digital systems (i.e., digital networks), three metrics are used to express the cost of manufacturing these systems: the size of the network and the minimal area or volume of a layout of the network in Thompson's 2-D or 3-D grid models[1] [T79, T80, L85, LR86]. The major aim of these metrics is to help the designers of digital systems evaluate the pros and cons of the many design alternatives so that they can focus on the most promising ones.

We argue that all these metrics poorly express the cost of large-scale, communication-intensive systems, and therefore provide little help to the designers of such systems.

---

[0] This work was supported in part by BSF Grant No. 94-266.

[1] In these models, the nodes of the network are embedded in distinct points of a two or three dimensional finite grid and the arcs of the network are realized by edge-disjoint paths of the grid. The *area* or *volume* of the layout is the total number of points of this grid.

O. Goldreich et al. (Eds.): Shimon Even Festschrift, LNCS 3895, pp. 129–142, 2006.

The first metric—the network size—plainly ignores the cost of the interconnection.

The second metric—the area—is useful only for small systems since large ones are invariably built in 3-D. Advocates of this metric argue that the large systems of today will be small systems tomorrow, but this is irrelevant; we need today to build the large systems of today, and we will need tomorrow to build the large systems of tomorrow.

The third metric—the volume—is inadequate since it ignores the cardinal engineering issues of the spatial partitioning of the system into physical subsystems and the packaging of these subsystems.

In contemporary packaging technology [B90, BMM78] the following hierarchy of package types are used: chip, Multi-Chip Module, Printed Circuit Board, box and cabinet. (Some levels of the hierarchy may be absent in any given system.) The volume metric is not only inadequate for contemporary packaging technology but seems inadequate for future ones as well since it ignores the following issues:

– Packages, wiring and cooling are expensive; volume is essentially free. (Note that this is not the case with the area metric; silicon area is expensive even if not used.)
– There is a great variation in the density of wires across the packaging hierarchy. In contemporary technology there is a gap of three orders of magnitude between the density of intra-chip wires and inter-cabinet ones.
– There is a level of the packaging hierarchy that serves for *Field Replaceable Units* (FRUs). It is mandatory that a failed FRU can be replaced without taking apart the entire system, but this is routinely violated by 3-D grid layouts.
– It is desirable that the system is made of few different types of *modules*, especially when these modules are fabricated by mass production techniques.

To summarize, the 3-D grid model is an abstraction that is very detached from the systems it aims to model; it is valid only for showing that some design alternatives are very bad, but not for showing that some alternatives are good.

We propose a new cost metric for large-scale, communication-intensive digital systems. The metric focuses on packaging and is based on the assumption that the input/output capability of the packages, rather than their internal capacity, is the dominant factor in the packaging of the system. Hence, the metric models the case where all the packages have unbounded internal capacity and the same limited input/output capability while the goal is to minimize the number of packages.

Assume that the input/output capability of the packages is k; i.e., a package is connected to the rest of the system (and to the external world) by k wires[2] and each wire implements a single edge of the network. This gives rise to the concept of a k-*parceling* of a network – a partition of the network into components such

---

[2] In contemporary packaging technology [B90, BMM78], a VLSI chip is connected to the rest of the system by a few hundred wires.

that each component is connected to the rest of the system (and the external world) by at most k edges. (Such a partition is of interest when the number of external edges of the network is much larger than k.) The k–*parceling number* of a network N, denoted $P(N, k)$, is the minimal number of components in a k–parceling.

In addition to being relevant to the engineering of large digital systems, the parceling metric has two significant advantages over the volume metric. Firstly, an optimal parceling of the network readily provides the designer with a top-level physical partitioning of the system. Secondly, although the parceling number does not reflect the issue of modularity – the desire to have a few different types of subsystems – the subsystems themselves are explicit in the parceling model while being absent in the 3-D grid model. Hence, the modularity of the composition can be addressed within the same parceling model.

The proposed metric has its own weaknesses, the major one of which is ignoring the intra-package wiring. The metric should be used as a first-order approximation of the system cost, and should be applied recursively within the packaging hierarchy.

Notable examples of communication-intensive subsystems are interconnection networks like the Butterfly and sorting networks like the Bitonic one.

In spite of the great theoretical breakthrough of the AKS sorting network [AKS83], the best practical networks are still the Batcher sorting networks invented in the 60's. Batcher [B68] has presented two sorting networks, the *Odd-Even sorting network* and the *Bitonic sorting network*. The depth of both networks is $\frac{1}{2} \log(n)(\log(n) + 1)$, where n is the number of inputs and is assumed henceforth to be a power of two. The former network has somewhat fewer nodes than the latter one; however, the latter network is more regular and therefore is preferred by designers of sorting and switching systems [HK84, WE85, GHMSL91]. Hence, of these two sorting networks, only the Bitonic one is studied in this paper.

Let B_n and S_n denote the Butterfly and the Bitonic sorting network, respectively, both with n external input edges. Let $|N|$ denote the size (number of nodes) of a network N. For our networks of interest, $|B_n| = \frac{1}{2} n \log(n)$ and $|S_n| = \frac{1}{4} n \log(n)(\log(n)+1)$. The paper establishes the following explicit (rather than asymptotic) upper and lower bounds on the parceling of these networks for any $1 < k < n$, both powers of 2:

$$\alpha(k)|B_n|/|B_k| \leq P(B_n, 2k) \leq |B_n|/|B_k| + n/k$$

and

$$\beta(\min(k, n/k))(|S_n| - 4|S_{k/4}|n/k)/|B_k| \leq P(S_n, 2k)$$
$$\leq (|S_n| - 4|S_{k/4}|n/k)/|B_k|.$$

The functions α, $\beta : \mathbb{N} \to \mathbb{R}$ of the lower bounds are defined in Section 4 and satisfy $\lim_{i \to \infty} \alpha(i) = \lim_{i \to \infty} \beta(i) = 1$ and, for all i, $\frac{1}{2} \leq \alpha(i)$ and $\frac{1}{8} \leq \beta(i)$.

No lower bounds on the parceling number of these networks were previously known. The above upper bound of the Butterfly is well-known. Our upper bound of the Bitonic sorting network is new and is lower (by about $\frac{1}{2} n \log n/k$) than the bound established by previous works [HK84, GHMSL91]. Our main contribution, however, has to do with modularity. Our parceling of S_n into $2k$-parcels uses only two modules. Namely, each parcel is either a copy of[3] B_k or a copy of another network, let's call it $X_{n,k}$. This improves previous parceling [HK84, GHMSL91] of this network which use $\log k$ modules.

A weakness of our parceling of S_n is that one of the modules, $X_{n,k}$, depends on n. This dependency can be removed, not for S_n itself, but for a variant of it. Namely, there is an n-input sorting network $\widehat{S}_{n,k}$ that is very similar to S_n and has a $2k$-parceling that is made of two modules which do not depend on n; one of these modules is B_k and the other is S_k; moreover, this parceling has the same number of parcels as the above parceling of S_n.

Other contributions of this paper are the techniques used to establish the above results. We consider the parceling of a wider family of networks—those having the extended buddy property (to be defined). The upper bound on members of this family is established via a straightforward parceling. The lower bound is established by transferring such a bound from one network to another via graph embedding.

2   Preliminaries

In this paper, a digital system (or subsystem) is represented by an *h-graph*—a directed[4] (multi-)graph having a designated node called *the host* [LS83] and denoted h. The host is not actually part of the system; it represents the external world; edges incident to it represent the input/output edges of the system.

Formally, a directed graph G is a pair $G = \langle V^G, E^G \rangle$, while an h-graph is a triplet $G = \langle V^G, E^G, h^G \rangle$. When no confusion can arise, we sometimes omit the graph superscript (G in this case) from such notations.

Let G be a directed graph (or h-graph) and s a subset of its nodes. An *input* (resp., *output*) of s is an edge whose head (resp., tail) is in s and whose tail (resp., head) is not in s. We denote by $\mathrm{I}^G(s)$ (resp., $\mathrm{O}^G(s)$) the number of inputs (resp., outputs) of s and define $\mathrm{IO}^G(s) \stackrel{\triangle}{=} \mathrm{I}^G(s) + \mathrm{O}^G(s)$.

Let G be an h-graph. A *parcel* of G is a subset of $V^G - \{h^G\}$. For a positive integer k, a *k-parcel* of G is a parcel p such that $\mathrm{IO}(p) \leq k$. A *k-parceling* of G is a partitioning of $V^G - \{h^G\}$ into (disjoint) k-parcels. The *k-parceling number of G*, denoted $\mathrm{P}(G, k)$, is the minimal number of parcels in a k-parceling of G. (In the special case where G has no k-parceling we define $\mathrm{P}(G, k) = \infty$.) Note that the designated host is mandatory in the definition of the parceling number; otherwise, any graph has a parceling number of 1.

---

[3] Each copy has its own association of the min/max functionality with the two edges emerging from each comparator; this association does not need to be wired-in and can be established during the initialization of the system.

[4] The fact that the graph is directed is clearly irrelevant as far as parceling is concern. However, we find the exposition to be more convenient in the context of directed graphs.

Typically, interconnection networks (e.g., the Butterfly) are represented by host-less graphs. In this case, we use the following uniform way to augment a graph by a host. For a positive integer j and a host-less directed graph G, the h-graph $h_j(G)$ is the graph generated from G by adding a host h and the minimal number of edges, all incident to h, such that the in-degree and out-degree of any node $v \neq h$ is at least j. (Note that h_2 reflects the usual usage of the Butterfly and the Bitonic sorting network.) For a host-less graph G, we sometimes use $P(G, k)$ to denote $P(h_2(G), k)$.

A (host-less) *directed layered graph* (*dil-graph* in short) is an acyclic directed graph G whose node-set is partitioned into sets called *layers* (or *levels*), denoted $L_1^G, L_2^G, \cdots, L_{d^G}^G$, and whose edge-set is partitioned into sets called *edge-levels*, denoted $E_1^G, E_2^G, \cdots, E_{d^G-1}^G$, such that any edge of E_i^G leads from a node of L_i^G to a node of L_{i+1}^G. The number of levels, d^G, is called the *depth* of G.

This paper studies the parceling of two well known dil-graphs: the Butterfly and the Bitonic sorting network. Henceforth, n is used exclusively to denote the number of input edges of these graphs and is always a power of two and greater than one.

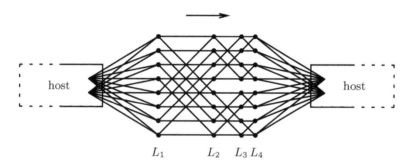

Fig. 1. $h_2(B_{16})$—the 16-input augmented Butterfly

For a definition of the Butterfly see, for example, [L92]. Our notation of the Butterfly somewhat differs from the common one as follows: Our Butterfly is a dil-graph, and we name the Butterfly by the number of its input edges. (This naming is chosen to be consistent with the Bitonic sorting network.) So, B_n is the Butterfly having $\log n$ levels each of $n/2$ nodes. Figure 1 depicts $h_2(B_{16})$—the 16-input augmented Butterfly.

The following three lemmas state several well known properties of the Butterfly that are used in our study. Proofs of these lemmas can be found, for example, in [EL97, GL04]. The first well known property is the Banyan property [LGGL73] — a directed graph is *Banyan* if it is acyclic and for any node u of in-degree 0 and any node v of out-degree 0 there is exactly one directed path from u to v.

Lemma 1. *The Butterfly is Banyan.*

Lemma 2. *Let v and v' be two nodes in the same level of a Butterfly B_n. Then there is an automorphism[5] π of B_n such that $\pi(v) = v'$. Moreover, let $(v{\to}u)$ and $(v'{\to}u')$ be two edges in the same edge-level of B_n. Then there is an automorphism π of B_n such that $\pi(v) = v'$ and $\pi(u) = u'$.*

For a dil-graph G and an interval of integers $I \subset [1, d^G]$, the I *segment of* G, denoted $G[I]$, is the subgraph of G induced by $\bigcup_{i \in I} L_i$. (That is, $G[I]$ is a dil-graph of depth $|I|$ whose levels are $L_1^{G[I]} = L_{\min(I)}^G$, $L_2^{G[I]} = L_{\min(I)+1}^G$, etc.) A *segment of G* is a subgraph of G of the form $G[I]$ for some interval I. For $1 \le i \le j \le d^G$, we shorten $G[[i, j]]$ to $G[i, j]$ and we use $G[i, \infty]$ to denote $G[i, d^G]$.

Lemma 3. *Let G' be a segment of B_n. Then each connected[6] component of G' is isomorphic to $B_{2^{d^{G'}}}$.*

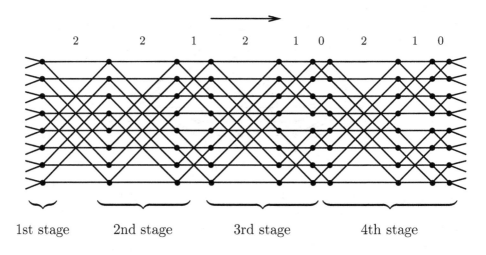

Fig. 2. S_{16}—the 16-input Bitonic sorting graph

Let us now consider the Bitonic sorting network. (For a definition see [B68].) The n–input Bitonic network has $\frac{1}{2} \log(n)(\log(n) + 1)$ levels of $n/2$ nodes each; it is conceptually partitioned into $\log n$ segment called *stages*, where the i–th stage is of depth i and is composed of 2^{n-i} disjoint $(2^{i-1} \times 2)$–mergers[7]. This

---

[5] Note that, since our Butterfly is directed, any automorphism of it is necessarily a level-preserving one.

[6] The term 'connected' means: connected in the base undirected graph.

[7] An $(i \times j)$–merger merges j sorted lists, of i elements each, into a single sorted list.

conceptual division and the functionality of the stages are related to the behavior of the network but not to its graph and are therefore largely irrelevant to our study.

Let S_n denote the dil-graph of the n–input Bitonic network. Figure 2 depicts S_{16} drawn in a "butterfly-style," i.e., each edge-level is composed of 4–edge 'butterflies' all of the same size. The only important attribute of the network which is absent in the graph is the distinction between the edge that carries the minimal key and the edge that carries the maximal one out of the two keys that emerge from each comparator; this distinction is, of course, irrelevant to the parceling of the network. (This distinction is critical for being a sorting network; however, such a differentiation of the edges can be established during the system initialization.)

3   Upper Bounds

This section establishes upper bounds on the parceling number of the Butterfly and the Bitonic sorting network. Rather than studying the parceling of only these networks, we consider the parceling of a wider family of graphs—those having the extended buddy property defined below. This family includes the Butterfly and the major part of any large Bitonic sorting network.

For an integer $j \geq 1$, a dil-graph G has the j–*buddy property* if for any G', a segment of G of depth at most j, each connected component of G' is isomorphic to $B_{2^{d^{G'}}}$—i.e., each connected component is a depth-$d^{G'}$ Butterfly[8].

Note that any dil-graph has the 1–buddy property, that the 2–buddy property is the plain buddy property [BFJM87], that if G has the j–buddy property then it has the j'–buddy property for any $j' \leq j$, that if G has the j–buddy property then so does any segment of G, and that if G has the d^G–buddy property then it has the j–buddy property for any j.

By Lemma 3, B_n has the j–buddy property for any j. As for S_n, Figure 2 reveals that S_{16} has the 2-buddy property (but not the 3-one), that $S_{16}[2, 10]$ has the 3–buddy property, and that $S_{16}[4, 10]$ has the 4–buddy property.

The next two lemmas extract the only properties of the Bitonic network that are relevant to our study. They establish that, for $j << \log n$, S_n has two overlapping segments; the smaller one, *the tail*, composed of the first $\frac{1}{2}j(j+1) = d^{S_{2^j}}$ levels has a compact parceling, while the larger one, *the head*, is composed of all but the first $\frac{1}{2}(j-2)(j-1) = d^{S_{2^{j-2}}}$ levels and has the j–buddy property. The first lemma is straightforward.

Lemma 4. *Let* $1 \leq j \leq \log n$. *Then:*
a. *Any connected component of* $S_n[1, \frac{1}{2}j(j+1)]$ *is isomorphic to* S_{2^j}.
b. $P(S_n[1, \frac{1}{2}j(j+1)], 2 \cdot 2^j) = n/2^j$.

---

[8] By a theorem of Bermond, Fourneau and Jean-Marie [BFJM87], G has the j–buddy property iff for any G' as above: each connected component of G' is Banyan and has exactly $2^{d^{G'}-1}$ nodes in each level.

The second lemma has been established in [GL04].

Lemma 5. *For* $1 \leq j \leq \log n$: $S_n[\frac{1}{2}(j-2)(j-1)+1, \infty]$ *has the* j*-buddy property.*

A parceling of a dil-graph is called a *straightforward parceling* if for any edge-level E_i: either all its edges are inter-parcel ones or all are intra-parcel ones (but not necessarily within a single parcel). We establish upper bounds on the parceling number of B_n and S_n via straightforward parceling.

A directed graph is l-bounded if the in-degree and the out-degree of any node are at most l. Note that the Butterfly, the Bitonic sorting network and any dil-graph having the 2-buddy property are 2-bounded, and that if G is l-bounded then $h_l(G)$ is a directed Euler graph.

For a directed graph or h-graph G and $p \subset V^G$, define $G(p)$ as the subgraph of G induced by p. The next lemma is immediate.

Lemma 6. *Let* G *be a host-less* l*-bounded graph and* $p \subset V^G$. *Then:*

$$IO^{h_l(G)}(p) = 2l|p| - 2|E^{G(p)}|.$$

Hence, the total number of inputs and outputs of a parcel p *in* $h_l(G)$ *depends on* l *and* $G(p)$ *but does not depend otherwise on* G.

Lemma 7. *Let* G_1 *and* G_2 *be subgraphs of a host-less directed* l*-bounded graph* G *such that* G_1 *and* G_2 *are node-disjoint and together cover all the nodes of* G. *Then*

$$P(h_l(G), k) \leq P(h_l(G_1), k) + P(h_l(G_2), k)$$

for any k.

Proof. By Lemma 6, any k-parcel of $h_l(G_1)$ or $h_l(G_2)$ is also a k-parcel of $h_l(G)$. Hence, any two k-parcelings of $h_l(G_1)$ and $h_l(G_2)$ combine into a k-parceling of $h_l(G)$. □

The next lemma is immediate.

Lemma 8. *Let* G *be a dil-graph with the* d^G*-buddy property,* $d^G \leq j$, *and* $m = 2|L_1|$. *Then* $P(G, 2 \cdot 2^j) \leq \lceil m/2^j \rceil$.

The following lemma establishes an upper bound on the parceling number of graphs having the extended buddy property.

Lemma 9. *Let* G *be a dil-graph with the* $\max(2, j)$*-buddy property,* $1 \leq j \leq d^G$, *and* $m = 2|L_1|$. *Then* $P(G, 2 \cdot 2^j) \leq \lceil d^G/j \rceil m/2^j$.

Proof. Partition the interval $[1, d^G]$ into $\lceil d^G/j \rceil$ intervals such that the cardinality of each is at most j. This induces a partitioning of G into segments of depth at most j. Since G has the 2-buddy property, it is 2-bounded. Apply Lemmas 7 and 8 and note that $\lceil m/2^j \rceil = m/2^j$. □

An immediate consequence of Lemma 9 is the following upper bound on the parceling number of B_n.

Theorem 1. *Let n and k be powers of 2 such that $2 \le k \le n$. Then*

$$P(B_n, 2k) \le |B_n|/|B_k| + n/k.$$

The next theorem establishes an upper bound on the parceling number of S_n.

Theorem 2. *Let n and k be powers of 2 such that $8 \le k \le n$. Then*

$$P(S_n, 2k) \le |S_n[d^{S_{k/4}} + 1, \infty]|/|B_k|.$$

Proof. The inequality clearly holds when $n = k$, so assume $k < n$. Let $k' = \log k$. Following Lemmas 4 and 5, S_n is covered by two overlapping segments: the tail $T = S_n[1, d^{S_k}]$ and the head $H = S_n[d^{S_{k/4}} + 1, \infty]$. The head H has the k'-buddy property, and the tail has an obvious optimal parceling which establishes $P(T, 2k) = n/k$.

To parcel S_n we partition it into a head and a tail (which may differ from the above H and T) and parcel each segment separately as follows. Let $\overline{H}$ be the minimal segment of S_n such that its depth is a multiple of k' and, together with T, it covers all of S_n. Let $\overline{T}$ be the rest of S_n; i.e., $\overline{T}$ is the segment of S_n such that $(\overline{T}, \overline{H})$ is a partition of S_n. The network $\overline{H}$ is a segment of H and, therefore, it has the k'-buddy property. By Lemma 9,

$$P(\overline{H}, 2k) \le |\overline{H}|/|B_k|. \tag{1}$$

Moreover, this bound is achieved by a parceling in which all the parcels are isomorphic to B_k. We have: $d^{S_k} - d^{S_{k/4}} = 2k - 1$. Hence,

$$d^{\overline{H}} \le d^{S_n} - d^{S_k} + k' - 1 = d^{S_n} - d^{S_{k/4}} - k'.$$

This implies:

$$|\overline{H}| \le |S_n[d^{S_{k/4}} + 1, \infty]| - |S_n[1, k']|. \tag{2}$$

By Lemma 4(a), T is the disjoint sum of n/k sub-networks, all isomorphic to each other, and each sub-network has k input and k output edges. The new tail, $\overline{T}$, is a segment of T. Both T and $\overline{T}$ are 2-regular; hence, the above statement holds also for $\overline{T}$; i.e., $\overline{T}$ is the disjoint sum of n/k sub-networks all isomorphic to each other and each sub-network has k input and k output edges. This implies:

$$P(\overline{T}, 2k) \le n/k = |S_n[1, k']|/|B_k|. \tag{3}$$

By Lemma 7,
$$P(S_n, 2k) \le P(\overline{T}, 2k) + P(\overline{H}, 2k). \tag{4}$$

Combining inequalities (1, 2, 3, 4) yields:

$$P(S_n, 2k) \le |S_n[d^{S_{k/4}} + 1, \infty]|/|B_k|.$$

$\square$

We conjecture that, for k a power of two, the parcelings implied by Lemma 9 and Theorem 2 are optimal for graphs having the j–buddy property and for the Bitonic network.

Our parceling of the Bitonic network is somewhat more compact then the parceling presented in [HK84] and [GHMSL91]. However, the main advantage of our parceling is modularity. Our parceling is made of only two modules while that of [HK84] and [GHMSL91] is made of $\log k + 1$ modules. One of our modules is B_k and the other depends on both k and n. This latter dependency on n is undesirable and can be removed via a slight modification of the network S_n as follow. The new network, let us call it $\widehat{S}_{n,k}$, is derived from S_n by extending the latter with at most $k - 1$ additional levels at its head so that the segment $\widehat{S}_{n,k}[d^{S_k} + 1, \infty]$ has the k-buddy property and its width is a multiple of k. (There are several networks having this property; see, for example, [GL04].) The extended network is, of course, a sorting network. In fact, all the additional comparators of this network are redundant — through one of their input edges they receive a key which is always greater than (or equal to) the key received through the other edge. Applying the procedure of this section to $\widehat{S}_{n,k}$, instead of S_n, produces a parceling having the same number of parcels and the same number of modules, but these modules do not depend on n; they are B_k and S_k.

4   Lower Bounds

This section establishes lower bounds on the parceling number of the Butterfly and the Bitonic sorting network. We accomplish this task in three steps as follows. Firstly, we establish a lower bound for l–bounded graphs having no parallel edges. Secondly, we employ a graph embedding [L92] to transfer the above lower bound to graphs having the extended buddy property. Finally, from the latter bound we infer lower bounds for the Butterfly and the Bitonic sorting network.

We begin by showing that, under certain conditions, the parceling number of a graph is no smaller than the parceling number of any of its subgraphs. A subgraph G' of a directed graph G is a *convex* subgraph of G if for any directed path t of G whose endpoints are both in G': all the nodes and edges of t are in G'. For example, any segment of a dil-graph is a convex subgraph.

Lemma 10. *Let G' be a convex subgraph of a host-less, directed, l–bounded graph G. Then:*
a. $IO^{h_l(G)}(V^{G'}) \le IO^{h_l(G)}(V^G)$.
b. $P(h_l(G'), k) \le P(h_l(G), k)$ for any k.

(Note that G is not necessarily acyclic.)

Proof. To establish (a) note that any directed simple cycle of $h_l(G)$ crosses at most one input of $V^{G'}$, and if it does cross an input, then it must pass through h.

Since $h_l(G)$ is a directed Euler graph, there is a set C of edge-disjoint directed simple cycles of $h_l(G)$ that cover all the edges of $h_l(G)$. By the above note on simple cycles,

$$I^{h_l(G)}(V^{G'}) = |\{c \in C : c \text{ crosses an input of } V^{G'}\}| \le I^{h_l(G)}(V^G).$$

Since $h_l(G)$ is an Euler graph, $\text{IO}^{h_l(G)} = 2I^{h_l(G)}$. This established (a).

To prove (b), it suffices to show that for any k–parcel p of $h_l(G)$, $p' \stackrel{\triangle}{=} p \cap V^{G'}$ is a k–parcel of $h_l(G')$. Let p be such a parcel. Define $Q \stackrel{\triangle}{=} G(p)$ and $Q' \stackrel{\triangle}{=} G(p')$. Note that Q' is a convex subgraph of Q. By Lemma 6 and claim (a) for the graphs Q and Q',

$$\text{IO}^{h_l(G')}(p') = \text{IO}^{h_l(Q)}(V^{Q'}) \le \text{IO}^{h_l(Q)}(V^Q) = \text{IO}^{h_l(G)}(p).$$

□

Lemma 11. *Let G be an l–bounded dil-graph with no parallel edges, and let $0 < k < l$. Then:*
a. $|p| \le lk/(l - k)$ for any $(2kl)$–parcel p of $h_l(G)$.
b. $|G|(l - k)/lk \le \mathrm{P}(h_l(G), 2kl)$.

Proof. To prove (a), let p be a $(2kl)$–parcel of $h_l(G)$, and define $Q = G(p)$. We claim that $I^Q(\{v\}) \le k$ for any node $v \in p$. To establish this claim, let Q' be the subgraph of Q induced by $\{u : (u \to v) \in E^Q\}$. Since G is a dil-graph, Q' is an edge-less convex subgraph of Q. By Lemmas 6 and 10(a),

$$2l|Q'| = \text{IO}^{h_l(Q)}(Q') \le \text{IO}^{h_l(Q)}(Q) \le 2kl.$$

Hence, $|Q'| \le k$; since there are no parallel edges, $I^Q(\{v\}) \le k$. This establishes our claim.

By this claim, each member of p contributes at least $(l - k)$ to $I^{h_l(G)}(p)$. By symmetry, the same holds for $O^{h_l(G)}(p)$. Hence, $2|p|(l - k) \le \text{IO}^{h_l(G)}(p) \le 2lk$. This establishes (a). Claim (b) follows immediately from (a). □

Next, we transfer the lower bound of Lemma 11 to graphs having the buddy property via graph embedding. An *embedding* of a directed[9] graph G (the *guest*) into a directed graph H (the *host*) is a function π defined on $V^G \bigcup E^G$ such that $\pi(V^G) \subset V^H$ and for any edge $(u \stackrel{e}{\to} v) \in E^G$, $\pi(e)$ is a directed path of H leading from $\pi(u)$ to $\pi(v)$.

The *dilation* of π is the maximal length of $\pi(e)$ paths; the *congestion* of an edge e of H is the number of edges of G whose paths cross e; the *congestion* of π is the maximal congestion of edges of H. For a function f and a set s, define $f^{-1}(s) \stackrel{\triangle}{=} \{x : f(x) \text{ is defined and } f(x) \in s\}$. Under this notation, the *load of* π is $\max\{|\pi^{-1}(\{u\})| : u \in V^H\}$.

The next lemma transfers a lower bound from a guest graph to a host graph.

Lemma 12. . *Let π be an embedding of an h-graph G into an h-graph H such that the congestion of π is c and $\pi^{-1}(\{h^H\}) = \{h^G\}$. Then $\mathrm{P}(G, kc) \le \mathrm{P}(H, k)$ for any k.*

---

[9] Graph embedding is usually defined for undirected graphs [L92]; it is more convenient here, however, to define it in the context of directed graphs.

Proof. Let P^H be a k–parceling of H with $|P^H| = P(H, k)$. Define the following parceling of G:

$$P^G \triangleq \{\pi^{-1}(p) : p \in P^H \text{ and } \pi^{-1}(p) \neq \emptyset\}.$$

By definition, P^G is a partitioning of $V^G - \{h^G\}$ and $|P^G| \leq |P^H|$. It remains to show that P^G is a (kc)–parceling. Consider a parcel $p \in P^G$. We have $p = \pi^{-1}(q)$ for some $q \in P^H$. Each of the inputs (outputs) of p is embedded into a path that crosses, at least once, an input (output) of q. Since $\text{IO}^H(q) \leq k$ and each edge of H is used by at most c paths, we must have $\text{IO}^G(p) \leq kc$.   □

Lemma 13. *Let $j \geq 1$ and let a dil-graph H have the $(j + 1)$–buddy property. Then there is a 2^j–bounded dil-graph G with no parallel edges and with $|G| \geq |H|/j$ and there is an embedding π of $h_{2^j}(G)$ into $h_2(H)$ whose congestion is 2^{j-1} and $\pi^{-1}(\{h^H\}) = \{h^G\}$.*

Proof. Define the dil-graph G of depth $d^G = \lceil d^H/j \rceil$ by:

and
$$L_i^G \triangleq L_{1+(i-1)j}^H$$

$E_i^G \triangleq \{u \to v : u \in L_i^G, v \in L_{i+1}^G, H \text{ has a directed path from } u \text{ to } v\}$.

Since H has the $(j + 1)$–buddy property, G is 2^j–bounded. In fact, since the Butterfly is Banyan (Lemma 1), the in-degree of all G nodes, except those of L_1^G, and the out-degree of all G nodes, except those of $L_{d^G}^G$, are exactly 2^j.

Construct the required embedding as follows. Each node of G is mapped to itself; the host of G is mapped to the host of H; each edge $(u \to v) \in E^G$ is mapped to the unique path in H from u to v; the 2^j parallel edges from h to a node v of G are mapped evenly onto the two parallel edges from h to v in H; finally, the 2^j parallel edges from a node v of G to h are mapped evenly onto the paths leading from v to h in H.

It remains to check that the congestion of the embedding is 2^{j-1}. We consider only edges of of $h_2(H)$ that reside in the first $(d^G - 1)j$ edge-levels of H. The analysis for the other edges is similar. Such an edge e belongs to a connected component C of $H[1 + (i - 1)j, 1 + ij]$ for some i. Due to the $(j + 1)$–buddy property, C is isomorphic to $B_{2^{j+1}}$. Let E' be the edge-level of C having e. A total of $(2^j)^2$ paths uses the $2 \cdot 2^j$ edges of E', and, by Lemma 2, the congestion of all these edges is the same. Hence, the congestion of e is 2^{j-1}.   □

Lemma 14. *Let H be a dil-graph having the $(j + 1)$–buddy property, $j \geq 1$, k a power of 2 and $2 \leq k \leq 2^j$. Then:*

$$((1 - k/2^{j+1}) \log k/j) \ |H|/|B_k| \leq P(H, 2k).$$

Proof. Let j, H and k be as above. Define $l \triangleq 2^j$. By Lemma 13, there is a dil-graph G such that G is l–bounded, has no parallel edges, $|G| \geq |H|/j$ and there is an embedding of $h_l(G)$ into $h_2(H)$ with congestion 2^{j-1}. By Lemma 12,

$$P(h_l(G), 2k2^{j-1}) \leq P(H, 2k).$$

Applying Lemma 11(b) on the graph G with $k' = \frac{1}{2}k$ yields:

$$[(2l - k)/lk] \ |G| = [(l - k')/lk'] \ |G| \leq P(h_l(G), 2k'l) = P(h_l(G), 2k2^{j-1}).$$

Combining the above inequalities yields:

$$\left((1 - k/2^{j+1}) \log k/j\right) |H|/|B_k| = [(2l - k)/lk]|H|/j$$
$$\leq [(2l - k)/lk]|G|$$
$$\leq \mathrm{P}(h_l(G), 2k2^{j-1})$$
$$\leq \mathrm{P}(H, 2k).$$

$\square$

Define the function $\alpha : (\mathbb{N} + 2) \to \mathbb{R}$ by

$$\alpha(k) \triangleq (\log(k) - \tfrac{1}{2})/(\log(k) + \lceil \log \log(k) \rceil).$$

By elementary calculus, $\lim_{k \to \infty} \alpha(k) = 1$ and $\alpha(k) \geq \tfrac{1}{2}$ for any k which is a power of 2 and is greater than 1. Moreover, for $j = k + \lceil \log \log(k) \rceil$ we have $\alpha(k) \leq (1 - k/2^{j+1}) \log(k)/j$. This, together with Lemma 14 and the fact that a Butterfly has the i-buddy property for any i, implies our first lower bound, concerning the Butterfly.

Theorem 3. *Let n and k be powers of 2 such that $2 \leq k \leq n$. Then*

$$\alpha(k)|B_n|/|B_k| \leq \mathrm{P}(B_n, 2k).$$

Define the function $\beta : (\mathbb{N} + 2) \to \mathbb{R}$ by

$$\beta(m) \triangleq \alpha(m)(\log(m) - \log \log(m))/(\log(m) + 2).$$

By elementary calculus, $\lim_{m \to \infty} \beta(m) = 1$ and $\beta(m) \geq \tfrac{1}{8}$ for any m that is a power of 2 and is greater than 1. The next theorem establishes a lower bound on the parceling number of the Bitonic network.

Theorem 4. *Let n and k be powers of 2 such that $8 \leq k < n$. Then*

$$\beta(\min(k, n/k))|S_n[d^{S_{k/4}} + 1, \infty]|/|B_k| \leq \mathrm{P}(S_n, 2k).$$

Proof. Let $m = \min(k, n/k)$, $j = \log(k) + \lceil \log \log(m) \rceil$ and let H be the final segment of S_n without the first $d^{S_{2j-1}}$ levels; namely, $H = S_n[d^{S_{2j-1}} + 1, \infty]$. By Lemma 10(b), $\mathrm{P}(H, 2k) \leq \mathrm{P}(S_n, 2k)$. By Lemma 5, H has the $(j+1)$–buddy property. By Lemma 14,

$$\left((1 - k/2^{j+1}) \log k/j\right) |H|/|B_k| \leq \mathrm{P}(H, 2k).$$

We have:

$$\alpha(m) = (1 - 1/(2 \log(m)) \log(m)/(\log(m) + \lceil \log \log(m) \rceil)$$
$$\leq (1 - 1/(2^{\lceil \log \log(m) \rceil + 1}) \log(k)/(\log(k) + \lceil \log \log(m) \rceil)$$
$$= (1 - k/2^{j+1}) \log k/j.$$

Recall that $d^{S_{2i}} = \tfrac{1}{2}i(i + 1)$. Hence,

$$|S_n[d^{S_{k/4}} + 1, \infty]|(\log(m) - \log \log(m))/(\log(m) + 2)$$
$$\leq |S_n[d^{S_{k/4}} + 1, \infty]|(\log(n) - \log(k) - \lceil \log \log(m) \rceil + 1)/(\log(n) - \log(k) + 2)$$
$$\leq |H|.$$

Combining the above inequalities yields:

$$\beta(\min(k, n/k))|S_n[d^{S_{k/4}} + 1, \infty]|/|B_k| \leq \mathrm{P}(H, 2k) \leq \mathrm{P}(S_n, 2k).$$

$\square$

Acknowledgments

I wish to thank Arny Rosenberg for many helpful comments.

References

[AKS83]   M. Ajtai, J. Komlós, E. Szemerédi (1983): Sorting in $c \log n$ parallel steps. *Combinatorica*, Vol. 3, pp. 1–19.

[B90]   H. B. Bakoglo (1990): *Circuits, interconnections and packaging for VLSI*, Addison-Wesley.

[B68]   K.E. Batcher (1968): Sorting networks and their application, *Proc. AFIPS 1968 Spring Joint Computer Conf.*, Vol. 32, pp. 307–314.

[BMM78]   C. G. Bell, J. C. Mudge and J. E. McNamara (1978): *Computer Engineering*, Digital Press.

[BFJM87]   J.C. Bermond, J.M. Fourneau and A. Jean-Marie (1987/88): Equivalence of multistage interconnection networks, *Information Processing Letters*, Vol. 26, pp. 45–50.

[BCR91]   S.N. Bhatt, F.R.K. Chung, A.L. Rosenberg (1991): Partitioning circuits for improved testability. *Algorithmica 6*, 37–48.

[EL97]   S. Even and A. Litman (1997): Layered Cross Product - a technique to construct interconnection networks. *Networks*, vol. 29, pp. 219-223.

[GHMSL91]   J.N. Giacopelli, J.J. Hickey, W.S. Marcus, W.D. Sincoskie and M. Littlewood (1991): Sunshine: a high-performance self-routing broadband packet switch architecture, *IEEE Journal on Selected Areas in Communications*, Vol. 9, pp. 1289–1298.

[GL04]   N. Golbandi and A. Litman (2004): Characterizations of generalized butterfly networks, *T.R. no. CS-2004-10, Dept. of Computer Science, Technion, Israel*, URL: http://www.cs.technion.ac.il/users/wwwb/cgi-bin/tr-info.cgi?2004/CS/CS-2004-10.

[LGGL73]   L. R. Goke and G. J. Lipovski (1973): Banyan Networks For Partitioning Multiprocessing Systems. *Proceedings of the 1st Annual Symposium on Computer Architecture*, pp. 21–28.

[HK84]   A. Huang and S. Knauer (1984): Starlite: a wideband digital switch, *Proceedings of GLOBECOME'84*, pp. 121–125.

[LR86]   F.T. Leighton and A.L. Rosenberg (1986): Three-dimensional circuit layouts. *SIAM J. Comput. 15*, pp. 793–813.

[L92]   F. T. Leighton (1992): *Introduction to parallel algorithms and architectures*, Morgan Kaufmann.

[L85]   C.E. Leiserson (1985): Fat-trees: universal networks for hardware-efficient supercomputing. *IEEE Trans. Comp., C-34*, 892–901.

[LS83]   C. E. Leiserson and J. B. Saxe (1983): Optimizing synchronous systems, *Journal of VLSI and Computer Systems*, Vol. 1, pp. 41–67.

[T79]   C.D. Thompson (1979): Area-time complexity for VLSI, *Proceeding of the 11th Annual ACM Symposium on Theory of Computing* pp. 81–88.

[T80]   C.D. Thompson (1980): *A complexity theory for VLSI*. Ph.D. Thesis, CMU.

[WE85]   N. H. E. Weste and K. Eshraghian (1985): *Principles of CMOS VLSI design,* 1st edition, Addison-Wesley.

An Application Intersection Marketing Ontology

Xuan Zhou[1], James Geller[1], Yehoshua Perl[1], and Michael Halper[2]

[1] CS Department, New Jersey Institute of Technology, Newark, NJ 07102*
{xxz1279, geller, perl}@oak.njit.edu
[2] Mathematics & Computer Science Department, Kean University, Union, NJ 07083
mhalper@kean.edu

Abstract. We consider the design of an ontology for marketing knowledge. Such an ontology contains two hierarchies, a customer hierarchy and a product hierarchy. The product hierarchy representation is straightforward, as in general each level consists of products that are more specific than the products on the previous level. However, the customer hierarchy is problematic, since it involves many independent dimensions such as age, gender, income, *etc.* A straightforward ordering of the different dimensions to create a tree hierarchy is ineffective. We present an innovative design for the customer hierarchy based on introducing intersections of options for various dimensions *on demand.* We call such an ontology an intersection ontology. The advantages of such a design are explored and evaluated using our Web marketing project.

1 Introduction

1.1 Motivation

The result of market research is *marketing knowledge* that is used as input for target marketing activities. However, marketing knowledge is usually complex, consisting of many detailed facts, which by themselves do not give any clear picture and in combination are often overwhelming. What is desirable is an organization of marketing knowledge in an ontology that allows for the explicit representation of interesting abstractions and generalizations.

Ontologies have become important resources in many application domains. However, in marketing, ontologies have been close to non-existent. In this paper, we develop a kind of ontology, called an intersection ontology, for a marketing application and explore its advantages.

1.2 What Are Ontologies?

We will start with a fairly non-technical summary of what ontologies are and what they are useful for. Ontology is known as the branch of philosophy concerned with the study of the nature of being. However, when computer scientists

---

* This research was funded in part by the New Jersey Commission for Science and Technology through the New Jersey Center for Software Engineering.

O. Goldreich et al. (Eds.): Shimon Even Festschrift, LNCS 3895, pp. 143–163, 2006.

are referring to "an ontology" they mean a computer implementation of human-like knowledge.

Ontologies are descendants of the semantic networks in Artificial Intelligence. Quillian's first semantic network in 1968 was a computer implementation of a dictionary [1]. Terms in dictionaries refer to other terms, and Quillian implemented these references by pointers. However, as one term could have different meanings, a distinction is made between terms and concepts. Concepts are the fundamental building blocks of all semantic networks and ontologies.

A concept is a basic unit of knowledge, and, as opposed to a term, a concept is unambiguous. Quillian used only a small number of kinds of links which have been extensively studied and greatly refined since then. The most fundamental of these links, describes a generalization/specialization relationship between two concepts. This relationship satisfies transitivity. It has been variously called IS-A, sub-concept, subclass, a-kind-of, *etc.* It allows property inheritance, as follows.

Humans have additional "local" information about concepts. For example, solid objects have color, size, *etc.* We call this kind of local information "attributes", "properties" or "slots". If a general concept has an attribute (vehicles have a weight), then a specific sub-concept will have the same property (cars have a weight). One can imagine that inheritance is the propagation of a property from the general concept to the more specific concept against the direction of the IS-A link. Besides the IS-A links, ontologies contain other links, e.g., likes, owns, connected-to, *etc.* Most of these additional links have no "built-in behavior". These links are variously called associative relationships, roles, semantic relationships, and are labeled by their name. Relationships are inherited down along IS-A links.

Because a concept cannot be more general than itself, and because of the transitivity of the IS-A links, there cannot be any cycles of IS-A links in a semantic network. Furthermore, it is practical to have one concept (often called THING) that is a generalization of every concept in an ontology. Thus, the concepts and IS-A links in an ontology form a hierarchy with a root. In other words, the hierarchy of an ontology is a rooted Directed Acyclic Graph (DAG), where the nodes represent the concepts and the links represent IS-A relationships. Furthermore, the concepts and the IS-A links together would form a weakly connected component.

The definition of an ontology as a graph results in a natural diagram representation for ontologies. Figure 1 shows an example of an ontology. This example is adapted from [2] by eliminating other relationships such as part-of. In this and later figures, every box stands for a concept. Bold arrows (typically pointing upwards) stand for IS-A relationships. Thin arrows stand for other relationships. The IS-A relationships in this example form a tree. Later we will see examples using DAGs. Family terms, such as *child, ancestor* and *descendant* are used. A number of other extensions exist for ontologies, e.g. rules or axioms. However, these are not used in our model of an intersection ontology and will be omitted.

Thus, we present the definition of an ontology as follows: An ontology is a directed graph of nodes, which represent concepts, and edges, which represent

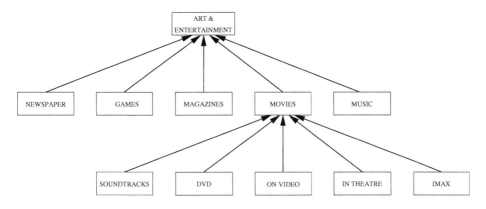

Fig. 1. A Partial Interest Hierarchy (Tree)

IS-A and semantic relationships between pairs of nodes. Concepts are labeled by unique terms. Concepts have additional (name, value) pairs, called attributes, where the attribute name needs to be unique for each concept. The set of all concepts together with the set of all IS-A links form a rooted, connected, Directed Acyclic Subgraph of the ontology. This subgraph is called the taxonomy of the ontology. Both attributes and semantic relationships may be inherited downwards, against the direction of the IS-A links, from more general concepts to more specific concepts.

Modern ontologies are attributed to Thomas Gruber [3] who built on a rich history which we briefly reviewed in [4]. Ontology building deals with modeling the world with shareable knowledge structures. With the emergence of the Semantic Web, the development of ontologies and ontology integration have become very important [5–8]. The Semantic Web is a vision, for a next generation Web, of Tim Berners-Lee, the inventor of the original Web, and colleagues. This vision is described in a figure called the "layer cake" of the Semantic Web [5]. This figure consists of nine functional layers of increasing technical complexity and abstraction. Each layer supports all the layers above it. Ontologies are flush in the middle of the layer cake. All the layers below Ontology, such as XML and RDF Schema are well developed. All the layers above ontologies, such as Rules and Proofs are well established within Artificial Intelligence (AI), but do not exist in widely applicable form outside of AI.

Ontologies will be used in the Semantic Web as follows. The current Web has shown that string matching by itself is often not sufficient for finding specific concepts. Rather, special programs are needed that search the Web for the concepts specified by a user. Such programs, which are activated once and traverse the Web without further supervision, are called agent programs.

Successful agent programs will search for concepts as opposed to words. Due to the well known homonym and synonym problems, it is difficult to select between different concepts expressed by the same word (e.g., Jaguar the animal, or Jaguar the car). However, having additional information about a concept, such as which concepts are related to it, makes it easier to solve this matching problem. For example, if that Jaguar that IS-A car is desired, then the agent knows which of the meanings to look for.

Ontologies provide a repository of this kind of relationship information. To make the creation of the Semantic Web easier, Web page authors will derive the terms of their pages from existing ontologies, or develop new ontologies for the Semantic Web.

Many technical problems remain for ontology developers, e.g. scalability. Yet, it is obvious that the Semantic Web will never become a reality if ontologies cannot be developed to the point of functionality, availability and reliability comparable to the existing components of the Web.

Some ontologies are used to represent the general world or word knowledge. Other ontologies have been used in a number of specialized areas. An overview of ontologies and their usages and properties can be found in [9]. For a comprehensive review of established ontologies see [10]. Two special issues on ontologies are [11, 8].

1.3 Ontologies in the Context of Web Marketing

Our work on marketing ontologies is part of a larger project that deals with the extraction of marketing knowledge from the World-Wide Web [12]. We have created a large database of customers. We extracted information from the home pages of individual Web users. Our database contains demographic information and interests of each customer.

We would prefer information about products that each of these customers has bought. However, this information is not publicly accessible on the Web. On the other hand, there are many very low-level interests with corresponding products. For example, the Yahoo interest hierarchy contained over 31,000 interests when we analyzed it. Many of these interests are as specific as the names of actresses or singers. If somebody has an interest in "Jennifer Lopez" then one may comfortably presume that this person might buy CDs or movies of Jennifer Lopez. Thus, information about interests can, to some degree, "stand in" for information about products.

We have processed the relational database of demographic and interest information with the WEKA data mining algorithm [13] and have found association rules between classifications of customers and interests. Thus, we needed an ontology that allows us to represent the resulting association rules of a data mining operation in a *succinct* format.

Note that we are not designing a marketing domain ontology which needs to represent all varied aspects of the marketing domain. We are creating an intersection ontology as an integral part of a marketing system. Our application deals with customer classifications needed for a marketing ontology. Our ontology is, in Sowa's terms, an application ontology [14], serving our marketing project [12] described above. As such, our marketing ontology concentrates only on representing purchasing knowledge, as described in detail in Section 2.

The straightforward representation of a customer classification is a tree hierarchy. The root represents the concept PERSON. The various demographic dimensions are ordered. At each of the levels we consider one different demographic dimension according to the above order and branch each node in the previous level to all possible options of this level's dimension. However, as we shall show there are problems with this representation.

To overcome these problems, we draw on Sowa's notion of representing conceptual knowledge using distinctions [15] and on Wille's use of intersections in Formal Concept Analysis [16, 17]. Due to the demands of the domain, realizing there is no natural order among the demographic dimensions and the need for an economical representation, we have developed an ontology that relies heavily on the use of "intersections" of concepts. To further economize, our ontology only contains those intersections about which we have marketing knowledge that needs to be represented. Thus, concepts are inserted into the ontology dynamically on demand. In an intersection hierarchy all the options of all dimensions are children of PERSON. All the relevant customer classifications appear in the next level, each classification as a child of all its options. Such a representation is called a three-level intersection hierarchy. Finally we represent a more economical solution where the customer classifications can be distributed over several levels – the multi-level intersection hierarchy.

Section 2 discusses in more detail why an ontology for marketing knowledge is useful. In Section 3, we will show the design of a customer hierarchy by ordered dimensions and the problems arising from it. Then, in Section 4, we will consider an alternative design for the customer hierarchy by creating "intersections" which results in an intersection ontology. In Section 5, we show the network design of a specific kind of intersection ontology, called multi-level intersection ontology.

The evaluation based on our Web marketing project is described in Section 6. In Section 7, we discuss how our marketing intersection ontology relates to Sowa's knowledge engineering by distinctions. Our conclusions appear in Section 8.

2 Representation of Marketing Knowledge

The essence of our marketing ontology is a collection of buy-relationships from customer classifications to product classifications. The basic facts we need to represent are of the form that a specific classification of customers tends to buy a given product or family of products. For example, "Married women with children buy toys." The challenge is to find a representation of this kind of knowledge in a convenient and economical way that fits into our ontology framework.

The marketing ontology needs to contain two hierarchies, a customer classification hierarchy, in short, *customer hierarchy*, and a product classification hierarchy, in short, *product hierarchy*. The group with the classification MARRIED WOMAN WITH CHILDREN (TOY) needs to be identifiable in the customer (product) hierarchy, either as a node or a group of nodes. To achieve the desired succinct representation, we prefer a single node for the customer classification concept and a second single node for the product classification concept. We connect those two nodes by a single relationship link with the label "buys", which is an economical representation capturing the desired marketing knowledge for an ontology.

Figure 2 shows a tiny ontology excerpt of four nodes with three "buys" connections. The node WOMAN WITH CHILDREN and its child MARRIED WOMAN WITH CHILDREN belong to the customer hierarchy. The node TOY and its child DOLL belong to the product hierarchy. The three connections are labeled "buys." The "buys" relationship to TOYs is inherited from WOMAN WITH CHILDREN to MARRIED WOMAN WITH CHILDREN. The inherited relationship is a dashed arrow, usually not shown in diagrams, since it can be inferred.

On the other hand, if the customer classification is represented by k nodes ($k > 0$) and the product classification is represented by l nodes ($l > 0$), then up to $k * l$ "buys" relationships are needed to represent the proper marketing knowledge, which is less desirable. Figure 3 represents a tiny part of a customer hierarchy and a product hierarchy. In Figure 3, two nodes are need to represent "men with children" or "electric toys". Thus, 4 arrows are needed to represent the fact that "men with children buy electric toys."

An alternative way with nodes representing "men with children" and "electric toys", respectively, with an arrow connecting them offers a more economical representation. However, if we represent ELECTRIC TOYS and NON-ELECTRIC TOYS at level two and the distinction between OUTDOOR and INDOOR at level three, then "men with children buy outdoor toys" will require 4 arrows. As we will discuss later, for each sequential ordering of the relevant dimensions, there are some marketing knowledge facts with an uneconomical representation.

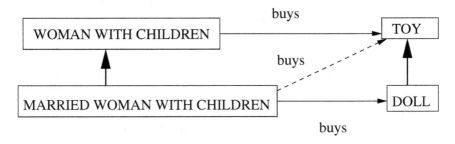

Fig. 2. Extract of a Marketing Ontology

We use the link with the label "buys" to mean "is likely to buy". Thus, "buys" is a statement strictly about a (meaningful) percentage of the population satisfying the demographic data.

For practical usability, a marketing knowledge representation should be as simple as possible. For example, if data mining tells us that married men with children buy diapers, and married women with children buy diapers, then an

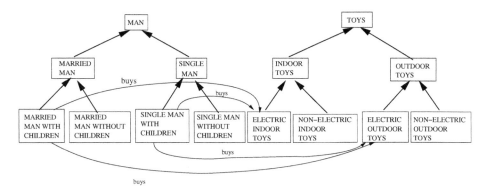

Fig. 3. We need $k * l$ arrows to express a simple Marketing Fact

assertion that married people with children buy diapers is better. Such information should be attached to exactly the concepts about which we are expressing knowledge. In our case, we would like to associate this knowledge with the concept *married people with children*, assuming such a concept *exists* in the ontology.

Finding a marketing ontology that enables the representation of *all the needed concepts* explicitly without creating a combinatorial explosion of concepts for customers is non-trivial. An intersection ontology achieves exactly this goal. However, first, we will describe the straightforward alternative, a tree representation with ordered dimensions, and explain why it is inappropriate for marketing knowledge.

3   Customer Tree Hierarchy with Ordered Dimensions

Following customary practice in marketing, as used, for instance, by MediaMark [18], we perform a classification of customers along various dimensions such as gender (man, woman), age (five age groups), marital status (single, married, separated), children status (with children, no child), *etc.*

Marketing research may reveal knowledge about buying habits of a customer classified according to several dimensions simultaneously. For example, consider the sentence: "Middle-aged married men with children buy books on early childhood development." We want in the customer hierarchy a node which corresponds exactly to the above customer classification.

Consider a tree hierarchy according to the four dimensions listed above, each dimension appears at a different level of the hierarchy. The tree hierarchy starts with the root node PERSON at level 1. The division into the classifications MAN and WOMAN happens at level 2. The division of men (and of women) according to five age groups happens at level 3. There is an obvious redundancy, as the same age choices are made twice, once below MAN and once below WOMAN. The next two levels follow the distinction according to marital status among three options, and children status, respectively. For a figure of a similar tree hierarchy see Figure 6(a) in Section 6.

In this tree hierarchy, which we will refer to as T, we are using a linear order of the various dimensions of a customer. In other words, we prioritize the different dimensions. The above order of dimensions was working well for the above given example, because the customer class (middle-aged married men with children) is represented by a unique leaf node which is the source for the "buys" relationship to the node representing the product BOOKs ON EARLY CHILDHOOD DEVELOPMENT.

Some marketing knowledge should be attached at a single non-leaf node in the tree hierarchy T. For example, "Men buy football tickets" would be expressed by a relationship that has the second level node MAN as its source and FOOTBALL TICKET as its target.

In the last examples, customer classification is represented as one node in T, from which one "buys" relationship link to a product node is emanating. In other situations, the description of a class of customers may not fit so neatly into the tree hierarchy T, as there might be a mismatch between this class and the order of dimensions in T. Consider, "People with children invest in Education IRAs." Even older people may have children, and people may also invest in IRAs for their grandchildren, so no age bracket applies here. To capture this class of customers, we need to refer to 30 leaf nodes in the tree hierarchy T, since the dimension considering children is at the lowest level in T. Furthermore, each of those nodes will require a "buys" relationship to an EDUCATION IRA node in the product hierarchy. The marketing knowledge "People with children invest in Education IRAs," expressed in a short sentence, requires 30 links in our marketing ontology. This is clearly an uneconomical representation of marketing knowledge. However, there is no inherent reason why we chose, for example, the distinction between MAN and WOMAN at the second level, above all the other dimensions. If, for example, the children status dimension would have been chosen as the top-level dimension in the hierarchy, then one node and one "buys" link would have been sufficient to represent this customer class and the associated marketing knowledge. Hence, for every ordering of the dimensions, the hierarchy will be well matched to some customer classes but ill fitting for others. Thus, we have identified a serious problem which may occur for *any* choice of ordering the dimensions, where many cases of marketing knowledge will require many links. The problem is inherent in the fact *that* an ordering is used.

Besides this problem of uneconomical representation of marketing knowledge, this straightforward representation has two secondary problems. One problem is the explosion of the total number of nodes. The number of just the leaves in T is the product of the numbers of options for all dimensions. In our tree hierarchy T of only four dimensions, each with few choices, there are 60 leaves. In the market research field, practitioners have identified many more dimensions. For example, the ten dimensions appearing in the MediaMark Web site [18] for customer classification are: Gender, Age, Household income, Education, Employment status/occupation, Race with region, Marital status, County size, Marketing region, and Household size. Since any combination of dimensions may appear in a customer classification, the tree hierarchy must be fully developed by expanding all dimensions. This need was demonstrated before with the example using the classification PERSON WITH CHILDREN.

The second problem with ordered dimensions is related to the explosion of nodes. Whole subtrees are repeated over and over. For example, the subtree with the marital choices is repeated for every age group. If a marketing executive decides to add a marital status "WIDOWED", then this update has to be performed in every subtree, leading to the well-known danger of inconsistencies (update anomalies).

4   Customer Intersection Hierarchy

The difficulties we encountered in designing a tree hierarchy customer ontology stem from the fact that there is no preferred order of the various dimensions. Thus, a possible solution is to avoid prioritizing the dimensions. To solve this problem we draw on Sowa's notion of representing conceptual knowledge using distinctions [15]. Sowa claims, for example, that there is no order between the distinctions Concrete/Abstract and Object/Process. All four concepts: Concrete, Abstract, Object, and Process are children of Thing. A concept such as PhysicalObject is an intersection of the concepts Concrete and Object. We call the result of consistently applying such distinctions for all dimensions *on demand* an *intersection ontology*. The significance of creating concepts for an ontology only on demand will be explained below.

We note that we may encounter some dimensions without a natural priority between them in the product hierarchy as well. Figure 3 demonstrates this situation between the location dimension (indoor, outdoor) and the operating mode dimension (electric, non-electric) of toys. Nevertheless, the situation in general is quite different from that of the customer hierarchy, where all dimensions are mutually independent. In the marketing field, there is an established practice of considering some dimensions of product classification prior to others. For example, Men's Wear and Women's Wear are typically in different departments and probably even on different floors of a department store. Each of these are further partitioned into various kinds of clothing, shoes, accessories *etc.* Furthermore, customers are used to this ordering of products and search accordingly for what they desire. Hence, while in the customer hierarchy, all dimensions are independent, some dimensions without natural priority between them exist for products. To handle these cases of independent dimensions for products, one could follow Sowa's [15] practice, where intersections appear only for these few mutually independent dimensions. In the balance of this paper, we will concentrate on the customer hierarchy.

The customer intersection hierarchy has a unique root node representing the concept PERSON at level 1. Each option of each dimension is now represented as a child of the root node at the second level of the hierarchy (see Figure 4). We call such a node an *option node*. For example, in Figure 4, we have the WOMAN option node and the MARRIED option node.

The next question is how to represent a customer classification involving several dimensions. For example, the classification MARRIED WOMAN WITH CHILDREN involves three dimensions: gender, marital status and children status. The solution is to define in the hierarchy a new kind of node that represents a combination of several options, one option for each of several dimensions (Figure 4). For example, a MARRIED WOMAN node represents the combination of the option WOMAN for the gender dimension and the MARRIED option for the marital status dimension. Another node represents WOMAN WITH CHILDREN, a combination of options for gender and children status. The more complicated classification MARRIED WOMAN WITH CHILDREN represents a combination of options for three dimensions: WOMAN for gender, MARRIED for marital status and WITH CHILDREN for children status.

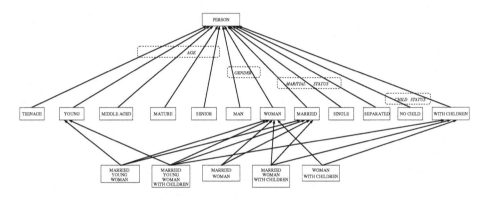

Fig. 4. A Sample Customer Intersection Hierarchy with Option Nodes in Level 2

We call a node that represents a combination of options of various dimensions an *intersection node*, since it represents the classification of a set of customers which is the mathematical intersection of several sets of customers, each with a one-dimensional classification. For example, the set of MARRIED WOMAN is the intersection of two sets MARRIED and WOMAN. Every intersection node is a child of each of the option nodes corresponding to the options involved in the intersection. For example, MARRIED WOMAN is a child of both MARRIED and WOMAN option nodes. Hence all intersection nodes appear in level 3 of the customer intersection hierarchy. Intersection classes of different kinds have appeared in various Object-Oriented Database (OODB) models of medical ontology representations [19–24].

Note that the representation of Figure 4 is superior to the tree hierarchy representation of Section 3, where neither of the classifications mentioned above in this section corresponds to a single node. For instance, MARRIED WOMAN WITH CHILDREN needs to be represented by several nodes in the tree hierarchy T, because the AGE dimension is not mentioned in this classification. In T, AGE is the second dimension, and both MARRIED and WITH CHILDREN are below AGE in the hierarchy. Thus, to incorporate MARRIED, all AGE choices are included, too. As a result, five nodes of T are needed due to the five options of the AGE dimension. Each of these nodes will have a link to DOLL, to capture the marketing knowledge "Married women with children buy dolls," represented by one link in Figure 2. Hence T is not an economical representation of this marketing knowledge.

As another example, fifteen nodes are needed to represent WOMAN WITH CHILDREN in T. This number corresponds to the multiplication of the number of options for the AGE and MARITAL STATUS dimensions, both not mentioned in this classification. Again, 15 links will be needed to represent the marketing knowledge "Women with children buy toys," represented by one link in Figure 2.

The reason for this large number of nodes of T for a classification is that in the tree hierarchy, for each dimension added to the classification, the number of relevant nodes is multiplied by the number of options for this dimension. This is because at each level the classification of each node of the previous level is further subdivided into nodes according to the options considered at this level. Thus, the representation in Figure 4, using intersection nodes, has the advantage that each classification (selecting one option for each dimension), independent of the number of dimensions involved, and independent of the number of their options, is represented by a single node. Hence, each "buys" link, starting at a customer class that is described by an intersection node, has a unique source. In contrast, in the customer tree hierarchy T, it is typical to require several nodes with "buys" relationships for such a classification.

Option nodes may have attributes and relationships. Intersection nodes inherit these properties from all their parents, enabling multiple inheritance of properties. The root node and option nodes may also be sources in "buys" relationships.

At first glance it might appear that with intersection nodes we will generate hierarchies that are even larger than with ordered dimensions, as we have a large number of nodes already at the second level. However, the opposite is the case. A crucial aspect of our definition of intersection ontologies is that concepts below the second level are only created on demand. That is, *only* nodes which represent a combination of dimensions needed for the marketing knowledge in our database are represented in the hierarchy. If no marketing knowledge about a specific combination of dimensions exists, then we do not create an intersection node for this combination!

More specifically, if we do not need a specific group of customers from our database as the source of a "buys" relationship then we do not need to create the corresponding concept node. Thus, if there is no marketing knowledge available in our database about a single man of Alaskan ethnic origin over seventy, then we will not create a corresponding general concept in our ontology. Intersection nodes are created only on demand if the need for them arises. Traditional general ontologies and domain ontologies typically attempt to represent everything that may exist. For our marketing application, this would result in an explosion of concepts. With the intersection hierarchy, the explosion of nodes is controlled.

Only concepts needed are created. In the ordered dimension representation, a node which is not a leaf cannot be omitted from the tree hierarchy, even if no marketing knowledge is available regarding this node, since marketing knowledge may exist about any of its descendants.

Definition: The size of an ontology is a pair (a, b) where a is the number of nodes and b is the number of relationships.

For instance, the size of the ontology of Figure 4 is (18, 26).

5 Multi-level Intersection Hierarchy

Now we will explicitly consider the network connecting all the nodes in the customer intersection hierarchy. We first describe formally the network of the intersection hierarchy, informally described in the previous section. This network will be denoted the three-level intersection hierarchy. Our discussion will show that the three-level intersection hierarchy is not a proper representation. We will then introduce an alternative network, the multi-level intersection hierarchy, overcoming the deficiencies of the three-level intersection hierarchy.

Consider an intersection node which represents the concept of a combination of k options O_{i_1}, O_{i_2},..., O_{i_k}, one for each of the corresponding k dimensions $(k \leq n)$ of the n existing dimensions. Such a concept (node) is more specific than (a child of) each of the option concepts (nodes) which represents one of the options O_{i_j}, $1 \leq j \leq k$, since the set of customers which satisfy all the options O_{i_1}, O_{i_2},..., O_{i_k} simultaneously, is a subset of each of the customer sets which satisfies one option O_{i_j}, where $1 \leq j \leq k$.

In the three-level intersection hierarchy, each intersection node is at the third level, since all its k option parents are the second level. Hence, the name of this network. (See Figure 4 for a sample of a three-level customer hierarchy.)

Now we will discuss in detail why the three-level intersection hierarchy is improper. In the three-level intersection hierarchy, only IS-A relationships between an intersection node and option node are presented. Consider two specific intersection nodes in Figure 4, MARRIED WOMAN and MARRIED WOMAN WITH CHILDREN. The second classification is more specialized than the first classification, since the set of customers, classified by MARRIED WOMAN WITH CHILDREN, is a subset of the set of customers classified by MARRIED WOMAN. To express this specialization, the intersection node MARRIED WOMAN WITH CHILDREN should have as a parent the intersection node MARRIED WOMAN.

In the three-level intersection hierarchy in Figure 4, the node MARRIED WOMAN WITH CHILDREN has three parents: WOMAN, MARRIED and WITH CHILDREN. Should those parent relationships also exist after adding the parent MARRIED WOMAN? The node MARRIED WOMAN itself has as parents the option nodes WOMAN and MARRIED. A relationship from MARRIED WOMAN WITH CHILDREN to WOMAN (or to MARRIED) is implied by the transitivity of the IS-A relationship.

Thus, we conclude that the three-level representation does not fulfill all our requirements for a proper representation, because it does not capture the specialization which exists between intersection nodes. We will now introduce an alternative representation, the more refined *multi-level intersection hierarchy* that allows expressing parent-child relationships between two intersection nodes, when one represents a more specific concept than the other. For a multi-level hierarchy representation of the nodes of Figure 4, see Figure 5.

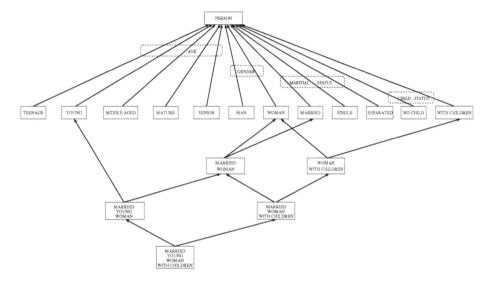

Fig. 5. A Sample Multi-Level Customer Hierarchy

In Figure 5, the node MARRIED WOMAN WITH CHILDREN has no parent relationship to the option nodes. On the other hand, the node MARRIED YOUNG WOMAN has a parent relationship to the option node YOUNG since the hierarchy contains neither the node YOUNG WOMAN nor the node YOUNG MARRIED which would have been parents of MARRIED YOUNG WOMAN and would have implied, as an intermediate node, the IS-A relationship to the option node YOUNG by transitivity.

Note that Figure 5 has 5 levels. The number of explicit parent relationships in Figure 5 is 22 versus 26 such relationships in Figure 4. Both figures have 18 nodes.

The definition for visual complexity of a diagram was introduced and used in [25–27]. We will now modify it for use with ontologies.

Definition: The *visual complexity* C of an ontology of size (a, b) is the ratio of the number of relationships (= links) to the number of nodes, $C = b/a$.

Hence the three-level intersection ontology of Figure 4 has size $(18, 26)$ and visual complexity $C = 26/18 = 1.44$. On the other hand, the multi-level intersection ontology of Figure 5 has size $(18, 22)$ and visual complexity $C = 22/18 = 1.22$.

In this example, the multi-level ontology has lower size and lower visual complexity in comparison with the corresponding three-level ontology. Note that our visual complexity measure is a global measure for an ontology, compared to the notions of "tangled" and "sparse" used in [28] to measure local properties of the top level hierarchy.

To summarize, the three-level intersection hierarchy representation is not proper, because it does not capture IS-A relationships between intersection nodes. Such relationships are captured by the multi-level intersection hierarchy which also has other advantages, as follows.

1. The multi-level representation allows to use inheritance between intersection nodes, which is not possible in the three-level hierarchy. For example, if we know that women with children buy toys, we inherit this fact to married women with children. In this way, the multi-level representation maintains one of the major advantages of ontologies, the economy brought about by inheritance-based reasoning.
2. The distribution of intersection nodes over several levels, due to the additional specialization IS-A relationship between such nodes, simplifies orientation of the user in such a hierarchy.
3. The number of explicit parent relationships is typically smaller than in the three-level intersection hierarchy. (This is not necessarily true, as one can intentionally design a counterexample.) This makes the multi-level intersection hierarchy diagram smaller in *size* and lower in *visual complexity* than the equivalent three-level intersection hierarchy.

6   Evaluation

We will use the customer ontology of our marketing project to evaluate the design of the multi-level ontology versus the other designs. In our Web marketing project, we have collected 301,109 valid records of person's information. A record of information is considered valid when it has a valid email address and at least one expressed interest. Some of the information is expressed in foreign characters, which we ignore. After filtering, we have 274,665 records. However, most people also provide more information such as their age, gender and marital status. Regarding these as three dimensions for PERSON, we constructed the customer ontology for our project and show how the *ordered dimensions* tree hierarchy, the *three-level intersection* hierarchy, and the *multi-level intersection* hierarchy representation will perform, respectively.

The dimensions of AGE, GENDER and MARITAL STATUS have 6, 2 and 6 options respectively. Each record is represented as an instance of a corresponding classification (node) in the ontology. However, some nodes only contain fewer than 100 records. For marketing purposes, we ignore such nodes which do not represent useful information.

Using the design of *ordered dimensions*, we have the ontology as in Figure 6(a). The blank boxes stand for nodes without enough instances, and are not created.

In this figure, each node represents a meaningful customer classification from a marketing point of view, with the corresponding number of persons in our database. For instance, there are 23709 records for those who are males whose ages are between 10 and 19, whose marital status is not specified.

The tree hierarchy in Figure 6(a) has 62 nodes and 61 IS-A links and the visual complexity of 0.98. However, using this hierarchy, when trying to represent all the customer concepts with marketing knowledge, some of the concepts are not represented by a single node. To represent such a concept, multiple nodes, distributed in different parts of the hierarchy of Figure 6(a), have to be collected. For example, due to the order of the dimensions, to represent the concept AGE 20-29, 11 nodes, structured in 2 subtrees in Figure 6(a), are needed, as shown in Figure 7(a). Moreover, to represent the concept MALE and DIVORCED, 4 nodes need to be collected, as shown in Figure 7(b).

The number of possible concepts with one dimension is $2+6+6 = 14$ and with two dimensions is $2 \times 6 + 2 \times 6 + 6 \times 6 = 60$. Hence the number of possible concepts with one or two dimensions is 74. We are not considering here the concepts with three dimensions, since they are properly represented in Figure 6(a) by a single node leaf. Among those 74 concepts, 14 can be found, in levels 2 and 3 in Figure 6(a), as corresponding single nodes. Since 48 of them do not have enough instances, there are $74 - 14 - 48 = 12$ concepts which are not represented by a single node. Figure 6(b) summarizes those 12 concepts needed in addition to Figure 6(a) to represent every needed marketing knowledge concept. Every one of these 12 concepts needs to be represented by a group of nodes, distributed in various parts of Figure 6(a), shown as its children as in the Figure 7. For each concept in Figure 6(b), the number of these nodes is listed, adding up to 76 nodes. Note that Figures 7(a) and 7(b) show only the expansions of the first node and the sixth node in Figure 6(b), respectively. Thus, the number of nodes representing all the relevant concepts in the customer tree hierarchy is $62 + 76 = 138$.

In the design of the *multi-level intersection* hierarchy, we get the ontology hierarchy in Figure 8. There are 14 option nodes. The third level has 21 intersection nodes, each of which has 2 IS-A links to option nodes. The fourth level has 47 intersection nodes combining three dimensions. Out of 72 possible intersection nodes, 25 contain fewer than 100 records and are not represented. Thus, this design has $1 + 14 + 21 + 47 = 83$ nodes and 150 IS-A links. The visual complexity of the *multi-level intersection* hierarchy is $150/83 = 1.81$.

For the *three-level intersection* hierarchy, the figure is too large to be shown here. However, the figure is a modification of Figure 8 for the *multi-level intersection* hierarchy. The only difference, is that all the 47 nodes in the fourth level are moved to level 3 and are directly connected to the option nodes. Thus, we have 68 intersection nodes at level 3. The total nodes number again is 83, but the number of IS-A links is 197. The extra 47 IS-A links are since each of the 47 nodes has 3 IS-A links. The visual complexity is $197/83 = 2.37$.

In summary, the usage of intersection nodes insures that every relevant customer concept is represented by one single node in the hierarchy. The three-level

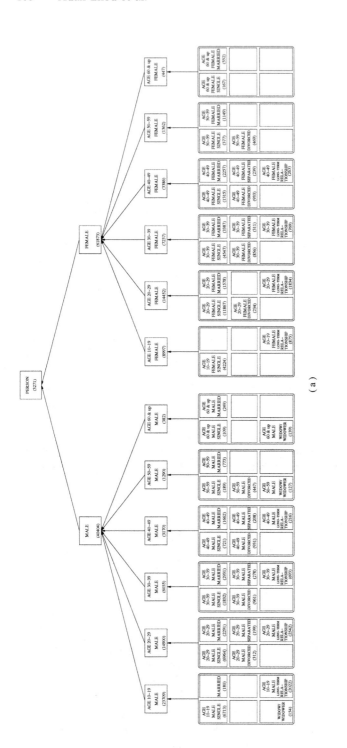

(a)

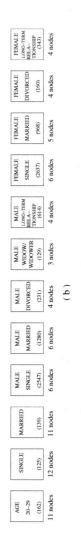

(b)

Fig. 6. Our Marketing Hierarchy with Ordered Dimensions

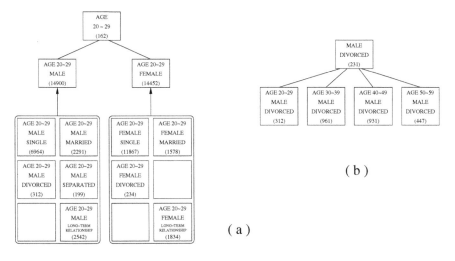

Fig. 7. The nodes collection samples in the Figure 6(b)

and multi-level intersection hierarchy have the same number of nodes. However, the *multi-level intersection* hierarchy has fewer links and lower visual complexity than the *three-level intersection* hierarchy.

In the design of the customer tree hierarchy, consisting of several trees (see Figure 6), there are 138 nodes. Comparing with the 83 nodes in the other two designs, the size is 66% bigger. This design has a lower visual complexity, since it is a forest of 13 trees. However, the measurement of visual complexity is secondary to the size, in this case.

7   Discussion

In principle, the mission of a general purpose ontology is to represent the real world and facilitate the exchange of information between heterogeneous systems. In this view, one would need to create every single intersection in the customer hierarchy, whether we have additional information about it or not, simply because it may be needed for information exchange. However, in our application ontology this would lead to an unreasonably large structure. As a matter of fact, Sowa [14], page 53, writes that "limited ontologies will always be useful for single applications in highly specialized domains. But to share knowledge with other applications, an ontology must be embedded within a more general framework." As a single application ontology, our marketing ontology has to serve its application as well as possible. Thus, it should be an economical representation that includes only nodes for which marketing knowledge is relevant. Also, for this specific application, there is no order whatsoever between any pair of dimensions. Thus, intersections need to be applied universally. This is not necessarily true for other application ontologies.

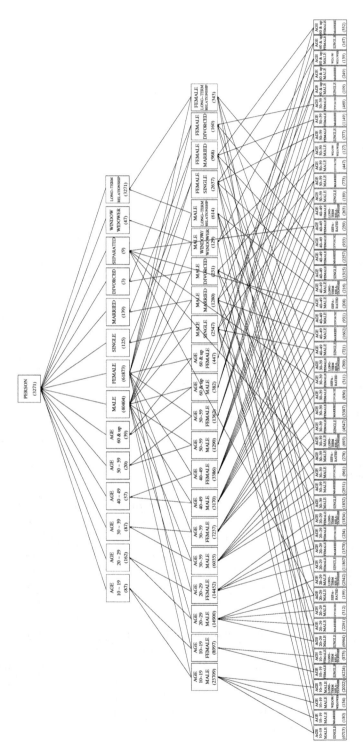

Fig. 8. Our Marketing Multi-Level Customer Hierarchy

Thus, we omit from our ontology information that may be "inferred." The principle of creating an ontology that is an economical representation goes back to the first semantic networks [1]. A major reason for storing attributes at a concept high up in a semantic network hierarchy was to eliminate duplication of information. Whenever any such attribute was needed at a lower level, it was inherited down.

The major difference in our case is that we are not omitting attributes at lower levels, we are omitting the lower levels altogether. Instead of using inheritance to instantiate the representation whenever needed, we are allowing the on demand creation of new intersection concepts which are children of two or more existing concepts. Similarly, [29] writes about dynamic additions in an ontology: "Note that according to this definition, an ontology includes not only the terms that are explicitly defined in it, but also terms that can be inferred using rules." Thus, one could view our approach as implementing a global inference rule which is triggered by existing data and infers new concepts.

One problem which exists in the design of ontologies is how to forbid the representation of an impossible combination. In our intersection ontology design, this translates into the question how to forbid impossible intersections. For example, we should not represent an intersection node TEENAGE MARRIED WOMAN since it is illegal for a teenager to get married.

How can we prevent impossible combinations in our intersection ontology? Note that our ontology's intersections are created on demand, based on available marketing knowledge. There should be no such impossible combinations in the available marketing knowledge, and thus no such intersection should be created. If, however, such an intersection is created, it comes from erroneous data and can be used for auditing errors in the given marketing knowledge, as we did in [23].

8 Conclusions

We have introduced an application-oriented ontology for marketing knowledge, based on the introduction of intersection concepts on demand. Instead of imposing an order on the classification dimensions which is satisfactory for some purposes (and users) but not for others, we completely eliminate ordered dimensions. Instead, we consistently use intersections of options for the various dimensions.

We described the development of an application ontology for customer classifications in a marketing knowledge base. This ontology needed to conform to a number of requirements. First and foremost, we wanted to make it easy to represent in the ontology, knowledge about likely buying behavior of classes of customers by a single (or a few) links from customer concepts to product concepts.

The intersection ontology representation satisfies this purpose because it allows the representation of "buys" relationships by single links whenever this is warranted by the marketing knowledge. Yet, our representation does not produce a combinatorial explosion of all possible intersection nodes. Rather, it only represents the concepts for customer classes which are necessary as sources for known "buys" relationships.

In the multi-level intersection ontology representation, intersection nodes of many option nodes may be placed at various levels. These nodes may have IS-A links to other intersection nodes as well as to option nodes. These IS-A links may be used for property inheritance, as in other concept hierarchies.

The multi-level hierarchy representation fulfills a secondary requirement that we have for ontologies, namely that they can be represented by diagrams of relatively low size and visual complexity. This representation typically requires lower visual complexity relative to the three-level intersection hierarchy. As described in Section 6, in the evaluation based on our marketing project, the multi-level representation has a 24% lower visual complexity than the three-level representation. Moreover, its size is 40% smaller than the size of the ordered dimensions representation. In conclusion, we showed that for the marketing domain an economical intersection ontology may be created by inserting intersections of options of the various classification dimensions on demand. Such a representation may also be proper for other applications.

References

1. Quillian, M.R.: Semantic memory. In Minsky, M.L., ed.: Semantic Information Processing. The MIT Press, Cambridge, MA (1968) 227–270
2. P. Bouquet, A. Dona, L.S., Zanobini, S.: Contextualized local ontology specification via ctxml. In: MeaN-02 AAAI Workshop on Meaning Negotiation, Edmonton, Alberta, Canada (2002)
3. Gruber, T.R.: A translation approach to portable ontology specifications. Knowledge Acquisition **5** (1993) 199–220
4. Geller, J., Perl, Y., Lee, J.: Guest editors' introduction to the special issue on ontologies: Ontology challenges: A thumbnail historical perspective. Knowledge and Information Systems **6** (2004) 375–379
5. Berners-Lee, T., Hendler, J., Lassila, O.: The Semantic Web. Scientific American **284** (2001) 34–43
6. Guarino, N., Poli, R.: Special issue on formal ontology in conceptual analysis and knowledge representation. Journal of Human-Computer Studies **43** (1995)
7. Hendler, J.: Special issue on agents and the semantic web. IEEE Intelligent Systems **16** (2001)
8. Gruninger, M., Lee, J.: Special issue on ontology applications and design. Communications of the ACM **45** (2002)
9. McGuinness, D.L.: Ontologies come of age. In Fensel, D., Hendler, J., Lieberman, H., Wahlster, W., eds.: Spinning the Semantic Web: Bringing the World Wide Web to Its Full Potential, Washington, D.C., MIT Press (2002)
10. Noy, N.F., Hafner, C.D.: The state of the art in ontology design. AI Magazine (Fall 1997) 53–74
11. Geller, J., Perl, Y., Lee, J.: Special issue on ontologies: Ontology challenges. Knowledge and Information Systems **6** (2004)
12. Geller, J., Scherl, R., Perl, Y.: Mining the web for target marketing information. In: Proceedings of CollECTer, Toulouse, France (April 20th, 2002)

13. Witten, I.H., Frank, E.: Data Mining. Morgan Kaufmann Publishers, San Francisco (2000)
14. Sowa, J.F.: Knowledge Representation. Brooks/Cole, Pacific Grove, CA (2000)
15. Sowa, J.: Distinctions, combinations and constraints. In: Proceedings of the Workshop on Basic Ontological Issues in Knowledge Sharing, Montreal, Canada (1995)
16. Wille, R.: Restructuring lattice theory: An approach based on hierarchies of concepts. In Rival, I., ed.: Ordered sets, Reidel, Dordrecht, Boston, MA (1982) 445–470
17. Wille, R.: Concept lattices and conceptual knowledge systems. Computers and Mathematics with Application **23** (1992) 493–515
18. MediaMark: MediaMark Web site (2001) http://www.mediamark.com.
19. Gu, H., Halper, M., Geller, J., Perl, Y.: Benefits of an OODB representation for controlled medical terminologies. JAMIA **6** (1999) 283–303
20. Gu, H., Perl, Y., Geller, J., Halper, M., Liu, L., Cimino, J.J.: Representing the UMLS as an OODB: Modeling issues and advantages. JAMIA **7** (2000) 66–80
21. Liu, L., Halper, M., Geller, J., Perl, Y.: Controlled vocabularies in OODBs: Modeling issues and implementation. Distributed and Parallel Databases **7** (1999) 37–65
22. Liu, L., Halper, M., Geller, J., Perl, Y.: Using OODB modeling to partition a vocabulary into structurally and semantically uniform concept groups. IEEE Transactions on Knowledge and Data Engineering **14** (2002) 850–866
23. Geller, J., Gu, H., Perl, Y., Halper, M.: Semantic refinement and error correction in large terminological knowledge bases. Data and Knowledge Engineering **45** (2003) 1–32
24. Gu, H., Perl, Y., Elhanan, G., Min, H., Zhang, L., Peng, Y.: Auditing concept categorizations in the UMLS. Artificial Intelligence in Medicine **31** (2004) 29–44
25. Gu, H., Perl, Y., Halper, M., Geller, J., Kuo, F., Cimino, J.J.: Partitioning an object-oriented terminology schema. Methods in Medical Informatics (2001) 204–212
26. Perl, Y., Geller, J., Gu, H.: Identifying a forest hierarchy in an OODB specialization hierarchy satisfying disciplined modeling. In: Proc. CoopIS'96, Brussels, Belgium (1996) 182–195
27. Gu, H., Perl, Y., Halper, M., Geller, J., Neuhold, E.J.: Contextual partitioning for comprehension of OODB schemas. Knowledge and Information Systems (KAIS) **6** (2004) 315–344
28. Campbell, A., Shapiro, S.: Ontologic mediation: An overview. In: Proceedings of the Workshop on Basic Ontological Issues in Knowledge Sharing, Montreal, Canada (1995) 16–25
29. Gómez-Pérez, A., Benjamins, V.R.: Overview of knowledge sharing and reuse components: Ontologies and problem-solving methods. In Benjamins, V., Chandrasekaran, B., Gómez-Pérez, A., Guarino, N., Uschold, M., eds.: Proceedings of the IJCAI-99 workshop on Ontologies and Problem-Solving Methods (KRR5), Stockholm, Sweden (1999) 1.1–1.15

How to Leak a Secret: Theory and Applications of Ring Signatures

Ronald L. Rivest[1], Adi Shamir[2], and Yael Tauman[1]

[1] Laboratory for Computer Science, Massachusetts Institute of Technology,
Cambridge, MA 02139
{rivest,tauman}@mit.edu,
http://theory.lcs.mit.edu/ rivest, http://www.mit.edu/ tauman
[2] Computer Science department, The Weizmann Institute, Rehovot 76100, Israel
adi.shamir@weizmann.ac.il

Abstract. In this work we formalize the notion of a *ring signature*, which makes it possible to specify a set of possible signers without revealing which member actually produced the signature. Unlike group signatures, ring signatures have no group managers, no setup procedures, no revocation procedures, and no coordination: any user can choose any set of possible signers that includes himself, and sign any message by using his secret key and the others' public keys, without getting their approval or assistance. Ring signatures provide an elegant way to leak authoritative secrets in an anonymous way, to sign casual email in a way that can only be verified by its intended recipient, and to solve other problems in multiparty computations.

Our main contribution lies in the presentation of efficient constructions of ring signatures; the general concept itself (under different terminology) was first introduced by Cramer et al. [CDS94]. Our constructions of such signatures are unconditionally signer-ambiguous, secure in the random oracle model, and exceptionally efficient: adding each ring member increases the cost of signing or verifying by a single modular multiplication and a single symmetric encryption. We also describe a large number of extensions, modifications and applications of ring signatures which were published after the original version of this work (in Asiacrypt 2001).

1 Introduction

The general notion of a *group signature scheme* was introduced in 1991 by Chaum and van Heyst [CV91]. In such a scheme, a trusted group manager predefines certain groups of users and distributes specially designed keys to their members. Individual members can then use these keys to anonymously sign messages on behalf of their group. The signatures produced by different group members look indistinguishable to their verifiers, but not to the group manager who can revoke the anonymity of misbehaving signers.

In this work we formalize the related notion of *ring signature schemes*. These are simplified group signature schemes that have only users and no managers (we call such signatures "ring signatures" instead of "group signatures" since rings

O. Goldreich et al. (Eds.): Shimon Even Festschrift, LNCS 3895, pp. 164–186, 2006.

are geometric regions with uniform periphery and no center).[3] Group signatures are useful when the members want to cooperate, while ring signatures are useful when the members do not want to cooperate. Both group signatures and ring signatures are *signer-ambiguous*, but in a ring signature scheme there are no prearranged groups of users, there are no procedures for setting, changing, or deleting groups, there is no way to distribute specialized keys, and there is no way to revoke the anonymity of the actual signer (unless he decides to expose himself). Our only assumption is that each member is already associated with the public key of some standard signature scheme such as RSA. To produce a ring signature, the *actual signer* declares an arbitrary set of *possible signers* that must include himself, and computes the signature entirely by himself using only his secret key and the others' public keys. In particular, the other possible signers could have chosen their RSA keys only in order to conduct e-commerce over the internet, and may be completely unaware that their public keys are used by a stranger to produce such a ring signature on a message they have never seen and would not wish to sign.

The notion of ring signatures is not completely new, but previous references do not crisply formalize the notion, and propose constructions that are less efficient and/or that have different, albeit related, objectives. They tend to describe this notion in the context of general group signatures or multiparty constructions, which are quite inefficient. For example, Chaum et al. [CV91]'s schemes three and four, and the two signature schemes in Definitions 2 and 3 of Camenisch's paper [Cam97] can be viewed as ring signature schemes. However the former schemes require zero-knowledge proofs with each signature, and the latter schemes require as many modular exponentiations as there are members in the ring. Cramer et al. [CDS94] show how to produce witness-indistinguishable interactive proofs. Such proofs could be combined with the Fiat-Shamir technique to produce ring signature schemes. Similarly, DeSantis et al. [SCPY94] show that interactive SZK for random self-reducible languages are closed under monotone boolean operations, and show the applicability of this result to the construction of a ring signature scheme (although they don't use this terminology).

The direct construction of ring signatures proposed in this paper is based on a completely different idea, and is exceptionally efficient for large rings (adding only one modular multiplication and one symmetric encryption per ring member both to generate and to verify such signatures). The resultant signatures are unconditionally signer-ambiguous and secure in the random oracle model. This model, formalized in [BR93], assumes that all parties have oracle access to a truly random function.

There have been several followup papers on the theory and applications of ring signatures. We summarize these results in Section 7.

---

[3] Hanatani and Ohta have pointed out that the idea of signing messages in the form of a ring dates back at least as far as 1756, the middle of Edo period, in Japan [HO05]. In that time, a group of farmers would sign their names in the form of a ring so as to conceal the identity of the group leader. (Had they signed sequentially, the leader would be expected to be the first on the list.)

2   Definitions and Applications

2.1   Ring Signatures

Terminology: We call a set of *possible signers* a *ring*. We call the ring member who produces the actual signature the *signer* and each of the other ring members a *non-signer*.

We assume that each possible signer is associated (via a PKI directory or certificate) with a public key P_k that defines his signature scheme and specifies his verification key. The corresponding secret key (which is used to generate regular signatures) is denoted by S_k. The general notion of a ring signature scheme does not require any special properties of these individual signing schemes, but our simplest construction assumes that they use trapdoor one-way permutations (such as the RSA functions) to generate and verify signatures.

A ring signature scheme is defined by two procedures:

- **ring-sign**$(m, P_1, P_2, \ldots, P_r, s, S_s)$ which produces a ring signature σ for the message m, given the public keys P_1, P_2, $\ldots$, P_r of the r ring members, together with the secret key S_s of the s-th member (who is the actual signer).
- **ring-verify**(m, σ) which accepts a message m and a signature σ (which includes the public keys of all the possible signers), and outputs either *true* or *false*.

A ring signature scheme is *set-up free*: The signer does not need the knowledge, consent, or assistance of the other ring members to put them in the ring; all he needs is knowledge of their regular public keys. Different members can use different independent public key signature schemes, with different key and signature sizes. Verification must satisfy the usual soundness and completeness conditions, but in addition we want the signatures to be *signer-ambiguous* in the sense that a signature should leek no information about the identity of the signer. This anonymity property can be either *computational* or *unconditional*. Our main construction provides unconditional anonymity in the sense that even an infinitely powerful adversary with access to an unbounded number of chosen-message signatures produced by the same ring member cannot guess his identity with any advantage, and cannot link additional signatures to the same signer.

Note that the size of any ring signature must grow linearly with the size of the ring, since it must list the ring members; this is an inherent disadvantage of ring signatures as compared to group signatures that use predefined groups.

2.2   Leaking Secrets

To motivate the title for this paper, suppose that Bob (also known as "Deep Throat") is a member of the cabinet of Lower Kryptonia, and that Bob wishes to leak a juicy fact to a journalist about the escapades of the Prime Minister, in such a way that Bob remains anonymous, yet such that the journalist is convinced that the leak was indeed from a cabinet member.

Bob cannot send to the journalist a standard digitally signed message, since such a message, although it convinces the journalist that it came from a cabinet member, does so by directly revealing Bob's identity.

It also doesn't work for Bob to send the journalist a message through a standard "anonymizer" [Ch81, Ch88, GRS99], since the anonymizer strips off all source identification and authentication: the journalist would have no reason to believe that the message really came from a cabinet member at all.

A standard group signature scheme does not solve the problem, since it requires the prior cooperation of the other group members to set up, and leaves Bob vulnerable to later identification by the group manager, who may be controlled by the Prime Minister.

The correct approach is for Bob to send the story to the journalist (through an anonymizer), signed with a ring signature scheme that names each cabinet member (including himself) as a ring member. The journalist can verify the ring signature on the message, and learn that it definitely came from a cabinet member. He can even post the ring signature in his paper or web page, to prove to his readers that the juicy story came from a reputable source. However, neither he nor his readers can determine the actual source of the leak, and thus the whistleblower has perfect protection even if the journalist is later forced by a judge to reveal his "source" (the signed document).

2.3   Designated Verifier Signature Schemes

A designated verifier signature scheme is a signature scheme in which signatures can only be verified by a single "designated verifier" chosen by the signer. It can be viewed as a "light signature scheme" which can authenticate messages to their intended recipients without having the nonrepudiation property. This concept was first introduced by Jakobsson, Sako and Impagliazzo at Eurocrypt 96 [JSI96].

A typical application is to enable users to authenticate casual emails without being legally bound to their contents. For example, two companies may exchange drafts of proposed contracts. They wish to add to each email an authenticator, but not a real signature which can be shown to a third party (immediately or years later) as proof that a particular draft was proposed by the other company.

One approach would be to use zero knowledge interactive proofs, which can only convince their verifiers. However, this requires interaction and is difficult to integrate with standard email systems and anonymizers. We can use noninteractive zero knowledge proofs, but then the authenticators become signatures which can be shown to third parties. Another approach is to agree on a shared secret symmetric key k, and to authenticate each contract draft by appending a message authentication code (MAC) for the draft computed with key k. A third party would have to be shown the secret key to validate a MAC, and even then he wouldn't know which of the two companies computed the MAC. However, this requires an initial set-up procedure to generate the secret symmetric key k.

A designated verifier scheme provides a simple solution to this problem: company A can sign each draft it sends, naming company B as the designated verifier.

This can be easily achieved by using a ring signature scheme with companies A and B as the ring members. Just as with a MAC, company B knows that the message came from company A (since no third party could have produced this ring signature), but company B cannot prove to anyone else that the draft of the contract was signed by company A, since company B could have produced this draft by itself. Unlike the case of MAC's, this scheme uses public key cryptography, and thus A can send unsolicited email to B signed with the ring signature without any preparations, interactions, or secret key exchanges. By using our proposed ring signature scheme, we can turn standard signature schemes into designated verifier schemes, which can be added at almost no cost as an extra option to any email system.

3   Efficiency of Our Ring Signature Scheme

When based on Rabin or RSA signatures, our ring signature scheme is particularly efficient:

- signing requires one modular exponentiation, plus one or two modular multiplications for each non-signer.
- verification requires one or two modular multiplications for each ring member.

In essence, generating or verifying a ring signature costs the same as generating or verifying a regular signature plus an extra multiplication or two for each non-signer, and thus the scheme is truly practical even when the ring contains hundreds of members. It is two to three orders of magnitude faster than Camenisch's scheme, whose claimed efficiency is based on the fact that it is 4 times faster than earlier known schemes (see bottom of page 476 in his paper [Cam97]). In addition, a Camenisch-like scheme uses linear algebra in the exponents, and thus requires all the members to use the same prime modulus p in their individual signature schemes. One of our design criteria is that the signer should be able to assemble an arbitrary ring without any coordination with the other ring members. In reality, if one wants to use other users' public keys, they are much more likely to be RSA keys, and even if they are based on discrete logs, different users are likely to have different moduli p. The only realistic way to arrange a Camenisch-like signature scheme is thus to have a group of consenting parties.

4   The Proposed Ring Signature Scheme (RSA Version)

Suppose that Alice wishes to sign a message m with a ring signature for the ring of r individuals $A_1, A_2, \ldots, A_r$, where the signer Alice is A_s, for some value of s, $1 \leq s \leq r$. To simplify the presentation and proof, we first describe a ring signature scheme in which all the ring members use RSA [RSA78] as their individual signature schemes. The same construction can be used for any other trapdoor one way permutation, but we have to modify it slightly in order to use trapdoor one way functions (as in, for example, Rabin's signature scheme [Rab79]).

4.1 RSA Trapdoor Permutations

Each ring member A_i has an RSA public key $P_i = (n_i, e_i)$ which specifies the trapdoor one-way permutation f_i of $\mathbf{Z}_{n_i}$:

$$f_i(x) = x^{e_i} \pmod{n_i} .$$

We assume that only A_i knows how to compute the inverse permutation f_i^{-1} efficiently, using trapdoor information (i.e., $f_i^{-1}(y) = y^{d_i} \pmod{n_i}$, where $d_i = e_i^{-1} \pmod{\phi(n_i)}$ is the trapdoor information). This is the original Diffie-Hellman model [DH76] for public-key cryptography.

Extending trapdoor permutations to a common domain

The trapdoor RSA permutations of the various ring members will have domains of different sizes (even if all the moduli n_i have the same number of bits). This makes it awkward to combine the individual signatures, and thus we extend all the trapdoor permutations to have as their common domain the same set $\{0, 1\}^b$, where 2^b is some power of two which is larger than all the moduli n_i's.

For each trapdoor permutation f over $\mathbf{Z}_n$, we define the extended trapdoor permutation g over $\{0, 1\}^b$ in the following way. For any b-bit input m define nonnegative integers q and r so that $m = qn + r$ and $0 \le r < n$. Then

$$g(m) = \begin{cases} qn + f(r) & \text{if } (q+1)n \le 2^b \\ m & \text{else.} \end{cases}$$

Intuitively, g is defined by using f to operate on the low-order digit of the n-ary representation of m, leaving the higher order digits unchanged. The exception is when this might cause a result larger than $2^b - 1$, in which case m is unchanged. If we choose a sufficiently large b (e.g. 160 bits larger than any of the n_i's), the chance that a randomly chosen m is unchanged by the extended g becomes negligible. (A stronger but more expensive approach, which we don't need, would use instead of $g(m)$ the function $g'(m) = g((2^b - 1) - g(m))$ which can modify all its inputs). The function g is clearly a permutation over $\{0, 1\}^b$, and it is a one-way trapdoor permutation since only someone who knows how to invert f can invert g efficiently on more than a negligible fraction of the possible inputs.

4.2 Symmetric Encryption

We assume the existence of a publicly defined symmetric encryption algorithm E such that for any key k of length l, the function E_k is a permutation over b-bit strings. Here we use the ideal cipher model which assumes that all the parties have access to an oracle that provides truly random answers to new queries of the form $E_k(x)$ and $E_k^{-1}(y)$, provided only that they are consistent with previous answers and with the requirement that E_k be a permutation. It was shown in [BSS02] that the ideal cipher model can be reduced to the random oracle model without almost any efficiency loss.[4] For simplicity we use the ideal cipher model in this presentation.

4.3   Hash Functions

We assume the existence of a publicly defined collision-resistant hash function h that maps arbitrary inputs to strings of length l, which are used as keys for E. We model h as a random oracle. (Since h need not be a permutation, different queries may have the same answer, and we do not consider "h^{-1}" queries.)

4.4   Combining Functions

We define a family of keyed "combining functions" $C_{k,v}(y_1, y_2, \ldots, y_r)$ which take as input a key k, an initialization value v, and arbitrary values y_1, y_2, $\ldots$, y_r in $\{0,1\}^b$. Each such combining function uses E_k as a sub-procedure, and produces as output a value z in $\{0,1\}^b$ such that given any fixed values for k and v, we have the following properties.

1. **Permutation on each input:** For each $s \in \{1, \ldots, r\}$, and for any fixed values of all the other inputs y_i, $i \neq s$, the function $C_{k,v}$ is a one-to-one mapping from y_s to the output z.
2. **Efficiently solvable for any single input:** For each $s \in \{1, \ldots, r\}$, given a b-bit value z and values for all inputs y_i except y_s, it is possible to efficiently find a b-bit value for y_s such that $C_{k,v}(y_1, y_2, \ldots, y_r) = z$.
3. **Infeasible to solve verification equation without trapdoors:** Given k, v, and z, it is infeasible for an adversary to solve the equation

$$C_{k,v}(g_1(x_1), g_2(x_2), \ldots, g_r(x_r)) = z \tag{1}$$

for $x_1, x_2, \ldots, x_r$, (given access to each g_i, and to E_k) if the adversary can't invert any of the trapdoor functions $g_1, g_2, \ldots, g_r$.

For example, the function

$$C_{k,v}(y_1, y_2, \ldots, y_r) = y_1 \oplus y_2 \oplus \cdots \oplus y_r$$

(where $\oplus$ is the exclusive-or operation on b-bit words) satisfies the first two of the above conditions, and can be kept in mind as a candidate combining function. Indeed, it was the first one we tried. But it fails the third condition since for any choice of trapdoor one-way permutations g_i, it is possible to use linear algebra when r is large enough to find a solution for $x_1, x_2, \ldots, x_r$ without inverting any of the g_i's. The basic idea of the attack is to choose a random value for each x_i, and to compute each $y_i = g_i(x_i)$ in the easy forward direction. If the number of values r exceeds the number of bits b, we can find with high probability a subset

---

[4] It was shown in [LR88] that the ideal cipher model can *always* be reduced to the random oracle model (with some efficiency loss).

of the y_i bit strings whose XOR is any desired b-bit target z. However, our goal is to represent z as the XOR of all the values $y_1, y_2, \ldots, y_r$ rather than as a XOR of a random subset of these values. To overcome this problem, we choose for each i *two* random values x_i' and x_i'', and compute their corresponding $y_i' = g_i(x_i')$ and $y_i'' = g_i(x_i'')$. We then define $y_i''' = y_i' \oplus y_i''$, and modify the target value to $z' = z \oplus y_1' \oplus y_2', \ldots \oplus y_r'$. We use the previous algorithm to represent z' as a XOR of a random subset of y_i''' values. After simplification, we get a representation of the original z as the XOR of a set of r values, with exactly one value chosen from each pair (y_i', y_i''). By choosing the corresponding value of either x_i' or x_i'', we can solve the verification equation without inverting any of the trapdoor one-way permutations g_i. (One approach to countering this attack, which we don't explore further here, is to let b grow with r.)

Even worse problems can be shown to exist in other natural combining functions such as addition mod 2^b. Assume that we use the RSA trapdoor functions $g_i(x_i) = x_i^3 (\bmod\ n_i)$ where all the moduli n_i have the same size b. It is known [HW79] that any nonnegative integer z can be efficiently represented as the sum of exactly nine nonnegative integer cubes $x_1^3 + x_2^3 + \ldots + x_9^3$. If z is a b-bit target value, we can expect each one of the x_i^3 to be slightly shorter than z, and thus their values are not likely to be affected by reducing each x_i^3 modulo the corresponding b-bit n_i. Consequently, we can solve the verification equation $(x_1^3 \bmod n_1) + (x_2^3 \bmod n_2) \ldots + (x_9^3 \bmod n_9) = z(\bmod\ 2^b)$ with nine RSA permutations without inverting any one of them.

Our proposed combining function utilizes the symmetric encryption function E_k as follows:

$$C_{k,v}(y_1, y_2, \ldots, y_r) = E_k(y_r \oplus E_k(y_{r-1} \oplus E_k(y_{r-2} \oplus E_k(\ldots \oplus E_k(y_1 \oplus v)\ldots)))) \ .$$

This function is applied to the sequence $(y_1, y_2, \ldots, y_r)$, where $y_i = g_i(x_i)$, as shown in Figure 1.

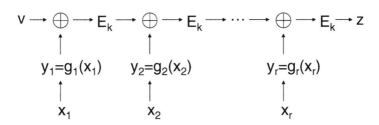

Fig. 1. An illustration of the proposed combining function

This function is clearly a permutation on each input, since the XOR and E_k functions are permutations. In addition, it is efficiently solvable for any single input since knowledge of k makes it possible to run the evaluation forwards from the initial v and backwards from the final z in order to uniquely compute any missing value y_i.

This function can be used to construct a signature scheme as follows: In order to sign a message m, set $k = h(m)$, where h is some predetermined hash function, and output $x_1, \ldots, x_r$ such that $C_{k,v}(g_1(x_1), g_2(x_2), \ldots, g_r(x_r)) = v$. Notice that forcing the output z to be equal to the input v, bends the line into the ring shape shown in Fig. 2.

A slightly more compact ring signature variant can be obtained by always selecting 0 as the "glue value" v. This variant is also secure, but we prefer the total ring symmetry of our main proposal.

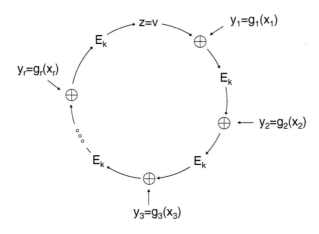

Fig. 2. Ring signatures

4.5   The Ring Signature Scheme

We now formally describe the signature generation and verification procedures:

Generating a ring signature:

   Given the message m to be signed, a sequence of public keys $P_1, P_2, \ldots, P_r$ of all the ring members (each public key P_i specifies a trapdoor permutation g_i), and a secret key S_s (which specifies the trapdoor information needed to compute g_s^{-1}), the signer computes a ring signature as follows.

1. **Determine the symmetric key:** The signer first computes the symmetric key k as the hash of the message m to be signed:

$$k = h(m)$$

   (a more complicated variant computes k as $h(m, P_1, \ldots, P_r)$; however, the simpler construction is also secure.)

2. **Pick a random glue value:** Second, the signer picks an initialization (or "glue") value v uniformly at random from $\{0,1\}^b$.
3. **Pick random x_i's:** Third, the signer picks random x_i for all the other ring members $1 \leq i \leq r$, where $i \neq s$, uniformly and independently from $\{0,1\}^b$, and computes

$$y_i = g_i(x_i) \ .$$

4. **Solve for y_s:** Fourth, the signer solves the following ring equation for y_s:

$$C_{k,v}(y_1, y_2, \ldots, y_r) = v \ .$$

By assumption, given arbitrary values for the other inputs, there is a unique value for y_s satisfying the equation, which can be computed efficiently.
5. **Invert the signer's trapdoor permutation:** Fifth, the signer uses his knowledge of his trapdoor in order to invert g_s on y_s, to obtain x_s:

$$x_s = g_s^{-1}(y_s) \ .$$

6. **Output the ring signature:** The signature on the message m is defined to be the $(2r+1)$-tuple:

$$(P_1, P_2, \ldots, P_r; v; x_1, x_2, \ldots, x_r) \ .$$

Verifying a ring signature:
 A verifier can verify an alleged signature

$$(P_1, P_2, \ldots, P_r; v; x_1, x_2, \ldots, x_r) \ .$$

on the message m as follows.

1. **Apply the trapdoor permutations:** First, for $i = 1, 2, \ldots, r$ the verifier computes

$$y_i = g_i(x_i) \ .$$

2. **Obtain k:** Second, the verifier hashes the message to compute the symmetric encryption key k:

$$k = h(m) \ .$$

3. **Verify the ring equation:** Finally, the verifier checks that the y_i's satisfy the fundamental equation:

$$C_{k,v}(y_1, y_2, \ldots, y_r) = v \ . \tag{2}$$

If the ring equation (2) is satisfied, the verifier accepts the signature as valid. Otherwise the verifier rejects.

4.6   Security

The identity of the signer is unconditionally protected with our ring signature scheme. To see this, note that for each k and v the ring equation has exactly $(2^b)^{(r-1)}$ solutions, and all of them can be chosen by the signature generation procedure with equal probability, regardless of the signer's identity. This argument does not depend on any complexity-theoretic assumptions or on the randomness of the oracle (which determines E_k).

The soundness of the ring signature scheme must be computational, since ring signatures cannot be stronger than the individual signature scheme used by the possible signers.

Theorem 1. *The above ring signature scheme is secure against adaptive chosen message attacks in the ideal cipher model (assuming each public key specifies a trapdoor one-way permutation).*

We need to prove that in the ideal cipher model, any forging algorithm A which on input $(P_1, \ldots, P_r)$ can generate with non-negligible probability a new ring signature for m^* by analyzing polynomially many ring signatures for other chosen messages $m \neq m^*$, can be turned into an algorithm B that inverts one of the trapdoor one-way permutations corresponding to $(P_1, \ldots, P_r)$ on a random input, with non-negligible probability.

The basic idea behind the proof is the following: We first show that the ring signing oracle "does not help" A in generating a new signature. This is done by showing that the ring signing oracle can be simulated by an efficient algorithm that has control over the oracles h, E and E^{-1}. We then show that any forgery algorithm (with no ring signing oracle) can be used to invert one of the trapdoor permutations $g_1, \ldots, g_r$ corresponding to the public keys $(P_1, \ldots, P_r)$, on a random input y. This is done by showing how to control the oracles h, E, and E^{-1}, so as to force the "gap" between the output and input values of two cyclically consecutive E_k's along the ring equation of the forgery to be equal to the value y. This forces the forger to close the gap by providing the corresponding $g_i^{-1}(y)$ in the generated signature (for some $i \in \{1, \ldots, r\}$). Since y is a random value which is not known to the forger, the forger cannot "recognize the trap" and refuse to sign the corresponding messages.

In what follows, we prove Theorem 1 by formalizing the above basic idea.

Proof of Theorem 1: Assume that there exists a forging algorithm A, that succeeds in creating a ring forgery with non-negligible probability. More specifically, algorithm A takes as input a set of random public keys $(P_1, P_2, \ldots, P_r)$ (but not any of the corresponding secret keys), where each P_i specifies a trapdoor one-way permutation g_i. Algorithm A is also given oracle access to h, E, E^{-1}, and to a ring signing oracle. It can work adaptively, querying the oracles at arguments that may depend on previous answers. Eventually, it produces a valid ring signature on a new message that was not presented to the ring signing oracle, with a non-negligible probability (over the random answers of the oracles and its own random coin tosses).

We show that A can be turned into an algorithm B, that takes as input a set of random trapdoor one-way permutations $g_1, \ldots, g_r$ and a random value $y \in \{0,1\}^b$, and outputs with non-negligible probability a value $g_i^{-1}(y)$ for some $i \in \{1, \ldots, r\}$.

Remark. Note that in order to get a contradiction, we actually need to construct an algorithm B' that takes as input a *single* random trapdoor one-way permutation g and a random element y and outputs with non-negligible probability $g^{-1}(y)$. Such an algorithm B' can be easily constructed from algorithm B as follows. Given a trapdoor one-way permutation g and a random element y, algorithm B' chooses at random an element $j \in \{1, \ldots, r\}$, and random trapdoor permutations $g_1, \ldots, g_r$ such that $g_j \triangleq g$, and runs algorithm B on input $(g_1, \ldots, g_r)$ and y. Algorithm B outputs with non-negligible probability a value of the form $g_i^{-1}(y)$ for some i. Since j is uniformly chosen in $\{1, \ldots, r\}$ and since r is polynomially bounded, it follows that B outputs with non-negligible probability the value $g_j^{-1}(y) = g^{-1}(y)$, as desired. Thus, it suffices to construct algorithm B.

Algorithm B will be constructed in two steps. We first show how to convert the (adaptive) forger A into an (oblivious) forger A' that does not make any queries to the ring signing oracle. We then show how A' can be converted into an inverter algorithm B.

The construction of A′. Algorithm A' uses A as a black-box, while simulating its ring signing oracle, as follows. Every time that A queries its ring signing oracle with a message m, algorithm A' will simulate the response by providing a random vector $(v, x_1, x_2, \ldots, x_r)$ as a ring signature of m. It then adjusts the random answers to queries of the form $E_{h(m)}$ and $E_{h(m)}^{-1}$, to support the correctness of the ring equation for these messages. Namely, A' chooses randomly $r - 1$ values $z_1, \ldots, z_{r-1}$, lets $z_0 = z_r = v$, and (mentally) sets $E_{h(m)}(z_i \oplus g_{i+1}(x_{i+1})) = z_{i+1}$ and $E_{h(m)}^{-1}(z_{i+1}) = z_i \oplus g_{i+1}(x_{i+1})$. Whenever A queries its encryption oracle with query of the form $E_{h(m)}(z_i \oplus g_{i+1}(x_{i+1}))$, algorithm A' does not forward this query to the encryption oracle; rather, it pretends that the answer to this query was z_{i+1} and feeds this value as a response to A. Similarly, whenever A queries its decryption oracle with a query of the form $E_{h(m)}^{-1}(z_{i+1})$, algorithm A' does not forward this query to the decryption oracle; rather, it pretends that the answer to this query was $z_i \oplus g_{i+1}(x_{i+1})$. Once A generates an output, A' copies it as its own output.

Lemma 1. A' *succeeds in forging a ring signature with non-negligible probability, assuming that A succeeds in forging a (new) ring signature with non-negligible probability.*

Proof of Lemma 1: We show that the probability that A' succeeds in forging a ring signature is essentially the same as probability in which A succeeds in forging a (new) ring signature. For this it suffices to prove that the view of A

when interacting with its original oracles is (statistically) indistinguishable from the view of A, when interacting with the following modified oracles:

1. The ring signature oracle is modified by replacing it with an oracle that on input any message m outputs a totally random vector $(v, x_1, x_2, \ldots, x_r)$.
2. The encryption and decryption oracles are modified by restricting $E_{h(m)}(z_i \oplus g_{i+1}(x_{i+1})) = z_{i+1}$ and restricting $E_{h(m)}^{-1}(z_{i+1}) = z_i \oplus g_{i+1}(x_{i+1})$, where m is any message that was signed by the (modified) ring signing oracle, and $(z_0, z_1, \ldots, z_r)$ are the associated random values chosen by A'.

The main thing to realize when arguing the above is that A cannot ask an oracle query of the form $E_{h(m)}(z_i \oplus g_{i+1}(x_{i+1}))$ or $E_{h(m)}^{-1}(z_{i+1})$, before providing m to the signing oracle (except with negligible probability). This is so since all the values $z_0 = v$ and $z_1, \ldots, z_{r-1}$ are chosen randomly by A' *after* A sends m as a query to the ring signing oracle, and thus cannot be guessed in advance by A.[5] $\square$

The construction of B. Algorithm B, on input $g_1, \ldots, g_r$ and a random value $y \in \{0, 1\}^b$, uses A' on input $(g_1, \ldots, g_r)$ as a black-box (while simulating its oracles), in order to find a value $g_i^{-1}(y)$, for some $i \in \{1, \ldots, r\}$.

We first note that A' must query the oracle h with the message that it is actually going to forge (otherwise the probability of satisfying the ring equation becomes negligible). Assume that, with non-negligible probability, A' forges the j'th message that it sends to the oracle h. We denote this message by m^*. Algorithm B begins by guessing randomly this index j. Note that B guesses the correct value with non-negligible probability (since A' makes in total at most polynomially many queries to the oracle h).

Recall that A' has access to three oracles: h, E, E^{-1}. Algorithm B simulates the oracle h in the straightforward manner: Whenever A makes a query to h, the query is answered by a uniformly chosen value (unless this query has previously appeared, in which case it is answered the same way as it was before, to ensure consistency).

Whenever A makes a query to E_k or E_k^{-1}, algorithm B first checks whether $k = h(m^*)$ (where m^* is the j'th query that A sends the oracle h). If $k \neq h(m^*)$ (or if A has not yet queried its j'th query to the oracle h), then B simulates these oracles in the straightforward manner. Namely, each query to E_k or E_k^{-1} is answered randomly, unless the value of this query has already been determined

---

[5] Note that A could have queried its random oracle h with the message m before sending m to the signing oracle, and thus could have queried $E_{h(m)}$ and $E_{h(m)}^{-1}$ with arbitrary messages of its choice. However, since A is polynomially bounded, it sends at most polynomially many queries. Thus, the probability that any one of these queries is of the form $E_{h(m)}(z_i \oplus g_{i+1}(x_{i+1}))$ or $E_{h(m)}^{-1}(z_{i+1})$ is negligible. Also, notice that A' never sends queries of the form $E_{h(m)}(z_i \oplus g_{i+1}(x_{i+1}))$ or $E_{h(m)}^{-1}(z_{i+1})$ to his oracles. This follows from the fact that A cannot send oracle queries of this form *before* providing m to the signing oracle (except with negligible probability), and if A sends queries of this form *after* sending m to the ring signing oracle, A' does not forward these queries to his oracles (but rather simulates the answers himself).

by B, in which case it is answered with the predetermined value. Note that so far, the simulated oracles are identically distributed to the real oracles, and thus in particular A' cannot distinguish between the real oracles and the simulated oracles.

It remains to simulate the oracles E_k and E_k^{-1}, for $k = h(m^*)$. Recall that the goal of algorithm B is to compute $x_i = g_i^{-1}(y)$, for some i. The basic idea is to slip this value y as the "gap" between the output and input values of two cyclically consecutive E_k's along the ring equation of the final forgery, which forces A' to close the gap by providing the corresponding x_i in the generated signature. This basic idea is carried out in the following way.

Let $k = h(m^*)$. The forger A' asks various E_k and E_k^{-1} queries, obtaining a disjoint set of pairs $\{(w_i, z_i)\}_{i=1}^{t}$, where $z_i = E_k(w_i)$. Finally A' presents a forgery of the form $(v; x_1, \ldots, x_r)$. Then with overwhelming probability there exist $\{z_{i_1}, \ldots, z_{i_r}\} \subseteq \{z_i\}_{i=1}^{t}$ such that the following holds:

1. $v = z_{i_r}$
2. $E_k(z_{i_{j-1}} \oplus g_j(x_j)) = z_{i_j}$ for every $j \in \{1, \ldots, r\}$, where $z_{i_0} \triangleq v$.

Without loss of generality, we may assume that for every $j = 1, \ldots, r$, either w_{i_j} is queried in the "clockwise" E_k direction, or z_{i_j} is queried in the "counterclockwise" E_k^{-1} direction, but not both (because this is redundant). We distinguish between the following three cases:

Case 1: For every $j = 1, \ldots, r$, the value w_{i_j} is queried in the "clockwise" E_k direction.

Case 2: For every $j = 1, \ldots, r$, the value z_{i_j} is queried in the "counterclockwise" E_k^{-1} direction.

Case 3: Some of these queries are in the "clockwise" E_k direction and some are in the "counterclockwise" E_k^{-1} direction.

We next show how in each of these cases, algorithm B can simulate answers to these queries in such a way that the ring signature of m^* generated by A' would yield the value $g_i^{-1}(y)$ for some $i \in \{1, \ldots, r\}$. We note that with overwhelming probability, E_k and E_k^{-1} are not constrained up to the point where A' queries the oracle h with query m^*. Thus, B will do the following immediately after A' queries the oracle h with query m^*.

Case 1: The structure of the ring implies that there must exist $j \in \{1, \ldots, r\}$ such that w_{i_j} was queried *before* $w_{i_{j-1}}$, where $w_{i_0} \triangleq w_{i_r}$.

Assume that w_{i_j} was queried before $w_{i_{j-1}}$. Then B will guess which query, out of *all* the queries $\{w_i\}$ that are sent to E_k by A', corresponds to w_{i_j} and which query corresponds to $w_{i_{j-1}}$ (there are only polynomially many possibilities and thus he will succeed with non-negligible probability). Next B will provide an answer to $w_{i_{j-1}}$ based on its knowledge of w_{i_j}. More precisely, B will set the output of $E_k(w_{i_{j-1}})$ to be $w_{i_j} \oplus y$ (so that the XOR of the values across the gap is the desired y). All other queries are answered randomly (unless the value of this query has already been determined by B, in which case it is answered with the predetermined value).

Case 2: This case is completely analogous to the previous case, and so B behaves accordingly.

Case 3: The structure of the ring implies that there must exist $j \in \{1, \ldots, r\}$ such that w_{i_j} was queried in the "clockwise" E_k direction whereas the proceeding z_{i_j+1} was queried in the "counterclockwise" E_k^{-1} direction (where $z_{i_{r+1}} \triangleq z_{i_1}$). Assume that w_{i_j} was queried in the "clockwise" E_k direction whereas $z_{i_{j+1}}$ was queried in the "counterclockwise" E_k^{-1} direction.

As in the previous two cases, B will guess which query, out of all the queries $\{w_i\}$ that are sent to E_k by A', corresponds to w_{i_j}, and which query, out of all the queries $\{z_i\}$ that are sent to E_k^{-1} by A', corresponds to $z_{i_{j+1}}$. Next B will answer the query corresponding to $E_k(w_{i_j})$ with a random value z and will answer the query corresponding to $E_k^{-1}(z_{i_{j+1}})$ with $z \oplus y$ (so that the XOR of the values across the gap is the desired y). All other queries are answered randomly (unless the value of this query has already been determined by B, in which case it is answered with the predetermined value).

Note that since y is a random value, the simulated oracles E_k and E_k^{-1} cannot be distinguished from the real oracles, and therefore, with non-negligible probability, A' will output a signature $(v; x_1, \ldots, x_r)$ to a message m^*. Moreover, with non-negligible probability there exists $i \in \{1, \ldots, r\}$ such that $g_i(x_i) = y$, as desired. $\qquad\square$

5   Our Ring Signature Scheme (Rabin Version)

Rabin's public-key cryptosystem [Rab79] has more efficient signature verification than RSA, since verification involves squaring rather than cubing, which reduces the number of modular multiplications from 2 to 1. However, we need to deal with the fact that the Rabin mapping $f_i(x_i) = x_i^2 \pmod{n_i}$ is not a permutation over $\mathbf{Z}_{n_i}^*$, and thus only one quarter of the messages can be signed, and those which can be signed have multiple signatures.

We note that Rabin's function, $f_N(x) = x^2 \pmod{N}$, is actually a permutation over $\{x : x < \frac{N}{2} \wedge (\frac{x}{N}) = 1\}$, assuming N is a Blum integer. Moreover, it can be easily extended to be a permutation over $\mathbb{Z}_N^*$ ([G04, Section C.1]). However this permutation is no longer so efficient, since in order to compute it on a value x, one first needs to compute $(\frac{x}{N})$, which is a relatively expensive computation. Moreover, both in the signing and verifying procedures, the number of times that a Jacobi symbol needs to be computed grows linearly with the size of the ring.

Rather than trying to convert Rabin's function to a permutation, we suggest the following natural operational fix: when signing, change your last random choice of x_{s-1} if $g_s^{-1}(y_s)$ is undefined. Since only one trapdoor one-way function has to be inverted, the signer should expect on average to try four times before succeeding in producing a ring signature. The complexity of this search is essentially the same as in the case of regular Rabin signatures, regardless of the size of the ring.

A more important difference is in the proof of unconditional anonymity, which relied on the fact that all the mappings were permutations. When the g_i's are not permutations, there can be noticeable differences between the distribution of the x_i's that are randomly chosen, and the distribution of x_s that is actually computed in a given ring signature. This could lead to the identification of the real signer among all the possible signers, and can be demonstrated to be a real problem in many concrete types of trapdoor one-way functions.

Consider, for example, a trapdoor one-way function family $\mathcal{F} = \{\mathcal{F}_n\}_{n \in \mathbb{N}}$ such that every $f \in \mathcal{F}_n$ is a function from $\{0,1\}^n$ to $\{0,1\}^n$, with the property that every y in the image of f has many pre-images, one which is of the form $x = x_1 \ldots x_n$ such that $x_1 = \ldots, x_{\log n} = 0$. Moreover, assume that the trapdoor information always finds this particular pre-image. In this case it will be easy to distinguish the real signer from the non-signers, as the x_s associated with the real signer will have the $\log n$ most significant bits equal to 0, whereas the x_i's associated with non-signers will be randomly chosen.

We overcome this difficulty in the case of Rabin signatures with the following simple observation:

Observation: Let S be a given finite set of "marbles" and let $B_1, B_2, \ldots, B_n$ be disjoint subsets of S (called "buckets") such that all non-empty buckets have the same number of marbles, and every marble in S is in exactly one bucket. Consider the following sampling procedure: pick a bucket at random until you find a non-empty bucket, and then pick a marble at random from that bucket. Then this procedure picks marbles from S with uniform probability distribution.

Rabin's functions $f_i(x_i) = x_i^2 \pmod{n_i}$ are extended to functions $g_i(x_i)$ over $\{0,1\}^b$ in the usual way. Let the set B_s of marbles be all the b-bit numbers $u = qn_s + r$ in which $r \in \mathbb{Z}_{n_s}^*$ and $(q+1)n_s \leq 2^b$. Each marble is placed in the bucket to which it is mapped by the extended Rabin mapping g_s. In the Rabin ring signature algorithm, each x_i that corresponds to a non-signer is chosen randomly in $\{0,1\}^b$, whereas x_s that corresponds to the signer is chosen by first choosing a random non-empty bucket and then choosing a random marble from that bucket. The fact that each bucket contains either zero or four marbles, together with the above observation, implies that x_s is uniformly distributed in B_s. The fact that the uniform distribution over $\{0,1\}^b$ is statistically close to the uniform distribution over B_s implies that the distribution of x_i's that correspond to non-signers is statistically close to the distribution of x_s that corresponds to the real signer. Consequently, even an infinitely powerful adversary cannot distinguish between signers and non-signers by analyzing actual ring signatures produced by one of the possible signers.

6   Generalizations and Special Cases

The notion of ring signatures has many interesting extensions and special cases. In particular, ring signatures with $r = 1$ can be viewed as a randomized version of Rabin's signature scheme (or RSA's signature scheme): As shown in Fig. 3, the verification condition can be written as $(x^2 \bmod n) = v \oplus E_{h(m)}^{-1}(v)$. The right hand side is essentially a hash of the message m, randomized by the choice of v.

Ring signatures with $r = 2$ have the ring equation:

$$E_{h(m)}(x_2^2 \oplus E_{h(m)}(x_1^2 \oplus v)) = v$$

(see Fig. 3). A simpler ring equation (which is not equivalent but has the same security properties) is:

$$(x_1^2 \bmod n_1) = E_{h(m)}(x_2^2 \bmod n_2)$$

where the modular squares are extended to $\{0,1\}^b$ in the usual way. This is our recommended method for implementing designated verifier signatures in email systems, where n_1 is the public key of the sender and n_2 is the public key of the recipient.

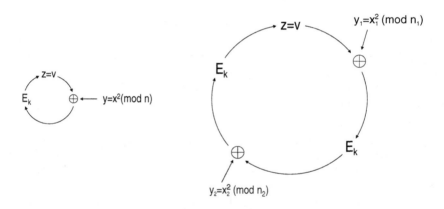

Fig. 3. Rabin-based Ring Signatures with $r = 1, 2$

In regular ring signatures it is impossible for an adversary to expose the signer's identity. However, there may be cases in which the signer himself wants to have the option of later proving his authorship of the anonymized email (e.g., if he is successful in toppling the disgraced Prime Minister). Yet another possibility is that the signer A wants to initially use {A,B,C} as the list of possible signers, but later prove that C is *not* the real signer. There is a simple way to implement these options, by choosing the x_i values for the non-signers in a pseudorandom rather than truly random way. To show that C is *not* the author, A publishes the seed which pseudorandomly generated the part of the signature associated with C. To prove that A *is* the signer, A can reveal a single seed which was used to generate all the non-signers' parts of the signature. The signer A cannot misuse this technique to prove that he is not the signer since his

x_i is computed by applying g^{-1} to a random value given to him by the oracle (where g is the trapdoor one-way permutation corresponding to his public key). Thus, his x_i is extremely unlikely to have a corresponding seed. Note that these modified versions can guarantee only computational anonymity, since a powerful adversary can search for such proofs of non-authorship and use them to expose the signer.

A different approach that guarantees unbounded anonymity is to choose the x_i value for each non-signer by choosing a random w_i and letting $x_i = f(w_i)$, where f is a one-way function with the additional property that each element in the range has a pre-image under f. By demonstrating w_i, the signer proves that the i'th ring member is not the signer. Notice that the fact that the signer (which corresponds to the s'th ring member) is computationally bounded, implies that he cannot produce $f^{-1}(x_s)$, and therefore he cannot prove that he himself is not the signer. Moreover, an adversary with unlimited computational power cannot figure out who the signer is since any x_i (including x_s) has a pre-image under f.

7   Followup Work

In this section we summarize the followup papers on the theory and applications of ring signatures.

Deniable Ring Signature Schemes. In [Na02] Naor defined the notion of *Deniable Ring Authentication*. This notion allows a member of an ad hoc subset of participants (a ring) to convince a verifier that a message m is authenticated by one of the members of the subset without revealing by which one, and the verifier cannot convince a third party that message m was indeed authenticated. Naor also provided an efficient protocol for deniable ring authentication based on any secure encryption scheme. The scheme is interactive. Susilo and Mu [SM03, SM04] constructed *non interactive* deniable ring authentication protocols. They first showed in [SM03] how to use any ring signature scheme and a chameleon hash family to construct a deniable ring signature scheme. In this construction the verifier is assumed to be associated with a pair of secret and public keys (corresponding to the chameleon hash family). They then showed in [SM04] how to use any ring signature scheme and an ID based chameleon hash family [AM04] to construct a deniable ring signature scheme. In this construction the verifier is only assumed to have his ID published.

Threshold and General Access Ring Signature Schemes. A t-threshold ring signature scheme is a ring signature scheme where each ring signature is a proof that at least t members of the ring are confirming the message. In a general access ring signature scheme, members of a set can freely choose any family of sets including their own set, and prove that all members of some set in the access structure have cooperated to compute the signature, without revealing any information about which set it is.

There have been many papers which considered these scenarios. The early work of [CDS94] has already considered this scenario, and showed (using different terminology) that a witness indistinguishable proof (with witnesses that correspond to some monotone access structure), can be combined with the Fiat-Shamir paradigm, to obtain a monotone access ring signature scheme. The work of Naor [Na02] also contains a construction of a general access (and in particular threshold) ring signature scheme. His scheme is interactive and its security is based only the existence of secure encryption schemes. There have been subsequent works which consider the general access scenario, such as [HS04a].

The work of Bresson et. al. [BSS02] contains a construction of a threshold ring signature scheme (proven secure in the Random Oracle Model under the RSA Assumption). Subsequent works which consider the threshold setting are [Wei04, KT03, WFLW03] (where security is proved in the Random Oracle Model).

Identity-based Ring Signature Schemes. Shamir introduced in 1984 the concept of Identity-based (ID-based) cryptography [Sha84]. The idea is that the public-key of a user can be publicly computed from his identity (for example, from a complete name, an email or an IP address). ID-based schemes avoid the necessity of certificates to authenticate public keys in a digital communication system. This is especially desirable in applications which involve a large number of public keys in each execution, such as ring signatures.

The first to construct an ID-based ring signature scheme were Zhang and Kim [ZK02]. Its security was analyzed in [Her03], based on bilinear pairings in the Random Oracle Model. Subsequent constructions of ID-based ring signatures appear in [HS04b, LW03a, AL03, TLW03, CYH04].

Identity-based Threshold Ring Signature Schemes. ID-based threshold ring signature schemes proven secure in the Random Oracle Model, under the bilinear pairings were constructed in [CHY04, HS04c]. This was extended in [HS04c], to a general access setting, where any subset of users S can cooperate to compute an anonymous signature on a message, on behalf of any family of users that includes S.

Separable Ring Signature Schemes. A ring signature scheme is said to be separable if all participants can choose their keys independently with different parameter domains and for different types of signature schemes. Abe et. al. [AOS02] were the first to address the problem of constructing a separable ring signature scheme. They show how to construct a ring signature scheme from a mixture of both trapdoor-type signature schemes (such as RSA based) and three-move-type signature schemes (such as Discrete Log based). This was extended in [LWW03] to the threshold setting.

Linkable Ring Signature Schemes. The notion of linkable ring signatures, introduced by Liu et al. [LWW04], allows anyone to determine if two ring signatures are signed by the same group member. In [LWW04] they also presented a linkable ring signature scheme that can be extended to the threshold setting.

Their construction was improved in [TWC+04], who presented a separable linkable threshold ring signature scheme.

Verifiable Ring Signature Schemes. Lv and Wang [LW03b] formalized the notion of verifiable ring signatures, which has the following additional property: if the actual signer is willing to prove to a recipient that he signed the signature, then the recipient can correctly determine whether this is the fact. We note that this additional property was considered in our (original) work, and as was mentioned in Section 6, we showed that this property can be obtained by choosing the x_i values for the non-signers in a pseudorandom rather than a truly random way.

Accountable Ring Signaure Schemes. An accountable ring signature scheme, a notion introduces by Xu and Yung [XY04], ensures the following: anyone can verify that the signature is generated by a user belonging to a set of possible signers (that may be chosen on-the-fly), whereas the actual signer can nevertheless be identified by a designated trusted entity. Xu and Yung [XY04] also presented a framework for constructing accountable ring signatures. The framework is based on a compiler that transforms a traditional ring signature scheme into an accountable one.

Short Ring Signature Schemes. Dodis et. al. [DKNS04] were the first to construct a ring signature scheme in which the length of an "actual signature" is independent of the size of the ad hoc group (where an "actual signature" does not include the group description). We note that in all other constructions that we are aware of, the size of an "actual signature" is at least linear in the size of the group. Their scheme was proven secure in the Random Oracle Model assuming the existence of accumulators with one-way domain (which in turn can be based on the Strong RSA Assumption).

Ring Authenticated Encryption. An authenticated encryption scheme allows the verifier to recover and verify the message simultaneously. Lv et al. [LRCK04] introduced a new type of authenticated encryption, called ring authenticated encryption, which loosely speaking, is an authenticated encryption scheme where the verifiability property holds with respect to a ring signature scheme.

References

[AL03] Amit K. Awasthi and Sunder Lal. ID-based Ring Signature and Proxy Ring Signature Schemes from Bilinear Pairings. In Cryptology ePrint Archive: Report 2004/184.

[AM04] G. Ateniese and B. de Medeiros. Identity-Based Chameleon Hash and Applications. In *Financial Cryptography 2004*, pages 164-180.

[AOS02] Masayuki Abe, Miyako Ohkubo, and Koutarou Suzuki. 1-out-of-n Signatures from a Variety of Keys. In *ASIACRYPT 2002, LNCS 2501*, pages 415-432.

[BGLS03] D. Boneh, C. Gentry, B. Lynn, and H. Shacham. Aggregate and Verfiably Encrypted Signatures from Bilinear Maps. In *E. Biham, ed., Proceedings of Eurocrypt 2003, vol. 2656 of LNCS*, pages 416-432.

[BR93] M. Bellare and P. Rogaway. Random Oracles are Practical: a Paradigm for Designing Efficient Protocols. In *ACM Conference on Computer and Communications Security 1993*, pages 62-73.

[BSS02] Emmanuel Bresson, Jacque Stern, and Michael Szydlo. Threshold Ring Signatures and Applications to Ad-Hoc Groups (Extended abstract). In *Advances in Cryptology - Proceedings of CRYPTO '02, LNCS 2442*, pages 465-480.

[Cam97] Jan Camenisch. Efficient and Generalzied Group Sigmatures. In *Advances in Cryptology - Eurocrypt 97*, pages 465-479.

[Ch81] David Chaum. Untraceable Electronic Mail, Return Addresses and Digital Pseudonyms. In *Communications of the ACM 24(2)*, pages 84-88, 1981.

[Ch88] David Chaum. The Dining Cryptographers Problem: Unconditional Sender and Recipient Untraceability. In *Journal of Cryptology 1(1)*, pages 65-75, 1988.

[CDS94] Ronald Cramer, Ivan Damgård, and Berry Schoenmakers. Proofs of partial knowledge and simplified design of witness hiding protocols. In Yvo Desmedt, editor, *Advances in Cryptology – CRYPTO '94*, pages 174–187, Berlin, 1994. Springer-Verlag. Lecture Notes in Computer Science Volume 839.

[CHY04] Sherman S.M. Chow, Lucas C.K.Hui, and S.M.Yiu. Identity Based Threshold Ring Signature. In *Cryptology ePrint Archive: Report 2004/179*.

[CV91] David Chaum and Eugène Van Heyst. Group signatures. In D.W. Davies, editor, *Advances in Cryptology — Eurocrypt '91*, pages 257–265, Berlin, 1991. Springer-Verlag. Lecture Notes in Computer Science No. 547.

[CYH04] Sherman S.M. Chow, S.M. Yiu, and Lucas C.K. Hui. Efficient Identity Based Ring Signature. In *Cryptology ePrint Archive: Report 2004/327*.

[DH76] W. Diffie and M. E. Hellman. New directions in cryptography. *IEEE Trans. Inform. Theory*, IT-22:644–654, November 1976.

[DKNS04] Y. Dodis, A. Kiayias, A. Nicolosi, and V. Shoup Anonymous Identification in Ad-Hoc Groups. In em EUROCRYPT 2004, LNCS 3027, pages 609-626. Springer-Verlag, 2004.

[G04] O. Goldreich. *Foundations of Cryptography: Volume 2 – Basic Applications.* Cambridge University Press, 2004.

[GRS99] David M. Goldschlag, Michael G. Reed, and Paul F. Syverson. Onion Routing for Anonymous and Private Internet Connections. In *Communications of the ACM 42(2)*, pages 39-41, 1999.

[Her03] Javier Herranz. A Formal Proof of Security of Zhang and Kim's ID-Based Ring Signature Scheme. In *WOSIS 2004*, pages 63-72.

[HO05] Y. Hanatani and K. Ohta. Two Stories of Ring Signatures. Crypto 2005 rump session talk. Available at http://www.iacr.org/conferences/crypto2005/r/38.ppt. A photo of the 1756 "ring signature" is available at http://www.nihonkoenmura.jp/theme3/takarabito07.htm.

[HS03] J. Herranz and G. Saez. Forking Lemmas in the Ring Signatures' Scenario. In *Cryptology ePrint Archive: Report 2003/067*.

[HS04a] J. Herranz and G. Saez. Ring Signature Schemes for General Ad-Hoc Access Structures. In *Proceedings of ESAS 2004*, pages 54-65.

[HS04b] J. Herranz and G. Saez. New Identity-Based Ring Signature Schemes. In *Information and Communications Security (ICICS 2004)*, pages 27-39. Springer-Verlag, 2004.

[HS04c] J. Herranz and G. Saez. Distributed Ring Signatures for Identity-Based Scenarios. In *Cryptology ePrint Archive: Report 2004/190*.

[HW79] G. H. Hardy and E. M. Wright. *An Introduction to the Theory of Numbers*. Oxford, fifth edition, 1979.

[JSI96] M. Jakobsson, K. Sako, and R. Impagliazzo. Designated verifier proofs and their applications. In Ueli Maurer, editor, *Advances in Cryptology - EuroCrypt '96*, pages 143–154, Berlin, 1996. Springer-Verlag. Lecture Notes in Computer Science Volume 1070.

[KT03] Hidenori Kuwakado, Hatsukazu Tanaka. Threshold Ring Signature Scheme Based on the Curve. In *IPSJ JOURNAL Abstract Vol.44*, pages 8-32.

[LR88] M. Luby and C. Rackoff. How to construct pseudorandom permutations from pseudorandom functions. *SIAM J. Computing*, 17(2):373–386, April 1988.

[LRCK04] J. Lv, K. Ren, X. Chen and K. Kim. Ring Authenticated Encryption: A New Type of Authenticated Encryption. In *The 2004 Symposium on Cryptography and Information Security, vol. 1/2*, pages 1179-1184.

[LW03a] C.Y. Lin and T. C. Wu. An Identity Based Ring Signature Scheme from Bilinear Pairings. In *Cryptology ePrint Archive, Report 2003/117, 2003*.

[LW03b] J. Lv, X. Wang. Verifiable Ring Signature. In *Proc. of DMS 2003 - The 9th International Conference on Distribted Multimedia Systems*, pages 663-667, 2003.

[LWW03] Joseph K. Liu, Victor K. Wei, and Duncan S. Wong. A Separable Threshold Ring Signature Scheme. In *ICISC 2003, LNCS 2971*, pages 12-26.

[LWW04] J. K. Liu, V. K. Wei, and D. S. Wong. Linkable Spontaneous Anonymous Group Signatures for Ad Hoc Groups (Extended Abstract). In *ACISP 2004 , volume 3108 of LNCS*, pages 325-335, Springer-Verlag, 2004.

[Na02] Moni Naor. Deniable Ring Authentication. In *CRYPTO 2002, LNCS 2442*, pages 481-498. Springer-Verlag, 2002.

[Rab79] M. Rabin. Digitalized signatures as intractable as factorization. Technical Report MIT/LCS/TR-212, MIT Laboratory for Computer Science, January 1979.

[RSA78] Ronald L. Rivest, Adi Shamir, and Leonard M. Adleman. A method for obtaining digital signatures and public-key cryptosystems. *Communications of the ACM*, 21(2):120–126, 1978.

[Sha84] Adi Shamir Identity Based Cryptosystems and Signature Schemes. In *Proceedings of CRYPTO 84, LNCS 196*, pages 47-53. Springer-Verlag, 1984.

[SCPY94] Alfredo De Santis, Giovanni Di Crescenzo, Giuseppe Persiano, and Moti Yung. On monotone formula closure of SZK. In *Proc. 35th FOCS*, pages 454–465. IEEE, 1994.

[SM03] Willy Susilo and Yi Mu. Non-Interactive Deniable Ring Authentication. In *the 6th International Conference on Information Security and Cryptology (ICISC 2003)*, pages 397-412, 2003.

[SM04] Willy Susilo and Yi Mu. Deniable Ring Authentication Revisited. In *Applied Cryptography and Network Security (ACNS 2004), LNCS 3089*, pages 149-163. Springer-Verlag, 2004.

[TLW03] Chunming Tang, Zhupjun Liu, and Mingsheng Wang. An Improved Identity-Based Ring Signature Scheme from Bilinear Pairings. In *NM Research Preprints*, pages 231-234. MMRC, AMSS, Academia, Sinica, Beijing, No. 22, December, 2003.

[TWC+04] P. P. Tsang, V. K. Wei, T. K. Chan, M. H. Au, J. K. Liu, and D. S. Wong. Separable Linkable Threshold Ring Signatures. In *INDOCRYPT 2004*, 384-398.

[Wei04] Victor K. Wei. A Bilinear Spontaneous Anonymous Threshold Signature for Ad Hoc Groups. In *Cryptology ePrint Archive: Report 2004/039*.

[WFLW03] Ducan S. Wong, Karyin Fung, Joseph K. Liu, and Victor K. Wei. On the RS-Code Construction of Ring Signature Schemes and a Threshold Setting of RST. In *Information and Communications Security (ICICS'03), LNCS 2836*, pages 34-46. Springer-Verlag, 2003.

[XY04] Shouhuai Xu and Moti Yung. Accountable Ring Signatures: A Smart Card Approach. In *Sixth Smart Card Research and Advanced Application IFIP Conference*, pages 271-286.

[ZK02] Fangguo Zhang and Kwangjo Kim. ID-Based Blind Signature and Ring Signature from Pairings. In *ASIACRYPT 2002, LNCS 2501*, pages 533-547. Springer-Verlag, 2002

A New Related Message Attack on RSA

Oded Yacobi[1] and Yacov Yacobi[2]

[1] Department of Mathematics, University of California San Diego,
9500 Gilman Drive, La Jolla, CA 92093, USA
`oyacobi@math.ucsd.edu`
[2] Microsoft Research, One Microsoft Way, Redmond, WA 98052, USA
`yacov@microsoft.com`

In memory of Professor Shimon Even.

Abstract. Coppersmith, Franklin, Patarin, and Reiter show that given two RSA cryptograms $x^e \bmod N$ and $(ax + b)^e \bmod N$ for known constants $a, b \in \mathbb{Z}_N$, one can usually compute x in $O(e \log^2 e)$ $\mathbb{Z}_N$-operations (there are $O(e^2)$ messages for which the method fails).
We show that given e cryptograms $c_i \equiv (a_i x + b_i)^e \bmod N$, $i = 0, 1, ...e - 1$, for any known constants $a_i, b_i \in \mathbb{Z}_N$, one can deterministically compute x in $O(e)$ $\mathbb{Z}_N$-operations that depend on the cryptograms, after a pre-processing that depends only on the constants. The complexity of the pre-processing is $O(e \log^2 e)$ $\mathbb{Z}_N$-operations, and can be amortized over many instances. We also consider a special case where the overall cost of the attack is $O(e)$ $\mathbb{Z}_N$-operations. Our tools are borrowed from numerical-analysis and adapted to handle formal polynomials over finite-rings. To the best of our knowledge their use in cryptanalysis is novel.

1 Introduction

RSA is the most popular public key cryptosystem in use today in commercial applications. It maps a message $m \in \mathbb{Z}_N$ into a cryptogram $c \in \mathbb{Z}_N$ using $c \equiv m^e \bmod N$, where (e, N) is the public key. The secret key is d, such that $ed \equiv 1 \bmod \varphi(N)$, where $\varphi(N)$ is the Euler totient function. Decryption is done using $m \equiv c^d \bmod N$. Usually N and d are at least 1000 bits long, however, e is usually short. Many applications still use 16 bit long e.

Messages with known relations may occur for example if an attacker pretends to be the recipient in a protocol that doesn't authenticate the recipient, and in addition the message is composed of the content concatenated with a serial number. In that case the attacker can claim that she didn't receive the transmission properly and ask that it be sent again. The next transmission will have the same content as the original but an incremented serial number. If the increment is known we have a known relation. Other examples appear in [4].

The protocol defines the relations among messages and the relations imply the attack method. For example, in the above example the relations are very simple: $c_i \equiv (x + i)^e \bmod N$. We later show extremely efficient attack for this case. More general linear relations require a different more complex attack.

O. Goldreich et al. (Eds.): Shimon Even Festschrift, LNCS 3895, pp. 187–195, 2006.

Related message attacks can be avoided altogether if before RSA-encryption the message M is transformed using e.g. the OAEP function ([3]; There are other methods and some issues are not settled yet, see [5]). This transformation destroys the relations between messages and increases the message length.

Nevertheless it is useful to know the ramifications in case for some reason one chooses not to use OAEP or similar methods (even though it is highly recommended). For example RFID tags may pose tough engineering challenges of creating very compact cryptosystems, and the trade-off must be known precisely.

In [4] it was shown that given two RSA cryptograms $x^e \bmod N$, and $(ax + b)^e \bmod N$ for any known constants $a, b \in \mathbb{Z}_N$ one can usually compute x in $O(e \log^2 e)$ $\mathbb{Z}_N$-operations (there are $O(e^2)$ messages for which the method fails, see footnote 1 in [4]).

We show that given e cryptograms $c_i \equiv (a_i x + b_i)^e \bmod N$, $i = 0, 1, ... e - 1$, for any known constants $a_i, b_i \in \mathbb{Z}_N$, such that the vectors (a_i, b_i), $i = 0, 1, ... e - 1$, are linearly independent, and $a_i \neq 0$, then one can deterministically compute x in $O(e)$ $\mathbb{Z}_N$-operations, after doing $O(e \log^2 e)$ pre-computations[3].

These pre-computations depend only on the constants a_i and b_i as determined by the encryption protocol. For example, for the encryption protocol suggested at the beginning of this section we have $a_i = 1$ and $b_i = i$. The cost of the pre-computations can be amortized over many instances of the problem.

Our problem could be solved by using the Newton expansion of $c_i \equiv (a_i x + b_i)^e \bmod N$, renaming $z_j = x^j$ and using linear algebra to find z_1. However, our method is more efficient.

We also show that in the special case where $c_i \equiv (ax + b \cdot i)^e \bmod N$, $i = 0, 1, ... e - 1$, for any known constants $a, b \in \mathbb{Z}_N$, where $\gcd(a, N) = \gcd(b, N) = \gcd(e!, N) = 1$[4], we show in section 3.2 how to deterministically compute x in overall $O(e)$ $\mathbb{Z}_N$-operations using

$$x \equiv a^{-1} b(b^e e!)^{-1} \left(\sum_{i=0}^{e-1} \binom{e-1}{i} \cdot c_i \cdot (-1)^{e-1+i} - \frac{e-1}{2} \right) \bmod N$$

It remains an open problem whether the new approach can improve the general case of implicit linear dependence, i.e., suppose for known constants a_i, $i = 0, 1, 2, ... k$, there is a known relation $\sum_{i=1}^{k} a_i x_i = a_0$ among messages $x_1, x_2, ... x_k$. The current complexity of attacking this problem is $O(e^{k/2} k^2)$ [4].

Our major attack-tools are divided-differences and finite-differences. These tools are borrowed from numerical-analysis, and adapted to handle formal poly-

---

[3] It appears noteworthy that while the on-line part of the attack requires just $O(e)$ multiplications, it would take the party doing the encryptions (the victim) $O(e \log(e))$ multiplications to compute all the needed ciphertexts. Thus, the latter can be argued to be the total run time (as opposed to "cost") of the on-line part of the attack. With this interpretation, the attack runs only $\log(e)$ times "faster" then that of Coppersmith et al., rather than the $\log^2(e)$.

[4] If any of the above gcd conditions do not hold then the system is already broken.

nomials over finite-rings. To the best of our knowledge their use in cryptanalysis is novel.

For a survey of the work on breaking RSA see [2]. An earlier version of this paper appeared in PKC'05.

2   Main Result

2.1   Divided Differences

We borrow the concept of *divided-differences* from numerical analysis and adapt it to handle formal polynomials over finite rings. This will allow us to extract the message from a string of e cryptograms whose underlying messages are linearly related. We specialize our definitions to the ring of integers modulo N, a product of two primes (the "RSA ring"). All the congruences in this paper are taken mudulo N.

Definition 1. *Let h be a polynomial defined over the ring of integers modulo N, and let $x_0, x_1, ...x_n$ be distinct elements of the ring such that $(x_0 - x_i)^{-1} \bmod N$ exist for $i = 0, 1, ...n$. The nth divided-difference of h relative to these elements is defined as follows:*

$[x_i] \equiv h(x_i),$

$[x_0, x_1] \equiv \frac{[x_0] - [x_1]}{x_0 - x_1},$

$[x_0, x_1, ...x_n] \equiv \frac{[x_0, x_1, ...x_{n-1}] - [x_1, x_2, ...x_n]}{x_0 - x_n}.$

Let x be an indeterminate variable, and for $i = 0, 1, ...n$, let $x_i \equiv x + b_i$ for some known constants b_i (these are the general explicit linear relations that we assume later). We can now view the above divided differences as univariate polynomials in x defined over $\mathbb{Z}_N$.

The following lemma is true for the divided difference of any polynomial mod N, but for our purposes it is enough to prove it for the RSA polynomial $x^e \bmod N$. Related results are stated in [8]. Before beginning the proof we introduce some notation borrowed from [7]. Let $\pi_k(y) \equiv \prod_{i=0}^{k} (y - x_i)$. Then taking the derivative of π_k with respect to y we have for $i \leq k$

$$\pi'_k(x_i) \equiv \prod_{\substack{0 \leq j \leq k \\ j \neq i}} (x_i - x_j)$$

By induction on k the following equality easily follows

$$[x_0, ..., x_k] \equiv \sum_{i=0}^{k} \frac{h(x_i)}{\pi'_k(x_i)} \tag{1}$$

Let $C_t(p)$ denote the t_{th} coefficient of the polynomial p, starting from the leading coefficients (the coefficients of the highest powers). We use $C_t[x_0, ..x_k]$ as a shorthand for $C_t([x_0, ..x_k])$.

Lemma 1. *Let $[x_0, ..., x_n]$ be the nth divided difference relative to the RSA poly-*
nomial $h(x) \equiv x^e \bmod N$, and let $x_0, x_1, ...x_n$ be distinct elements of the ring
such that $(x_0 - x_i)^{-1} \bmod N$ exist for $i = 0, 1, ...n$. Then (i) for $0 \leq n \leq e$, if
$\binom{e}{e-n} \neq 0 \bmod N$ *then* $\deg[x_0, ..., x_n] = e - n$. *(ii)* $C_{e-n}[x_0, x_1, .., x_n] \equiv \binom{e}{e-n}$
(an important special case is $C_1[x_0, x_1, .., x_{e-1}] \equiv e \bmod N$).

Comment: In practice the condition in claim (i) always holds, since $e \ll N$.

Proof. The claim is trivial for $n = 0$. For $n \geq 1$ we prove the equivalent
proposition that $C_t[x_0, ..., x_n] = 0$ for $t = e, e-1, ..., e-n+1$ and $C_{e-n}[x_0, ..., x_n]$
is independent of the b_i and is not congruent to 0. We use the notations $1/b$
and b^{-1} interchangeably. We induct on n. When $n = 1$

$$[x_0, x_1] \equiv \frac{(x + b_0)^e - (x + b_1)^e}{b_0 - b_1} \equiv \frac{\sum_{i=0}^{e} \binom{e}{i} x^i [b_0^{e-i} - b_1^{e-i}]}{b_0 - b_1}$$

Note that by our assumption $(b_0 - b_1)^{-1} \bmod N$ exist. So $C_e[x_0, x_1] \equiv 0$
and $C_{e-1}[x_0, x_1] \equiv e$ and indeed our claim is true for $n = 1$. For the inductive
hypothesis let $n = k - 1$ and assume that $C_t[x_0, ..., x_{k-1}] \equiv 0$ for $t = e, e -$
$1, ..., e - (k-1) + 1$ and $C_{e-(k-1)}[x_0, ..., x_{k-1}]$ is independent of the b_i and is not
congruent to 0. We want to show that when $n = k$, $C_t[x_0, ..., x_k] \equiv 0$ for
$t = e, e - 1, ..., e - k + 1$ and $C_{e-k}[x_0, ..., x_k]$ is independent of the b_i and is not
congruent to 0.

The fact that $C_t[x_0, ..., x_k] \equiv 0$ for $t = e, e-1, ..., e-k+1$ follows immediately
from the inductive hypothesis and Definition 1. It takes a little more work to
show that $C_{e-k}[x_0, ..., x_k]$ is independent of the b_i.

Using (1):

$$[x_0, x_1, ..., x_k] \equiv \sum_{i=0}^{k} \frac{(x + b_i)^e}{\pi'_k(x_i)} \equiv \sum_{j=0}^{e} \binom{e}{j} x^j [\frac{b_0^{e-j}}{\pi'_k(x_0)} + \frac{b_1^{e-j}}{\pi'_k(x_1)} + ... + \frac{b_k^{e-j}}{\pi'_k(x_k)}]$$

We want to show that $C_{e-k}[x_0, x_1, ..., x_k]$ is independent of the b_i.

$$C_{e-k}[x_0, x_1, .., x_k] \equiv \binom{e}{e-k} [\frac{b_0^k}{\pi'_k(x_0)} + \frac{b_1^k}{\pi'_k(x_1)} + ... + \frac{b_k^k}{\pi'_k(x_k)}] \quad (2)$$

So now it is sufficient to show that

$$\sum_{i=0}^{n} (-1)^i \frac{b_i^n}{(b_0 - b_i)...(b_{i-1} - b_i)(b_i - b_{i+1})...(b_i - b_n)} \quad (3)$$

is independent of the b_i.

We first multiply (3) by the necessary terms to get a common denominator.
We introduce some compact notation that will simplify the process. For a given
set of constants $b_0, b_1, ...b_k$ define

$$\delta(h, i) \equiv (b_h - b_i)$$
$$\delta(h, i, j) \equiv (b_h - b_i)(b_h - b_j)\delta(i, j)$$
$$\vdots$$
$$\delta(i_0, ..., i_k) \equiv (b_{i_0} - b_{i_1})(b_{i_0} - b_{i_2}) \cdots (b_{i_0} - b_{i_k})\delta(i_1, ..., i_k)$$

Similarly we can also define $\delta_j \equiv \delta(0, 1, ..., \bar{j}, ..., k)$ where the bar denotes that the index is missing (so if $k = 4$ then $\delta_3 = \delta(0, 1, 2, 4,)$). Then (3) becomes:

$$\frac{b_0^k \delta_0 - b_1^k \delta_1 + \cdots + (-1)^k b_k^k \delta_k}{\delta(0, 1, ..., k)} \tag{4}$$

We want to show that (4) is independent of the b_i. In fact it equals 1. To see this consider the Vandermonde matrix:

$$V \equiv \begin{bmatrix} 1 & b_0 & b_0^2 & \cdots & b_0^k \\ 1 & b_1 & b_1^2 & \cdots & b_1^k \\ \vdots & \vdots & \vdots & \ddots & \vdots \\ 1 & b_k & b_k^2 & \cdots & b_k^k \end{bmatrix}$$

The denominator in (4) is the well-known formula for $\det(V)$. The numerator is the determinant of V gotten by expanding along the last row. We conclude from (2) that $C_{e-k}[x_0, x_1, .., x_k] \equiv \binom{e}{e-k}$, which is certainly independent of the b_i. This also implies that $C_{e-k}[x_0, x_1, .., x_k]$ is not congruent to 0 when $k \leq e$. By induction we are done.

2.2   Related-Messages Attack

Here we consider the general case where for $i = 0, 1, ...e - 1$, $x_i \equiv a_i x + b_i \bmod N$. $N = pq$ is an RSA composite (p and q are large primes, with some additional restrictions which are irrelevant in the current discussion), and the constants a_i, b_i are known. Of course it is sufficient to consider just the case where $x_i \equiv x + b_i$. We now show how to deterministically compute x in $O(e)$ $\mathbb{Z}_N$-operations after some pre-computation that depends only on the known constants. If the constants b_i hold for many unknown values of cryptograms x^e then the cost of pre-computations can be amortized and discarded. We show that the cost of the additional computations that depend on the value of x is $O(e)$.

Specifically, $\pi'_n(x_k)$ is independent of y and of x, hence for all k these coefficients can be computed in advance. In that case the cost of computing $[x_0, x_1, ...x_{e-1}] \equiv ux + v \equiv w(x)$ is $O(e)$.

For each particular value x we know how to compute the value $w(x)$ without knowing x using Lemma 1 and Formula (1). More explicitly, Let $c_i \equiv (x + b_i)^e \bmod N$, $i = 0, 1, 2, ...e - 1$, be the given cryptograms, whose underlying messages are linearly related, and let $\pi'_{e-1}(x_k) \equiv \prod_{\substack{i=0 \\ i \neq k}}^{e-1} (b_k - b_i)$. We

use p_k as a shorthand for $\pi'_{e-1}(x_k)$. Then

$$w(x) \equiv \sum_{k=0}^{e-1} \frac{[x_k]}{\pi'_{e-1}(x_k)} \equiv \sum_{k=0}^{e-1} \frac{c_k}{p_k}.$$

Note that the condition that the pairs (a_i, b_i), $i = 0, 1, ... e - 1$, are distinct and $a_i \neq 0$, imply that we can replace them with equivalent relations where $a_i = 1$, and where for all i, j, $b_k - b_i \neq 0$. Hence $\pi'_{e-1}(x_k)^{-1}$ exist and $w(x)$ is well defined.

From Lemma 1 (ii) we know that $u = e$. Note also that $w(0) \equiv v \equiv \sum_{k=0}^{e-1} b_k^e \cdot p_k^{-1} \bmod N$, and we can compute it in the pre-computation phase (before intercepting the cryptograms). So we can find $x \equiv (w(x) - v)e^{-1} \bmod N$.

The following algorithm summarizes the above discussion:

Algorithm 1:

Given cryptograms $c_i \equiv (x + b_i)^e \bmod N$, $i = 0, 1, 2, ... e - 1$, with known constants b_i, find x.

Method:

1. Pre computation:

For $k = 0, ... e - 1$, compute $p_k^{-1} \equiv \prod_{\substack{i=0 \\ i \neq k}}^{e-1} (b_k - b_i)^{-1}$;

$v \equiv \sum_{k=0}^{e-1} b_k^e \cdot p_k^{-1} \bmod N$;

2. Real-time computation: $x \equiv e^{-1} \cdot ((\sum_{k=0}^{e-1} c_k p_k^{-1}) - v) \bmod N$.

The complexity of the pre-computation is $O(e \log^2(e))$ (see Appendix), and the complexity of the real time computations is $O(e)$.

3   Special Case

3.1   Finite Differences

We now consider the special case where the e cryptograms are of the form $c_i \equiv (ax + b \cdot i)^e \bmod N$, $i = 0, 1, ... e - 1$, for any known constants $a, b \in \mathbb{Z}_N$, where $\gcd(a, N) = \gcd(b, N) = \gcd(e!, N) = 1$. The special linear relations among these cryptograms allows us to deterministically compute x in overall $O(e)$ $\mathbb{Z}_N$-operations. As before x denotes an indeterminate variable.

Definition 2. *For h a polynomial over any ring let $\Delta^{(0)}(x) \equiv h(x)$, and let*

$$\Delta^{(i)}(x) \equiv \Delta^{(i-1)}(x + 1) - \Delta^{(i-1)}(x), i = 1, 2, ...$$

It is easy to see that the degree of the polynomials resulting from this simpler process keep decreasing as in the case of divided-differences. More precisely:

Lemma 2. *In the special case where $x_i \equiv x + i$, and $\gcd(n!, N) = 1$, $[x_0, x_1, ... x_n] \equiv \Delta^{(n)}(x)/n!$*

A similar relation can be derived when $x_i \equiv ax + ib$, for known constants a, b. The next two lemmas are stated for general polynomials $h(x)$, although eventually we use them for $h(x) \equiv x^e \bmod N$. Let $m = \deg(h)$, and $0 \le k \le m$. By induction on k:

Lemma 3. $\Delta^{(k)}(x) \equiv \sum_{i=0}^{k} \binom{k}{i} \cdot h(x+i) \cdot (-1)^{k-i} \bmod N$.

For the algorithm we will need explicit formulas for the two leading terms of $\Delta^{(k)}(x)$. Let $h(x) = \sum_{i=0}^{m} a_i x^i$ and let $T^{(k)}_{a_m, a_{m-1}}(x)$ denote the two leading terms of $\Delta^{(k)}(x)$.

Lemma 4. $T^{(k)}_{a_m, a_{m-1}}(x) \equiv \frac{(m-1)!}{(m-k)!} x^{m-k-1} (a_m m(x + k(m-k)/2) + a_{m-1}(m-k))$.

Proof. We induct on k. The basis step is trivial. We verify one more step that is needed later.

$$T^{(1)}_{a_m, a_{m-1}}(x) \equiv x^{m-2}(a_m m(x + \frac{m-1}{2}) + a_{m-1}(m-1)) \tag{5}$$

$\Delta^{(1)}(x) \equiv h(x+1) - h(x)$, whose two leading terms are indeed equal to $T^{(1)}_{a_m, a_{m-1}}(x)$ above. Now assume that the two leading terms of $\Delta^{(k-1)}(x)$ are $T^{(k-1)}_{a_m, a_{m-1}}(x) \equiv \alpha x^{m-k+1} + \beta x^{m-k}$, where $\alpha \equiv \frac{(m-1)!}{(m-k)!} a_m m$, and $\beta \equiv \frac{(m-1)!}{(m-k)!} [a_m m k(m-k)/2 + a_{m-1}(m-k)]$. The proof can be completed by showing that $T^{(1)}_{\alpha, \beta}(x) \equiv T^{(k)}_{a_m, a_{m-1}}(x)$. This can be done by computing the first difference of $T^{(k-1)}_{a_m, a_{m-1}}(x)$, substituting α for a_m and β for a_{m-1} in equation (5) to get the claim.

3.2 Related-Messages Attack with Lowered Complexity

Using the results of section 3.1 we consider the special case where $x_i \equiv x + i$ (or likewise $x_i \equiv ax + bi$, for known a, b) and use the simpler finite-differences to yield overall complexity $O(e)$.

In lemmas 3 and 4 let $h(x) \equiv x^e \bmod N$, where $e \ge 3$. Thus $a_n \equiv 1, a_{n-1} \equiv 0$, and $T^{(e-1)}_{1,0} \equiv e!(x + (e-1)/2)(\bmod N)$. Lemmas 1 and 2 imply that after the $e - 1$ finite difference we have a linear congruence $ux + v \equiv w$. Then lemma 4 gives us the values of u and v, and lemma 3 tells us how to compute w given the e cryptograms.

Specifically $u \equiv e!$, $v \equiv e!(e-1)/2$ and $w \equiv \sum_{i=0}^{e-1} \binom{e-1}{i} \cdot c_i \cdot (-1)^{e-1+i}$ where $c_i \equiv (x+i)^e$ (all the congruences are taken $\bmod N$). This equation is solvable iff $e!^{-1} \bmod N$ exists, which holds for practical (small) values of e. The computation of w dominates, and takes $O(e)$ operations in $\mathbb{Z}_N$ (since $\binom{e-1}{i}$ can be computed from $\binom{e-1}{i-1}$ using one multiplication and one division).

If $x_i \equiv ax + bi \bmod N$, $i = 0, 1, 2...e - 1$, for known a and b, with $\gcd(a, N) = \gcd(b, N) = 1$, we can likewise compute x. Given cryptogram

$c_i \equiv (ax+b\cdot i\ )^e \bmod N$ we can transform it into $c_i' \equiv c_i \cdot b^{-e} \equiv (z+i)^e \bmod N$, where $z \equiv xab^{-1} \bmod N$.   So

$$x \equiv a^{-1}b(b^e e!)^{-1}[\sum_{i=0}^{e-1} \binom{e-1}{i} \cdot c_i \cdot (-1)^{e-1+i} - \frac{e-1}{2}]\ \bmod N.$$

which is computable in $O(e)$ $\mathbb{Z}_N$ operations.

4   Conclusions

We have shown new attacks on RSA-encryption assuming known explicit linear relations between the messages. Our attacks require more information (i.e., intercepting more cryptograms), but they run faster than previously published attacks. When the public exponent is 32 bits long (as recommended in [4]) our attacks run three orders of magnitudes faster than previous attacks.

This should be taken into consideration when designing very compact cryptosystems (e.g., for RFID tags), although the default should be using some form of protection like OAEP+ to destroy such known relations. Our attack tools are borrowed from numerical analysis and adapted to handle formal polynomials defined over finite rings.

Open problems: Can these or similar tools be used to attack other cases of known relations, such as implicit linear relations or explicit non-linear relations?

A non-linear case. If $c_i = (a_i x + b_i)^{f_i}$ where f_i is a known constant, then we can find x in time $O(ef)$, where $f = \operatorname{lcm}\{f_i\}$, by first converting $c_i = (a_i x + b_i)^{f_i e}$ to $c_i' = c_i^{f/f_i} = (a_i x + b_i)^{fe}$, then solving for x using the previous method with ef playing the role of e.

Acknowledgements

Special thanks go to Gideon Yuval who suggested looking into divided differences, and to Peter Montgomery who made numerous valuable suggestions and corrections (in particular to the appendix and to lemma 4). We also thank Don Coppersmith, Kamal Jain, Adi Shamir, and Venkie (Ramarathnam Venkatesan), for helpful discussions on earlier applications of the finite difference technique. Finally, we thank Arjen Lenstra, PKC'05 reviewers, and reviewers of this book who made valuable suggestions that improved this paper.

References

1. Aho Hopcroft and Ullman: "The Design and Analysis of Computer Algorithms", Addison Wesley, 1974, ISBN 0-201-00029-6.
2. D. Boneh: "Twenty Years of Attacks on the RSA Cryptosystem", in Notices of the American Mathematical Society (AMS), Vol. 46, No. 2, pp. 203–213, 1999.

3. M. Bellare and P. Rogaway: "Optimal asymmetric encryption", Eurocrypt'94: 92-111.
4. Don Coppersmith, Matthew Franklin, Jacques Patarin, Michael Reiter: "Low-Exponent RSA with related Messages", Proc. of Eurocrypt'96, LNCS 1070, pp. 1-9.
5. E. Fujisaki, T. Okamoto, D. Pointcheval, J. Stern: "RSA-OAEP Is Secure Under the RSA Assumption", J. Crypt. Vo. 17, No.2, March'04, pp. 81-104 (Springer Verlag).
6. Ronald Rivest, Adi Shamir, Leonard M. Adleman: "A Method for Obtaining Digital Signatures and Public-Key Cryptosystems", CACM 21(2): 120-126 (1978).
7. Volkov, E.A., "Numerical Methods". New York: Hemisphere Publishing Corporation, pp.48, 1987.
8. Whittaker, E. T. and Robinson, "The Calculus of Observations: A Treatise on Numerical Mathematics", 4th ed. New York: Dover, pp. 20-24, 1967.

Appendix: The Complexity of the Pre-processing

The following algorithm, due to Peter Montgomery, computes the pre-processing phase of Algorithm 1 in $O(e \log^2 e)$ time. We currently do not know of a better algorithm for the general case.

For $k = 0, ...e - 1$, we need to compute $p_k = \pi'_k(y) \equiv \prod_{\substack{i=0 \\ i \neq k}}^{e-1} (b_k - b_i)$. We use

the observation stated before Formula (1). The algorithm proceeds as follows (time complexity for each step is included in the brackets):

1. Expand the formal polynomial $\pi(y) \equiv \prod_{i=0}^{e-1} (y - x_i)$ in indeterminate variable
   y ($O(e \log^2 e)$,as explained below).
2. Compute the formal derivative of $\pi(y)$ ($O(e)$).
3. Simultaneously evaluate the value of the derivative in the given points b_i, $i = 0, 1, ...e - 1$ ($O(e \log^2 e)$, see [1] pp. 294, Corollary 2).

Expanding step (1) above:
Suppose we have a polynomial multiplication algorithm that works in time $O(n \log n)$, where n is the degree of the polynomials. Multiply pairs (there are $n/2$ many pairs). Then multiply the resulting $n/4$ pairs at cost $O(2 \log 2)$ each. And so on. There are $\log e$ many levels. Let $e = 2^k$. The total cost is $e \sum_{i=0}^{k} i = O(e \log^2 e)$.

Note that if the b_i happen to be some powers of one primitive n_{th} root of unity, $w \in Z_N$, then we can use DFT in $O(n \log n)$. However, for arbitrary $b'_i s$ chances to have this condition with $n = O(e)$ are negligible.

A Tale of Two Methods

Reuven Bar-Yehuda[1] and Dror Rawitz[2]

[1] Department of Computer Science, Technion, Haifa 32000, Israel
reuven@cs.technion.ac.il
[2] Caesarea Rothschild Institute, University of Haifa, Haifa 31905, Israel
rawitz@cri.haifa.ac.il

Abstract. We describe two widely used methods for the design and analysis of approximation algorithms, the *primal-dual schema* and the *local ratio technique*. We focus on the creation of both methods by revisiting two results by Bar-Yehuda and Even—the linear time primal-dual approximation algorithm for set cover, and its local ratio interpretation. We also follow the evolution of the two methods by discussing more recent studies.

1 Introduction

We describe two approximation methods for solving combinatorial optimization problems, the *primal-dual schema* and the *local ratio technique*. We specifically focus on the contribution of two papers written by Reuven Bar-Yehuda and Shimon Even in the early 1980's. In their first paper [8] Bar-Yehuda and Even presented a linear programming (LP) based approximation algorithm for the *set cover problem*, and for its the well known special case, the *vertex cover problem*. The idea of using linear programming for approximating set cover was not new—it was used before by Chvátal [17] and Hochbaum [24]. However, the specific way in which linear programming was used was new. Bar-Yehuda and Even's algorithm [8] constructs simultaneously a primal integral solution and a dual feasible solution without solving either the primal or dual programs. Their algorithm was the first to operate in this way, which later became known as the primal-dual schema. The local ratio technique was first used about a couple of years later in a second paper by Bar-Yahuda and Even [9] that deals with the set cover problem. In this paper they presented a local ratio analysis of the algorithm from [8]. They also developed a $(2 - \frac{\log \log n}{2 \log n})$-approximation algorithm for the vertex cover problem, which is partially based on the local-ratio technique. Over the years the two methods have become immensely popular. Numerous algorithms which use either the primal-dual schema or the local ratio technique were published. Almost two decades later, Bar-Yehuda and Rawitz [13] proved that the two methods are actually equivalent.

O. Goldreich et al. (Eds.): Shimon Even Festschrift, LNCS 3895, pp. 196–217, 2006.

Before going any further, we present some basic concepts relating to *combinatorial optimization* and *approximation algorithms*. An optimization problem is a problem consisting of a set of possible instances. Each possible instance has a set of candidate solutions, called *feasible solutions*, each of which is associated with a weight. In a minimization (resp., maximization) problem our goal is to find a feasible solution of minimum (resp., maximum) weight. Such a solution is called an *optimal solution*. The weight of an optimal solution is call the *optimum*. For example, in the *vertex cover problem*, an instance consists of a simple graph $G = (V, E)$, and a weight function w on the vertices. A solution is a set of vertices, and a feasible solution is a subset $U \subseteq V$ such that each edge in E has at least one end-point in U. Such a feasible solution is called a *vertex cover*. The weight of a vertex cover U is the total weight of the vertices in U. In the vertex cover problem our goal is to obtain a minimum weight vertex cover. The special case in which $w(u) = 1$ for every $u \in V$ is referred to as the *unweighted vertex cover problem*. In this problem, our goal is to find a vertex cover of minimum cardinality.

Since the vertex cover problem and many other optimization problems are NP-hard, we are forced to compromise. Instead of seeking algorithms that compute optimal solutions in polynomial time, we are willing to settle for efficient algorithm that compute near optimal solutions, or *approximate solutions*. A solution whose weight is within a factor of r of the optimum is called *r-approximate*. An *r-approximation algorithm* is an algorithm that computes r-approximate solutions.

For example, consider Algorithm **UnweightedVC** which is a 2-approximation algorithm for unweighted vertex cover due to Gavril (see [20]).

Algorithm 1 - UnweightedVC(G): a 2-approximation algorithm for vertex cover

---
1: $U \leftarrow \emptyset$
2: **while** there exists an uncovered edge **do**
3:    Let (u, v) be an uncovered edge
4:    $U \leftarrow U \cup \{u, v\}$
5: **end while**
6: Return U

Clearly, this algorithm runs in linear time. Also, U is a vertex cover because this is the termination condition of the algorithm. However, how close is the size of the solution U to the size of an optimal vertex cover? We show that the size of U is quite close to the optimum by proving that it is not more than twice the optimum. Denote by M the set of edges that are considered in Line 3. Clearly, M is a *maximal matching*. Since there are no two edges in M that share a common vertex, any vertex cover must be at least as large as M. Hence, if U^* is an optimal vertex cover, then $|U| = 2|M| \leq 2|U^*|$.

Consider the analysis of Algorithm **UnweightedVC**. It is based on the following simple idea. First, we find a lower bound on the optimum value and then we show that the size of the solution computed by the algorithm is bounded by r times the lower bound, where r is the approximation ratio of the algorithm. The bound in our case is the size of M. This theme is widely used in the field of approximation algorithms, especially in approximation algorithm that are based on linear programming—many combinatorial optimization problems can be expressed as linear integer programs, and the value of an optimal solution to their *LP-relaxation* provides the desired bound.

As we shall see in the sequel, algorithms that fall within the scope of either the primal-dual schema or the local ratio technique use a variation on the lower bound idea (or, upper bound, in the maximization case). Let W denote the weight of the solution computed by the algorithm. Instead of finding directly some lower bound B on the optimum such that $W \leq r \cdot B$, we break down the weight of the solution into a sum of partial weights $W = W_1 + \ldots + W_k$. Then, for each such partial weight W_i we find a "partial" lower bound B_i such that $W_i \leq r \cdot B_i$. Our solution is r-approximate since the sum of the partial lower bounds is not greater than the optimum. In both methods the breakdown of W is determined by the manner in which the solution is constructed by the algorithm. In fact, the algorithm constructs the solution in such a manner as to ensure that such a breakdown exists. The breakdown is done in steps, where in the ith step, the algorithm determines the ith partial weight, and the ith lower bound B_i. In the primal-dual schema the partial weight and bound are induced by an increase of the dual solution, while in the local ratio technique they are determined by the construction of a weight function.

The remainder of this essay is organized as follows. In Section 2 we present several basic results in the area of linear programming, and formally define the problems that we consider in this essay. Bar-Yehuda and Even's [8] primal-dual approximation algorithm for set cover is presented, in *hitting set terms*, in Section 3. Afterwards, we give a general description of the schema, and demonstrate it on an extension of the hitting set problem called *generalized hitting set*. The local ratio version of the approximation algorithm for set cover [9] is given in Section 4. This section also contains a general description of the local ratio technique, and a local ratio algorithm for generalized hitting set. Finally, in Section 5 we survey results that were obtained in both methods during the last two decades, and discuss the connection between the two methods.

2   Preliminaries

2.1   Linear Programming

In this section we state several basic facts from the theory of linear programming. Note that the section is written in terms of minimization problems. (Similar arguments can be made in the maximization case.) For more details about linear programming the reader is referred to, e.g., [28, 29, 31].

Consider the following linear integer program:

$$
\begin{aligned}
\min \ & \sum_{j=1}^{n} c_j x_j \\
\text{s.t.} \ & \sum_{j=1}^{n} a_{ij} x_j \geq b_i && \forall i \in \{1, \ldots, m\} \\
& x_j \in \mathbb{N} && \forall j \in \{1, \ldots, n\}
\end{aligned}
\tag{IP}
$$

The LP-relaxation of IP is obtained by removing the integrality constraints:

$$
\begin{aligned}
\min \ & \sum_{j=1}^{n} c_j x_j \\
\text{s.t.} \ & \sum_{j=1}^{n} a_{ij} x_j \geq b_i && \forall i \in \{1, \ldots, m\} \\
& x_j \geq 0 && \forall j \in \{1, \ldots, n\}
\end{aligned}
\tag{P}
$$

Let OPT(IP) and OPT(P) denote the optimum of IP and P, respectively. Notice that any feasible solution of IP is also feasible with respect to P. Hence,

Observation 1 $\text{OPT}(P) \leq \text{OPT}(IP)$.

We refer to P as the *primal* linear program. The following linear program is the *dual* of P:

$$
\begin{aligned}
\max \ & \sum_{i=1}^{m} b_i y_i \\
\text{s.t.} \ & \sum_{i=1}^{m} a_{ij} y_i \leq c_j && \forall j \in \{1, \ldots, n\} \\
& y_i \geq 0 && \forall i \in \{1, \ldots, m\}
\end{aligned}
\tag{D}
$$

A solution of P is called a *primal solution*, and a solution of D is called a *dual solution*. A solution of IP is referred to as an *integral primal solution*.

The connection between the primal and dual optima is given by the following two theorems (the second is given without proof):

Theorem 2 (Weak Duality). *Let x and y be a pair of primal and dual solutions. Then, $b^T y \leq c^T x$.*

Proof.

$$
\sum_{j=1}^{n} c_j x_j \geq \sum_{j=1}^{n} \left(\sum_{i=1}^{m} a_{ij} y_i \right) x_j = \sum_{i=1}^{m} \left(\sum_{j=1}^{n} a_{ij} x_j \right) y_i \geq \sum_{i=1}^{m} b_i y_i
\tag{1}
$$

where the first inequality follows from a summation of the dual constraints, and the second follows from a summation of the primal constraints. ☐

Theorem 3 (Strong Duality). *Let x^* and y^* be a pair of optimal primal and dual solutions. Then, $b^T y^* = c^T x^*$.*

It follows that

Observation 4 *Let x be an integral primal solution, and let y be a dual solution. Then, $b^T y \leq \text{OPT}(D) = \text{OPT}(P) \leq \text{OPT}(IP) \leq c^T x$.*

The Strong Duality Theorem provides us with a way to characterize a primal-dual pair of optimal solutions.

Theorem 5 (Complementary Slackness Conditions). *Let x and y be a pair of primal and dual solutions. Then, x and y are optimal if and only if the following conditions, called the* complementary slackness conditions, *are satisfied:*

$$\text{Primal conditions: } \forall j, \ x_j > 0 \Rightarrow \sum_{i=1}^{m} a_{ij} y_i = c_j$$

$$\text{Dual conditions: } \quad \forall i, \ y_i > 0 \Rightarrow \sum_{j=1}^{n} a_{ij} x_j = b_i$$

Proof. First, assume x and y are optimal. By the Strong Duality Theorem it follows that $c^T x = b^T y$. Therefore, the inequalities in Equation (1) become equalities:

$$\sum_{j=1}^{n} c_j x_j = \sum_{j=1}^{n} \left(\sum_{i=1}^{m} a_{ij} y_i \right) x_j = \sum_{i=1}^{m} \left(\sum_{j=1}^{n} a_{ij} x_j \right) y_i = \sum_{i=1}^{m} b_i y_i \qquad (2)$$

The primal complementary slackness conditions are implied by the first equality, and the dual conditions are implied by the third equality.

For the other direction assume that the complementary slackness conditions are satisfied. In this case Equality (2) is satisfied as well, and therefore $c^T x = b^T y$. x and y are optimal by the Weak Duality Theorem. $\square$

2.2   The Problems

Recall that, in the *vertex cover problem*, an instance consists of a simple graph $G = (V, E)$, and a weight function w on the vertices, and our goal is to obtain a minimum weight vertex cover. Hence, the vertex cover problem can be formulated by the following linear integer program:

$$
\begin{aligned}
\min \ &\sum_{u \in V} w(u) x_u \\
\text{s.t. } \ &x_u + x_v \geq 1 \qquad \forall (u, v) \in E \\
&x_u \in \{0, 1\} \qquad \forall u \in V
\end{aligned}
\qquad \text{(VC)}
$$

where $x_u = 1$ if and only if u is in the vertex cover. The LP-relaxation of VC is obtained by replacing the integrality constraints by: $x_u \geq 0$ for every $u \in V$.

(Notice that the possible inequalities of the form $x_u \leq 1$ are redundant.) We denote the LP relaxation of VC by VC-P. (Henceforth, a -P suffix denotes the LP-relaxation of a linear integer program, while a -D suffix denotes the dual of the LP-relaxation.)

The *hitting set problem* is defined as follows. The input consists of a collection of subsets $\mathcal{S} = \{S_1, \ldots, S_m\}$ of the ground set U of size n. Each element $u \in U$ is associated with a positive weight $w(u)$. A set H is said to *hit* a subset S if $H \cap S \neq \emptyset$. A *hitting set* is a set $H \subseteq U$ that hits every subset $S \in \mathcal{S}$. In the hitting set problem our goal is find a hitting set of minimum total weight. Given a hitting set instance, we denote by $\mathcal{S}(u)$ the collection of sets that contain u, i.e., $\mathcal{S}(u) \triangleq \{S : u \in S\}$. We define $s_{\max} \triangleq \max_{S \in \mathcal{S}} |S|$. Note that the vertex cover problem is a special case of hitting set in which all sets are of size two, and hence $s_{\max} = 2$ in this special case. We also note that some of the results in this survey were originally written in terms of the *set cover problem*. In the *set cover* problem we are given a collection of sets $\mathcal{S}$ of the ground set U, and a weight function on the subsets. The objective is to find a minimum weight collection of sets that covers all elements, or a minimum weight *set cover*. It is easy to see that set cover and hitting set are equivalent problems in the sense that each is obtained from the other by switching the roles of sets and elements.

The hitting set problem can be formulated by the following linear integer program:

$$\min \sum_{u \in U} w(u) x_u$$
$$\text{s.t. } \sum_{u \in S} x_u \geq 1 \qquad \forall S \in \mathcal{S} \qquad\qquad \text{(HS)}$$
$$x_u \in \{0, 1\} \qquad \forall u \in U$$

where $x_u = 1$ if and only if u is in the hitting set. The LP-relaxation of HS is obtained by replacing the integrality constraints by: $x_u \geq 0$ for every $u \in U$. We denote that LP relaxation by HS-P. The dual of HS-P is:

$$\max \sum_{S \in \mathcal{S}} y_S$$
$$\text{s.t. } \sum_{S \in \mathcal{S}(u)} y_S \leq w(u) \qquad \forall u \in U \qquad\qquad \text{(HS-D)}$$
$$y_S \geq 0 \qquad \forall S \in \mathcal{S}$$

In the *generalized hitting set problem* we are also given of a collection of subsets $\mathcal{S}$ of the ground set U, and our goal is to hit the sets in $\mathcal{S}$ by using elements from U. However, in this case, we are allowed not to hit a set S, provided that we pay a tax $w(S)$. Hence, the weight function w is define on both the elements and the subsets. Formally, the input is a collection of sets $\mathcal{S} = \{S_1, \ldots, S_m\}$ of the ground set $U = \{1, \ldots, n\}$, and a weight function on the elements and subsets, and our goal is to find a minimum-weight set $H \subseteq U$, where the weight of H is the weight of the elements in H and the weight of the sets that are not hit by H. The hitting set problem is the special case where the tax $w(S)$ is infinite for every set $S \in \mathcal{S}$.

In order to formulate generalized hitting set using a linear program it would be convenient to slightly change the problem definition. Instead of simply searching for a set of elements H, we shall search for a set of elements H and a subcollection of sets $\mathcal{T}$ such that for every set S either S is hit by H, or it is contained in $\mathcal{T}$, i.e., we seek a pair $(H, \mathcal{T})$ where $H \subseteq U$, $\mathcal{T} \subseteq \mathcal{S}$, and for all $S \in \mathcal{S}$, either $H \cap S \neq \emptyset$ or $S \in \mathcal{T}$. This means that we allow a set S to be both hit by H and contained in $\mathcal{T}$. Clearly, for a given generalized hitting set instance, the optima of both problems are the same. Moreover, any solution for the second definition can be easily turned into a solution for the first. Hence, the generalized hitting set problem can be formalized as follows:

$$\min \sum_{u \in U} w(u)x_u + \sum_{S \in \mathcal{S}} w(S)x_S$$
$$\text{s.t.} \sum_{u \in S} x_u + x_S \geq 1 \qquad \forall S \in \mathcal{S} \qquad \text{(GHS)}$$
$$x_u \in \{0, 1\} \qquad \forall u \in U$$
$$x_S \in \{0, 1\} \qquad \forall S \in \mathcal{S}$$

where $x_u = 1$ if and only if the element u is contained in H, and $x_S = 1$ if and only if the subset S is contained in $\mathcal{T}$. As usual, the LP-relaxation is obtained by replacing the integrality constraints by: $x_u \geq 0$ for every $u \in U$ and $x_S \geq 0$ for every $S \in \mathcal{S}$. We denote that LP relaxation by GHS-P. The dual is:

$$\max \sum_{S \in \mathcal{S}} y_S$$
$$\text{s.t.} \sum_{S \in \mathcal{S}(u)} y_S \leq w(u) \qquad \forall u \in U \qquad \text{(GHS-D)}$$
$$y_S \leq w(S) \qquad \forall S \in \mathcal{S}$$
$$y_S \geq 0 \qquad \forall S \in \mathcal{S}$$

Notice that a generalized hitting set instance can be viewed as a hitting set instance in which each set S contains a unique element u_S whose weight is $w(u_S) \triangleq w(S)$. This way, we can pay $w(S)$ for the element u_S instead of paying the tax $w(S)$ for not hitting S. In the sequel we present several $s_{\max}$-approximation algorithms for hitting set. It follows that these algorithm can be used to obtain $(s_{\max} + 1)$-approximate solutions for generalized hitting set. However, we also show how to obtain $s_{\max}$-approximate solutions for generalized hitting set.

An important notion in the design of approximation algorithms using primal-dual or local ratio is the notion of *minimal solutions*. A feasible solution is said to be minimal with respect to set inclusion (or minimal for short) if all its proper subsets are not feasible. Minimal solutions arise naturally in the context of *covering problems*, which are the problems for which feasible solutions have the property of being monotone inclusion-wise, that is, the property that adding items to a feasible solution cannot render it infeasible. For example, adding an element to a hitting set yields a hitting set, so *hitting set* is a covering problem. In contrast, adding an edge to a spanning tree does not yield a tree, so *minimum spanning tree* is not a covering problem. However, if instead of focusing only on trees, we consider all sets of edges that intersect all non trivial cuts in the given graph the problem becomes a covering problem.

It is easy to see that generalized hitting set (under the second definition) is a covering problem. The following observation formalizes the fact that it makes no sense to add a set S to T if $H \cap S \neq \emptyset$.

Observation 6 *Let (H, T) be a minimal solution, and let x be the incidence vector of (H, T). Then, (i) $x_S = 1$ if and only if $\sum_{u \in S} x_u = 0$, and (ii) $\sum_{u \in S} x_u + x_S \leq s_{\max}$.*

We note that the use of minimality in the context of the generalized hitting set problem is somewhat artificial. However, it will assist us in demonstrating the use of minimality in the design of primal-dual and local ratio approximation algorithms for covering problems.

3   The Primal-Dual Schema

In this section we present the primal-dual $s_{\max}$-approximation algorithm for hitting set from [8]. Afterwards we give a general description of the primal-dual schema, and demonstrate it on the generalized hitting set problem.

An r-approximation algorithm for a minimization problem that is based on a primal-dual analysis produces an integral primal solution x and a dual solution y such that the weight of the primal solution is not more than r times the value of dual solution. Namely, it produces an integral primal solution x and a dual solution y such that

$$c^T x \leq r \cdot b^T y \tag{3}$$

The integral primal solution x is r-approximate due to Observation 4.

There are several ways to find such a pair of primal and dual solutions. The first one to do so was Chvátal [17], who proved that the greedy algorithm for hitting set computes $\mathcal{H}_m$-approximate solutions, where $\mathcal{H}_m$ is the mth harmonic number. (Recall that, in hitting set terms, m is the number of sets.) In his analysis he obtained an *infeasible* dual solution whose value as not less than the weight of the integral primal solution that was computed by the algorithm. Then, he showed that if the dual solution is divided by $\mathcal{H}_m$ it becomes feasible. This method was later called *dual fitting*, and the feasible solution was referred to as *shrunk dual*. (See [32, 26] for more details.)

Hochbaum [24] presented several $s_{\max}$-approximation algorithms for hitting set that require the solution of a linear program. The first algorithm is as follows: (i) compute an optimal solution y^* of HS-D, and (ii) return $H_D = \{u : \sum_{S \in \mathcal{S}(u)} y_S^* = w(u)\}$. We show that H_D is a hitting set. Assume by contraposition that H_D is not a hitting set. Then, there exists a set S such that $S \cap H_D = \emptyset$. Let $\varepsilon = \min_{u \in S}\{w(u) - \sum_{S \in \mathcal{S}(u)} y_S^*\}$. Clearly, $\varepsilon > 0$. Furthermore, if we raise y_S^* by ε it remains feasible in contradiction to the optimality y^*. Next, we show that H_D is $s_{\max}$-approximate. Let x_D be the incidence vector of H_D. Then,

$$
\begin{aligned}
w(H_D) &= \sum_{u \in U} w(u) x_u \\
&= \sum_{u \in U} x_u \sum_{S \in \mathcal{S}(u)} y_S^* \\
&= \sum_{S \in \mathcal{S}} y_S^* \sum_{u \in S} x_u \\
&\leq s_{\max} \sum_{S \in \mathcal{S}} y_S^* \\
&\leq s_{\max} \cdot \text{OPT}
\end{aligned}
$$

and we are done.

An $s_{\max}$-approximate hitting set can also be found by solving the primal linear program HS-P. Consider the following algorithm: (i) compute an optimal solution x^* of HS-P, and (ii) return $H_P = \{u : x_u^* > 0\}$. H_P must be a hitting set, since otherwise, x^* is not feasible. Moreover, by the complementary slackness conditions $H_P \subseteq H_D$ (where H_D is defined as above). Hence, H_P is $s_{\max}$-approximate as well. Note that it is even enough to consider only elements whose primal variable is at least $1/s_{\max}$. That is, the set $H_P' = \{u : x_u^* \geq 1/s_{\max}\}$ is also an $s_{\max}$-approximate solution. H_P' is feasible since there exists $u \in S$ such that $x_u^* \geq 1/s_{\max}$ for every subset $S \in \mathcal{S}$, and H_P' is $s_{\max}$-approximate since $H_P' \subseteq H_P$.

Following the work of Hochbaum [24], Bar-Yehuda and Even [8] presented another $s_{\max}$-approximation algorithm for hitting set that uses primal-dual arguments. As opposed to Hochbaum's algorithm, this algorithm is not based on finding an *optimal* dual (or primal) solution, and therefore it is more efficient. The key observation that was made by Bar-Yehuda and Even [8] is that the dual solution, y^*, used in Hochbaum's analysis does not have to be optimal. A dual solution y is called *maximal* if an increase in y_i makes y infeasible, for any i. Clearly, an optimal dual solution is also maximal. It is not hard to verify that $H_D' = \{u : \sum_{S \in \mathcal{S}(u)} y_S = w(u)\}$ is a hitting set for any maximal (and not necessarily optimal) dual solution y. Hence, Hochbaum's analysis stays intact when a maximal dual solution is used (and H_D is replaced by H_D'). The improved running time is due to the fact that a simple greedy algorithm can compute a maximal dual solution in linear time. Algorithm **PD-HS** is the algorithm from [8] given in terms of hitting set.

It is not hard to verify that the running time of Algorithm **PD-HS** is $O(\sum_{S \in \mathcal{S}} |S|)$, which means that it runs in linear time. Observe that, in every iteration, y_S is raised as much as possible while maintaining feasibility, hence y is a maximal dual solution. Hence, the set of elements whose corresponding dual constraint is tight (i.e., H_D') constitute an $s_{\max}$-approximate hitting set. Algorithm **PD-HS** does not return the set of elements whose corresponding

---

Algorithm 2 - PD-HS$(U, \mathcal{S}, w)$: a primal-dual $s_{\max}$-approximation algorithm for hitting set

---

1: $H \leftarrow \emptyset$
2: $y \leftarrow 0$
3: **while** $\mathcal{S} \neq \emptyset$ **do**
4:    Let $S \in \mathcal{S}$
5:    $v \leftarrow \operatorname{argmin}_{u \in S} \left\{ w(u) - \sum_{S' \in \mathcal{S}(u)} y_{S'} \right\}$
6:    $y_S \leftarrow w(v) - \sum_{S' \in \mathcal{S}(v)} y_{S'}$
7:    $H \leftarrow H \cup \{v\}$
8:    $\mathcal{S} \leftarrow \mathcal{S} \setminus \mathcal{S}(v)$
9: **end while**
10: Return H

---

dual constraint is tight. However, an element v may be added to H only if its corresponding dual constraint is tight (i.e., $H \subseteq H'_D$). Moreover, every subset S contains at least one such element. Hence, H is feasible and therefore also $s_{\max}$-approximate.

It is important to notice that since the choice of v (in Line 6) is made according to the tightness of the dual constraints, and not according to the values of the dual variables, it is enough to compute, in every iteration, the tightness of the dual constraints, instead of maintaining the dual solution y. The actual values of the dual variables are needed only for purposes of analysis. Hence, Lines 5–6 of Algorithm **PD-HS** can be replaced with the following two lines:

5: $v \leftarrow \operatorname{argmin}_{u \in S} \{w(u)\}$
6: For every $u \in S$ do: $w(u) \leftarrow w(u) - w(v)$

In fact, the original algorithm was presented in this way in [8].

Algorithm **PD-HS** computes an integral primal solution x, the incidence vector of H, and a dual solution y such that the weight of x is bounded by $s_{\max}$ times the value of y. This seems like a neat trick, but can we use this idea in order to approximate other problems? We shall see that the connection between x and y is somewhat more complicated than what is implied by the analysis of Algorithm **PD-HS**. Clearly, Algorithm **PD-HS** picks only elements whose corresponding dual constraint is tight. Hence, if $x_u = 1$ (i.e., $u \in H$) then $\sum_{S \in \mathcal{S}(u)} y_S = w(u)$. Now, consider a set $S \in \mathcal{S}$, and the corresponding constraint $\sum_{u \in S} x_u \geq 1$. Clearly, $\sum_{u \in S} x_u \leq s_{\max}$ for any S such that $y_S > 0$ (or, for any other S). Putting it all together we get that x and y satisfy the following conditions:

$$\forall u \in U, \ x_u > 0 \Rightarrow \sum_{S \in \mathcal{S}(u)} y_S = w(u)$$

$$\forall S \in \mathcal{S}, \ y_S > 0 \Rightarrow 1 \leq \sum_{u \in S} x_u \leq s_{\max}$$

The first set of conditions are exactly the primal complementary slackness conditions, while the second is a relaxation of the dual conditions. Moreover, the relaxation factor is exactly the approximation ratio of Algorithm **PD-HS**. As we shall see this idea is not limited to the hitting set problem.

Let x be an integral primal solution, and let y be a dual solution. Also, assume that x and y satisfy the following *relaxed complementary slackness conditions*:

$$\text{Primal conditions:} \qquad \forall j, \ x_j > 0 \Rightarrow \sum_{i=1}^{m} a_{ij} y_i = c_j$$

$$\text{Relaxed dual conditions:} \ \forall i, \ y_i > 0 \ \Rightarrow b_i \leq \sum_{j=1}^{n} a_{ij} x_j \leq r \cdot b_i$$

Then,

$$\sum_{j=1}^{n} c_j x_j = \sum_{j=1}^{n} \left(\sum_{i=1}^{m} a_{ij} y_i \right) x_j = \sum_{i=1}^{m} \left(\sum_{j=1}^{n} a_{ij} x_j \right) y_i \leq r \cdot \sum_{i=1}^{m} b_i y_i$$

which means that x is r-approximate. Hence, we have found a way to compute a pair of integral primal and dual solutions that satisfy Inequality (3).

Indeed, a typical primal-dual algorithm computes a primal-dual pair (x, y) that satisfies the relaxed complementary slackness conditions. Moreover, a primal-dual algorithm usually constructs the primal-dual pair in such a way that the relaxed complementary slackness conditions are satisfied throughout its execution. It starts with an infeasible primal solution and a feasible dual solution (usually, $x = 0$ and $y = 0$). It iteratively raises the dual solution, and improves the feasibility of the primal solution while maintaining the following two invariants: (i) a primal variable is increased only if its corresponding primal condition is satisfied, and (ii) a dual variable is increased only if its corresponding relaxed dual condition is satisfied. (We note that many primal-dual algorithms change several dual variables in each iteration. However, it can be shown that it is enough to raise only a single dual variable in each iteration [15, 13].) Hence, an iteration of a primal-dual r-approximation algorithm (for a covering problem) can be informally described as follows:

1. Find a primal constraint, $\alpha x \geq \beta$, such that $\alpha x \leq r \cdot \beta$ for every feasible solution x.
2. Raise the dual variable that corresponds to the above primal constraint until some dual constraint becomes tight.
3. Add an element whose corresponding dual constraint is tight to the primal solution.

Steps (1) and (2) ensure that the relaxed dual conditions are satisfied, while Step (3) ensures that the primal conditions are satisfied. The reader is referred to [15, 13] for a formal description.

In many primal-dual algorithms the Step (1) is slightly modified. Instead of finding a primal constraint for which $\alpha x \leq r \cdot \beta$ for every feasible solution x, it is enough to find a primal constraint $\alpha x \geq \beta$, for which $\alpha x \leq r \cdot \beta$ for every minimal solution x (or, sometimes, for every feasible solution x that satisfies some other property $\mathcal{P}$). Such a constraint is called *r-effective* (or *r-effective with respect to* $\mathcal{P}$). In this case, the algorithm must compute a minimal solution x (or a solution x that satisfies $\mathcal{P}$), since otherwise the dual solution y do not satisfy the relaxed dual complementary slackness conditions at termination. Primal-dual algorithms that compute minimal solutions usually use a primal pruning procedure that is sometimes referred to as *reverse deletion*. Algorithms that use a property $\mathcal{P}$ other than minimality use some sort of solution correction procedure that depend on $\mathcal{P}$. (See [15, 13] for more details.)

We demonstrate the above ideas on the generalized hitting set problem. Algorithm **PD-GHS** is an $s_{\max}$-approximation algorithm for generalized hitting set that was presented by Bar-Yehuda and Rawitz [13].

Algorithm 3 - PD-GHS$(U, \mathcal{S}, w)$: a primal-dual $s_{\max}$-approximation algorithm for generalized hitting set

---
1: $H \leftarrow \emptyset, \mathcal{T} \leftarrow \emptyset$
2: $y \leftarrow 0$
3: **for all** $S \in \mathcal{S}$ **do**
4:     Raise y_S until some dual constraint becomes tight
5:     **if** there exists an element $u \in S$ whose dual constraint became tight **then**
6:         $H \leftarrow H \cup \{u\}$
7:     **else**
8:         $\mathcal{T} \leftarrow \mathcal{T} \cup \{S\}$
9:     **end if**
10: **end for**
11: $\mathcal{T} \leftarrow \mathcal{T} \setminus \bigcup_{u \in H} \mathcal{S}(u)$.
12: Return $(H, \mathcal{T})$

We show that Algorithm **PD-GHS** computes a pair of minimal primal solution and dual solution that satisfies the relaxed complementary slackness conditions. First, $(H, \mathcal{T})$ is minimal due to Line 11. Also, the primal conditions are satisfied by the construction of the primal solution. Now, consider a set $S \in \mathcal{S}$ and its corresponding primal constraint: $\sum_{u \in S} x_u + x_S \geq 1$. According to Observation 6 we know that $\sum_{u \in S} x_u + x_S \leq s_{\max}$ for every minimal solution x. Hence, any minimal solution satisfies the relaxed dual slackness condition corresponding to S. It follows that Algorithm **PD-GHS** computes $s_{\max}$-approximate solutions.

We note that in some cases proving that the relaxed dual slackness conditions are satisfied for some r may be a difficult task, e.g., the 2-approximation algorithms for the *feedback vertex set problem* [16]).

We also remark that most primal-dual algorithms in the literature are based on a predetermined LP formulation, and therefore several dual variable are changed in each iteration of the algorithm. Hence, most primal-dual algorithms do not refer explicitly to the relaxed complementary slackness conditions. As mentioned above, such algorithms can be altered such that only a single dual variable is changed in each iteration (see [15, 13]). After doing so these analyses can be easily explained using the relaxed conditions. It is important to note that the combinatorial properties of the problem that were used in the analysis are usually presented in a much clearer fashion when the analysis change only a single dual variable in each iteration and is based on the relaxed complementary slackness conditions. Hence, such analyses tend to be simpler and more elegant.

4   Local Ratio

We re-consider Gavril's 2-approximation algorithm for unweighted vertex cover (Algorithm **UnweightedVC**). In each iteration the algorithm picks two vertices u and v that cover the uncovered edge (u, v). Since this edge must be covered, any vertex cover must contain at least one of the vertices. Hence, if we take both u and v we decrease the optimum by at least one, while adding not more than two vertices to the solution. Notice that this argument is local in the sense that it refers separately to any edge in M. (Recall that M is the maximal matching constructed by the algorithm.) This simple idea is at the heart of the local ratio technique.

We show how to extend this algorithm to an $s_{\max}$-approximation algorithm for the weighted hitting set problem. Imagine that we have to actually purchase the elements we select as our solution. Rather than somehow deciding on which elements to buy and then paying for them, we adopt the following strategy. We repeatedly select an element and pay for it. However, the amount we pay need not cover its entire cost; we may return to the same vertex later and pay some more. In order to keep track of the payments, whenever we pay ε for a vertex, we lower its marked price by ε. When the marked price of an element drops to zero, we are free to take it, as it has been fully paid for. The heart of the matter is the rule by which we select the element and decide on the amount to pay for it. Actually, we select up to $s_{\max}$ elements each time and pay ε for each, in the following manner. We select any subset S whose elements have non-zero weight, and pay $\varepsilon = \min_{u \in S} w(u)$ for every element in S. As a result, the weight of at least one of the elements drops to zero. After $O(n)$ rounds, prices drop sufficiently so that every set contains an element of zero weight. Hence, the set of all zero-weight elements is a hitting set.

We formalize the above discussion by Algorithm **LR-HS** which is a linear time $s_{\max}$-approximation algorithm that was presented by Bar-Yehuda and Even [9]. (The original algorithm was presented in set cover terms.) We say that a set is *positive* if all its elements have strictly positive weights. Notice that on instances of unweighted vertex cover it is identical to Gavril's 2-approximation algorithm.

To formally analyze the algorithm consider the ith iteration. Let S_i be the set that was selected in this iteration, and let ε_i be the weight of the minimum weight element in S_i. Since every hitting set must contain at least one element in S_i, decreasing the weight of the elements in S_i by ε_i lowers the weight of every hitting set by at least ε_i. Hence, the optimum must also decrease by at

---

Algorithm 4 - LR-HS$(U, \mathcal{S}, w)$: a local ratio s_{max}-approximation algorithm for hitting set

---

1: **while** there exists a positive set S **do**
2:    $\varepsilon \leftarrow \min_{u \in S} \{w(u)\}$
3:    For every $u \in S$ do: $w(u) \leftarrow w(u) - \varepsilon$
4: **end while**
5: Return the set $H = \{u : w(u) = 0\}$

---

least ε_i. Thus, in the ith round we pay $s_{max} \cdot \varepsilon_i$ and lower the optimum by at least ε_i. Since H is a zero weight hitting set (with respect to the final weights), the optimum has decreased to zero. Hence, OPT $\geq \sum_i \varepsilon_i$. One the other hand, since our payments fully cover H, its weight is bounded by $\sum_i s_{max} \cdot \varepsilon_i$. H is s_{max}-approximate, because $w(H) \leq \sum_i s_{max} \cdot \varepsilon_i \leq s_{max} \cdot$ OPT.

It is interesting to note that the proof that the solution found is s_{max}-approximate does not depend on the actual value of ε in any given iteration. In fact, any value between 0 and $\min_{u \in S} \{w(u)\}$ would yield the same result (by the same arguments). We chose $\min_{u \in S} \{w(u)\}$ for the sake of efficiency. This choice ensures that the number of elements with positive weight strictly decreases with each iteration.

In Algorithm **LR-HS** we have paid $s_{max} \cdot \varepsilon$ for lowering OPT by at least ε in each round. Other local ratio algorithms can be explained similarly—one pays in each round at most $r \cdot \varepsilon$, for some r, while lowering OPT by at least ε. If the same r is used in all rounds, the solution computed by the algorithm is r-approximate. This idea works well for several problems. However, it is not hard to see that this idea works mainly because we make a down payment on several elements, and we are able to argue that OPT must drop by a proportional amount because every solution must contain one of these elements. This localization of the payments is at the root of the simplicity and elegance of the analysis, but it is also the source of its weakness: how can we design algorithms for problems in which no single element (or set of elements) is necessarily involved in every optimal solution? For example, consider the *feedback vertex set* problem, in which we are given a graph and a weight function on the vertices, and our goal is to remove a minimum weight set of vertices such that the remaining graph contains no cycles. Clearly, it is not always possible to find a constant number of vertices such that at least one of them is part of every optimal solution!

It helps to view the payments made by the algorithm as the subtraction of a new weight function w_1 from the current weight function w. For example, examine an iteration of Algorithm **LR-HS**. The action taken in Line 3 is equivalent to defining a new weight function:

$$w_1(u) = \begin{cases} \varepsilon & u \in S, \\ 0 & u \notin S, \end{cases}$$

and subtracting it from w. The analysis above implies that:

Observation 7 *Every hitting set is s_{max}-approximate with respect to w_1.*

Hence, any $s_{\max}$-approximate hitting set H with respect to $w - w_1$ is also $s_{\max}$-approximate with respect to w_1. By the following theorem H is also $s_{\max}$-approximate with respect to w.

Theorem 8 (Local Ratio Theorem [9,6]). *Let $(\mathcal{F}, w)$ be a minimization problem, where $\mathcal{F}$ is a set of constraints on $x \in \mathbb{R}^n$, and $w \in \mathbb{R}^n$ is a weight function. Also, let w_1 and w_2 be weight functions such that $w = w_1 + w_2$. Then, if x is r-approximate with respect to $(\mathcal{F}, w_1)$ and with respect to $(\mathcal{F}, w_2)$, then x is r-approximate with respect to $(\mathcal{F}, w)$.*

Proof. Let x^*, x_1^*, x_2^* be optimal solutions with respect to $(\mathcal{F}, w), (\mathcal{F}, w_1)$, and $(\mathcal{F}, w_2)$, respectively. Then, $wx = w_1 x + w_2 x \le r \cdot w_1 x_1^* + r \cdot w_2 x_2^* \le r \cdot (w_1 x^* + w_2 x^*) = r \cdot wx^*$. □

Note that $\mathcal{F}$ can include arbitrary feasibility constraints and not just linear constraints. Nevertheless, all successful applications of the local ratio technique to date involve problems in which the constraints are linear.

This idea of weight decomposition leads us to the to Algorithm **Recursive-LR-HS** which is a recursive version of Algorithm **LR-HS**.

Algorithm 5 - Recursive-LR-HS$(U, \mathcal{S}, w)$: a local ratio $s_{\max}$-approximation algorithm for hitting set

---

1: **if** $\mathcal{S} = \emptyset$ **then**
2:     Return $\emptyset$
3: **end if**
4: Let $S \in \mathcal{S}$
5: $v \leftarrow \operatorname{argmin}_{u \in S} \{w(u)\}$
6: $\varepsilon \leftarrow w(v)$
7: Define $w_1(u) = \begin{cases} \varepsilon & u \in S, \\ 0 & u \notin S, \end{cases}$
8: $H \leftarrow \{v\} \cup \textbf{Recursive-LR-HS}(U \setminus \{v\}, \mathcal{S} \setminus \mathcal{S}(v), w - w_1)$
9: Return H

We first note that this algorithm is slightly different from Algorithm **LR-HS**, since not all the vertices that have zero weight at the recursive base are necessarily taken into the solution. (A similar pruning procedure can be added to Algorithm **LR-HS** as well.)

Since Algorithm **Recursive-LR-HS** is recursive, it is natural to use induction in its analysis. First, it is not hard to show that the solution returned is a hitting set by induction on the number of recursive calls. (Note that this number is bounded by the number of elements.) We prove that the solution is $s_{\max}$-approximate by induction. In the base case, $\emptyset$ is an optimal solution. For the inductive step, let H be the solution returned, and denote $w_2 = w - w_1$. By the induction hypothesis $H \setminus \{v\}$ is $s_{\max}$-approximate with respect to $(U \setminus \{v\}, \mathcal{S} \setminus \mathcal{S}(v), w_2)$. Since $w_2(v) = 0$, the optima of $(\mathcal{S}, U, w_2)$

and of $(U \setminus \{v\}, \mathcal{S} \setminus \mathcal{S}(v), w_2)$ are the same. Hence, H is $s_{\max}$-approximate with respect to $(\mathcal{S}, U, w_2)$. Due to Observation 7 any hitting set with respect to the instance $(\mathcal{S}, U)$ is $s_{\max}$-approximate with respect to w_1, therefore H is $s_{\max}$-approximate with respect to $(\mathcal{S}, U, w_1)$. Finally, H is $s_{\max}$-approximate with respect to $(\mathcal{S}, U, w)$ as well due to the Local Ratio Theorem.

Observation 7 states that the weight function w_1 is *well behaved*. That is, from its view point all hitting sets weigh roughly the same (up to a multiplitive factor of $s_{\max}$). We formalize the notion of well behaves weight functions.

Definition 9 *A weight function w is said to be r-effective with respect to property $\mathcal{P}$ if there exists a number b such that $b \leq wx \leq r \cdot b$ for all feasible solutions x that satisfy $\mathcal{P}$.*

In Algorithm **Recursive-LR-HS** the property $\mathcal{P}$ uses is simply *feasibility*. However, in many local ratio algorithms the property $\mathcal{P}$ is *minimality*, and in this case w is simply called *r-effective*. When $\mathcal{P}$ is satisfies by every solutions w is sometimes called *fully r-effective*.

It turns out that in many cases it is convenient to use algorithms that are based on weight decomposition. This is especially true when the local ratio advancement step includes more than a constant number of elements, or when w_1 is well behaved for solutions that satisfy a certain property (usually, minimal solutions) and not for every solution.

A typical local-ratio r-approximation algorithm for a covering problem is recursive, and works as follows. Given a problem instance with a weight function w, we find a non-negative weight function $w_1 \leq w$ such that (i) every minimal solution is r-approximate with respect to w_1, and (ii) there exists some index j for which $w(j) = w_1(j)$. We subtract w_1 from w, and remove some zero weight element from the problem instance. Then, we recursively solve the new problem instance. If the solution returned is infeasible the above mentioned element is added to it. This way we make sure that the solution is minimal with respect to the current instance. Since the solution for the current instance is r-approximate with respect to both w_1 and $w - w_1$, it is also r-approximate with respect to w by the Local-Ratio Theorem. The base of the recursion occurs when the problem instance has degenerated into an empty instance.

We demonstrate these ideas by presenting an $s_{\max}$-approximation algorithm for the generalized hitting set problem. The algorithm is taken from [11] and is called Algorithm **LR-GHS**. For purposes of conciseness we represent each set S by an element u_S that is contained in S and whose weight is $w(u_S) \triangleq w(S)$. Recall that, this way, we can pay $w(S)$ for the element u_S instead of paying the tax $w(S)$ for not hitting S.

Note that the main difference between Algorithms **Recursive-LR-HS** and **LR-GHS** is the fact that first simply adds an element to the solution found by the recursive call (Line 8), while the latter adds the element only in case the solution returned by the recursive call is infeasible without it (Lines 8-12). As we shall see this modification makes sure that the solution returned is not only feasible but also minimal.

Algorithm 6 - LR-GHS(U, S, w): a local ratio s_{max}-approximation algorithm for generalized hitting set

---

1: **if** $S = \emptyset$ **then**
2:    Return $\emptyset$
3: **end if**
4: Let $S \in S$
5: $v \leftarrow \mathrm{argmin}_{u \in S} \{w(u)\}$
6: $\varepsilon \leftarrow w(v)$
7: Define $w_1(u) = \begin{cases} \varepsilon & u \in S, \\ 0 & \text{otherwise.} \end{cases}$
8: $H' \leftarrow$ **LR-GHS**$(U \setminus \{v\}, S \setminus S(v), w - w_1)$
9: $H \leftarrow H'$
10: **if** H is not feasible **then**
11:    $H \leftarrow H \cup \{v\}$
12: **end if**
13: Return H

We show that Algorithm **LR-GHS** computes minimal solutions. The proof is by induction on the recursion. At the recursion basis the solution returned is the empty set, which is both feasible and minimal. For the inductive step, we show that $H \setminus \{u\}$ is not feasible for every $u \in H$. First, if $H = H'$, then H is minimal since H' is minimal with respect to $(U \setminus \{v\}, S \setminus S(v))$ by the inductive hypothesis. Consider the case where $H = H' \cup \{v\}$. If $u \neq v$ and $H \setminus \{u\}$ is feasible, then $H' \setminus \{u\}$ is feasible with respect to $(U \setminus \{v\}, S \setminus S(v))$, and therefore H' is not minimal in contradiction to the inductive hypothesis. Also, observe that v is added to H only if H' is not feasible.

It remains to show that Algorithm **LR-GHS** returns s_{max}-approximate solutions. The proof is by induction on the recursion. In the base case the solution returned is the empty set, which is optimal. For the inductive step, H' is s_{max}-approximate with respect to $(U \setminus \{v\}, S \setminus S(v))$, and w_2 by the inductive hypothesis. Since $w_2(v) = 0$, H is also s_{max}-approximate with respect to $(U \setminus \{v\}, S \setminus S(v))$, and w_2. Due to Observation 6, and the fact that H is minimal, it is also s_{max}-approximate with respect to w_1. Thus by the Local Ratio Theorem H is s_{max}-approximate with respect to w as well.

5   The Evolution of Both Methods

In this section we follow the evolution of both primal-dual and local ratio.

5.1   Applications to Various Problems

Covering problems and minimal solutions. During the early 1990's the primal-dual schema was used extensively to design and analyze approximation algorithms for *network design problems*, such as the *Steiner tree problem* (see, e.g., [30, 1, 21]). In fact, this line of research has introduced the idea of using minimal solutions to the primal-dual schema. Subsequently, several primal-dual approximation frameworks were proposed. Goemans and Williamson [22] presented a generic primal-dual approximation algorithm based on the hitting set problem. They showed that it can be used to explain many classical (exact and approximation) algorithms for special cases of the *hitting set* problem, such as *shortest path*, *minimum spanning tree*, and *minimum Steiner forest*. Following [21], Bertsimas and Teo [15] proposed a primal-dual framework for covering problems. As in [22] this framework enforces the primal complementary slackness conditions while relaxing the dual conditions. However, in contrast to previous studies, Bertsimas and Teo [15] express each advancement step as the construction of a single valid inequality, and an increase of the corresponding dual variable (as opposed to an increase of several dual variables).

About ten years after the birth of the local ratio technique [9], Bafna et al. [3] extended the technique in order to construct a 2-approximation algorithm for the *feedback vertex set problem*. Their algorithm was the first local ratio algorithm that used the notion of minimal solutions. Following Bafna et al. [3], Fujito [19] presented a generic local ratio algorithm for a certain family of node deletion problems. Later, Bar-Yehuda [6] presented a local ratio framework for covering problems, which extends the one in [19] and can be used to explain many known optimization and approximation algorithms for covering problems.

Other minimization problems. Both primal-dual and local ratio were also applied to non-covering minimization problems. For example, in [11] the local ratio technique was used in the design of a 2-approximation algorithm for bounded integer programs with two variables per constraint. Recently, Guha et al. [23] presented a primal-dual 2-approximation algorithm to the *capacitated vertex cover problem*.

Jain and Vazirani [27] presented a 3-approximation algorithm for the *metric uncapacitated facility location problem*. Their algorithm was the first primal-dual algorithm that approximated a problem whose LP formulation contains inequalities with negative coefficients. However, this algorithm deviates from the primal-dual schema. Their algorithm does not employ the usual mechanism of relaxing the dual complementary slackness conditions, but rather it relaxes the *primal* conditions. (Note that this algorithm has a non-LP interpretation in the spirit of local ratio [18].)

Bar-Yehuda and Rawitz [12] presented local ratio interpretations of known algorithms for *minimum s-t cut* and the *assignment problem*. These algorithms are the first applications of local ratio to use negative weights. The corresponding primal-dual analyses are based on new IP formulations of these fundamental problems that contain negative coefficients.

Maximization problems. By the turn of the 20th century both methods were used extensively in the context of minimization algorithms. However, there was no application of either method that approximated a maximization problem. The first study to present a local-ratio and primal-dual approximation algorithm for a maximization problem was by Bar-Noy et al. [4]. In this paper the authors used the local-ratio technique to develop an approximation framework for resource allocation and scheduling problems. A primal-dual interpretation was also presented.

Bar-Noy et al.'s [4] result paved the way for other studies dealing with maximization problems. In [5] Bar-Noy et al. developed approximation algorithms for two variants of the problem of scheduling on identical machines with *batching*. Akcoglu at al. [2] presented approximation algorithms for several types of *combinatorial auctions*.

5.2   Equivalence Between the Two Methods

It has often been observed that primal-dual algorithms have local ratio interpretations, and vice versa. Bar-Yehuda and Even's primal-dual algorithm for *hitting set* [8] was analyzed using local ratio in [9]. The local ratio 2-approximation algorithm for *feedback vertex set* by Bafna et al. [3] was interpreted within the primal-dual schema [16]. The 2-approximation of a family of *network design* problems by Goemans and Williamson [21] was explained using local ratio in [6] (see also [7]). And, finally, Bar-Noy et al.'s [4] approximation framework for resource allocation and scheduling problems was developed initially using the local-ratio approach, and then explained it (in the same paper) in primal-dual terms. Thus, over the years there was a growing sense that the two seemingly distinct approaches share a common ground, but the exact nature of the connection between them remained unclear (see, e.g., [33], where this was posed as an open question). The issue was resolved in a paper by Bar-Yehuda and Rawitz [13], in which two approximation frameworks are defined, one encompassing the primal-dual schema, and the other encompassing the local ratio technique, and showed that these two frameworks are equivalent.

The equivalence between the paradigms is based on the simple fact that increasing a dual variable by ε is equivalent to subtracting the weight function obtained by multiplying the coefficients of the corresponding primal constraint by ε from the primal objective function. In other words, this means that an r-effective inequality can be viewed as an r-effective weight function and vice versa. For example, the coefficients of the generalized vertex cover constraint $\sum_{u \in S} x_u + x_S \geq 1$ are the same as the coefficients of the weight function w_1 from Algorithm **LR-GHS** up to a multiplitive factor of ε. Furthermore, both primal-dual analysis of Algorithm **PD-GHS** and local ratio analysis of Algorithm **LR-GHS** are based on Observation 6. The equivalence between the methods is constructive, meaning that an algorithm formulated within one paradigm can be translated quite mechanically to the other paradigm.

5.3   Fractional Local Ratio and Fractional Primal-Dual

The latest important development in the context of local ratio is a new variant of the local ratio technique called *fractional local ratio* [10]. As we have seen, a typical local ratio algorithm is recursive, and it constructs, in each recursive call, a new weight function w_1. In essence, a local ratio analysis consists of comparing, at each level of the recursion, the solution found in that level to an optimal solution for the problem instance passed to that level, where the comparison is made with respect to w_1. Thus, different optima are used at different recursion levels. The superposition of these "local optima" may be significantly worse than the "global optimum," i.e., the optimum of the original problem instance. Conceivably, we could obtain a better bound if at each level of the recursion we approximated the weight of a solution that is optimal with respect to the original weight function. This is the idea behind the fractional local ratio approach. More specifically, a fractional local ratio algorithm uses a single solution x^* to the original problem instance as the yardstick against which all intermediate solutions (at all levels of the recursion) are compared. In fact, x^* is not even feasible for the original problem instance but rather for a relaxation of it. Typically, x^* will be an optimal fractional solution to an LP relaxation of the problem.

Recently, Bar-Yehuda and Rawitz [14] have shown that the fractional approach extends to the primal-dual schema as well. As in fractional primal-dual the first step in a *fractional primal-dual* r-approximation algorithm is to compute an optimal solution to an LP relaxation of the problem. Let P be the LP relaxation, and let x^* be an optimal solution of P. Next, as usual in primal-dual algorithms, the algorithm produces an integral primal solution x and a dual solution y, such that r times the value of y bounds the weight of x (we use minimization terms). However, in contrast to other primal-dual algorithms, y is not a solution to the dual of P. The algorithm induces a new LP, denoted by P', that has the same objective function as P, but contains inequalities that may not be valid with respect to the original problem. Nevertheless, we make sure that x^* is a feasible solution of P'. The dual solution y is a feasible solution of the dual of P'. The primal solution x is r-approximate, since the optimum value of P' is not greater than the optimum value of P.

5.4   Further Reading

A survey that describes the primal-dual schema and several recent extensions of the primal-dual approach is given in [33]. A detailed survey on the local ratio technique (including fractional local ratio) is given in [7].

Acknowledgment

We thank Oded Goldreich and Jonathan Laserson for their proof reading and suggestions.

References

1. A. Agrawal, P. Klein, and R. Ravi. When trees collide: An approximation algorithm for the generalized Steiner problem on networks. *SIAM Journal on Computing*, 24(3):440–456, 1995.

2. K. Akcoglu, J. Aspnes, B. DasGupta, and M.-Y. Kao. Opportunity cost algorithms for combinatorial auctions. In E. J. Kontoghiorghes, B. Rustem, and S. Siokos, editors, *Applied Optimization: Computational Methods in Decision-Making Economics and Finance*. Kluwer Academic Publishers, 2002.

3. V. Bafna, P. Berman, and T. Fujito. A 2-approximation algorithm for the undirected feedback vertex set problem. *SIAM Journal on Discrete Mathematics*, 12(3):289–297, 1999.

4. A. Bar-Noy, R. Bar-Yehuda, A. Freund, J. Naor, and B. Shieber. A unified approach to approximating resource allocation and schedualing. *Journal of the ACM*, 48(5):1069–1090, 2001.

5. A. Bar-Noy, S. Guha, Y. Katz, J. Naor, B. Schieber, and H. Shachnai. Throughput maximization of real-time scheduling with batching. In *13th Annual ACM-SIAM Symposium on Discrete Algorithms*, pages 742–751, 2002.

6. R. Bar-Yehuda. One for the price of two: A unified approach for approximating covering problems. *Algorithmica*, 27(2):131–144, 2000.

7. R. Bar-Yehuda, K. Bendel, A. Freund, and D. Rawitz. Local ratio: a unified framework for approximation algorithms. *ACM Computing Surveys*, 36(4):422–463, 2004.

8. R. Bar-Yehuda and S. Even. A linear time approximation algorithm for the weighted vertex cover problem. *Journal of Algorithms*, 2(2):198–203, 1981.

9. R. Bar-Yehuda and S. Even. A local-ratio theorem for approximating the weighted vertex cover problem. *Annals of Discrete Mathematics*, 25:27–46, 1985.

10. R. Bar-Yehuda, M. M. Halldórsson, J. Naor, H. Shachnai, and I. Shapira. Scheduling split intervals. In *13th Annual ACM-SIAM Symposium on Discrete Algorithms*, pages 732–741, 2002.

11. R. Bar-Yehuda and D. Rawitz. Efficient algorithms for bounded integer programs with two variables per constraint. *Algorithmica*, 29(4):595–609, 2001.

12. R. Bar-Yehuda and D. Rawitz. Local ratio with negative weights. *Operations Research Letters*, 32(6):540–546, 2004.

13. R. Bar-Yehuda and D. Rawitz. On the equivalence between the primal-dual schema and the local ratio technique. *SIAM Journal on Discrete Mathematics*, 2005. To appear.

14. R. Bar-Yehuda and D. Rawitz. Using fractional primal-dual to schedule split intervals with demands. In *13th Annual European Symposium on Algorithms*, 2005. To appear.

15. D. Bertsimas and C. Teo. From valid inequalities to heuristics: A unified view of primal-dual approximation algorithms in covering problems. *Operations Research*, 46(4):503–514, 1998.

16. F. A. Chudak, M. X. Goemans, D. S. Hochbaum, and D. P. Williamson. A primal-dual interpretation of recent 2-approximation algorithms for the feedback vertex set problem in undirected graphs. *Operations Research Letters*, 22:111–118, 1998.

17. V. Chvátal. A greedy heuristic for the set-covering problem. *Mathematics of Operations Research*, 4(3):233–235, 1979.

18. A. Freund and D. Rawitz. Combinatorial interpretations of dual fitting and primal fitting. In *1st International Workshop on Approximation and Online Algorithms*, volume 2909 of *LNCS*, pages 137–150. Springer-Verlag, 2003.

19. T. Fujito. A unified approximation algorithm for node-deletion problems. *Discrete Applied Mathematics and Combinatorial Operations Research and Computer Science*, 86:213–231, 1998.

20. M. R. Garey and D. S. Johnson. *Computers and Intractability; A Guide to the Theory of NP-Completeness*. W.H. Freeman and Company, 1979.

21. M. X. Goemans and D. P. Williamson. A general approximation technique for constrained forest problems. *SIAM Journal on Computing*, 24(2):296–317, 1995.

22. M. X. Goemans and D. P. Williamson. The primal-dual method for approximation algorithms and its application to network design problems. In Hochbaum [25], chapter 4, pages 144–191.

23. S. Guha, R. Hassin, S. Khuller, and E. Or. Capacitated vertex covering. *Journal of Algorithms*, 48(1):257–270, 2003.

24. D. S. Hochbaum. Approximation algorithms for the set covering and vertex cover problems. *SIAM Journal on Computing*, 11(3):555–556, 1982.

25. D. S. Hochbaum, editor. *Approximation Algorithms for NP-Hard Problem*. PWS Publishing Company, 1997.

26. K. Jain, M. Mahdian, E. Markakis, A. Saberi, and V. Vazirani. Greedy facility location algorithms analyzed using dual-fitting with factor-revealing LP. *Journal of the ACM*, 50(6):795–824, 2003.

27. K. Jain and V. V. Vazirani. Approximation algorithms for metric facility location and k-median problems using the primal-dual schema and Lagrangian relaxation. *Journal of the ACM*, 48(2):274–296, 2001.

28. H. Karloff. *Linear Programming*. Progress in Theoretical Computer Science. Birkhäuser Boston, 1991.

29. C. H. Papadimitriou and K. Steiglitz. *Combinatorial Optimization; Algorithms and Complexity*. Prentice-Hall, Inc., 5th edition, 1982.

30. R. Ravi and P. Klein. When cycles collapse: A general approximation technique for constrained two-connectivity problems. In *3rd Conference on Integer Programming and Combinatorial Optimization*, pages 39–56, 1993.

31. A. Schrijver. *Theory of linear and integer programming*. Wiley Publishers, 1998.

32. V. V. Vazirani. *Approximation algorithms*. Springer-Verlag, Berlin, 2nd edition, 2001.

33. D. P. Williamson. The primal dual method for approximation algorithms. *Mathematical Programming (Series B)*, 91(3):447–478, 2002.

Dinitz' Algorithm:
The Original Version and Even's Version

Yefim Dinitz

Dept. of Computer Science, Ben-Gurion University of the Negev,
Beer-Sheva 84105, Israel
dinitz@cs.bgu.ac.il

Abstract. This paper is devoted to the max-flow algorithm of the author: to its original version, which turned out to be unknown to non-Russian readers, and to its modification created by Shimon Even and Alon Itai; the latter became known worldwide as "Dinic's algorithm". It also presents the origins of the Soviet school of algorithms, which remain unknown to the Western Computer Science community, and the substantial influence of Shimon Even on the fortune of this algorithm.

1 Introduction

The reader may be aware of the so called "Dinic's algorithm" [4][1], which is one of the first (strongly) polynomial max-flow algorithms, while being both one of the easiest to implement and one of the fastest in practice. This introduction discusses two essential influences on its fortune: the first supported its invention, and the other furthered its publicity, though changing it partly.

The impact of the late Shimon Even on Dinic's algorithm dates back to 1975. You may ask how a person in Israel could influence something in the former USSR, when it was almost impossible both to travel abroad from the USSR and to publish abroad or even to communicate with the West? This question will clear up after first considering the impact made by the early Soviet school of computing and algorithms.

The following anecdote sheds some light on how things were done in the USSR. Shortly after the "iron curtain" fell in 1990, an American and a Russian, who had both worked on the development of weapons, met. The American asked: "When you developed the Bomb, how were you able to perform such an enormous amount of computing with your weak computers?". The Russian responded: "We used better algorithms."

This was really so. Russia had a long tradition of excellence in Mathematics. In addition, the usual Soviet method for attacking hard problems was to combine pressure from the authorities with people's enthusiasm. When Stalin decided to develop the Bomb, many bright mathematicians, e.g., Izrail Gelfand and my first Math teacher, Alexander Kronrod, put aside their mathematical studies and delved deeply into the novel area of computing. They have assembled teams

---

[1] In this paper, the references are given in chronological order.

O. Goldreich et al. (Eds.): Shimon Even Festschrift, LNCS 3895, pp. 218–240, 2006.

of talented people, and succeeded. The teams continued to grow and work on the theory and practice of computing.

The supervisor of my M.Sc. thesis was George Adel'son-Vel'sky, one of the fathers of Computer Science. Among the students in his group at that time were M. Kronrod (one of the future "Four Russians", i.e. the four authors of [3]), A. Karzanov (the future author of the $O(n^3)$ network flow algorithm [9]) and other talented school pupils of A. Kronrod. This was in 1968, long after the Bomb project had been completed. The work on the foundations of the chess program "Kaissa", created by members of A. Kronrod's team under the guidance of Adel'son-Vel'sky, was almost finished; "Kaissa" won the first world championship in 1974. Adel'son-Vel'sky's new passion became discrete algorithms, which he felt had a great future.

The fundamental contribution of Adel'son-Vel'sky to computer science was AVL-trees. He (AV) and Eugene Landis (L) published the paper [1] about AVL-trees, which consists of just a few pages. Besides solving an important problem, it presented a bright approach to data structure maintenance. While this approach became standard in the USSR, it was still unknown in the West. No reaction immediately followed their publication, until several years later another paper, 15 pages long, was published by a researcher (unknown to me), who understood how AVL-trees work and explained this to the Western community, in its own language. Since then, AVL-trees and the data structure maintenance approach became corner-stones of computer science.

We, Adel'son-Vel'sky's students, absorbed the whole paradigm of the Soviet computing school from his lectures. This paradigm consisted of eagerness to develop economical algorithms based on the deep investigation of a problem and on the use of smart data structure maintenance and amortized running time analysis as necessary components. All this became quite natural for us in 1968, seventeen years before the first publication in the West on amortized analysis by R. Tarjan [21]. Hence, it was not surprising that my network flow algorithm, invented in January 1969, improved the Ford&Fulkerson algorithm by using and maintaining a layered network data structure and employing a delicate amortized analysis of the running time.

However, at that time, such an approach was very unusual in the West. Shimon Even and (his then Ph.D. student) Alon Itai at the Technion (Haifa) were very curious and intrigued by the two new network flow algorithms: mine [4] and that of Alexander Karzanov [9] in 1974. It was very difficult for them to decipher these two papers (each compressed into four pages, to meet the page restriction of the prestigious journal Doklady). However, Shimon Even was not used to giving up. After a three-day long effort, Even and Itai understood both papers, except for the layered network maintenance issue. The gaps were spanned by using Karzanov's concept of blocking flow (which was implicit in my paper), and by a beautiful application of DFS for finding each augmenting path.

It is well known that Shimon Even was an excellent lecturer. During the next couple of years, Even presented "Dinic's algorithm" in lectures, which he gave in many leading research universities of the West. The result was important, the idea was fresh, the algorithm was very nice, and the combination of BFS for constructing the layered network and DFS for operating it was fascinating. "Dinic's algorithm" was a great success and gained a place in the annals of the computer science community. Hardly anyone was aware that the algorithm, taught in many universities since then, is not the original version, and that a considerable part of the beauty of its known version—combining BFS and DFS—was due to the contribution of Even and Itai. Also, its name was rendered incorrectly as [dinik] instead of [dinits].

The original paper, published in a Soviet journal, was not understood also by others in the West. This algorithm and many other achievements of the Moscow school of algorithms in the network flow area were published as a book [10]. This book was well known all over the former USSR, and students of leading Soviet universities studied its contents as an advanced course. After more than 15 years, the book appeared in the West (in Russian) and was reviewed in English by A. Goldberg and D. Gusfield [25]. A few years later, I finally explained the original version of my algorithm, as published in [4], to Shimon Even.

The rest of the paper is devoted to the description of my algorithm and its various versions and to the development of max-flow algorithms following it. The presentation is a bit didactic; it shows that serious and devoted work towards maximal understanding of a phenomenon and the best implementation of algorithms may bring important, unexpected fruits.

2   The Original Dinitz' Algorithm

2.1   The Max-Flow Problem and the Ford&Fulkerson Algorithm

Max-Flow Problem Let us recall the max-flow problem definition. A capacitated directed graph $G = (V, E, c)$, where c is a non-negative capacity function on edges, with vertices s distinguished as the source and t as the sink is given. A flow is defined as a function on the (directed) edges, f, satisfying the following two laws:

- *Capacity constraint*: $\forall e \in E : 0 \leq f(e) \leq c(e)$, and
- *Flow conservation*: $\forall v \in V \setminus \{s, t\} : \sum_{(v,u)\in E} f(v, u) - \sum_{(u,v)\in E} f(u, v) = 0$.

The quantity $\sum_{(v,u)\in E} f(v, u) - \sum_{(u,v)\in E} f(u, v)$ is called the net flow from v. The value of a flow f is defined as the net flow from s (which is equal to the negated net flow from t). The task is to find a flow of the maximal value, called "maximum flow" (see Figure 1a for illustration). Initially, some feasible flow function, f_0, is given, which is by default the zero function.

Another equivalent is formed by considering a flow as a *skew-symmetric* function on the pairs of vertices connected by at least one edge in the given graph. For such a pair v, u, if one of the edges between them is absent, in E, we add it to E with capacity and flow zero (clearly, the problem remains equivalent). Given any feasible flow function, f, we define the new function $\bar{f}$ by setting

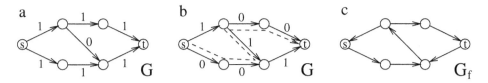

Fig. 1. (a) A flow network G and a maximum flow in it, of value 2 (the numbers denote flows in edges; all edge capacities are 1). (b) Another flow, f, in same network, of value 1. In terms of Sections 3 and 5, f is "maximal" and "blocking", resp. The dashed line shows the unique augmenting path, P, which *decreases* the flow on the diagonal edge. The residual capacity $c_f(P)$ is 1, and pushing the flow along P results in the above maximum flow. (c) The residual network G_f, where the residual capacity of all edges is 1. The path as in (b) is the unique path from s to t in G_f.

$\bar{f}(v, u) = f(v, u) - f(u, v)$ for any edge (v, u) in the (extended) E. This represents the total flow from v to u on the pair of edges (v, u) and (u, v) together. It is easy to see that $\bar{f}$ satisfies the following three constraints:

- *Capacity constraint*: $\forall (v, u) \in E : \bar{f}(v, u) \leq c(v, u)$,
- *Skew symmetry*: $\forall (v, u) \in E : \bar{f}(u, v) = -\bar{f}(v, u)$, and
- *Flow conservation*: $\forall v \in V \setminus \{s, t\} : \sum_{(v, u) \in E} \bar{f}(v, u) = 0$.

The net flow from v is defined then as $\sum_{(v, u) \in E} \bar{f}(v, u)$. To retrieve f from $\bar{f}$ use the formula $f(v, u) = \max(0, \bar{f}(v, u))$; it results in a feasible flow function according to the original definition. Both transformations keep the flow value. Hence, using any one of them may be considered legal.

We say that an edge (v, u) is *saturated*, if $\bar{f}(v, u)$ is equal to $c(v, u)$. We call the difference $c(v, u) - \bar{f}(v, u)$ the *current* or *residual* capacity of the edge (v, u), and denote it by $c_f(v, u)$. In the original flow definition terms, the property of being saturated is equivalent to the combined property: $f(v, u) = c(v, u)$ & $f(u, v) = 0$, while the residual capacity is equal to $c(v, u) - f(v, u) + f(u, v)$. The meaning of residual capacity is "how much may be added to the flow from v to u on edges (v, u) and (u, v) together"; the formulae use the essential equivalence between increasing the flow in (v, u) and decreasing it in (u, v). Both flow representations are widely known; while Ford and Fulkerson introduced the first one in [2], some of the popular textbooks, e.g. [23], use the second one.

In what follows, we use the skew-symmetric form of flows, where the concepts of saturated edge and residual capacity are easier to handle. Nevertheless, for simplicity, we denote flows by regular Latin letters, e.g. f, not $\bar{f}$, and in explanations, we refer $f(v, u)$ as the flow on edge (v, u).

Ford&Fulkerson's Algorithm The classic Ford&Fulkerson algorithm [2] finds a maximum flow using the following natural idea: augmenting the current flow iteratively, by pushing an additional amount of flow on a single source-sink path at each iteration, as long as possible. Let f denote the current flow. A path from

s to t is sought, such that none of its edges is saturated by f; such a path is called "augmenting". When an augmenting path, P, is found, its current capacity $c_f(P) = \min_{e \in P} c_f(e) > 0$ is computed. Then, the flow on every edge of P is increased by $c_f(P)$ (of course, the corresponding update of the skew-symmetric flow values at the opposite edges—decreasing them by $c_f(P)$—is made as well; we will not mention this, in what follows). As a result of this iteration, the flow value grows by $c_f(P)$ and at least one edge of P becomes saturated. For illustration see Figure 1b.

For convenience sake, let us define the *residual network* $G_f = (V, E_f)$, where E_f consists of all unsaturated edges in E. Now, an augmenting path is simply any path in G_f from s to t (see Figure 1c). Clearly, it is easy to find an augmenting path, if exists, given G and f: just run any search algorithm (e.g. BFS or DFS) on G_f, from s. It is easy to see that, after each iteration, using an augmented path P, the change of the residual network is as follows: it loses all the edges of P which became saturated and acquires all the edges *opposite* to the edges of P that were saturated before the iteration.

The famous theorem of Ford and Fulkerson states that when no augmenting path exists (i.e. t is disconnected from s in G_f), the current flow is maximum. Clearly, if the initial data—capacities and the initial flow—is integer, the current flow remains integer and eventually becomes maximum. That is, the Ford&Fulkerson algorithm (henceforth, *FF*) is finite and pseudo-polynomial in the integer case. For the general case, Ford and Fulkerson provided an example of a network with an execution of FF in it which runs infinitely; moreover, the flow value converges at a *quarter* of the maximal possible one. So, the question of the existence of a polynomial, or even of a finite or converging algorithm, for the general case, remained open. This was settled affirmatively by J. Edmonds and R.M. Karp [5] and by Y. Dinitz [4], independently. The algorithms suggested in these papers are modifications of FF.

2.2 Layered Network Data Structure for Accelerating Iterations

Layered Network The initial intention of the author of [4] was just to accelerate iterations of FF by means of a smart data structure. Recall that all the parts of an iteration of FF, except for finding an augmenting path, P, cost time proportional to the length of P, that is $O(|P|) = O(|V|)$. Indeed, all computations and updates are made along P, while the remainder of the data is not touched. However, the search running on G costs $O(|E|)$; one may even say $\Theta(|E|)$, according to the search algorithms. Notice that $\Theta(|E|)$ is, in general, substantially more than $O(|P|)$ or $O(|V|)$, and is $O(|V|^2)$ in the general case. So, finding an augmenting path is the computational time bottleneck of an iteration of FF.

We will use BFS from s as the network scanning and path search algorithm. Recall that BFS assigns to each vertex reachable from s its distance (the number of edges in the shortest path) from s, and builds a tree rooted at s, where the single edge incoming each vertex, except for s itself, comes from a vertex at the distance less by one. A shortest path from s to t is restored as the chain of such incoming edges, beginning from t. Let us try to save the information achieved at a BFS run for the following iterations. Notice that at least one edge of the BFS tree, lying on the path from s to t, is saturated at the iteration, so it must be erased from the tree. Thus, s and t become disconnected in this tree, and there are no easy means to connect them again using unsaturated edges. For illustration see Figure 2a,b.

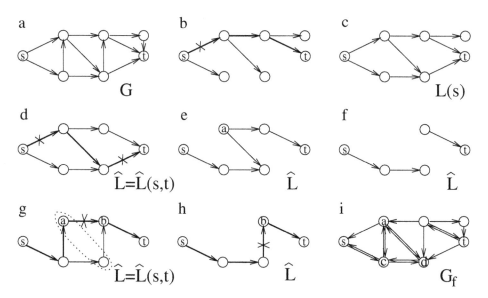

Fig. 2. (a) A network G (edge capacities are not shown). (b) A BFS tree from s in G, with the path from s to t in it shown in bold. The edge that would be saturated as a result of the flow augmentation along it is shown crossed. (c) The network $L(s)$, w.r.t. the zero initial flow. (d) The layered network $\hat{L} = \hat{L}(s,t)$, of length 3, and a path from s to t in it. Two its crossed edges are saturated by the flow augmentation. (e) The network $\hat{L}$ with the saturated edges removed. (f) *RightPass* began with removing dead-end a and the edges outgoing from it. (g) After vanishing of $\hat{L}$, the new $\hat{L} = \hat{L}(s,t)$, w.r.t. the current flow, is of length 4 (its second layer is encircled by a dotted line). The flow augmentation along the path shown in bold saturates the crossed edge. Vertex a will be cleaned from $\hat{L}$. (h) The updated layered network; the augmentation along its single path from s to t saturates the crossed edge. (i) The residual network G_f after vanishing of $\hat{L}$ of length 4. BFS execution in G_f marks as reachable from s vertices s, a, c, and d, but not t. Hence, the current flow is maximum.

Let us try to enrich the data structure while building it. Notice that BFS arranges vertices into layers, according to their distance from s. The BFS tree includes only the *first edge found*, leading to each vertex from the previous layer; it ignores all other edges coming to it from that layer, though they may be no less useful than that first edge. The idea is to keep *all those edges* in our data structure.

We begin with some definitions for an arbitrary digraph H. For an edge (v, u), the vertex v is called its *tail* and u its *head*. Let a source vertex s be given. Let $dist(v, u)$ denote the distance to a vertex u from vertex v; we denote $dist(v) = dist(s, v)$. Let the *ith vertex layer* V_i be the set of vertices with $dist(v) = i$ (where $V_0 = \{s\}$), and the *ith edge layer* E_i be the set of all the edges of H going from V_{i-1} to V_i. We define $L(s)$ as the digraph $(\cup V_i, \cup E_i)$. Notice that a straightforward extension of BFS builds $L(s)$, with the same running time $O(|E|)$ as that of the regular BFS; in what follows we call it "the extended BFS". It is easy to see, by the properties of BFS, that $L(s)$ *is the union of all the vertices and edges of all shortest paths from s in H*. For illustration see Figure 2c.

Since we are interested only in the paths from s to another given vertex, t, let us prune $L(s)$ to $\hat{L}(s, t)$, as in [4]. The network $\hat{L}(s, t)$ contains ℓ layers, where $\ell = dist(s, t)$ is called its *length*. The vertices of its ith layer, $\hat{V}_i$, are characterized by a double property: they are at distance i from s, while t is at distance $\ell - i$ from them. The ith layer of its edges, $\hat{E}_i$, consists of all the edges of H going from $\hat{V}_{i-1}$ to $\hat{V}_i$. The pruning of $L(s)$ to $\hat{L}(s, t)$ is easy: we just run the extended BFS once more, but *on $L(s)$ from t, using the opposite edge direction*. For illustration see Figure 2d.

Claim. The layered sub-graph $\hat{L}(s, t)$ is the union of all the vertices and edges residing on all the shortest paths from s to t.

Proof. For any shortest path from s to t, its ith vertex is at distance i from s, while t is at distance $\ell - i$ from it. Conversely, for any vertex v in $\hat{V}_i$, there is a path from s to v of length i and a path from v to t of length $\ell - i$; their concatenation is a shortest path from s to t going via v. Similar reasons suffice for the edges of $\hat{E}_i$. $\square$

An easy property of $L(s)$ and $\hat{L}(s, t)$ is that they have no "dead-ends": vertices without any incoming edge, except for s (for both $L(s)$ and $\hat{L}(s, t)$) and those with no outgoing edge, except for t (for $\hat{L}(s, t)$ only). Due to the layered structure and to this property, finding a shortest path from s to t, given any one of $L(s)$ and $\hat{L}(s, t)$, is almost as simple and exactly as fast (in time linear in its length) as using the BFS tree, by means of the following procedure:

PathFinding: Starting from t, repeatedly choose one of the incoming edges and pass to its tail. Necessarily, after ℓ steps the layer 0, that is s, is reached. The path consisting of the chosen edges (in the reverse order), is a shortest path from s to t.

Remark: Notice that given $\hat{L}(s, t)$, a similar procedure executed from s, in the direction of edges, also constructs a path from s to t, but in its natural order.

Layered Network Maintenance Suppose that, given a flow network G and a flow f in it, we had built the sub-graph $\hat{L}(s, t)$ of the residual network G_f (by two extended BFSs). Having found an augmenting path, by *PathFinding* in $\hat{L}(s, t)$, we pushed some amount of flow along it (as in FF). A natural idea is to find the next augmenting path, somehow using the existing $\hat{L}(s,t)$. We accomplish this by adjusting it to the new flow.

Throughout the algorithm, we use a *layered network* data structure. It is a general digraph, whose vertex set consists of consequent layers, so that the leftmost one is $\{s\}$, while each its edge goes from some layer to the next one; $L(s)$ and $\hat{L}(s,t)$ are particular examples of a layered network. We begin the algorithm by initializing our data structure $\hat{L}$ as $\hat{L}(s,t)$ of the residual network w.r.t. the initial flow. The maintenance of $\hat{L}$, after an iteration, using an augmenting path P, is as follows. We first remove from $\hat{L}$ all the edges of P that became saturated; clearly, we can extend the flow changing procedure of FF to provide a list of such edges, Sat, keeping its running time $O(\ell)$. Notice that updated $\hat{L}$ is contained in the current residual network, since only the edges, such as above, disappeared from the residual network as a result of the flow change. Therefore, any path from s to t found in the updated $\hat{L}$ would be an augmenting path, w.r.t. the current flow. For illustration see Figure 2e.

Observe further that applying *PathFinding* to updated $\hat{L}$ might be stuck at a dead-end vertex with no incoming edges. In order to restore the original property of the layered network, to have no dead-ends, let us apply to it the following procedure **Cleaning**. We initialize the left queue of edges Q^l and right queue of edges Q^r by the list of saturated edges Sat. The main loop of *Cleaning* consists of the right and left passes, processing edges in Q^r and Q^l, respectively, one by one (in an arbitrary order), as follows:

RightPass For each edge $e \in Q^r$, if its right end-vertex has no incoming edges, then it is deleted from $\hat{L}$, together with all its outgoing edges, while inserting those edges into Q^r (for illustration see Figure 2e,f.)

LeftPass For each edge $e \in Q^l$, if its left end-vertex has no outgoing edges, then it is deleted from $\hat{L}$, together with all its incoming edges, while inserting those edges into Q^l (for illustration see Figure 2g,h.)

Notice that any deleted element of $\hat{L}$ is absent in all paths from s to t in the current $\hat{L}$; in another words, the above procedure does not destroy any path from s to t in $\hat{L}$. If *Cleaning* empties $\hat{L}$, it reports on its *vanishing*.

It is easy to see that the cleaned $\hat{L}$ contains no dead-ends (for illustration see Figure 2g,h.) Indeed, a vertex becomes a dead-end when its last incoming or last outgoing edge is removed from $\hat{L}$; thus, any such event should be detected when processing that edge and that vertex should be removed from $\hat{L}$ during *Cleaning*. Now, assuming that *Cleaning* has not caused $\hat{L}$ to vanish, *PathFinding* in the cleaned $\hat{L}$ should be executed without any problem. This allows us to find an augmenting path and to execute the entire next iteration of FF once more in $O(\ell)$ time. Notice that this time includes neither the time of building $\hat{L}$, nor of its cleaning. This is intentional; it is quite usual to count the cost of initializing and maintaining a data structure separately.

Such accelerated iterations are executed until $\hat{L}$ vanishes. The part of the algorithm, beginning from the building of a layered network and ending at its vanishing point, is called a *phase*. The algorithm consists of consequent phases, repeated until the next layered network construction reveals that it cannot be built. This happens when the current residual network G_f contains no path from s to t. In this case, by the Ford&Fulkerson theorem, the current flow f is maximum, i.e. the problem is solved. For illustration see Figure 2.

In what follows, we refer to the suggested algorithm as the *Dinitz algorithm*, or *DA*. Its general description is as follows (the iteration invariant is discussed and proved in Section 2.3):

The Original Dinitz Algorithm

Input:
    a flow network $G = (V, E, c, s, t)$,
    a feasible flow f, in G (equal to zero, by default).

```
/* Phase Loop: */
    dowhile
    begin
        Build L̂(s,t) in G_f, using the extended BFS;
        if L̂(s,t) = ∅ then return f
        else L̂ ← L̂(s,t);
        /* Iteration Loop: */
        while L̂ is not empty do
        /* Iteration Invariant: L̂ is the union of all shortest augmenting paths
*/
        begin
            P ← PathFinding(L̂);
            Sat ← FlowChange(P);
            /* Cleaning(L̂): */
            begin
                Removal of edges in Sat;
                Q^r, Q^l ← Sat;
                RightPass(Q^r);
                LeftPass(Q^l);
            end;
        end;
    end;
```

2.3   Algorithm Analysis

Two aspects of DA have to be analyzed:

- How much does it cost to maintain the layered network?
- How many iterations are contained in a phase? How many phases are contained in the algorithm?

Layered Network Maintenance Cost It is easy to construct an extremal example, where the entire layered network $\hat{L} = \hat{L}(s,t)$ of size $O(|E|)$ vanishes after a single iteration. For example, if the first edge layer consists of just a single edge from s, of the minimal capacity, then the first executed iteration saturates this edge and thus disconnects t from s.

We use the amortized analysis for bounding the total cost of *all* cleanings during a single phase. Let us charge the cost of all relevant maintenance operations to *removed elements* of $\hat{L}$, as follows. A removed *edge* pays for the operations applied to it, to the total of $O(1)$, and for checking its two end-vertices; note that checking whether a list it is empty or not costs $O(1)$, as well. A removed *vertex* pays for the operations applied to itself, except for checking as above, and for the arrangement of the loop to remove its incident edges (but not for iterations of that loop), which also totals $O(1)$. Therefore, the overall maintenance cost during a phase is $O(|E| + |V|)$, which is $O(|E|)$, since $\hat{L}$ is connected.

Running Time w.r.t. Iterations and Phases Recall that the construction of the layered network, when initializing a phase, costs $O(|E|)$. Hence, the total cost of all layered network data structure operations, except for *PathFinding*, is $O(|E|)$ per phase. Recall that the total cost of an accelerated iteration is $O(\ell) = O(|V|)$, where ℓ is the length of the layered network.

Let us denote the total number of iterations of DA by $\#it$ and the number of phases by $\#ph$ (it might be infinite, in general). Then, the total running time of DA is bounded as $O(\#it \cdot |V| + \#ph \cdot |E|)$. First, this is better than the bound $O(\#it \cdot |E|)$ of FF. Indeed, DA is much faster than FF, even if FF uses only shortest augmenting paths (see in [13] the results of an experiment comparing them, among other max-flow algorithms). Second, the essential structure of FF, when accelerated, remains exactly the same, so neither finiteness proofs, nor bounds for the number of iterations ever proved for FF suffer from the acceleration.

However, we still have no provable reason to consider DA be *theoretically* faster than FF. Indeed, there might be just a single iteration during a phase, in the worst case, and there is no provable evidence that the average number of iterations per phase $\frac{\#it}{\#ph}$ is higher than $\Omega(1)$. To summarize for the time being, DA seems to be just helpful heuristics for FF.

The Maintenance of the Layered Network Is Perfect Let us continue analyzing the data structure, presenting its purity and beauty. Let us call a method of data structure maintenance *perfect* if, before every iteration, the data structure is as if built from scratch based on the current data. The method chosen in the original paper [4] and described above is such. The following proposition implies straightforwardly that $\hat{L}$ *coincides with* $\hat{L}(s,t)$ *of the current residual network before every iteration of DA.*

Proposition 1. *After any iteration in the phase of DA with the layered network of length ℓ, the updated layered network is the union of all augmenting paths of length ℓ, while there is no shorter augmenting path (w.r.t. the current flow).*

Proof. Let us consider an arbitrary iteration in the phase with the layered network of length ℓ. We relate to the function $d(v)$ on vertices as the distance from

s to v in the residual network *at the beginning of the phase*. In other words, $d(v)$ is the number of the layer to which v was related by the first extended BFS, during the process of the building of $L(s)$.

Recall that, during this phase, the flow on an edge (v, u) is increased only if $d(u) = d(v) + 1$; only those edges may be saturated and thus *removed* from G_f. Also, the flow on an edge (u, v) is decreased only if $d(v) = d(u) - 1$; only those edges may be unsaturated and thus *added* to G_f; we call the added edges "new". An immediate consequence is that the value of function d is defined for any vertex reachable from s in the current residual network G_f (equivalently, any such vertex was reachable from s also at the beginning of the phase). Another easy consequence is that no edge saturated during a certain phase can be unsaturated later during the same phase. Thus, no edge removed from G_f can be restored to it during the same phase.

First, let us consider augmenting paths of length ℓ without new edges (henceforth "old paths"), and prove that they are the same in the updated $\hat{L}$ and in the current G_f (we use the generic notion G_f for the residual network, assuming f changes from iteration to iteration).

Let us consider an arbitrary old path, P, of length ℓ; by construction, it was contained in the layered network $\hat{L}$ at the beginning of the phase. Assume that P is contained in G_f after some iteration. This means that none of its edges were saturated during the phase. So, none of its edges were removed from $\hat{L}$ as saturated. Hence, the existence of P in $\hat{L}$ is self-supporting, since it prevents its nodes from ever being removed from $\hat{L}$ as dead-ends. Therefore, P is contained in the current layered network $\hat{L}$. Now assume that P is contained in the current $\hat{L}$. This means that none of its edges were saturated during the phase. Thus, P is contained in G_f. Summarizing, the layered network and the residual network contain the same old paths of length ℓ. Recall that there exist no old paths of length less than ℓ.

We now prove that the length of any "new" (not old) path, P', is at least $\ell + 2$. Let P' contain k new edges; since P is new, $k \geq 1$. Let us concentrate on the change of function d along P' (recall that d is defined at all its vertices). Its total increase, from $d(s) = 0$ to $d(t) = \ell$, is exactly ℓ. Along any one of its old edges, d increases by at most 1; along any one of its new edges, d decreases by exactly 1 (for illustration see Figure 3). Hence, the number of old edges should be at least $\ell + k$. That is, the length of P' is at least $\ell + 2k \geq \ell + 2$, as required.

Thus, we see that after any iteration of the phase, there is no augmenting path of length less than ℓ, while all the augmenting paths of length ℓ, if any, (are old and) are *contained* in the current $\hat{L}$.

Moreover, since cleaning restores the property of $\hat{L}$ of having no dead-ends, reasoning, as in the proof of Claim in Section 2.2, shows that all the vertices and edges of $\hat{L}$, if any, belong to the paths from s to t of length ℓ. This suffices.   □

At this point, consider the moment after the last iteration of some phase. At that moment, $\hat{L}$ had vanished, so it contains no path from s to t of length ℓ. By Proposition 1, there is no augmenting path of length ℓ or shorter, w.r.t. the current flow. Therefore, the distance from s to t in G_f is at least $\ell + 1$,

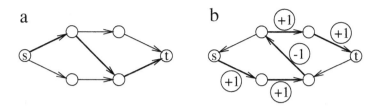

Fig. 3. (a) A layered network of length 3, and an augmenting path in it, shown in bold. (b) A part of the residual network after pushing the flow along this path, and a path from s to t in it, shown bold. The encircled number near an edge shows the change of d along it.

and so is the length of the layered network built at the beginning of the next phase (if the flow is not yet maximum). Thus, as a by-product of the perfect method of maintaining the layered network data structure, we have arrived at the remarkable property of DA that *the length of the layered network grows strictly from phase to phase.*

The Polynomial Time Bound of DA An easy consequence of the foregoing property is that there are at most $|V| - 1$ phases in DA, since the distance from s to t cannot exceed $|V| - 1$.

It is easy to see that the number of iterations during a single phase is at most $|E|$, since at least one out of at most $|E|$ edges of $\hat{L}$ is removed from it at every iteration. Therefore, the total running time at each phase, which consists of the accelerated iteration times and the layered network building and maintenance time, is $O(|E| \cdot |V| + |E|) = O(|V||E|)$. Thus, DA is finite and its total running time is $O(|V|^2|E|)$. We arrive at our main result:

Theorem 1. *The Dinitz algorithm builds a maximum flow in time $O(|V|^2|E|)$.*

Summary Let us summarize what is established on the behavior of DA.

- DA consists of phases. Each one contains iterations changing the flow using shortest augmenting paths of a fixed length, ℓ.
- At the beginning of each phase, the extended BFS builds in $O(|E|)$ time a layered network data structure, $\hat{L}$, of length ℓ. The layered network is constantly maintained during the phase as the union of all shortest augmenting paths of length ℓ, until it vanishes, in total time $O(|E|)$.
- The layered structure of $\hat{L}$ and absence of dead-ends in it allow for the execution of every iteration of FF in $O(\ell) = O(|V|)$ time.
- The layered network is strictly pruned after each iteration of FF. Therefore, the number of iterations at each phase is bounded by $|E|$.
- When the layered network vanishes, there is no augmenting path of length lesser than or equal to ℓ, w.r.t the current flow. Hence, the length of the next layered network, equal to the length of the currently shortest augmenting path, is strictly greater than ℓ.

- Since the length of $\hat{L}$ grows from phase to phase, there are at most $|V| - 1$ phases.
- When DA stops, the current flow is maximum.
- The running time of DA is $O(|V|^2|E|)$.

A Historical Remark In Adel'son-Vel'sky's Algorithms class, the lecturer had a habit of giving the problem to be discussed at the next meeting as an exercise to students. The DA was invented in response to such an exercise. At that time, the author was not aware of the basic facts regarding FF, in particular, of the idea of decreasing the current flow on opposite edges during a flow push along a path. As a consequence from the then naive view, Proposition 1 turned out to be quite natural, following from *just exhausting* the set of all paths of length ℓ, since the issue of "new" augmenting paths had not arisen at all. So, the entire effort of the inventor was devoted to suggesting the best accelerating data structure.

Because of the above gap in knowledge, the first time bound was $O(|V||E|)$, since the total number of iterations had been erroneously considered to be $|E|$, according to the maximal possible number of saturated edges. After learning about the inverse edges of Ford and Fulkerson, the author completed the proof of Proposition 1 and corrected the time bound.

Ignorance sometimes has its merits. Very probably, DA would not have been invented then, if the idea of possible saturated edge desaturation had been known to the author.

3   The Version of Shimon Even and Alon Itai

There are various approaches for using a data structure for a sequence of iterations of an algorithm. The perfect way, as chosen in [4] and described above, is not the goal from the point of view of the algorithm itself. The updated data structure may not be like one built from scratch, but it should work; the only requirement is that the desired bound for the running time can be proved. This was the approach chosen by Shimon Even and Alon Itai. In what follows, we first relax some requirements for the layered network data structure, using the notation defined in Section 2. Then we briefly describe the version of DA of Even and Itai, as it appears in the Even's textbook [14].

Note that using the two-terminal layered network $\hat{L}(s,t)$ is a luxury; the one-terminal layered network $L(s)$ is quite sufficient. Indeed, in order for *PathFinding* to work, a layered network should have no dead-ends in the direction of s (i.e. without incoming edges), while the existence of dead-ends in the direction of t does no harm. Hence, DA may switch to the layered network L, which is initialized by $L(s)$. In fact, the building of $L(s)$ may be stopped upon finishing the ℓth layer.

Moreover, during the update of L, DA may remove dead-ends in the direction of s only; that is, at *Cleaning*, it is sufficient to execute $RightPass(Q^r)$ only. Then, L vanishes when t is removed from it. Notice that the original property of $L(s)$, to be the union of all the shortest paths from s to all the vertices reachable from it, is *not* preserved by such maintenance. However, this is not essential to DA, since it uses the shortest paths to t only. The right invariant of L is to *contain* all the augmenting paths of length ℓ (not to be their union), while there is no shorter augmenting path. Obviously, this property holds after the initializing L at the beginning of a phase. Its maintenance during the phase, from iteration to iteration, is proved like the proof of Proposition 1. Notice that the weaker invariant above implies the property of increasing the layered network's length from phase to phase, as well. Therefore, the analysis of the number of phases, the number of iterations, and the running time of DA, for the version where the one-terminal layered network is used, is the same as in Section 2.3.

Even and Itai go much farther while changing DA. They admit dead-ends in L in both directions; thus, they cancel *Cleaning* and are urged to give up using *PathFinding*. They even "burn their bridges behind them" by finding a path from s to t in L beginning from s. They suggest another way to find an augmenting path: a search of the DFS type, combined with *encountered* dead-ends removal. Each iteration of their version of DA is as follows.

Their DFS begins to build a path from s, incrementing it edge by edge, from one layer to the next. Such path building stops either by arrival at t (a success), or at another dead-end. In the last case, DFS backtracks to the tail of the edge leading to that dead-end, while removing that edge from L as useless (i.e. not contained in any path from s to t in L). After that, DFS continues as above, incrementing the current path when possible or backtracking and removing the last edge from L, otherwise. This process ends either when DFS is at s and has no edge to leave it, indicating that the phase is finished, or when it arrives at t. In the latter case, an augmenting path is built. Then, a flow change is executed along that path, as at FF, while removing from L all its edges which have become saturated; recall that there is at least one such edge.

The running time of a phase, after executing the extended BFS to initialize L, is counted by Even and Itai as divided into intervals between *arrivals at dead-ends*. First, each such event causes the removal of an edge from L, so there may be at most $|E|$ such events. Secondly, there may be at most ℓ forward steps of DFS between neighboring events, which costs $O(\ell)$. Except for those steps, there may be either a single backtracking and edge removal, which costs $O(1)$, or a flow change, together with saturated edge removals, which costs $O(\ell)$. In any case, every interval between consequent events costs $O(\ell)$, which totals $O(\ell \cdot |E|) = O(|V||E|)$ per phase, as desired.

Summarizing our discussion on the version of DA suggested by Even and Itai, we see that their use of a non-cleaned layered network, where ad-hoc cleaning is made "on route", is quite successful: not only the same $O(\cdot)$ time bounds are acquired, but also the practical effectiveness (where constants are concerned as well) is not lost. Needless to say, their version not only has its own real beauty, but is somewhat "sexy" running DFS on the layered network constructed by (extended) BFS.

Regarding proof of the crucial property of DA, that the length of the layered network increases from phase to phase, Even and Itai use *no* data structure invariant, maintained from iteration to iteration, in contrast with the original paper [4]. Instead, they explicitly prove the above property, by analyzing the situation when a layered network vanishes. They characterize such a situation by the property of the flow accumulated *in* L during the phase being "maximal" (as distinguished from "maximum"), defined as a flow such that any path from s to t in L contains at least one edge saturated by it (a "blocking" flow, in notation of Karzanov [9]). Lemma 5.4 in [14] proves that after arriving at a maximal flow in the layered network of length ℓ, the currently shortest augmenting path is longer than ℓ.

It is interesting that the arguments in the proof of that Lemma are similar to those in the proof of Proposition 1. However, the reader may see clearly that the overall emphases, in the presentation of Dinic's algorithm at [14] are quite different from those of the original presentation of DA in [4], in both the algorithm iteration and the algorithm analysis. The origin of this difference is apparently the rejection of the data structure maintenance approach, occurring everywhere in the version of Even and Itai. See Section 1 for a possible explanation of this phenomenon.

4   Implementation of DA by Cherkassky

At the implementation stage, nothing extra should be left over. The program designer should have removed all the non-essential parts, not only to decrease the volume of programming work and/or decrease constant factors at the running time, but also to reduce the probability of bugs: a simpler program has less bugs, and thus causes less risk of arriving, sometimes, at a sudden malfunction, or what is the worst, at a wrong result. The only restriction, while choosing an implementation, is the equivalence of the designed program to the original algorithm, which is required for the validity of the program.

From the 1970s, Boris Cherkassky worked on the quality implementation of various flow algorithms and on the experimental evaluation of their performance, see [13, 27]. His recommendations for the best implementation of DA are as follows:

- No layered network—neither $L(s)$ nor $\hat{L}(s,t)$—should be built. It is sufficient to compute just the layer number ("rank") $dist(v,t)$ for every vertex up to the distance $\ell = dist(s,t)$. This may be done by a single run of the *usual* BFS from t on the unsaturated edges, in the inverse edge direction.[2]
- The entire phase is conducted by a single DFS from s. Any saturated edge or edge not going from a vertex of rank i to a vertex of rank $i + 1$ is just skipped (instead of using the list of edges in the layered network).
- No edge removal is needed. If DFS backtracks on some edge, that edge will not participate in the remaining part of DFS automatically.

---

[2] We could retain the natural direction of edges and the ranks $dist(s,v)$, for consistency with the preceding discussion. Switching to the ranks $dist(v,t)$ of Cherkassky is done for better intuition and for consistency with the push-relabel technique, discussed in Section 5.

- When the outgoing edges from the current vertex, $v \neq s, t$, are exhausted (v becomes a dead-end), DFS backtracks from v. If s becomes a dead-end, DFS (and the phase) is completed.

- When arriving at t, the current DFS path is an augmenting path. The usual flow change is made along it. After this, DFS continues from the tail of the edge closest to s which has been saturated during this flow change.

- The only network data needed during the phases is the residual capacity c_f for all the edges. After the last phase, the flow is restored as the difference of capacities c and c_f.

The resulting implementation is as follows:

Implementation of DA by Cherkassky

```
/* Input: */
    a flow network G = (V, E, c, s, t),
    a feasible flow f, in G (equal to zero, by default).
Initialization:
    compute ∀e ∈ E : cf(e) = c(e) − f(e);
/* Phase Loop: */
    dowhile
    begin
        compute ∀v ∈ V : rank(v) = dist(v, t), by BFS from t on edges with
cf > 0,
            in the inverse edge direction;
        if rank(s) = ∞ then begin  f ← c − cf;  return f; end;
        while DFS from s do
        begin
            /* P denotes the current path and x the current vertex of DFS */
            any edge (x, y) s.t. cf(x, y) = 0 or rank(y) ≠ rank(x) − 1 is skipped;
            if x = t then
            begin
                ε ← min{cf(e) : e ∈ P};
                for edges (v, u) of P, from t downto s  do
                begin
                    cf(v, u) ← cf(v, u) − ε;  cf(u, v) ← cf(u, v) + ε;
                    if cf(u, v) = 0  then x ← u;
                end;
                /* continue DFS from x */
            end;
        end;
    end;
```

In Section 5, we will see that such a simplification of DA, especially using vertex ranks instead of layered networks, has a deep influence on the following research on max-flow algorithms.

5   On the Further Progress of Max-Flow Finding

In this section, we consider some issues related to DA itself and the development of the max-flow algorithms area from DA and on, on the level of ideas. Note that many interesting results and even directions in the max-flow area remain untouched or are mentioned quite briefly. We emphasize using layered networks and/or distance ranks of vertices for max-flow finding.

Edmonds' and Karp's Version of FF Edmonds and Karp proved the finiteness and polynomial time of FF, if only the shortest augmenting paths are used in it. The achieved time bound is $O(|V||E|^2) = O(|V|^5)$, which is higher than $O(|V|^2|E|) = O(|V|^4)$ of DA, since iterations are made in $O(|E|)$ time each, as usual in FF. A talk on their work was given at a conference held in the middle of 1968, just half a year before the invention of DA. However, their result became known in Moscow and was told to the author only at the end of 1972, after the journal version of their paper [5] had been published.

The Running Time of DA for Special Cases The "iron curtain" worked in both directions. In January 1971, at a conference in Moscow, the author and Alexander Karzanov reported on new time bounds of DA for specific network types [6,7]. In [6], it is shown that if DA is executed on a network with unit capacities, the running time of each phase is bounded by $O(|E|)$ (since each edge of $\hat{L}$ is saturated after participating in a one or two augmenting paths). Hence, the total running time of DA is $O(|V||E|)$ for such a network. In [7], the flow network modeling of the bipartite $n \times n$ matching is considered. The number of phases of DA in such a modeling network is shown to be $O(\sqrt{n})$. Using the result of [6], the running time of DA for solving the bipartite $n \times n$ matching is found to be $O(\sqrt{n}m) = O(n^{5/2})$, where m is the number of pairs allowed for matching.

An algorithm for the bipartite $n \times n$ matching, similar to DA, with the same time bound of $O(\sqrt{n}m) = O(n^{5/2})$, was suggested by Hopkroft and Karp [8]. Their algorithm is the only one cited in the West. Various time bounds for DA working in networks of special types, similar to those in [6,7], were suggested by Even and Tarjan in [11]; in particular, they showed that the number of phases of DA for a network with unit capacities is bounded as $O(\min\{|E|^{1/2}, |V|^{2/3}\})$, and thus its running time is $O(\min\{|E|^{1/2}, |V|^{2/3}\}|E|)$. Still, the textbook [14] published in 1979 does not mention the results of [6,7], when reviewing these topics.

Karzanov's Algorithm Karzanov was the first to leave the augmenting paths idea, in the research on max-flow finding. He suggested the "preflow-push" approach, as follows. He observed that at each phase of DA, it is sufficient to *somehow* find a flow in a layered network $\hat{L}(s,t)$, such that any path from s to t contains an edge saturated by it; he calls such a flow "blocking". His algorithm (henceforth, *KA*) finds a blocking flow in $O(|V|^2)$ time, thus arriving at an $O(|V|^3)$ max-flow algorithm [9].

Here, we provide an outline of KA. Each finding of a blocking flow begins from *Pushing*. It scans vertices of $\hat{L}(s,t)$ in their BFS order, starting from s, and pushes flow from the current vertex, v, on its outgoing edges, as much as possible. If the edges outgoing from v, $v \neq s$, have enough capacity to remove from it all the flow brought to it on incoming edges, then a flow balance is created at v; otherwise, a "flow excess" at v is created, to be fixed farther on. After finishing *Pushing*, the flow function is infeasible, in general, with a flow excess at some vertices; such a function is called a *preflow*. Then, KA begins *Balancing*, which scans vertices with a flow excess but without outgoing unsaturated edges. *Balancing* pushes flow from such a vertex *back* on its incoming edges. Then, a flow excess is created at its preceding vertices, where outgoing unsaturated edges may exist. Then, a new *Pushing* is executed, followed by a new *Balancing*, and so on until balancing the preflow at all the vertices, except for s. The result is a desired blocking flow.

It is worth to mention that the time bounds $O(|V|^2|E|)$ and $O(|V|^4)$ for DA and $O(|V|^3)$ for KA are tight; the corresponding problem instances and executions of algorithms are provided in [10, 17].

Improvements upon DA and KA   Much effort was made, based on DA and KA, to lower the running time bound of max-flow finding; the aim was to acquire an $O(|V||E|)$ max-flow algorithm. Following is a brief review, cut before the push-relabel algorithm.

Cherkassky first simplified KA and later suggested a "hybrid" of DA and KA running in time $O(|V|^2\sqrt{|E|})$ [12]. Galil accelerated his algorithm to run in time $O(|V|^{5/3}|E|^{2/3})$ [15]. Galil and Naaman suggested an acceleration of augmenting path finding in DA by means of storing parts of previously found augmenting paths [16]; Shiloach discovered a similar algorithm independently a bit later. Their algorithms run in time $O(|E||V|\log^2|V|)$, which differs from the "goal time" only in a poly-logarithmic factor. Sleator and Tarjan improved the algorithm from [16] to run in time $O(|E||V|\log|V|)$ by using trees instead of just paths [19]. These low running times were achieved by using smart, but heavy, data structures.

Towards the Push-Relabel Algorithm   The other max-flow research movement was slow, its explicit direction was to simplify the pushing-balancing technique of Karzanov. As seen a-posteriory, its implicit goal was the "push-relabel" approach, suggested finally in [20, 22]. Let us show an example of such a development by analyzing the algorithm of Boris Cherkassky [13].[3]

One of the crucial steps taken in [13] is to cancel the previous goal of each phase, to construct a feasible flow in the layered network, and hence to eliminate *Balancing*. Instead, the new requirement is to achieve a flow balance at every vertex only by the end of the algorithm's execution. The main technical observation, while transforming KA, is that *Balancing* at some vertex is equivalent

---

[3] The $O(|V|^3)$ max-flow algorithm described by Shiloach and Vishkin in [18] may also be considered as a precursor of one of the push-relabel algorithms.

to *Pushing* from it in future, when it will appear in some layered network at a greater distance from the sink t. As in his implementation of DA (see Section 4), Cherkassky does not build layered networks, but just computes the vertex ranks $d(v) = dist(v, t)$.

There are steps of only the following two types in his algorithm:

Push For a vertex with a flow excess, pushing from it as much flow as possible to the vertices of a rank smaller by one.

Relabel For vertices without unsaturated edges outgoing to vertices of a rank smaller by one, recomputing their ranks.

For the reader familiar with the push-relabel approach, the remarkable similarity of the above steps with those of the generic push-relabel algorithm is obvious. (The names of the types are taken from [22], not from [13].)

Moreover, Cherkassky never computes new vertex ranks from scratch, except at the beginning of the algorithm—he *maintains* them. That is, he just increases the current vertex ranks as needed. This maintenance is done as that of the *global layered network*, which was sketched in [4] and given later in detail in [10] (see the next subsection). So, also the maintenance of vertex ranks is equivalent to the relabeling method of [22]. Summarizing, the algorithm of [13] is one of the possible implementations of the generic push-relabel algorithm [22].

The essential difference between the algorithm [13] and the approach of [22] is that the former is rigid, while the latter is generic, i.e. leaving freedom to choose any order for processing vertices with a flow excess. Leftovers of the phase structure of DA and KA prevented Cherkassky from canceling the division of his algorithm into phases, inherited from DA. At each phase of his algorithm, all the vertices are processed in the BFS order from farther to closer to the sink, and after that, all vertex ranks are updated. He uses the technique of the global layered network for accelerating the algorithm only. He does not notice that it is able to maintain vertex ranks *after each elementary push* at the same total cost, so that after any push, the algorithm is ready and free to push at any other vertex requiring it. This oversight prevented the algorithm [13] from becoming generalized to the generic push-relabel algorithm.

Usually, in the modern research community, relaxations of suggested methods, which get rid of various leftovers, are made very quickly after the publication of an interesting result. However, because of the "iron curtain", the results of [13] were unknown in the West, preventing further development by Western researchers[4]. The push-relabel approach was invented by a Ph.D. student, Andrew Goldberg, in 1985 [20].

We may hope that all our efforts to the right direction are not in vain. The following remarkable observation, heard by the author from his teacher Alexander Kronrod, may be related to the invention of the push-relabel approach, to some extent: "A solution to an open problem is always found much more easily, if that problem has already been solved by somebody else, even in a case where the current solver is completely unaware of the existing solution".

---

[4] Ahuja et al., in [26], relates even seminal *introducing distance labels*, (*i.e. vertex ranks*) *instead of layered networks*, to [20].

Perfect Maintenance of the One-Terminal Layered Network For completeness, let us briefly describe the method of perfect maintaining of the one-terminal layered network $L(s)$ [10]. Note that using it may eliminate the phase borders from both the original DA and its implementation given in Section 4.

The layered network L is initialized by building $L(s)$ w.r.t. the initial flow by the extended BFS in time $O(|E|)$. After a flow change, we take care of the vertices of L that lose the last incoming edge (called "dead-ends"). At DA, such a vertex is only removed from the data structure, and now we must put it into its correct layer. After removing the saturated edges, we set the *temporary distance* from s to each dead-end v at $d'(v) = d(v) + 1$. Additionally, every outgoing edge (v, u) is removed from the list of edges incoming to u. If, as a result, u becomes a dead-end, we apply the same operation to it.

We process the dead-ends in the non-decreasing order of temporary distances d'. For each dead-end v, we check whether there is an incoming edge from the layer $d'(v) - 1$. If so, we assign $d(v) = d'(v)$ and scan its incident edges: we insert into L the edges going to v from the layer $L_{d(v)-1}$ and those going from v to the layer $L_{d(v)+1}$. Otherwise, we increase $d'(v)$ by one, to be processed once more. It may be easily shown that this processing updates L to become $L(s)$ w.r.t. the new flow. In total, each vertex is moved to the next layer at most $|V| - 2$ times, which implies that the total cost of updating L is $O(|V||E|)$ during the entire layered network maintenance.

Maintaining vertex ranks, as in [13, 22], differs from the above method in beginning from t, reversing edge directions, and removing all operations with edges.

After Inventing the Generic Push-Relabel Algorithm The previously best time bound for max-flow finding was improved by Goldberg and Tarjan in [22]; their algorithm runs in time $O(|V||E| \cdot \log(|V|^2/|E|))$. The algorithm combines the push-relabel approach, choosing a vertex of the maximal rank at each push step, with the dynamic tree technique of [19] (changed a bit). After that, more (slight) improvements of running time for max-flow finding were made; we do not list them, referring to [28] for a review.

Interestingly, after the invention of the push-relabel approach, the development of max-flow algorithms returned to the *blocking flow techniques*, that is to dividing an algorithm execution into phases, each computing a blocking flow in a certain auxiliary network. In a couple of years, Goldberg and Tarjan [24] achieved a result finer than that of [22], by suggesting an $O(|E| \cdot \log(|V|^2/|E|))$ algorithm for finding a blocking flow in an arbitrary *acyclic* graph (the need to use acyclic graphs, which generalize layered, arisen when building a fast min-cost max-flow algorithm). Thus, the above time bound $O(|V||E| \cdot \log(|V|^2/|E|))$ for max-flow finding was achieved by a blocking flow algorithm too.

The above-mentioned results leave a polynomial gap between the general and the unit capacity cases: the bounds on the number of blocking flow computations in the former case is $O(|V|)$ and in the latter, $O(\min\{|E|^{1/2}, |V|^{2/3}\})$. For the case when capacities are integers bounded by U, Goldberg and Rao [29] suggest a blocking flow algorithm with $O(\min\{|E|^{1/2}, |V|^{2/3}\} \cdot \log U)$ blocking flow computations.

They combine the idea of increasing the distance from the source to the sink after each phase with another interesting requirement for the result of a phase. Let us denote by Δ_f the residual flow value, i.e. the value of the maximum flow in G_f. Edmonds and Karp considered in [5] the "thickest" augmenting path strategy of FF, that is such that a path P maximizing the value of $c_f(P)$ is chosen at each iteration. They show that Δ_f decreases by at least the fraction $\frac{1}{|E|}$, i.e. becomes at most $(1 - \frac{1}{|E|})\Delta_f$, after each iteration. Hence, the number of iterations of this version of FF in the integer case is $O(|E| \cdot \log U)$.

In the algorithm of Goldberg and Rao [29], the auxiliary network for each phase is acyclic, instead of layered. They choose a certain threshold depending on Δ_f and assign to each edge a length of one or zero, depending on whether its residual capacity is below or above that threshold, respectively. They use the distances w.r.t. these edge lengths to construct an auxiliary graph; in order to make it acyclic, they contract all the zero length cycles. After each phase, either *the distance from s to t increases*, or Δ_f *decreases by a certain fraction*. As a consequence, the number of phases is bounded by $O(\min\{|E|^{1/2}, |V|^{2/3}\} \cdot \log U)$. A blocking flow at each phase is found by the algorithm from [24], thus arriving at the $O(\min\{|E|^{1/2}, |V|^{2/3}\}|E| \cdot \log U \cdot \log(|V|^2/|E|))$ total running time. Thus, the time bound for the unit capacity case is extended to the integral capacity case at the expense of a factor of $\log U \cdot \log(|V|^2/|E|)$. Note that unless U is very big, the time bound achieved in [29] for general flow networks is better than $O(|V||E|)$.

Acknowledgments

The author is grateful to Avraham Melkman for sharing the effort to clarify the explanation of Dinitz' algorithm, to Boris Cherkassky and Andrew Goldberg for useful information and comments, to Oded Goldreich for his patience and help in improving presentation of this paper, and to Ethelea Katzenell for her English editing.

References

1. G.M. Adel'son-Vel'sky and E.M. Landis. An algorithm for the organization of information. *Soviet Mathematics Doklady* **3** (1962), 1259–1263).
2. L.R. Ford and D.R. Fulkerson. Flows in networks. Princeton University Press, Princeton, NJ, 1962.
3. V.L. Arlazarov, E.A. Dinic, M.A. Kronrod, and I.A. Faradjev. On economical finding of transitive closure of a graph. *Doklady Akademii Nauk SSSR* **194** (1970), no. 3 (in Russian; English transl.: *Soviet Mathematics Doklady* **11** (1970), 1270–1272).
4. E.A. Dinic. An algorithm for the solution of the max-flow problem with the polynomial estimation. *Doklady Akademii Nauk SSSR* **194** (1970), no. 4 (in Russian; English transl.: *Soviet Mathematics Doklady* **11** (1970), 1277–1280).

5. J. Edmonds and R.M. Karp. Theoretical improvements in algorithmic efficiency for network flow problems. *J. of ACM* **19** (1972), 248–264.

6. E.A. Dinic. An efficient algorithm for the solution of the generalized set representatives problem. In: *Voprosy Kibernetiki. Proc. of the Seminar on Combinatorial Mathematics (Moscow, 1971).* Scientific Council on the Complex Problem "Kibernetika", Akad. Nauk SSSR, 1973, 49–54 (in Russian).

7. A.V. Karzanov. An exact time bound for a max-flow finding algorithm applied to the "representatives" problem. In: *Voprosy Kibernetiki. Proc. of the Seminar on Combinatorial Mathematics (Moscow, 1971).* Scientific Council on the Complex Problem "Kibernetika", Akad. Nauk SSSR, 1973, 66–70 (in Russian).

8. J. Hopkroft and R.M. Karp. An $n^{5/2}$ algorithm for maximum matchings in bipartite graphs. *SIAM J. on Computing* **2** (1973), 225–231.

9. A.V. Karzanov. Determining the maximum flow in the network by the method of preflows. *Doklady Akademii Nauk SSSR* **215** (1974), no. 1 (in Russian; English transl.: *Soviet Mathematics Doklady* **15** (1974), 434–437).

10. G.M. Adel'son-Vel'sky, E.A. Dinic, and A.V. Karzanov. *Network flow algorithms.* "Nauka", Moscow, 1975, 119 p. (in Russian; a review in English see in [25]).

11. S. Even and R.E. Tarjan. Network flow and testing graph connectivity. *SIAM J. on Computing* **4** (1975), 507–518.

12. B.V. Cherkassky. An algorithm for building a max-flow in a network, running in time $O(n^2\sqrt{p})$. In: *Mathematical Methods for Solving Economic Problems*, issue **7** (1977), "Nauka", Moscow, 117–126 (in Russian).

13. B.V. Cherkassky. A fast algorithm for constructing a maximum flow through a network. In *Combinatorial Methods in Network Flow Problems*. VNIISI, Moscow, 1979, 90–96 (in Russian; English transl.: *Amer. Math. Soc. Transl.* **158** (1994), no. 2, 23–30).

14. S. Even. *Graph Algorithms*. Computer Science Press, Rockville, MD, 1979.

15. Z. Galil. An $O(V^{5/3}E^{2/3})$ algorithm for the maximal flow problem. *Acta Inf.* **14** (1980), 221-242.

16. Z. Galil and A. Naaman. An $O(EV \log^2 V)$ algorithm for the maximal flow problem. *J. Comput. Syst. Sci.* **21** (1980), no. 2, 203–217.

17. Z. Galil. On the theoretical efficiency of various network flow algorithms. *Theor. Comp. Sci.* **14** (1981), 103–111.

18. Y. Shiloach and U. Vishkin. An $O(n^2 \log n)$ parallel max-flow algorithm. *J. of Algorithms* **3** (1982), 128–146.

19. D.D. Sleator and R.E. Tarjan. A data structure for dynamic trees. *J. Comput. Syst. Sci.* **24** (1983), 362–391.

20. A.V. Goldberg. A new max-flow algorithm. TR MIT/LCS/TM-291, Laboratory for Comp. Sci., MIT, Cambridge, MA, 1985.

21. R.E. Tarjan. Amortized computational complexity. *SIAM J. Alg. Disc. Meth.* **6** (1985), no.2, 306–318.

22. A.V. Goldberg and R.E. Tarjan. A new approach to the maximum flow problem. In: *Proc. of the 18th ACM Symp. on the Theory of Computing*, 136–146. Full paper in *J. of ACM* **35** (1988), 921–940.

23. T. Cormen, C. Leiserson, and R. Rivest. *Introduction to Algorithms*. McGraw-Hill, New York, NY, 1990.

24. A.V. Goldberg and R.E. Tarjan. Finding minimum-cost circulations by successive approximation. *Mathematics of Operations Research*, **15** (1990), issue 3, 430–466.

25. A.V. Goldberg and D. Gusfield. Book Review: Flow algorithms by G.M. Adel'son-Vel'sky, E.A. Dinits, and A.V. Karzanov. *SIAM Reviews* **33** (1991), no. 2, 306–314.

26. R.K. Ahuja, T.L. Magnanti, and J.B. Orlin. *Network Flows: Theory, Algorithms, and Applications*. Prentice Hall, Upper Saddle River, NJ, 1993.

27. B.V. Cherkassky and A.V. Goldberg. On implementing the push-relabel method for the maximum flow problem. *Algorithmica* **19** (1997), 390–410.

28. A.V. Goldberg. Recent developments in maximum flow algorithms (invited lecture). In: *Proc. of the 6th Scandinavian Workshop on Algorithm Theory (SWAT'98)*, LNCS 1432, Springer-Verlag, London, UK, 1998, 1–10.

29. A.V. Goldberg and S. Rao. Beyond the flow decomposition barrier. *Journal of ACM* **45** (1998), 753–782.

Survey of Disjoint NP-pairs and Relations to Propositional Proof Systems

Christian Glaßer[1], Alan L. Selman[2*], and Liyu Zhang[2]

[1] Theoretische Informatik, Universität Würzburg,
Am Hubland, 97074 Würzburg, Germany
`glasser@informatik.uni-wuerzburg.de`
[2] Department of Computer Science and Engineering,
University at Buffalo, Buffalo, NY 14260
{`selman,lzhang7`}`@cse.buffalo.edu`

Abstract. We survey recent results on disjoint NP-pairs. In particular, we survey the relationship of disjoint NP-pairs to the theory of proof systems for propositional calculus.

1 Introduction

A *disjoint* NP-*pair* is a pair (A, B) of nonempty, disjoint sets A and B such that both A and B belong to the complexity class NP.[3] We let DisjNP denote the collection of all disjoint NP-pairs. A *separator* of a disjoint NP-pair (A, B) is a set S such that $A \subseteq S$ and $B \subseteq \overline{S}$ (Figure 1). A fundamental question is whether (A, B) has a separator belonging to P. In this case the pair is P-*separable*; otherwise, it is P-*inseparable*.

To state this fundamental question differently, we want to know whether there is an efficient algorithm whose set of yes-instances includes the set A and whose set of no-instances includes the set B. The algorithm behaves arbitrarily on instances in the complement of $A \cup B$. That is, a disjoint NP-pair is a promise problem. To learn about promise problems we refer to Goldreich's survey paper [8] in this volume. The second author first became interested in promise problems, and specifically, in disjoint NP-pairs, in 1982 while working with Shimon Even and Yacov Yacobi. At that time they formulated the problem of cracking a public-key cryptosystem as a promise problem and observed that secure public-key cryptosystems do not exist unless P-*inseparable* pairs exist [4].

Disjoint NP-pairs also relate naturally to the theory of proof systems for propositional calculus [21, 20] and that is the connection we will explore here.

2 Preliminaries

The notations $\leq_m^p$ and $\leq_T^p$ denote polynomial-time-bounded many-one and Turing reducibility, respectively. Thus, we write $A \leq_m^p B$ if there is a function f computable in polynomial time, such that for all instances x, $x \in A \Leftrightarrow f(x) \in B$ and

---

* Research partially supported by NSF grant CCR-0307077.
[3] Nonemptyness ensures $A \nsubseteq B$ and $B \nsubseteq A$, which simplifies several proofs.

O. Goldreich et al. (Eds.): Shimon Even Festschrift, LNCS 3895, pp. 241–253, 2006.
© Springer-Verlag Berlin Heidelberg 2006

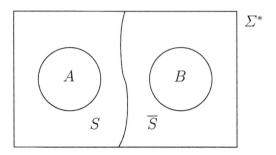

Fig. 1. An NP-pair (A, B) that is separated by S.

we write $A \leq_T^p B$ if there is an oracle Turing machine M such that $A = L(M, B)$ is the language accepted by M using B as the oracle.

We let PF denote the class of all polynomial-time-computable functions. A function f is *honest* if there is a polynomial q such that for every $y \in range(f)$ there exists $x \in dom(f)$ such that $f(x) = y$ and $|x| \leq q(|y|)$. If $f \in$ PF is an honest function and $A \in$ NP, then $f(A) \in$ NP.

To review the definition of standard exponential-time complexity classes,

$$E = \bigcup \{\text{DTIME}(k^n) \mid k \geq 1\}$$

and

$$NE = \bigcup \{\text{NTIME}(k^n) \mid k \geq 1\}.$$

For any complexity class $\mathcal{C}$, $co\mathcal{C} = \{L \mid \overline{L} \in \mathcal{C}\}$.

A nondeterministic Turing machine that has at most one accepting computation for any input is called by Valiant [27] an *unambiguous* Turing machine. Let UP denote the set of languages accepted by unambiguous Turing machines in polynomial time. Obviously, $P \subseteq UP \subseteq NP$, and it is not known whether either inclusion is proper. One reason UP is an interesting complexity class is because there exists a one-to-one, honest function $f \in$ PF whose inverse f^{-1} is not computable in polynomial time if and only if $P \neq UP$ [9].

A set A is *sparse* if there is a polynomial p such that for all n, A contains at most $p(n)$ strings of length n. We let SPARSE denote the collection of all sparse sets.

3   Propositional Proof Systems

Resolution calculus is one well-known example of a propositional proof system. A propositional formula ϕ in disjunctive normal form is a tautology if and only if there exists a proof w in the resolution system showing that $\neg\phi$ is not satisfiable. (Recall that a formula ϕ is a tautology if and only if its negation $\neg\phi$ is not satisfiable.) All propositional proof systems have three properties in common:

1. **Correctness:** If there is a proof in the system, then the formula is indeed a tautology.
2. **Completeness:** Every tautology can be proved within the system.
3. **Verifiability:** The validity of a proof can be easily verified.

Cook and Reckhow [3] formalized the intuitive notion of a proof system as follows: A *propositional proof system* (proof system for short) is a total function $f : \Sigma^* \to$ TAUT such that f is onto and polynomial-time-computable. (TAUT denotes the set of tautologies.) For any tautology ϕ, if $f(w) = \phi$, then w is a proof in the system f (f-*proof*) showing that ϕ is a tautology.

That the validity of proofs is easy to verify is formalized by the requirement that f is polynomial-time-computable. Furthermore, the definition requires that only tautologies have proofs (correctness) and that every tautology has a proof (completeness).

The following function shows that the resolution calculus can be interpreted as a propositional proof system in the formal sense.

$$
f(w) = \begin{cases}
\phi & : & \text{if } w = (\phi, u) \text{ and } u \text{ is a resolution refutation of } \neg\phi \\
\phi & : & \text{if } w = (\phi, u) \text{ and } |u| \geq 2^{|\phi|} \text{ and } \phi \text{ is a tautology} \\
\text{true} & : & \text{otherwise}
\end{cases}
$$

Note that in this definition, both lines, the first and the second one, are necessary. The first line makes sure that tautologies in disjunctive normal form have f-proofs not much longer than the corresponding resolution proofs. The second line takes care of all the tautologies that are not in disjunctive normal form. Note that in this line, the test for tautology can be done by exhaustive search in polynomial time, since $2^{|\phi|} \leq |w|$.

A propositional proof system f is not necessarily honest; it is possible that a formula $\phi \in$ TAUT has only exponentially long proofs w, i.e., $f(w) = \phi$ and $|w| = 2^{\Omega(|\phi|)}$. A proof system f is *polynomially bounded* if the function f is honest. Cook and Reckhow demonstrated that NP = coNP if and only if there exists a polynomially-bounded proof system, and they proposed attacking the question of whether NP equals coNP by studying propositional proof systems.

Let f and f' be two proof systems. We say that f *simulates* f' if there is a polynomial p and a function $h : \Sigma^* \to \Sigma^*$ such that for every $w \in \Sigma^*$, $f(h(w)) = f'(w)$ and $|h(w)| \leq p(|w|)$. So for every f'-proof w, $h(w)$ is an f-proof of the same tautology. If f simulates f', then f-proofs are not much longer than f'-proofs. If additionally $h \in$ PF, then we say that f *p-simulates* f'.

A proof system is *optimal* (resp., *p-optimal*) if it simulates (resp., p-simulates) every other proof system. The notion of simulation between proof systems is analogous to the notion of reducibility between problems. Using that analogy, optimal proof systems correspond to complete problems.

It is not known whether there exist optimal propositional proof systems, but the question is interesting, because if there is an optimal proof system f, then there is a polynomially-bounded proof system if and only if f is polynomially-bounded. The question of whether optimal propositional proof systems exist has been studied in detail. Krajíček and Pudlák [19, 13] showed that NE = coNE implies the existence of optimal proof systems. Ben-David and Gringauze [1] and Köbler, Meßner, and Torán [12] obtained the same conclusion under weaker assumptions. On the other hand, Meßner and Torán [18] and Köbler, Meßner, and Torán [12] proved that existence of optimal proof systems results in the existence of $\leq_m^p$-complete sets for the promise class NP∩SPARSE. In the same paper, they showed that there exist p-optimal proof systems only if the complexity class UP has a many-one complete set. These results hold relative to all oracles. Therefore, optimal proof systems exist relative to any oracle in which NE = coNE holds. Krajíček and Pudlák [13], Ben-David and Gringauze [1], and Buhrman et al. [2] constructed oracles relative to which optimal proof systems do not exist. In addition, NP∩SPARSE does not have complete sets relative to the latter oracle.

Razborov [21] related the study of propositional proof systems to disjoint NP-pairs. For every propositional proof system f, he associated a canonical disjoint NP-pair. Furthermore, he showed that if f is an optimal proof system, then the canonical pair for f is a complete disjoint NP-pair. We will explain these results in Section 5, but first, in order to define the notion of completeness for disjoint NP-pairs, it is necessary to describe reducibilities between disjoint NP-pairs.

4   Reductions Between Disjoint NP-pairs

Since disjoint pairs are simply an equivalent formulation of promise problems, disjoint pairs easily inherit the natural notions of reducibilities that exist between promise problems [4, 26, 9]. Hence, completeness and hardness notions follow naturally also. We review these here.

Definition 1 Let (A, B) and (C, D) be disjoint pairs.

1. (A, B) is many-one reducible in polynomial-time to (C, D), $(A, B) \leq_m^{pp} (C, D)$, if for every separator T of (C, D), there exists a separator S of (A, B) such that $S \leq_m^p T$.
2. (A, B) is Turing reducible in polynomial-time to (C, D), $(A, B) \leq_T^{pp} (C, D)$, if for every separator T of (C, D), there exists a separator S of (A, B) such that $S \leq_T^p T$.

The definitions tell us that for every separator of (C, D), there is a separator of (A, B) that is no more complex. In particular, if (C, D) is P-separable, then it follows immediately that (A, B) is P-separable. On the other hand, these definitions are nonuniform. Looking at $(A, B) \leq_T^{pp} (C, D)$, for example, if S_1 is a separator of (C, D), then there is an oracle Turing machine M_1 such that the set $L(M_1, S_1)$ is a separator of (A, B). However, for a different separator S_2 of (C, D), there might be a different Turing machine M_2 so that $L(M_2, S_2)$ is a separator of (A, B). This nonuniformity makes these definitions difficult to work with. Fortunately, they have the following equivalent formulations [9, 6]. Observe that the formulation for many-one reducibility simplifies enormously.

Theorem 2 (uniform reductions for pairs). *Let (A, B) and (C, D) be disjoint pairs.*

1. *(A, B) is many-one reducible in polynomial-time to (C, D) if and only if there exists a polynomial-time computable function f such that $f(A) \subseteq C$ and $f(B) \subseteq D$.*
2. *(A, B) is Turing reducible in polynomial-time to (C, D) if and only if there exists a polynomial-time oracle Turing machine M such that for every separator T of (C, D), there exists a separator S of (A, B) such that $S \leq_T^p T$ via M. That is, $S = L(M, T)$.*

Now we clearly have uniformity. The same oracle Turing machine M is used for all separators T.

The abbreviation 'pp' in $\leq_T^{pp}$, for example, stands for *polynomial-time-bounded promise reduction*. We retain the promise problem notation in order to distinguish reductions between disjoint NP-pairs from reducibilities between sets.

If $(A, B) \leq_m^{pp} (C, D)$ and $(C, D) \leq_m^{pp} (A, B)$, then we write $(A, B) \equiv_m^{pp} (C, D)$; if $(A, B) \leq_T^{pp} (C, D)$ and $(C, D) \leq_T^{pp} (A, B)$, then we write $(A, B) \equiv_T^{pp} (C, D)$. Obviously, $\equiv_m^{pp}$ and $\equiv_T^{pp}$ are equivalence relations.

Keeping with common terminology, a disjoint pair (A, B) is $\leq_m^{pp}$-complete ($\leq_T^{pp}$-complete) for the class DisjNP if $(A, B) \in$ DisjNP and for every disjoint pair $(C, D) \in$ DisjNP, $(C, D) \leq_m^{pp} (A, B)$ ($(C, D) \leq_T^{pp} (A, B)$), respectively).

Razborov raised the question of whether DisjNP contains complete pairs (i.e., complete disjoint NP-pairs). Although we are primarily interested in the question of whether there exist many-one complete pairs, let's pause for a moment to consider the question of whether there exist Turing-complete pairs. Even, Selman, and Yacobi [4] conjectured that DisjNP does not contain a disjoint pair all of whose separators are NP-hard (i.e., $\leq_T^p$-hard for NP.) The conjecture has strong consequences, for it implies that NP $\neq$ coNP, NP $\neq$ UP, and no public-key cryptosystem is NP-hard to crack [4,9]. For example, if NP $=$ coNP, then for every NP-complete S, the pair $(S, \bar{S})$ is in DisjNP and all of its separators are NP-hard (since S is the only separator). We conjecture that DisjNP does not contain Turing-complete pairs, but it would be difficult to prove this, because the the latter conjecture implies the former conjecture (which in turn implies NP $\neq$ coNP).

Proposition 3 *If there do not exist $\leq_T^{pp}$-complete pairs for the class DisjNP, then DisjNP does not contain a disjoint pair all of whose separators are NP-hard.*

Proof. Suppose there is a disjoint pair $(A, B) \in$ DisjNP such that all separators are NP-hard. We claim that (A, B) is $\leq_T^{pp}$-complete for DisjNP. Let (C, D) belong to DisjNP. Let S be an arbitrary separator of (A, B). Note that S is NP-hard and $C \in$ NP. So $C \leq_T^p S$. Since C is a separator of (C, D), this demonstrates that $(C, D) \leq_T^{pp} (A, B)$. $\square$

Glaßer et al. [6] constructed an oracle relative to which Turing-complete pairs do not exist for DisjNP.

5   Canonical Disjoint NP-pairs

The canonical pair of a propositional proof system f [21] is the disjoint NP-pair $(\mathrm{SAT}^*, \mathrm{REF}_f)$ where

$$\mathrm{SAT}^* = \{(x, 0^n) \mid x \in \mathrm{SAT} \text{ and } n \in \mathbb{N}\} \quad \text{and}$$
$$\mathrm{REF}_f = \{(x, 0^n) \mid \neg x \in \mathrm{TAUT} \text{ and } \exists y[|y| \leq n \text{ and } f(y) = \neg x]\}.$$

Informally, SAT^* is the set of satisfiable formulas (i.e., formulas whose negations are not tautologies), and REF_f is the set of easily refutable formulas (i.e., formulas whose negations have short proofs). It is straightforward to see that SAT^* and REF_f are disjoint and that they belong to NP.

The following easy to prove proposition states a strong connection between proof systems and disjoint NP-pairs.

Proposition 4 *Let f and g be propositional proof systems. If g simulates f, then $(\mathrm{SAT}^*, \mathrm{REF}_f) \leq_m^{pp} (\mathrm{SAT}^*, \mathrm{REF}_g)$.*

Proof. By assumption there exists a total function $h : \Sigma^* \to \Sigma^*$ and a polynomial p such that for all y, $g(h(y)) = f(y)$ and $|h(y)| \leq p(|y|)$. We claim that $(\mathrm{SAT}^*, \mathrm{REF}_f) \leq_m^{pp} (\mathrm{SAT}^*, \mathrm{REF}_g)$ via reduction r where $r(x, 0^n) \stackrel{df}{=} (x, 0^{p(n)})$. Clearly, if $(x, 0^n) \in \mathrm{SAT}^*$, then $(x, 0^{p(n)}) \in \mathrm{SAT}^*$ as well. Let $(x, 0^n) \in \mathrm{REF}_f$, i.e., $\neg x$ is a tautology and there exists y such that $|y| \leq n$ and $f(y) = \neg x$. So for $y' \stackrel{df}{=} h(y)$ it holds that $|y'| \leq p(n)$ and $g(y') = \neg x$ which shows $(x, 0^{p(n)}) \in \mathrm{REF}_g$. $\square$

Razborov's result (Corollary 8 below) states that if f is an optimal proof system, then $(\mathrm{SAT}^*, \mathrm{REF}_f)$ is a $\leq_m^{pp}$-complete NP-pair. This result is an immediate consequence of Proposition 4 and the following new result [7]. The latter states that every disjoint NP-pair is many-one equivalent to the canonical NP-pair of some propositional proof system.

Theorem 5. *For every disjoint NP-pair (A, B) there exists a proof system f such that $(\mathrm{SAT}^*, \mathrm{REF}_f) \equiv_m^{pp} (A, B)$.*

Proof. Let $\langle \cdot, \cdot \rangle$ be a polynomial-time computable, polynomial-time invertible pairing function such that $|\langle v, w \rangle| = 2|vw|$. Choose g that is polynomial-time computable *and* polynomial-time invertible such that $A \leq_m^p \mathrm{SAT}$ via g (such a g exists, since SAT is a paddable NP-complete set). Let M be an NP-machine that accepts B in time p. Define the following function f.

$$f(z) \stackrel{df}{=} \begin{cases} \neg g(x) & : \quad \text{if } z = \langle x, w \rangle, \ |w| = p(|x|), \ M(x) \text{ accepts along path } w \\ x & : \quad \text{if } z = \langle x, w \rangle, \ |w| \neq p(|x|), \ |z| \geq 2^{|x|}, \ x \in \mathrm{TAUT} \\ \text{true} & : \quad \text{otherwise} \end{cases}$$

The function is polynomial-time computable, since in the second case, $|z|$ is large enough so that $x \in \text{TAUT}$ can be decided in deterministic time $O(|z|^2)$. In the first case of f's definition, $x \in B$ and so $g(x) \notin \text{SAT}$. It follows that $f : \Sigma^* \to \text{TAUT}$. The mapping is onto, since for every tautology x,

$$f(\langle x, 0^{2^{|x|}} \rangle) = x.$$

Therefore, f is a propositional proof system.

Claim 6 $(\text{SAT}^*, \text{REF}_f) \leq_m^{pp} (A, B)$.

Choose arbitrary elements $a \in A$ and $b \in B$. The reduction function h is as follows.

```
1    input (x, 0ⁿ)
2    if n ≥ 2^|x| then
3        if x ∈ SAT then output a else output b
4    endif
5    if g⁻¹(x) exists then output g⁻¹(x)
6    output a
```

The exhaustive search in Line 3 is possible in quadratic time in $2^{|x|} \leq n$. So $h \in \text{PF}$.

Assume $(x, 0^n) \in \text{SAT}^*$. If we reach Line 3, then we output $a \in A$. Otherwise we reach Line 5. If $g^{-1}(x)$ exists, then it belongs to A. Therefore, in either case (output in Line 5 or in Line 6) we output an element from A.

Assume $(x, 0^n) \in \text{REF}_f$ (in particular $\neg x \in \text{TAUT}$). So there exists y such that $|y| \leq n$ and $f(y) = \neg x$. If we reach Line 3, then we output b. Otherwise we reach Line 5 and so it holds that $|y| \leq n < 2^{|x|}$ and $\neg x$ differs from the expression true (since the expression true does not start with the symbol $\neg$). Therefore, $f(y) = \neg x$ must be due to the first case in the definition of f. It follows that $g^{-1}(x)$ exists. So we output $g^{-1}(x)$ which belongs to B (again by the first case of f's definition). This shows Claim 6.

Claim 7 $(A, B) \leq_m^{pp} (\text{SAT}^*, \text{REF}_f)$.

The reduction function is $h'(x) \overset{df}{=} (g(x), 0^{2(|x|+p(|x|))})$. If $x \in A$, then $g(x) \in \text{SAT}$ and therefore, $h'(x) \in \text{SAT}^*$. Otherwise, let $x \in B$. Let w be an accepting path of $M(x)$ and define $z \overset{df}{=} \langle x, w \rangle$. So $|w| = p(|x|)$ and $|z| = 2(|x| + p(|x|))$. By the first case of f's definition, $f(z) = \neg g(x)$. Therefore, $h'(x) \in \text{REF}_f$. This proves Claim 7 and finishes the proof of Theorem 5. □

Corollary 8 (Razborov) *If there exists an optimal propositional proof system f, then $(\text{SAT}^*, \text{REF}_f)$ is a $\leq_m^{pp}$-complete NP-pair.*

Proof. Suppose that f is an optimal proof system. Let (A, B) be an arbitrary disjoint NP-pair. By Theorem 5, let g be a proof system such that

$$(A, B) \equiv_m^{pp} (\text{SAT}^*, \text{REF}_g).$$

We only use $(A, B){\leq^{pp}_m}(\text{SAT}^*, \text{REF}_g)$ and the fact that $(\text{SAT}^*, \text{REF}_g) \in \text{DisjNP}$. Since f is optimal, f simulates g. Thus, by Proposition 4,

$$(\text{SAT}^*, \text{REF}_g){\leq^{pp}_m}(\text{SAT}^*, \text{REF}_f).$$

Then, $(A, B){\leq^{pp}_m}(\text{SAT}^*, \text{REF}_f)$, from which it follows that $(\text{SAT}^*, \text{REF}_f)$ is ${\leq^{pp}_m}$-complete for DisjNP.   □

Also, we state the following corollary. It is convenient for us to define the Turing-degree of a pair $(A, B) \in \text{DisjNP}$ as follows:

$$\mathbf{d}(A, B) = \{(C, D) \in \text{DisjNP} \mid (A, B) \equiv^{pp}_T (C, D)\}.$$

So the Turing-degree of (A, B) is the class of pairs that are equivalent to (A, B) with respect to Turing reductions. In a canonical way, Turing reductions extend from pairs to Turing-degrees: $\mathbf{d}(A, B){\leq^{pp}_T}\mathbf{d}(C, D)$ if $(A, B){\leq^{pp}_T}(C, D)$. The *degree structure* of disjoint NP-pairs is the structure of the partial ordering $(\{\mathbf{d}(A, B) \mid (A, B) \in \text{DisjNP}\}, {\leq^{pp}_T})$.

Corollary 9 *Disjoint* NP-*pairs and canonical pairs for proof systems have identical degree structure.*

Every disjoint NP-pair we believe to be P-inseparable is many-one equivalent to some canonical pair that is also P-inseparable. We cannot prove that P-inseparable pairs exist, but there is evidence for their existence, for example, if $P \neq \text{UP}$ or if $P \neq \text{NP} \cap \text{coNP}$. On the other hand, the hypothesis that $P \neq \text{NP}$ does not seem to be sufficient to obtain P-inseparable disjoint NP-pairs. Homer and Selman [11] constructed an oracle relative to which $P \neq \text{NP}$ and all disjoint NP-pairs are P-separable.

Glaßer et al. [6] constructed an oracle O_1 relative to which optimal proof systems exist, and therefore, relative to which many-one complete disjoint NP-pairs exist. Also, they constructed an oracle O_2 relative to which many-one complete disjoint NP-pairs exist, but optimal proof systems do not exist. So relative to this oracle, the converse of Corollary 8 does not hold. Relative to O_2, there is a propositional proof system f whose canonical pair is complete, but f is not optimal. Hence, there is a propositional proof system g such that the canonical pair of g many-one reduces to the canonical pair of f, but f does not simulate g. The results of this section (Proposition 4, Theorem 5, and Corollary 9) present tight connections between disjoint NP-pairs and propositional proof systems. Nevertheless, relative to this oracle, the relationship is not as tight as one might hope for.

In light of Corollary 9, we should try to understand the degree structure of DisjNP. Glaßer, Selman, and Zhang [7] prove that between any two comparable and inequivalent disjoint NP-pairs (A, B) and (C, D) there exist P-inseparable, incomparable NP-pairs (E, F) and (G, H) whose degrees lie strictly between (A, B) and (C, D). Their result is an analogue of Ladner's result for NP [14]. The proof is based on Schöning's formulation [25] and uses techniques of Regan

[22, 23]. Thus, assuming that P-inseparable disjoint NP-pairs exist, the class DisjNP has a rich, dense, degree structure—and each of these degrees contains a canonical pair.

Observe that the premise of the following theorem is true as long as there exist P-inseparable disjoint NP-pairs.

Theorem 10. *Suppose there exist disjoint* NP-*pairs* (A, B) *and* (C, D) *such that* A, B, C, *and* D *are infinite,* $(A, B) \leq_T^{pp} (C, D)$, *and* $(C, D) \not\leq_T^{pp} (A, B)$. *Then there exist incomparable, strictly intermediate disjoint* NP-*pairs* (E, F) *and* (G, H) *between* (A, B) *and* (C, D) *such that* E, F, G, *and* H *are infinite. Precisely, the following properties hold:*

- $(A, B) \leq_m^{pp} (E, F) \leq_T^{pp} (C, D)$ *and* $(C, D) \not\leq_T^{pp} (E, F) \not\leq_T^{pp} (A, B)$;
- $(A, B) \leq_m^{pp} (G, H) \leq_T^{pp} (C, D)$ *and* $(C, D) \not\leq_T^{pp} (G, H) \not\leq_T^{pp} (A, B)$;
- $(E, F) \not\leq_T^{pp} (G, H)$ *and* $(G, H) \not\leq_T^{pp} (E, F)$.

Messner [16, 17] unconditionally proved the existence of propositional proof systems f and g such that f does not simulate g and g does not simulate f. Further he shows that the simulation order of propositional proof systems is dense. However, from this we cannot conclude a dense degree structure for disjoint NP-pairs. There exist infinite, strictly increasing chains of propositional proof systems (using simulation as the order relation $\leq$) such that all canonical pairs of these proofs systems belong to the same many-one degree of disjoint NP-pairs.

6   Uniform Enumerability

In this section we describe some recent results of Glaßer, Selman, and Sengupta [5] on reductions between disjoint NP-pairs. The main result is a list of equivalent statements to the assertion that there exists a many-one complete disjoint NP-pair, which, taken together, strongly suggests that the assertion does not hold.

We begin our exposition with the following definition of strongly many-one reductions, as defined by Köbler, Meßner, and Torán [12].

Definition 11 ([12]) (C, D) *strongly many-one reduces to* (A, B) *in polynomial time,* $(C, D) \leq_{sm}^{pp} (A, B)$, *if there is a polynomial-time computable function* f *such that* $f(C) \subseteq A$, $f(D) \subseteq B$, *and* $f(\overline{C \cup D}) \subseteq \overline{A \cup B}$.

Clearly, the added condition $f(\overline{C \cup D}) \subseteq \overline{A \cup B}$ states that instances violating the promise of (C, D) are mapped into instances that violate the promise of (A, B) (Figure 2). Equivalently, $f^{-1}(A) \subseteq C$ and $f^{-1}(B) \subseteq D$. Therefore, if $(C, D) \leq_{sm}^{pp} (A, B)$ via f, then $C \leq_m^p A$ via f, and $D \leq_m^p B$ via f.

Whereas Razborov proved that existence of an optimal proof system implies existence of a many-one complete disjoint NP-pair, Köbler, Meßner, and Torán proved with the same hypothesis existence of a complete disjoint NP-pair with respect to strongly many-one reductions. In particular, the result of Glaßer,

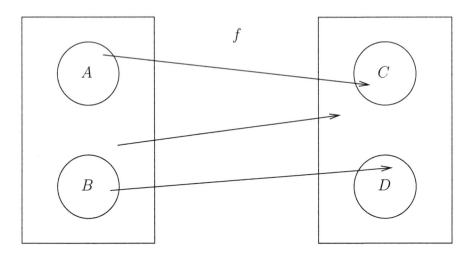

Fig. 2. A strong many-one reduction f from (A, B) to (C, D).

Selman, and Sengupta shows that these results of Razborov and Köbler, Meßner, and Torán are equivalent. That is, there exists a many-one complete disjoint NP-pair if and only if there exists a complete disjoint NP-pair with respect to strongly many-one reductions. Nevertheless, it is apparently true that the "stronger reduction" really is stronger. This is easy to see if we permit disjoint NP-pairs whose components are finite sets. However, for pairs whose components are infinite and coinfinite, strongly many-one reductions are identical to many-one reductions if and only if P = NP. We show this result now:

Theorem 12. *The following are equivalent:*

1. P $\neq$ NP.
2. *There are disjoint NP-pairs (A, B) and (C, D) such that A, B, C, D, $\overline{A \cup B}$, and $\overline{C \cup D}$ are infinite, and $(A, B) \leq_m^{pp} (C, D)$ but $(A, B) \not\leq_{sm}^{pp} (C, D)$.*

Proof. If P = NP, then given disjoint NP-pairs (A, B) and (C, D), A, B, C, and D are all in P. Given any string x, it can be determined whether $x \in A$, $x \in B$, or $x \in \overline{A \cup B}$, and x can be mapped appropriately to some fixed string in C, D, or $\overline{C \cup D}$. Therefore, $(A, B) \leq_{sm}^{pp} (C, D)$.

For the other direction, consider the clique-coloring pair (C_1, C_2) such that

$$C_1 = \{\langle G, k \rangle \mid G \text{ has a clique of size } k\}, \tag{1}$$

and

$$C_2 = \{\langle G, k \rangle \mid G \text{ has a coloring with } k - 1 \text{ colors}\}. \tag{2}$$

This is a disjoint NP-pair, and is known to be P-separable [15, 20]. Let S be the separator that is in P. Note that $(C_1, C_2) \leq_m^{pp} (S, \overline{S})$ via the identity function. (Note that this reduction is also invertible.) Let

$$C = \{\langle G, 3 \rangle \mid G \text{ is a cycle of odd length with at least 5 vertices}\}.$$

Let $S_1 = S - C$ and $S_2 = \overline{S} - C$. Both S_1 and S_2 are in P. Since any odd cycle with at least 5 vertices is not 2-colorable, and does not contain any clique of size 3, $C \cap C_1 = \emptyset$, and $C \cap C_2 = \emptyset$. Therefore, $(C_1, C_2) \leq_m^{pp} (S_1, S_2)$ via the identity function. Assume that $(C_1, C_2) \leq_{sm}^{pp} (S_1, S_2)$. Then $C_1 \leq_m^p S_1$, and $C_2 \leq_m^p S_2$. Hence C_1 and C_2 are in P. This is impossible, since NP $\neq$ P, and C_1 and C_2 are NP-complete. Thus, $(C_1, C_2) \not\leq_{sm}^{pp} (S_1, S_2)$.   □

Next we mention smart reductions. Grollmann and Selman [9] defined smart reductions in order to analyze the conjecture of Even, Selman, and Yacobi [4] that we discussed earlier.

Definition 13 ([9]) *A* smart reduction *from* (C, D) *to* (A, B) *is a Turing reduction from* (C, D) *to* (A, B) *such that if the input belongs to* $C \cup D$, *then all queries belong to* $A \cup B$.

A disjoint pair $(A, B) \in$ DisjNP is *smart* $\leq_T^{pp}$-*complete* for DisjNP if for every (C, D) in DisjNP there is a smart reduction from (C, D) to (A, B). Note that if (A, B) is $\leq_m^{pp}$-complete for DisjNP, then (A, B) is smart $\leq_T^{pp}$-complete for DisjNP as well.

Let $\{N_i\}_i$ be an effective enumeration of nondeterministic, polynomial-time bounded Turing machines. Now we define the central concept of this section.

Definition 14 DisjNP *is* uniformly enumerable *if there is a total computable function* $f : \Sigma^* \to \Sigma^* \times \Sigma^*$ *such that*

1. $\forall (i, j) \in \text{range}(f)[(L(N_i), L(N_j)) \in \text{DisjNP}]$.
2. $\forall (C, D) \in \text{DisjNP} \; \exists (i, j)[(i, j) \in \text{range}(f) \wedge C = L(N_i) \wedge D = L(N_j)]$.

The following theorem is a slight simplification of the main result of Glaßer, Selman, and Sengupta [5].

Theorem 15. *The following are equivalent.*

1. *There is a* $\leq_m^{pp}$-*complete disjoint NP-pair.*
2. *There is a* $\leq_{sm}^{pp}$-*complete disjoint NP-pair.*
3. *There is a smart* $\leq_T^{pp}$-*complete disjoint NP-pair.*
4. *DisjNP is uniformly enumerable.*

There is a long history of equating having complete sets with uniform enumerations. Hartmanis and Hemachandra [10], for example, proved this for the class UP, and it holds as well for NP ∩ co-NP and BPP. More recently, Sadowski [24] proved that there exists an optimal propositional proof system if and only if the class of all easy subsets of TAUT is uniformly enumerable.[4] It seems inconceivable that there would exist a total computable function that lists exactly the disjoint NP-pairs, and that is why we don't believe that many-one

---

[4] By Corollary 8, if DisjNP is not uniformly enumerable, then the class of all easy subsets of TAUT is also not uniformly enumerable.

complete disjoint NP-pairs exist, and hence, don't believe that optimal proof systems exist.

The most interesting direction of the proof is to show that if there exists a many-one complete disjoint NP-pair, then DisjNP is uniformly enumerable. We sketch this direction now:

Let (A, B) be a $\leq_m^{pp}$-complete disjoint pair. Let N_A and N_B be NP-machines that accept A and B, respectively. Let $\{f_i\}_i$ be an effective enumeration of polynomial-time computable functions. Input to the enumerator is a number encoding a triple $\langle i, j, k \rangle$. Output is a pair $\langle a, b \rangle$ to be described.

Given $\langle i, j, k \rangle$, we define nondeterministic Turing machines N_1' and N_2' as follows. On input x, N_1' computes $f_i(x) = q$ and then simulates both $N_A(q)$ and $N_B(q)$. At most one of these accepts. N_1' accepts x if $x \in L(N_j)$ and $q \in L(N_A)$. N_2' is defined similarly, except that N_2' accepts x if $x \in L(N_k)$ and $q \in L(N_B)$.

Let a and b be the indices of N_1' and N_2', respectively, and define $f(\langle i, j, k \rangle) = \langle a, b \rangle$. It is easy to see that $L(N_a)$ and $L(N_b)$ are disjoint. So for all i, j, and k, $(L(N_a), L(N_b)) \in \text{DisjNP}$, where $f(\langle i, j, k \rangle) = \langle a, b \rangle$.

Now let (C, D) be a disjoint NP-pair. For some indices j and k, $C = L(N_j)$ and $D = L(N_k)$. Then $(C, D) \leq_m^{pp} (A, B)$ by f_i, for some i. Consider, $\langle a, b \rangle = f(\langle i, j, k \rangle)$. The remainder of the proof, which is easy, shows that $C = L(N_a)$ and $D = L(N_b)$.

References

1. S. Ben-David and A. Gringauze. On the existence of propositional proof systems and oracle-relativized propositional logic. Technical Report 5, Electronic Colloquium on Computational Complexity, 1998.
2. H. Buhrman, S. Fenner, L. Fortnow, and D. van Melkebeek. Optimal proof systems and sparse sets. In *Proceedings 17th Symposium on Theoretical Aspects of Computer Science*, volume 1770 of *Lecture Notes in Computer Science*, pages 407–418. Springer Verlag, 2000.
3. S. Cook and R. Reckhow. The relative efficiency of propositional proof systems. *Journal of Symbolic Logic*, 44:36–50, 1979.
4. S. Even, A. Selman, and J. Yacobi. The complexity of promise problems with applications to public-key cryptography. *Information and Control*, 61:159–173, 1984.
5. C. Glaßer, A. Selman, and S. Sengupta. Reductions between disjoint NP-pairs. In *Proceedings 19th IEEE Conference on Computational Complexity*, pages 42–53. IEEE Computer Society, 2004.
6. C. Glaßer, A. Selman, S. Sengupta, and L. Zhang. Disjoint NP-pairs. *SIAM Journal on Computing*, 33(6):1369–1416, 2004.
7. C. Glaßer, A. Selman, and L. Zhang. Canonical disjoint NP-pairs of proposional proof systems. Technical Report 04-106, Electronic Colloquium on Computational Complexity, 2004.
8. O. Goldreich. On promise problems, in memory of Shimon Even (1935–2004). *This volume*, 2005.
9. J. Grollmann and A. Selman. Complexity measures for public-key cryptosystems. *SIAM Journal on Computing*, 17(2):309–335, 1988.

10. J. Hartmanis and L. A. Hemachandra. Complexity classes without machines: On complete languages for UP. *Theoretical Computer Science*, 58:129–142, 1988.
11. S. Homer and A. Selman. Oracles for structural properties: The isomorphism problem and public-key cryptography. *Journal of Computer and System Sciences*, 44(2):287–301, 1992.
12. J. Köbler, J. Messner, and J. Torán. Optimal proof systems imply complete sets for promise classes. *Information and Computation*, 184(1):71–92, 2003.
13. J. Krajíček and P. Pudlák. Propositional proof systems, the consistency of first order theories and the complexity of computations. *Journal of Symbolic Logic*, 54:1063–1079, 1989.
14. R. Ladner. On the structure of polynomial-time reducibility. *Journal of the ACM*, 22:155–171, 1975.
15. L. Lovász. On the shannon capacity of graphs. *IEEE Transactions on Information Theory*, 25:1–7, 1979.
16. J. Messner. *On the Simulation Order of Proof Systems*. PhD thesis, Universität Ulm, 2000.
17. J. Messner. On the structure of the simulation order of proof systems. In *Proceedings 27rd Mathematical Foundations of Computer Science*, Lecture Notes in Computer Science 1450, pages 581–592. Springer-Verlag, 2002.
18. J. Meßner and J. Torán. Optimal proof systems for propositional logic and complete sets. In *Proceedings 15th Symposium on Theoretical Aspects of Computer Science*, Lecture Notes in Computer Science, pages 477–487. Springer Verlag, 1998.
19. P. Pudlák. On the length of proofs of finitistic consistency statements in first order theories. In J. B. Paris et al., editor, *Logic Colloquium '84*, pages 165–196. North-Holland Amsterdam, 1986.
20. P. Pudlák. On reducibility and symmetry of disjoint NP-pairs. In *Proceedings 26th International Symposium on Mathematical Foundations of Computer Science*, volume 2136 of *Lecture Notes in Computer Science*, pages 621–632. Springer-Verlag, Berlin, 2001.
21. A. Razborov. On provably disjoint NP-pairs. Technical Report TR94-006, Electronic Colloquium on Computational Complexity, 1994.
22. K. Regan. On diagonalization methods and the structure of language classes. In *Proceedings Foundations of Computation Theory*, volume 158 of *Lecture Notes in Computer Science*, pages 368–380. Springer Verlag, 1983.
23. K. Regan. The topology of provability in complexity theory. *Journal of Computer and System Sciences*, 36:384–432, 1988.
24. Z. Sadowski. On an optimal propositional proof system and the structure of easy subsets of TAUT. *Theoretical Computer Science*, 288(1):181–193, 2002.
25. U. Schöning. A uniform approach to obtain diagonal sets in complexity classes. *Theoretical Computer Science*, 18:95–103, 1982.
26. A. Selman. Promise problems complete for complexity classes. *Information and Computation*, 78:87–98, 1988.
27. L. G. Valiant. Relative complexity of checking and evaluation. *Information Processing Letters*, 5:20–23, 1976.

On Promise Problems: A Survey[*]

Oded Goldreich

Department of Computer Science, Weizmann Institute of Science, Rehovot, Israel
oded.goldreich@weizmann.ac.il

Abstract. The notion of promise problems was introduced and initially studied by Even, Selman and Yacobi (*Inform. and Control*, Vol. 61, pages 159–173, 1984). In this article we survey some of the applications that this notion has found in the twenty years that elapsed. These include the notion of "unique solutions", the formulation of "gap problems" as capturing various approximation tasks, the identification of complete problems (especially for the class $\mathcal{SZK}$), the indication of separations between certain computational resources, and the enabling of presentations that better distill the essence of various proofs.

1 Introduction

The Theory of Computation excels in identifying fundamental questions and formulating them at the right level of abstraction. Unfortunately, the field's preoccupation with innovation comes sometimes at the expense of paying relatively modest attention to the proper presentation of these fundamental questions and the corresponding notions and results. One striking example is the way the basics are being taught.[1]

For example, in typical *Theory of Computation* classes, the focus is on "language recognition" devices, and fundamental questions like "P versus NP" are presented in these terms (e.g., do deterministic polynomial-time machines accept the same languages as non-deterministic polynomial-time machines). In my opinion, such a formulation diminishes the importance of the problem in the eyes of non-bright students, and hides the fundamental nature of the question (which is evident when formulated in terms of "solving problems versus checking the correctness of solutions"). Similarly, one typically takes the students through the proof of Cook's Theorem before communicating to them the striking message that "universal" problems exist at all (let alone that many natural problems like SAT are universal). Furthermore, in some cases, this message is not communicated explicitly at all.

---

[*] This survey started as a private communication, calling an expert's attention to numerous applications of promise problems; specifically, to capturing various notions of approximation and to the study of statistical zero-knowledge. The current volume provided the immediate incentive to turn these sporadic notes into a more comprehensive survey.

[1] The interested reader is referred to the author's article "On Teaching the Basics of Complexity Theory" (this volume), which focuses on the following two examples.

O. Goldreich et al. (Eds.): Shimon Even Festschrift, LNCS 3895, pp. 254–290, 2006.

This article focuses on a less dramatic case of a bad perspective, but still one that deserves considerable attention: I refer to the notion of *promise problems*, and to its presentation in theory of computation classes. Let me start by posing the following rhetorical question:

How many readers have learned about *promise problems* in an undergraduate "theory of computation" course or even in a graduate course on complexity theory?

Scant few? And yet I contend that almost all readers refer to this notion when thinking about computational problems, although they may be often unaware of this fact.

1.1   What Are Promise Problems

My view is that any decision problem is a promise problem, although in some cases the promise is trivial or tractable (and is thus possible to overlook or ignore). Formally, a promise problem is a partition of the set of all strings into three subsets:

1. The set of strings representing YES-instances.
2. The set of strings representing NO-instances.
3. The set of disallowed strings (which represent neither YES-instances nor NO-instances).

The algorithm (or process) that is supposed to solve the promise problem is required to distinguish YES-instances from NO-instances, and is allowed arbitrary behavior on inputs that are neither YES-instances nor NO-instances. Intuitively, this algorithm (or rather its designer) is "promised" that the input is either a YES-instance or a NO-instance, and is only required to distinguish these two cases. Thus, the union of the first two sets (i.e., the set of all YES-instances and NO-instances) is called the promise.

In contrary to the common perception, in my opinion, promise problems are no offshoot for *abnormal* situations, but *are rather the norm*: Indeed, the standard and natural presentation of natural decision problems is actually in terms of promise problems, although the presentation rarely refers explicitly to the terminology of promise problems. Consider a standard entry in [17] (or any similar compendium) reading something like "given a planar graph, determine whether or not ..." A more formal statement will refer to strings that represent planar graphs. Either way, one may wonder what should the decision procedure do when the input is *not* a (string representing a) planar graph. One common formalistic answer is that all strings are interpreted as representations of planar graphs (typically, by using a decoding convention by which every "non-canonical" representation is interpreted as a representation of some fixed planar graph). Another (even more) formalistic "solution" is to discuss the problem of distinguishing YES-instances from anything else (i.e., effectively viewing strings that violate the promise as NO-instances). Both conventions miss the true nature of the original computational problem, which is concerned with distinguishing planar graphs of one type from planar graphs of another type (i.e., the complementary type). That is, the conceptually correct perspective is that the aforementioned problem is a promise problem in which the promise itself is an easily recognizable set.

But, as observed by Even, Selman and Yacobi [13], the promise need not be an easily recognizable set, and in such a case the issue cannot be pushed under the carpet. Indeed, consider a computational problem that, analogously to the one above, reads "given a Hamiltonian graph, determine whether or not ..." In this case, the two formalistic conventions mentioned above fail: The first one cannot be implemented, whereas the second one may drastically affect the complexity of the problem.

Jumping ahead, we mention that the formulation of promise problems is avoided not without reason. Firstly, it is slightly more cumbersome than the formulation of ordinary decision problems (having a trivial promise that consists of the set of all strings). More importantly, as observed by Even, Selman and Yacobi [13], in some cases "well-known" *structural* relations (which refer to standard decision problems) need not hold for promise problems (in which the promise itself is hard to test for membership). For example, the existence of a promise problem in $\mathcal{NP} \cap \text{co}\mathcal{NP}$ that is $\mathcal{NP}$-hard (under Cook-reduction) does not seem to imply that $\mathcal{NP} = \text{co}\mathcal{NP}$. Still, the benefits of formulating computational problems in terms of promise problems is often more than worth the aforementioned costs.

1.2   Some Definitions

In accordance with the above discussion, promise problems are defined as follows.

Definition 1 (promise problems): *A promise problem Π is a pair of non-intersecting sets, denoted $(\Pi_{\text{YES}}, \Pi_{\text{NO}})$; that is, $\Pi_{\text{YES}}, \Pi_{\text{NO}} \subseteq \{0,1\}^*$ and $\Pi_{\text{YES}} \cap \Pi_{\text{NO}} = \emptyset$. The set $\Pi_{\text{YES}} \cup \Pi_{\text{NO}}$ is called the* promise.

An alternative formulation, used in the original paper [13], is that a promise problem is a pair (P, Q), where P is the promise and Q is a super-set of the YES-instances. Indeed, in some cases, it is more natural to use the original formulation (e.g., let P be the set of Hamiltonian graphs and Q be the set of 3-colorable graphs), but Definition 1 refers more explicitly to the actual computational problem at hand (i.e., distinguishes inputs in $\Pi_{\text{YES}} = P \cap Q$ from inputs in $\Pi_{\text{NO}} = P \setminus Q$).

Standard "language recognition" problems are cast as the special case in which the promise is the set of all strings (i.e., $\Pi_{\text{YES}} \cup \Pi_{\text{NO}} = \{0,1\}^*$). In this case we say that the promise is trivial. The standard definitions of complexity classes (i.e., classes of languages) extend naturally to promise problems. In formulating such an extension, rather than thinking on the standard definition as referring to the set of YES-instances and its complement, one better think of it as referring to two (non-intersecting) sets: the set of YES-instances and the set of NO-instances. We thus have definitions of the following form.

Definition 2 (three classes of promise problems):[2]

$\mathcal{P}$ is the class of promise problems that are solvable in (deterministic) polynomial-time. That is, the promise problem $\Pi = (\Pi_{\text{YES}}, \Pi_{\text{NO}})$ is in $\mathcal{P}$ if there exists a polynomial-time algorithm A such that:
  - For every $x \in \Pi_{\text{YES}}$ it holds that $A(x) = 1$.
  - For every $x \in \Pi_{\text{NO}}$ it holds that $A(x) = 0$.

$\mathcal{NP}$ is the class of promise problems that have polynomially long proofs of membership that are verifiable in (deterministic) polynomial-time. That is, the promise problem $\Pi = (\Pi_{\text{YES}}, \Pi_{\text{NO}})$ is in $\mathcal{NP}$ if there exists a polynomially bounded binary relation R that is recognized by a polynomial-time algorithm such that:
  - For every $x \in \Pi_{\text{YES}}$ there exists y such that $(x, y) \in R$.
  - For every $x \in \Pi_{\text{NO}}$ and every y it holds that $(x, y) \notin R$.

  We say that $R \subseteq \{0, 1\}^* \times \{0, 1\}^*$ is polynomially bounded if there exists a polynomial p such that for every $(x, y) \in R$ it holds that $|y| \leq p(|x|)$, and R is recognized by algorithm A if $A(x, y) = 1$ if and only if $(x, y) \in R$.

$\mathcal{BPP}$ is the class of promise problems that are solvable in probabilistic polynomial-time. That is, the promise problem $\Pi = (\Pi_{\text{YES}}, \Pi_{\text{NO}})$ is in $\mathcal{BPP}$ if there exists a probabilistic polynomial-time algorithm A such that:
  - For every $x \in \Pi_{\text{YES}}$ it holds that $\Pr[A(x) = 1] \geq 2/3$.
  - For every $x \in \Pi_{\text{NO}}$ it holds that $\Pr[A(x) = 0] \geq 2/3$.

That is, in each case, the conditions used in the standard definition (of language recognition) are applied to the partition of the promise (i.e., $\Pi_{\text{YES}} \cup \Pi_{\text{NO}}$), and nothing is required with respect to inputs that violate the promise.

The notion of a reduction among computational problems also extends naturally to promise problems. The next definition extends the most basic type of reductions (i.e., Karp and Cook reductions).

Definition 3 (reductions among promise problems): *The promise problem $\Pi = (\Pi_{\text{YES}}, \Pi_{\text{NO}})$ is Karp-reducible to the promise problem $\Pi' = (\Pi'_{\text{YES}}, \Pi'_{\text{NO}})$ if there exists a polynomial-time computable function f such that:*

- *For every $x \in \Pi_{\text{YES}}$ it holds that $f(x) \in \Pi'_{\text{YES}}$.*
- *For every $x \in \Pi_{\text{NO}}$ it holds that $f(x) \in \Pi'_{\text{NO}}$.*

The promise problem $\Pi = (\Pi_{\text{YES}}, \Pi_{\text{NO}})$ is Cook-reducible to the promise problem $\Pi' = (\Pi'_{\text{YES}}, \Pi'_{\text{NO}})$ if there exists a polynomial-time oracle machine M such that:

- *For every $x \in \Pi_{\text{YES}}$ it holds that $M^{\Pi'}(x) = 1$.*
- *For every $x \in \Pi_{\text{NO}}$ it holds that $M^{\Pi'}(x) = 0$.*

---

[2] Indeed, the following classes "absorb" the standard language classes. When we wish to refer to the latter, we will use Roman font.

where query q to Π' is answered by 1 if $q \in \Pi'_{YES}$, by 0 if $q \in \Pi'_{NO}$, and arbitrarily otherwise. Alternatively, we may consider the computation of M when given access to any total function $\sigma : \{0,1\}^* \to \{0,1,\perp\}$ that satisfies $\sigma(x) = 1$ if $x \in \Pi'_{YES}$ and $\sigma(x) = 0$ if $x \in \Pi'_{NO}$, where for $x \notin \Pi'_{YES} \cup \Pi'_{NO}$ the value of $\sigma(x)$ may be anything (in $\{0,1,\perp\}$). Such a function σ is said to conform with Π'. We then require that there exists a polynomial-time oracle machine M such that for every total function $\sigma : \{0,1\}^* \to \{0,1,\perp\}$ that conforms with Π' the following holds:

  - *For every $x \in \Pi_{YES}$ it holds that $M^\sigma(x) = 1$.*
  - *For every $x \in \Pi_{NO}$ it holds that $M^\sigma(x) = 0$.*

Randomized reductions are defined analogously.

We stress that the convention by which queries that do not satisfy the promise may be answered arbitrarily is consistent with the notion of solving a promise problem. Recall that solving the latter means providing correct answers to instances that satisfy the promise, whereas nothing is required of the "solver" in case it is given an instance that violates the promise. In particular, such a potential "solver" (represented by σ in the alternative formulation) may either provide wrong answers to instances that violate the promise or provide no answer at all (as captured by the case $\sigma(x) = \perp$). On the other hand, reductions are supposed to capture what can be done when given access to a device (represented by σ) that solves the problem at the target of the reduction. Thus, a reduction to a promise problem should yield the correct answer regardless of how one answers queries that violate the promise. We stress that the standard meaning of a reduction is preserved: *if Π is Cook-reducible to a promise problem in $\mathcal{P}$ (or in $\mathcal{BPP}$) then Π is in $\mathcal{P}$ (resp., in $\mathcal{BPP}$).*

The foregoing natural convention (regarding oracle calls to a promise problem) is the source of technical problems. In particular, unlike in the case of languages, a Cook-reduction to a promise problem in $\mathcal{NP} \cap \mathrm{co}\mathcal{NP}$ does not guarantee that the reduced problem is in $\mathcal{NP}$. (For further discussion, see Section 5.1. We stress, again, that a Cook-reduction to a promise problem in $\mathcal{P}$ does guarantee that the reduced problem is in $\mathcal{P}$.)

1.3   Some Indispensable Uses of Promise Problems

As argued in Section 1.1, promise problems are actually more natural than language recognition problems, and the latter are preferred mainly for sake of technical convenience (i.e., using less cumbersome formulations). However, in many cases, promise problems are indispensable for capturing important computational relations. For example, the notion of one computational problem being a special case (or a restriction) of another problem is best captured this way: The promise problem $\Pi = (\Pi_{YES}, \Pi_{NO})$ is a special case of $\Pi' = (\Pi'_{YES}, \Pi'_{NO})$ if both $\Pi_{YES} \subseteq \Pi'_{YES}$ and $\Pi_{NO} \subseteq \Pi'_{NO}$.

The above paragraph refers to the importance of promise problems in providing the nicest presentation of simple ideas, where by a nice presentation we mean one in which conceptual issues are explicitly represented (rather than hidden by technical conventions). We note that when simple ideas are concerned one may survive ugly presentations, but this becomes more difficult when the issues at hand are less simple. Furthermore, in some cases the notion of a promise problem is essential to the main results themselves. Most of this article will be devoted to surveying some of these cases, and a brief overview of some of them follows.

1. The study of the complexity of problems with unique solution must be formally cast in terms of promise problems. For example, unique-SAT is the promise problem having as YES-instances Boolean formulas that have a unique satisfying assignment and having as NO-instances unsatisfiable Boolean formulas. (See Section 2 for further discussion.)
2. The study of the hardness of approximation problems may be formally cast in terms of promise problems. This is especially appealing when one wants to establish the hardness of obtaining an approximation of the optimal value. Specifically, one often refers to "gap problems" which are promise problems having as YES-instances objects that have a relatively high (resp., low) optimum value and NO-instances that are objects with relatively low (resp., high) optimum value. (See Section 3 for further discussion.)
3. Promise problem allow to introduce complete problems for classes that are not known to have complete languages. A notable example is the class $\mathcal{BPP}$, and another important one is $\mathcal{SZK}$ (i.e., the class of problems having statistical zero-knowledge proof systems). Indeed, promise problems have played a key role in the study of the latter class. (See Section 4 for further discussion.)
4. Promise problem were used to indicate separations between certain computational devices with certain resource bounds. Examples appeared in the study of circuit complexity, derandomization, PCPs, and zero-knowledge. (See Section 5.2 for further discussion.)

Finally, we wish to call attention also to the expositional benefits of promise problems, further discussed in Sections 6 and 7. In particular, Section 6.1 discusses their application for proving various complexity lower-bounds, while in Section 6.2 they are used to distill the essence of a known result (i.e., BPP ⊆ PH). In Section 7 we present a suggestion for casting various "modified" complexity classes (i.e., "computations that take advice" and "infinitely often" classes) in terms of the classes themselves where the latter are understood as classes of promise problems.

1.4   Relation to Shimon Even (A Personal Comment)

As hinted above, promise problems were explicitly introduced by Even, Selman and Yacobi [13], and their study was initiated in [13]. In my opinion, the powerful combination of the natural notion that promise problems capture, their simple definition, and their wide applicability is one of Shimon Even's trade-marks. I vividly recall him telling me in one of our first meetings:

> *The very simple facts and the basic approaches are the ones that have most impact; they are the ones that get disseminated across the disciplines and even influence other disciplines. A work's most influential contribution may be introducing a good notation.*

Needless to say, science progresses by coping with difficult problems. Most scientific works are too complicated to have a far-reaching impact by themselves, but at times they lead to paradigm shifts that do have far-reaching impact, as argued by Kuhn [41]. These paradigm shifts, which are the most important contributions of science, are typically simple from a technical point of view. Thus, both Even and Kuhn viewed simplicity (at the frontier of science) as positively correlated with impact and importance.

In view of the above, I believe that in surveying the notion of promise problems and its wide applicability, I am surveying a central theme in Shimon's research, a theme that is prominently present also in other works of his.

1.5   A Comment About the Organization

In addition to the main sections mentioned above, the survey contains three appendices that provide further details regarding some of the results mention in the main text. These appendices may be ignored with no loss to the conceptual message of this survey. Among the three appendices, Appendices B and C are most relevant to the main message, because they offer a closer look at the role of promise problems in the surveyed results. In contrast, Appendix A demonstrates that reductions to promise problems may cleverly utilize queries that violate the promise (an issue further addressed in Section 5.1).

2   Unique Solutions and Approximate Counting of Solutions

In this section, we review the use of promise problems in stating central results regarding the complexity of finding unique solutions and the complexity of approximating the number of solutions to NP-problems. We call the reader's attention to the indispensable role of promise problems in the definition of "problems with unique solutions" and their role in formulating a decision version of the problem of "approximate counting". The latter theme will reappear in Section 3.

2.1   The Complexity of Finding Unique Solutions

The widely believed intractability of SAT cannot be due to instances that have a "noticeable fraction" of satisfying assignments. For example, given an n-variable formula that has at least $2^n/n$ satisfying assignments, it is easy to find a satisfying assignment (by trying $O(n)$ assignments at random). Going to the other extreme, one may ask whether or not it is easy to find satisfying assignments to SAT instances that have very few satisfying assignments (e.g., a unique satisfying assignment). As shown by Valiant and Vazirani [58], the answer is negative: the ability to find satisfying assignments to such instances yields the ability to find satisfying assignments to arbitrary instances. Actually, they showed that distinguishing *uniquely* satisfiable formulae from unsatisfied ones is not easier than distinguishing satisfiable formulae from unsatisfied ones.

In order to formulate the above discussion, we refer to the notion of promise problems. Specifically, we refer to the promise problem of distinguishing instances with a unique solution from instances with no solution. For example, unique-SAT (or uSAT) is the promise problem with YES-instances being formulae having a unique satisfying assignment and NO-instances being formulae having no satisfying assignment.

Theorem 4 [58]: SAT *is randomly reducible to* uSAT. *That is, there exists a randomized Cook-reduction of* SAT *to* uSAT.

A proof sketch is presented in Appendix A. The same result holds for any known NP-complete problem; in some cases this can be proven directly and in other cases by using suitable parsimonious reductions.[3]

2.2   The Complexity of Approximately Counting the Number of Solutions

A natural computational problem associated with an NP-relation R is to determine the number of solutions for a given instance; that is, given x, determine the cardinality of $R(x) \stackrel{\text{def}}{=} \{y : (x, y) \in R\}$. Certainly, the aforementioned counting problem associated with R is not easier than the problem of deciding membership in $L_R = \{x : \exists y \text{ s.t. } (x, y) \in R\}$, which can be cast as determining, for a given x, whether $|R(x)|$ is positive or zero.

We focus on the problem of approximating $|R(x)|$, when given x, up to a factor of $f(|x|)$, for some function $f : \mathsf{N} \to \{r \in \mathsf{R} : r > 1\}$ (which is bounded away from 1). Formulating this problem in terms of decision problems has several advantages (see analogous discussion at the end of Section 3.1), and can be done via promise problems. Specifically, the problem of approximating $|R(x)|$ can be cast as a promise problem, denoted $\#R^f$, such that the YES-instances are pairs (x, N) satisfying $|R(x)| \geq N$ whereas the NO-instances are pairs (x, N) satisfying $|R(x)| < N/f(|x|)$. Indeed, for every $f : \mathsf{N} \to \mathsf{R}$ such that $f(n) > 1 + (1/\mathrm{poly}(n))$, approximating $|R(x)|$ up to a factor of $f(|x|)$ is Cook-reducible to deciding $\#R^f$.[4]

---

[3] A parsimonious reduction (between NP-sets) is a Karp-reduction that preserves the number of solutions (i.e., NP-witnesses). That is, for NP-sets $L_R = \{x : (\exists y)(x, y) \in R\}$ and $L_{R'} = \{x : (\exists y)(x, y) \in R'\}$, the mapping f is a parsimonious reduction from L_R to $L_{R'}$ if for every x it holds that $|R'(f(x))| = |R(x)|$, where $R(x) \stackrel{\text{def}}{=} \{y : (x, y) \in R\}$ and $R'(x') \stackrel{\text{def}}{=} \{y' : (x', y') \in R'\}$.

[4] On input x, the Cook-reduction issues the queries $(x, f(|x|)^i)$, for $i = 0, 1, ..., \ell$, where $\ell = \mathrm{poly}(|x|)/\log_2(f(|x|))$. The oracle machine returns 0 if the first query was answered by 0, and $f(|x|)^i$ if i is the largest integer such that $(x, f(|x|)^i)$ was answered by 1.

Clearly, for every $f : \mathsf{N} \to \{r \in \mathsf{R} : r \geq 1\}$, deciding $\#R^f$ is at least as hard as deciding L_R. Interestingly, for any f that is bounded away from 1 and for any known NP-relation R, deciding $\#R^f$ is not harder than deciding L_R. We state this fact for the witness relation of SAT, denoted R_{SAT}.

Theorem 5 [53]: *For every $f : \mathsf{N} \to \mathsf{R}$ such that $f(n) > 1 + (1/\mathrm{poly}(n))$, the counting problem $\#R^f_{\mathsf{SAT}}$ is randomly Karp-reducible to* SAT.

A proof sketch is presented in Appendix A. The same result holds for any known NP-complete problem; in some cases this can be proven directly and in others by using suitable parsimonious reductions.

3   Gap Problems – Representing Notions of Approximation

Gap problems are a special type of promise problems in which instances are partitioned according to some metric leaving a "gap" between YES-instances and NO-instances. We consider two such metrics: in the first metric instances are positioned according to the value of the best corresponding "solution" (with respect to some predetermined objective function), whereas in the second metric instances are positioned according to their distance from the set of objects that satisfy some predetermined property.

3.1   Approximating the Value of an Optimal Solution

When constructing efficient approximation algorithms, one typically presents algorithms that given an instance find an almost-optimal solution, with respect to some desired objective function, rather than merely the value of such a solution. After all, in many settings, one seeks a solution rather than merely its value, and typically the value is easy to determine from the solution itself, thus making the positive result stronger. However, when proving negative results (i.e., hardness of approximation results), it is natural to consider the possibly easier task of approximating the value of an optimal solution (rather than finding the solution itself). This makes the negative result stronger, and typically makes the proof more clear.

Promise problems are the natural vehicle for casting computational problems that refer to approximating the value of an optimal solution. Specifically, one often refers to "gap problems" that are promise problem having as YES-instances objects that have a relatively high (resp., low) optimum value and NO-instances that are objects with relatively low (resp., high) optimum value. Indeed, this has been the standard practice since [7].

Let us demonstrate this approach by considering the known results regarding several famous approximation problems. For example, the complexity of Max-Clique is captured by the gap problem $\mathtt{gapClique}_{b,s}$, where b and s are functions of the number of vertices in the instance graph. The problem $\mathtt{gapClique}_{b,s}$ is a promise problem consisting of YES-instances that are N-vertex graphs containing a clique of size $b(N)$ and NO-instances that are N-vertex graphs containing no clique of size $s(N)$. Hastad's celebrated result asserts that, for every $\epsilon \in (0, 1/2)$, the promise problem $\texttt{gapClique}_{b_\epsilon, s_\epsilon}$ is NP-hard (under probabilistic Karp-reductions) [35], where $b_\epsilon(N) = N^{1-\epsilon}$ and $s_\epsilon(N) = N^\epsilon$.

Another famous approximation problem is Max3SAT. For any constant $s \in (0, 1)$, consider the gap problem $\texttt{gap3SAT}_s$ that consists of YES-instances that are satisfiable 3CNF formulae and NO-instances that are 3CNF formulae in which every truth assignment satisfies less than an s fraction of the clauses.[5] Note that the gap problem $\texttt{gap3SAT}_{7/8}$ is trivial, because every 3CNF formula has a truth assignment that satisfies at least a 7/8 fraction of its clauses. On the other hand, Hastad showed that, for every $\epsilon \in (0, 1/8)$, the promise problem $\texttt{gap3SAT}_{(7/8)+\epsilon}$ is NP-hard (under Karp-reductions) [36].

On the benefits of the framework of gap problems. The reader may wonder how essential is the use of gap problems in stating results of the aforementioned type. Indeed, one often states the Max-Clique result by saying that, for every $\epsilon > 0$, it is NP-hard to approximate the size of the maximum-clique in an N-vertex graph to within a factor of $N^{1-\epsilon}$. Firstly, we comment that the latter is merely a corollary of Hastad's result [35], which is actually a (randomized) Karp-reduction of $\mathcal{NP}$ to $\texttt{gapClique}_{N^{1-\epsilon}, N^\epsilon}$. The same holds with respect to all hardness of approximation results that are obtained through PCPs: They are obtained by Karp-reductions of PCPs with certain parameters to gap problems, where the former PCPs are shown to exist for $\mathcal{NP}$. In our opinion, it is nicer to present these results as hardness of certain gap problems (which reflects what is actually proved), and their meaning is at least as clear when stated in this way. More importantly, in some cases information is lost when using the "approximation factor" formulation. Consider for example the assertion that, for every $\epsilon \in (0, 1/8)$, it is NP-hard to approximate Max3SAT to within a factor of $(7/8) + \epsilon$. The latter assertion does not rule out the possibility that, given a satisfiable 3CNF formula, one can find an assignment that satisfies 90% of the clauses. This possibility is ruled out by the fact that $\texttt{gap3SAT}_{9/10}$ is NP-hard, and we comment that proving the latter result seems to require more work than proving the former [36].[6] Lastly, the formulation of promise problems seems essential to "reversing the PCP to approximation" connection [7, Sec. 8] (i.e., showing that certain NP-hardness results regarding approximation yield PCP systems with certain parameters).

---

[5] By a 3CNF formula we mean a conjunction of clauses, each consisting of exactly three different literals.

[6] Specifically, proving that $\texttt{gap3SAT}_{(7/8)+\epsilon}$ is NP-hard seems to require using a PCP with "perfect completeness" (as constructed in [36, Thm. 3.4]), whereas Hastad's initial construction [36, Thm. 2.3] does not have perfect completeness (and establishes the NP-hardness of distinguishing 3CNF formulae having a truth assignment that satisfies at least $1 - \epsilon$ fraction of the clauses from 3CNF formulae in which every truth assignment satisfies less than a $(7/8) + \epsilon$ fraction of the clauses [36, Thm. 3.1]).

3.2 Property Testing – The Distance Between YES and NO-Instances

In some sense, all research regarding property testing (cf. [50, 23]) can be cast in terms of promise problems, although this is typically not done – for reasons discussed below.

Property testing is a relaxation of decision problems, where the (typically sub-linear time) algorithm is required to accept (with high probability) any instance having the property (i.e., any instance in some predetermined set) and reject (with high probability) any instance that is "far from having the property" (i.e., being at large distance from any instance in the set). The algorithm, called a tester, may run in sub-linear time because it is given oracle access to the tested object, and thus need not read it entirely. We comment that, in all interesting cases, this algorithm needs to be probabilistic.

Typically, the distance parameter is given as input to the tester (rather than being fixed as in Section 3.1)[7], which makes the positive results stronger and more appealing (especially in light of a separation recently shown in [5]). In contrast, negative results typically refer to a fixed value of the distance parameter. Thus, for any *distance function* (e.g., Hamming distance between bit strings) and any property P, two natural types of promise problems emerge:

1. *Testing w.r.t variable distance*: Here instances are pairs (x, δ), where x is a description of an object and δ is a distance parameter. The YES-instances are pairs (x, δ) such that x has property P, whereas (x, δ) is a NO-instance if x is δ-far from any x' that has property P.
2. *Testing w.r.t a fixed distance*: Here we fix the distance parameter δ, and so the instances are merely descriptions of objects, and the partition to YES and NO instances is as above.

For example, for some fixed integer d, consider the following promise problem, denoted BPG$_d$, regarding bipartiteness of bounded-degree graphs. The YES-instance are pairs (G, δ) such that G is a bipartite graph of maximum degree d, whereas (G, δ) is a NO-instance if G is an N-vertex graph of maximum degree d such that more than $\delta \cdot dN/2$ edges must be omitted from G in order to obtain a bipartite graph. Similarly, for fixed integer d and $\delta > 0$, the promise problem BPG$_{d,\delta}$ has YES-instances that are bipartite graphs of maximum degree d and NO-instances that are N-vertex graphs of maximum degree d such that more than $\delta \cdot dN/2$ edges must be omitted from the graph in order to obtain a bipartite graph. In [26] it was shown that any tester for BPG$_{3, 0.01}$ must make $\Omega(\sqrt{N})$ queries (to the description of the graph, given as an oracle). In contrast, for every d and δ, the tester presented in [27] decides BPG$_{d,\delta}$ in time $\widetilde{O}(\sqrt{N}/\mathrm{poly}(\delta))$. In fact, this algorithm decides BPG$_d$ in time $\widetilde{O}(\sqrt{N}/\mathrm{poly}(\delta))$, where N and δ are explicitly given parameters.

---

[7] In fact, an analogous treatment applies to approximation problems as briefly surveyed in Section 3.1. Indeed a formulation of approximation problems in which the approximation factor is part of the input corresponds to the notion of an approximation scheme (which is not surveyed here).

The formulation typically used in the literature. Indeed, all research on property testing refers to the two aforementioned types of promise problems, where typically positive results refer to the first type and negative results refer to the second type. However, most works do not provide a strictly formal statement of their results (see further discussion below), because the formulation is rather cumbersome and straightforward. Furthermore, in light of the greater focus on positive results (and in accordance with the traditions of algorithmic research), such a formal statement is believed to be unnecessary.[8] Let us consider what is required for a formal statement of property testing results. The starting point is a specification of a property and a distance function, the combination of which yields a promise problem (of the first type), although the latter fact is never stated. The first step is to postulate that the potential "solvers" (i.e., property testers) are probabilistic oracle machines that are given oracle access to the "primary" input (i.e., the object in the aforementioned problem types). Indeed, this step need to be taken and is taken in all works in the area. Secondly, for a formal asymptotic complexity statement, one needs to specify the "secondary" (explicit) inputs, which consist of various problem-dependent parameters (e.g., N in the above examples) and the distance parameter δ (in case of BPG_d and any other problem of the first aforementioned type). This step is rarely done explicitly in the literature. Finally, one should state the complexity of the tester in terms of these explicit inputs.

4   Promise Problems Provide Complete Problems

Most of this section is devote to the key role that promise problems have played in the study of Statistical Zero-Knowledge proof systems. However, we start by reviewing the situation in the seemingly lower complexity class $\mathcal{BPP}$.

4.1   A Complete Problem for BPP

In terms of language recognition, finding a complete problem for BPP is a long-standing challenge. The same hold for establishing hierarchy theorems for BP-time (cf. [6, 16]). However, in terms of promise problems, both challenges are rather easy (as is the case for analogous questions regarding $\mathcal{P}$). Indeed, the following promise problem is complete (under deterministic Karp-reductions) for the (promise problem) class $\mathcal{BPP}$: The YES-instances are Boolean circuits that evaluate to 1 on at least a 2/3 fraction of their inputs, whereas the NO-instances are Boolean circuits that evaluate to 0 on at least a 2/3 fraction of their inputs. (Thus, the promise "rules out" circuits that evaluate to 1 on a p fraction of their

---

[8] Needless to say, a higher level of rigor is typically required in negative statements. Indeed, property testing is positioned between algorithmic research and complexity theory, and seems to be more influenced by the mind-frame of algorithmic research. (We comment that the positioning of a discipline is determined both by its contents and by sociology-of-science factors.)

input, where $p \in (1/3, 2/3)$.) A reduction from $\Pi \in \mathcal{BPP}$ to the aforementioned promise problem merely maps x to C_x, where C_x is a circuit that on input r emulates the computation of M on input x and random-tape r, where M is a probabilistic polynomial-time machine deciding Π.

Needless to say, the above also holds with respect to other complexity classes that are aimed to capture efficient randomized computation (e.g., $\mathcal{RP}$ and $\mathcal{ZPP}$).

4.2   Complete Problems for Statistical Zero-Knowledge

Statistical zero-knowledge (SZK) is a subclass of standard zero-knowledge (ZK, aka computational zero-knowledge), where the simulation requirement is more strict (i.e., requiring simulation that is statistically close to the true interaction rather than only computationally indistinguishable from it). For background see either [19, Chap. 4] or [20]. Typically (as is the case in all results reviewed below), the study of SZK is carried out *without* referring to any intractability assumptions (in contrast to the study of standard ZK, which is usually based on one-way functions; cf. [25] but see [56] for a recent exception).

Promise problem have played a key role in the comprehensive study of statistical zero-knowledge. (This study was carried out in the late 1990's and is nicely summarized in Vadhan's PhD Thesis [55].) This study of statistical zero-knowledge (SZK) was conducted by presenting and extensively studying two complete (promise) problems for the (promise problem) class $\mathcal{SZK}$. Specifically, these promise problems facilitate the establishment of various important properties of the class $\mathcal{SZK}$, because the definition of these promise problems is very simple in comparison to the actual definition of the class $\mathcal{SZK}$. Furthermore, the fact that the class has natural complete problems is of independent interest.

The two aforementioned complete problems are gapSD and gapENT, introduced and shown complete for $\mathcal{SZK}$ in [51] and [31], respectively. Both problems refer to pairs of distributions, where each distribution is represented by a "sampling circuit" (i.e., a circuit C represents the distribution seen at its output wires when feeding the input wires with uniformly distributed values). The YES-instances of gapSD are distributions that are at (statistical) distance at most $1/3$ apart, and the NO-instances are distributions that are at distance at least $2/3$ apart. The YES-instances of gapENT are pairs of distributions in which the first distribution has entropy greater by one unit than the entropy of the second distribution, and in the NO-instances the first distribution has entropy that is smaller by one unit from the entropy of the second distribution.

To demonstrate the power of the complete problem approach to the study of $\mathcal{SZK}$, note that the fact that gapENT is complete (under Karp-reductions) for $\mathcal{SZK}$ immediately implies that $\mathcal{SZK}$ is closed under complementation, which is a highly non-trivial result. For a more detailed presentation, which highlights the role of promise problems in the study of $\mathcal{SZK}$, the interested reader is referred to Appendix B.

5   Promise Problems as Indicators of Complexity: Pros and Cons

Given the common desire to appeal to traditional notions, one typically tries to avoid promise problems and formulate the assertions in terms of language recognition problems. As we have seen in previous sections, in some cases this desire can not be satisfied due to inherent (or seemingly inherent) reasons. In other cases, one turns to promise problems after failing to prove an analogous result for language recognition problems, although there seems to be no inherent reason to justify the failure (see examples in Section 5.2). The question, however, is whether we lose something important when working with promise problems (rather than with language recognition problems). Since we have already seen some of the benefit of promise problems, we start by considering the dark side (i.e., the latter question).

5.1   Con: The Failure of Some Structural Consequences

The problem with results regarding promise problems is that sometimes they do not have the same *structural* consequences as analogous results regarding language recognition. The most notorious example is that the existence of an $\mathcal{NP}$-hard (under Cook reductions) promise problem in $\mathcal{NP} \cap \mathrm{co}\mathcal{NP}$ does not seem to have any structural consequences, whereas an analogous result for a language recognition problem implies that $\mathcal{NP} = \mathrm{co}\mathcal{NP}$ (see Theorem 7 below). This fact was observed by Even, Selman and Yacobi [13], who presented the following $\mathcal{NP}$-complete problem, denoted xSAT: The YES-instances are pairs (ϕ_1, ϕ_2) such that $\phi_1 \in \mathsf{SAT}$ and $\phi_2 \notin \mathsf{SAT}$, whereas the NO-instances are pairs (ϕ_1, ϕ_2) such that $\phi_1 \notin \mathsf{SAT}$ and $\phi_2 \in \mathsf{SAT}$.

Theorem 6 [13, Thm. 4]: *$\mathcal{NP}$ is Cook-reducible to xSAT, which in turn is in $\mathcal{NP} \cap \mathrm{co}\mathcal{NP}$.*

Proof sketch: To see that xSAT is in $\mathcal{NP}$, consider the witness relation $R_1 = \{((\phi_1, \phi_2), \tau) : (\phi_1, \tau) \in R_{\mathsf{SAT}}\}$, whereas xSAT is in $\mathrm{co}\mathcal{NP}$ by virtue of the witness relation $R_2 = \{((\phi_1, \phi_2), \tau) : (\phi_2, \tau) \in R_{\mathsf{SAT}}\}$. A Cook-reduction of SAT to xSAT may consist of the following oracle machine that, on input a formula ϕ, tries to find a satisfying assignment to ϕ, and accepts if and only if it succeeds. On input ϕ and oracle access to xSAT, the machine proceeds as follows, starting with $\phi_\lambda \stackrel{\mathrm{def}}{=} \phi$ and $\tau = \lambda$ (the empty prefix of a potential satisfying assignment), and continuing as long as ϕ_τ has free variables.

1. Let $\phi_{\tau\sigma}$ be the formula obtained from ϕ_τ by setting the $|\tau| + 1^{\mathrm{st}}$ variable to σ.
2. Invoke the oracle on query $(\phi_{\tau 1}, \phi_{\tau 0})$. If the answer is 1 then let $\tau \leftarrow \tau 1$, otherwise $\tau \leftarrow \tau 0$.

Note that if ϕ_τ is satisfiable and the query $(\phi_{\tau 1}, \phi_{\tau 0})$ is answered with σ then $\phi_{\tau\sigma}$ is satisfiable, because the claim holds trivially if both $\phi_{\tau 1}$ and $\phi_{\tau 0}$ are satisfiable, and the oracle answer is definitely correct if exactly one of these formulae is satisfiable (since the promise is satisfied in this case). Thus, the above process finds a satisfying assignment to ϕ if and only if one exists. ■

What happened? We stress that a Cook-reduction to a promise problem does maintain the standard meaning of the concept; that is, if the target (promise) problem is tractable (i.e., is in $\mathcal{P}$ or $\mathcal{BPP}$) then so is the reduced problem. The issue is that if the target problem is in $\mathcal{NP} \cap \mathrm{co}\mathcal{NP}$ then (unlike in the case of trivial promises (i.e., language recognition problems)) it does not necessarily follow that the reduced problem is in $\mathcal{NP} \cap \mathrm{co}\mathcal{NP}$. This fact will be clarified by looking at the proof of Theorem 7, which refers to "smart reductions" to promise problems.

Note that the reduction used in the proof of Theorem 6 may make queries that violate the promise. Still, we have shown that the reduction remains valid regardless of the answers given to these queries (i.e., to queries that violate the promise). However, these queries fail the aforementioned structural consequences. One may eliminate the problems arising from such queries by requiring that the reduction does not make them (i.e., does not make queries that violate the promise). Such a reduction is called smart [34] (probably because it is smart to avoid making queries that violate the promise, although one may argue that it is even more clever to be able to use answers to such queries). Note that any Karp-reduction is smart. Smart reductions maintain the structural consequences established in the case of language recognition problems.

Theorem 7 [34, Thm. 2]: *Suppose that the promise problem Π' is reducible to the promise problem $\Pi = (\Pi_{\mathrm{YES}}, \Pi_{\mathrm{NO}})$ via a smart reduction, and that $\Pi \in \mathcal{NP} \cap \mathrm{co}\mathcal{NP}$. Then $\Pi' \in \mathcal{NP} \cap \mathrm{co}\mathcal{NP}$.*

Proof sketch: We prove that $\Pi' \in \mathcal{NP}$ and the proof that $\Pi' \in \mathrm{co}\mathcal{NP}$ is similar. Let M be the polynomial-time oracle machine guaranteed by the hypothesis. The transcript of the execution of $M^{\Pi}(x)$ contains the sequence of queries and answers to the oracle as well as the final decision of M, but the transcript itself (as a string) does not guarantee the correctness of the answers and thus the authenticity of the execution. The key observation is that the said answers can be augmented by corresponding NP-witnesses that guarantee the correctness of the answers, and thus the authenticity of the execution.

Specifically, on any input x (which satisfies the promise of Π'), machine M makes queries that are either in Π_{YES} or in Π_{NO}, and in each of these cases there is an NP-witness guaranteeing the correctness of the answer (because $\Pi \in \mathcal{NP}$ and $\Pi \in \mathrm{co}\mathcal{NP}$). Thus, an NP-witness for x may consist of the sequence of (answers and) corresponding NP-witnesses, each proving either that the query is in Π_{YES} or that the query is in Π_{NO}, thus certifying the correctness of the answers. Indeed, these NP-witnesses are all correct, because it is guaranteed that each query satisfies the promise (since the reduction is smart). Note that this sequence of NP-witnesses uniquely determines the execution of M, on input x

and oracle access to Π, and thus vouches for the correctness of the outcome of this computation. ∎

In contrast to the proof of Theorem 7, note that a query that violates the promise does not necessarily have an NP-witness (e.g., asserting that it violates the promise, or anything else). Thus, we cannot insist on having NP-witnesses for all queries, and once we allow "uncertified answers" (i.e., answers not backed by NP-witnesses) all bets are off.

Another look. Indeed, smart reduction salvage the structural consequences of reductions to language recognition problems, but this comes at the cost of restricting the consequences to smart reductions. That is, for $\Pi \in \mathcal{NP} \cap \mathrm{co}\mathcal{NP}$, rather than saying "if Π is NP-hard then $\mathcal{NP} = \mathrm{co}\mathcal{NP}$" one may only say "if Π is NP-hard under smart reductions then $\mathcal{NP} = \mathrm{co}\mathcal{NP}$". However, there is another way out, provided we know more about the promise problem $\Pi = (\Pi_{\mathrm{YES}}, \Pi_{\mathrm{NO}})$. For example, suppose that in addition to knowing that $\Pi \in \mathcal{NP} \cap \mathrm{co}\mathcal{NP}$, we know that the set Π_{NO} is in $\mathrm{co}\mathcal{NP}$ (i.e., $(\{0,1\}^* \setminus \Pi_{\mathrm{NO}}, \Pi_{\mathrm{NO}}) \in \mathcal{NP}$). Then, we can ask for NP-witnesses asserting either membership in $\{0,1\}^* \setminus \Pi_{\mathrm{NO}}$ or membership in Π_{NO}.

Theorem 8 (implicit in [11], see [21]):[9] *Suppose that the promise problem Π' is reducible to the promise problem $\Pi = (\Pi_{\mathrm{YES}}, \Pi_{\mathrm{NO}}) \in \mathrm{co}\mathcal{NP}$ and that $(\{0,1\}^* \setminus \Pi_{\mathrm{NO}}, \Pi_{\mathrm{NO}}) \in \mathcal{NP}$. Then $\Pi' \in \mathcal{NP} \cap \mathrm{co}\mathcal{NP}$.*

Note that $(\{0,1\}^* \setminus \Pi_{\mathrm{NO}}, \Pi_{\mathrm{NO}}) \in \mathcal{NP}$ implies that $\Pi = (\Pi_{\mathrm{YES}}, \Pi_{\mathrm{NO}}) \in \mathcal{NP}$, and thus the latter was not stated as a hypothesis in Theorem 8. To demonstrate the applicability of Theorem 8, we mention that it was recently shown (cf. [2] improving upon [22]) that certain promise problems (i.e., gap problems) regarding lattices are in $\mathcal{NP} \cap \mathrm{co}\mathcal{NP}$. It is actually obvious that the set of the corresponding NO-instances is in $\mathrm{co}\mathcal{NP}$. Applying Theorem 8, it follows that these (gap) problems are unlikely to be $\mathcal{NP}$-hard (rather than restricting the claim to smart reductions).

Proof sketch: Following the proof of Theorem 7, an NP-witness for x may consist of the sequence of (answers and) corresponding NP-witnesses, each "proving" either that the query is in $\{0,1\}^* \setminus \Pi_{\mathrm{NO}}$ or that the query is in Π_{NO}. Note that these witnesses exist for every query, but indeed, in case the query violates the

---

[9] This theorem is implicit in [11], which observes an oversight of [22]. In [22] certain gap problems regarding lattices were shown to be in $\mathcal{NP} \cap \mathrm{co}\mathcal{AM}$, and it was inferred that these (gap) problems are unlikely to be $\mathcal{NP}$-hard under smart reductions (because such a reduction will imply that $\mathcal{AM} = \mathrm{co}\mathcal{AM}$, which in turn will cause collapse of the Polynomial-time Hierarchy). In [11] it was observed that these problems are unlikely to be $\mathcal{NP}$-hard (under any Cook-reduction). Specifically, they showed that, for these gap problems, the argument of Theorem 7 can be extended using NP-witnesses that exist for the corresponding set $\{0,1\}^* \setminus \Pi_{\mathrm{NO}}$. This argument was abstracted in [21], where a theorem analogous to Theorem 8 is presented (referring to $\mathcal{AM}$ rather than to $\mathcal{NP}$).

promise, witnesses may exist to both claims. Still, the witnesses do guarantee the correctness of all answers to queries that satisfy the promise (although they do not indicate which queries satisfy the promise). However, guaranteeing the correctness of all queries that satisfy the promise suffices for guaranteeing the correctness of the outcome of the computation. Thus, although the sequence of witnesses does not determine (uniquely) the execution of M on input x and oracle access to Π, it does vouch for the correctness of the outcome of the computation. ∎

Generalization of Theorem 8. The following elegant generalization of Theorem 8 was suggested to us by Salil Vadhan. It considers two sets, S_Y and S_N, such that S_Y (resp., S_N) contains all YES-instances (resp., NO-instances) of Π but none of the NO-instances (resp., YES-instances).

Theorem 9 (Vadhan [priv. comm.]): *Let $\Pi = (\Pi_{YES}, \Pi_{NO})$ be a promise problem, and S_Y and S_N be sets such that $S_Y \cup S_N = \{0,1\}^*$, $\Pi_{YES} \subseteq S_Y \subseteq \{0,1\}^* \setminus \Pi_{NO}$ and $\Pi_{NO} \subseteq S_N \subseteq \{0,1\}^* \setminus \Pi_{YES}$. Suppose that the promise problems (S_Y, Π_{NO}) and (S_N, Π_{YES}) are both in $\mathcal{NP}$. Then, every promise problem that is Cook-reducible to Π, is in $\mathcal{NP} \cap co\mathcal{NP}$.*

We stress that S_Y and S_N cover the set of all strings but are not necessarily a partition of it (i.e., $S_Y \cup S_N = \{0,1\}^*$ but $S_Y \cap S_N$ may be non-empty). Theorem 8 is obtained as a special case by considering $S_Y = \{0,1\}^* \setminus \Pi_{NO}$ and $S_N = \Pi_{NO}$. The proof of Theorem 9 generalizes the proof of Theorem 8: the answer to each query is augmented by a corresponding NP-witness (asserting either membership in S_Y or membership in S_N). Again, "witnesses" exist for each query, and they are guaranteed to be correct in case the query satisfies the promise.

5.2   Pro: Shedding Light on Questions Concerning Complexity Classes

Recall that working with promise problems (rather than with language recognition problems) may result in the loss of some structural consequence. We stress, however, that the most fundamental feature of general reductions is maintained: if a problem is reducible to a tractable problem, then the former is also tractable. Here we address the issue of tractability, and discuss promise problems that are not about "gaps" or "unique solutions" nor complete for any natural class, at least not obviously. Still, they are important for the study of some natural complexity classes. Specifically, they indicate (or provide evidence to) separations between complexity classes that represent the computing power of certain computational devices with certain resource bounds.

Separating monotone and non-monotone circuit complexities. Several researchers have observed that Razborov's celebrated super-polynomial lower-bound on the monotone circuit complexity of Max-Clique [47, Thm. 2] actually establishes a lower-bound on a promise problem that is in $\mathcal{P}$.[10] Thus, this result actually establishes a super-polynomial separation between the monotone and non-monotone circuit complexities (of a monotone problem). Actually, a less-known result in the same paper [47, Thm. 3] asserted a similar lower-bound for Perfect Matching (cf. [48]), and so the said separation could have been established by a language recognition problem (but not by the more famous result of [47]). Interestingly, the clique lower-bound was improved to exponential in [4], but a similar result was not known for Perfect Matching. Thus, at that time, an *exponential separation of the monotone and non-monotone circuit complexities required referring to a promise problem* (i.e., the one mentioned in Footnote 10). Subsequently, an exponential separation for languages was shown by providing an exponential lower-bound on the monotone complexity of some other polynomial-time computable (monotone) function [54].

The derandomization of BPP versus the derandomization of MA. One obvious fact, rarely noted, is that results about derandomization of $\mathcal{BPP}$ imply results on the derandomization of $\mathcal{MA}$, where $\mathcal{MA}$ is the class of problems having a "randomized verification procedure" (i.e., the analogue of $\mathcal{NP}$ in which the validity of witnesses is determined by a probabilistic polynomial-time algorithm rather than by a deterministic polynomial-time algorithm). This observation holds provided that the former derandomization results relate to $\mathcal{BPP}$ as a class of promise problems (as in Definition 2) rather than to the corresponding class of language recognition problems. We note that all known derandomization results have this property. In any case, *in terms of promise problem classes*, we have that $\mathcal{BPP} \subseteq \mathcal{DTIME}(t)$ implies $\mathcal{MA} \subseteq \mathcal{NTIME}(\mathrm{poly}(t))$, provided that the function t is "nice". Specifically, $\mathcal{BPP} = \mathcal{P}$ implies $\mathcal{MA} = \mathcal{NP}$. For details see [32, Sec. 5.4] (or Appendix C).

Disjoint NP-pairs and proof complexity. Disjoint NP-pairs are promise problems such that both the set of YES-instances and the set of NO-instances are NP-sets. Such pairs are related to propositional proof systems in the sense that each such proof system gives rise to a ("canonical") disjoint NP-pair, and every disjoint NP-pair is computationally equivalent to a canonical pair associated with some propositional proof system. The existence of "optimal" propositional proof systems is thus equivalent to the existence of complete NP-pairs.[11] Hence, the study of a natural question regarding propositional proof systems is equivalent

---

[10] For $k \approx N^{2/3}$, this promise problem has YES-instances that are N-vertex graphs having a clique of size k, and NO-instances that are complete $(k-1)$-partite N-vertex graphs. This problem can be easily solved by a greedy attempt to construct a $(k-1)$-partition of the input graph. Needless to say, this greedy approach takes advantage of the promise.

[11] See [18] for definitions of the "canonical pair" associated with a propositional proof system, the "optimality" of propositional proof systems, and "complete NP-pairs" (which are merely promise problems that are complete for the class of Disjoint NP-pairs).

to the study of the reducibility properties of a class of promise problem. For details see [18].

Relations among PCP classes. Some of the appealing transformations among PCP classes are only known when these classes are defined in terms of promise problems (see, e.g., [7, Sec. 11] and [30, Sec. 4]). For example, the intuitive meaning of [7, Prop. 11.2] is that the randomness in a PCP can be reduced to be logarithmic in the length of the proof oracle, but the actual result is a randomized Karp reduction of any problem having a PCP to a *promise problem* having a PCP with the same query (and/or free-bit) complexity and proof-length but with logarithmic randomness. Similarly, the main PCP result of [30, Sec. 4] is a almost-linear length PCP not for SAT but rather for a promise problem to which SAT can be randomly Karp-reduced (by an almost length preserving reduction). We mention that the latter random reduction was eliminated by the subsequent work of [9].

Supporting the conjectured non-triviality of statistical zero-knowledge. Seeking to provide further evidence to the conjectured non-triviality of statistical zero-knowledge (i.e., the conjecture that $\mathcal{SZK}$ extends beyond $\mathcal{BPP}$), researchers tried to show statistical zero-knowledge proof systems for "hard" (language recognition) problems. At the time (i.e., late 1980's), it was known that Quadratic Residuosity and Graph Isomorphism are in $\mathcal{SZK}$ (cf., [33] and [25], respectively), but the belief that these problems are hard seems weaker than the belief that factoring integers or the Discrete Logarithm Problem are hard. So the goal was to present a statistical zero-knowledge proof system for a language recognition problem that is computationally equivalent to any of these search problems. This was almost done in [24], who showed an *analogous result for a promise problem.* Specifically, they presented a statistical zero-knowledge proof for a promise problem that is computationally equivalent to the Discrete Logarithm Problem. Needless to say, the "gap" between YES and NO instances in this promise problem plays a key role in showing that this problem is in $\mathcal{SZK}$. Thus, *based on this promise problem, the non-triviality of $\mathcal{SZK}$ is supported by the conjectured intractability of the Discrete Logarithm Problem.*

Following a great tradition. The last example follows a central tradition in the closely related field of Cryptography, where one often considers promise problems. These problems are often search problems that refer to inputs of a special form (although computationally equivalent decision (promise) problems are sometimes stated too). Typical examples include "cryptanalyzing" a sequence of ciphertexts that are "promised" to have been produced using the same encryption-key, and factoring an integer that in the product of two primes of approximately the same size. Indeed, these examples were among the concrete motivations to the definition of promise problems introduced by Even, Selman and Yacobi [13], following prior work of Even and Yacobi [14]. The latter paper (combined with [43]) has also demonstrated that NP-hardness (i.e., worst-case hardness) of the "cryptanalysis" task is a poor evidence for crypto-graphic security. Indeed, subsequent works in cryptography typically relate to the average-case complexity of "cryptanalysis", and the "promise problem na-ture" of the task is incorporated (implicitly) in the formulation by assigning zero (probability) weight to instances that violate the promise.

And something completely different. Finally, we mention the role of promise problems in the study of Quantum Computation and Communication. I am re-ferring to two elegant mathematical models of controversial relevance to the theory of computation, and admit that I do not understand the real meaning of these models. Still, I am told that the very definition of "Quantum NP-completeness" refers to promise problems, and that the known complete prob-lems are all promise problems (see, e.g., [39, 40] where the names QBNP and QMA are used). As for Quantum Communication, the only super-polynomial separations known between the power of classical and quantum communication complexities are for promise problems (see, e.g., Raz's paper [46]).

6 Promise Problems as Facilitators of Nicer Presentation

In previous sections, we have discussed the role of promise problems in providing a framework for several natural studies and in enabling several appealing results (e.g., complete problems for $\mathcal{SZK}$). In the current section we focus on their role as facilitators of nicer presentation of various results. We believe than an explicit use of promise problems in such cases clarifies the argument as well as reveals its real essence. We start with a rather generic discussion, and later turn to one concrete example (i.e., the well-known result BPP $\subseteq$ PH).

6.1 Presenting Lower-Bound Arguments

Numerous lower-bound arguments proceed by focusing on special cases of the original decision problem. As stated in Section 1.3, these special cases are promise problems. To be concrete, we refer to an example mentioned in Section 5.2: Razborov's lower-bound on the monotone circuit complexity of Max-Clique [47, Thm. 2] is commonly presented as a lower-bound on a promise problem that, for $k \approx N^{2/3}$, has YES-instances that are N-vertex graphs having a clique of size k, and NO-instances that are complete $(k-1)$-partite N-vertex graphs.

A similar strategy is adopted in numerous works (which are too numerous to be cited here). The benefit of this strategy is that it introduces additional structure that facilitates the argument. In some cases the act of restricting at-tention to special cases is even repeated several times. Needless to say, a proper formulation of this process involves the introduction of promise problems (which correspond to these special cases). It also relies on the trivial fact that any "solver" of a problem also solves its special cases (i.e., if some device solves the promise problem $(\Pi_{\text{YES}}, \Pi_{\text{NO}})$ then it also solves any $(\Pi'_{\text{YES}}, \Pi'_{\text{NO}})$ that satisfies both $\Pi'_{\text{YES}} \subseteq \Pi_{\text{YES}}$ and $\Pi'_{\text{NO}} \subseteq \Pi_{\text{NO}}$).

The use of promise problems becomes almost essential when one proves a lower-bound by a reduction from a known lower-bound for a promise problem, and the reduction uses the promise in an essential way. Consider, for example, the separation between rank and communication complexity proven by Nisan and Wigderson [44]. Their communication complexity lower bound is by a reduction of "unique disjointness" to their communication problem, while noting that the linear lower-bound on disjointness established by Razborov [49] holds also for the promise problem "unique disjointness" (where the sets are either disjoint or have an single element in their intersection). We stress that their reduction of unique disjointness uses the promise in an essential way (and may fail for instances that violate the promise).

6.2 BPP Is in the Polynomial-Time Hierarchy, Revisited

It is well-known that BPP is in the Polynomial-time Hierarchy (see proofs by Lautemann [42] and Sipser [52]). However, the known proofs actually establish stronger results. In my opinion, both the strength and the essence of the proof comes out best via the terminology of promise problem. Specifically, we consider the extension of the language classes RP and BPP to promise problems, and show that $\mathcal{BPP} = \mathcal{RP}^{\mathcal{RP}}$.

Following Definition 2, we define $\mathcal{RP}$ and $\mathrm{co}\mathcal{RP}$ as classes of promise problems that are solvable by *one-sided* error (rather than two-sided error) probabilistic polynomial-time algorithms. Specifically, $\Pi \in \mathcal{RP}$ (resp., $\Pi \in \mathrm{co}\mathcal{RP}$) if there exists a *probabilistic* polynomial-time algorithm A such that:

- For every $x \in \Pi_{\mathrm{YES}}$ it holds that $\Pr[A(x) = 1] \geq 1/2$ (resp., $\Pr[A(x) = 1] = 1$).
- For every $x \in \Pi_{\mathrm{NO}}$ it holds that $\Pr[A(x) = 0] = 1$ (resp., $\Pr[A(x) = 0] \geq 1/2$).

It is evident that $\mathcal{RP}^{\mathcal{RP}} \subseteq \mathcal{BPP}^{\mathcal{BPP}} = \mathcal{BPP}$ (where the last equality utilizes standard "error reduction"). Thus, we focus on the other direction (i.e., $\mathcal{BPP} \subseteq \mathcal{RP}^{\mathcal{RP}}$), following the proof ideas of Lautemann [42].

Theorem 10 ([10], following [42]): *Any problem in $\mathcal{BPP}$ is reducible by a one-sided error randomized Karp-reduction to $\mathrm{co}\mathcal{RP}$.*

Proof: Consider any BPP-problem with a characteristic function χ (which, in case of a promise problem, is a partial function, defined only over the promise). That is, for some probabilistic polynomial-time algorithm A and for every x on which χ is defined it holds that $\Pr[A(x) \neq \chi(x)] \leq 1/3$. Thus, for some polynomial p_0 and some polynomial-time recognizable relation $R_0 \subseteq \cup_{n \in \mathbb{N}}(\{0,1\}^n \times \{0,1\}^{p_0(n)})$ and for every x on which χ is defined it holds that

$$\Pr_{r \in \{0,1\}^{p_0(|x|)}}[R_0(x,r) \neq \chi(x)] \leq \frac{1}{3} \qquad (1)$$

where $R_0(x,y) = 1$ if $(x,y) \in R_0$ and $R_0(x,y) = 0$ otherwise. By straightforward "error reduction" we have that, for some other polynomial p and polynomial-time recognizable relation R,

$$\left|\{r \in \{0,1\}^{p(|x|)} : R(x,r) \neq \chi(x)\}\right| < \frac{2^{p(|x|)}}{2p(|x|)} \tag{2}$$

We show a randomized one-sided error (Karp) reduction of χ to co$\mathcal{RP}$. We start by stating the simple reduction, and next define the target promise problem.

THE REDUCTION: On input $x \in \{0,1\}^n$, the randomized polynomial-time mapping uniformly selects $s_1, ..., s_m \in \{0,1\}^m$, and outputs the pair $(x, \overline{s})$, where $m = p(|x|)$ and $\overline{s} = (s_1, ..., s_m)$.

THE PROMISE PROBLEM: We define the following co$\mathcal{RP}$ promise problem, denoted $\Pi = (\Pi_{\text{YES}}, \Pi_{\text{NO}})$.

- The YES-instances are pairs $(x, \overline{s})$ such that for every $r \in \{0,1\}^m$ there exists an i satisfying $R(x, r \oplus s_i) = 1$, where $\overline{s} = (s_1, ..., s_m)$ and $m = p(|x|)$.
- The NO-instances are pairs $(x, \overline{s})$ such that for at least half of the possible $r \in \{0,1\}^m$, it holds that $R(x, r \oplus s_i) = 0$ for every i, where again $\overline{s} = (s_1, ..., s_m)$ and $m = p(|x|)$.

To see that Π is indeed a co$\mathcal{RP}$ promise problem, we consider the following randomized algorithm. On input $(x, (s_1, ..., s_m))$, where $m = p(|x|) = |s_1| = \cdots = |s_m|$, the algorithm uniformly selects $r \in \{0,1\}^m$, and accepts if and only if $R(x, r \oplus s_i) = 1$ for some $i \in \{1, ..., m\}$. Indeed, YES-instances of Π are accepted with probability 1, whereas NO-instances are rejected with probability at least $1/2$.

ANALYZING THE REDUCTION: We claim that the above randomized mapping, denoted by M, reduces χ to Π. Specifically, we will prove:

Claim 1: If x is a YES-instance (i.e., $\chi(x) = 1$) then $\Pr[M(x) \in \Pi_{\text{YES}}] > 1/2$.
Claim 2: If x is a NO-instance (i.e., $\chi(x) = 0$) then $\Pr[M(x) \in \Pi_{\text{NO}}] = 1$.

We start with Claim 2, which refers to $\chi(x) = 0$ (and is easier to establish). Recall that $M(x) = (x, (s_1, ..., s_m))$, where $s_1, ..., s_m$ are uniformly and independently distributed in $\{0,1\}^m$. Observe that (by Eq. (2)), for every possible choice of $s_1, ..., s_m \in \{0,1\}^m$ and every $i \in \{1, ..., m\}$, the fraction of r's that satisfy $R(x, r \oplus s_i) = 1$ is at most $\frac{1}{2m}$. Thus, for every possible choice of $s_1, ..., s_m \in \{0,1\}^m$, the fraction of r's for which there exists an i such that $R(x, r \oplus s_i) = 1$ holds is at most $m \cdot \frac{1}{2m} = \frac{1}{2}$. Hence, the reduction always maps such an x to a NO-instance of Π (i.e., an element of Π_{NO}).

Turning to Claim 1 (which refers to $\chi(x) = 1$), we will show shortly that in this case, with very high probability, the reduction maps x to a YES-instance of Π. We upper-bound the probability that the reduction fails (in case $\chi(x) = 1$):

$$\Pr[M(x) \notin \Pi_{\mathrm{YES}}] = \Pr_{s_1,\ldots,s_m}[\exists r \in \{0,1\}^m \text{ s.t. } (\forall i) \ R(x, r \oplus s_i) = 0]$$

$$\leq \sum_{r \in \{0,1\}^m} \Pr_{s_1,\ldots,s_m}[(\forall i) \ R(x, r \oplus s_i) = 0]$$

$$\leq 2^m \cdot \left(\frac{1}{2m}\right)^m \ll \frac{1}{2}$$

Thus, the randomized mapping M reduces χ to Π, with one-sided error on YES-instances. Recalling that $\Pi \in \mathrm{co}\mathcal{RP}$, the theorem follows. ∎

Comment: The traditional presentation uses the above reduction to show that $\mathcal{BPP}$ is in the Polynomial-Time Hierarchy. One defines the polynomial-time computable predicate $\varphi(x, \overline{s}, r) \stackrel{\text{def}}{=} \bigvee_{i=1}^m (R(x, s_i \oplus r) = 1)$, and observes that

$$\chi(x) = 1 \Rightarrow \exists \overline{s} \, \forall r \ \varphi(x, \overline{s}, r) \tag{3}$$

$$\chi(x) = 0 \Rightarrow \forall \overline{s} \, \exists r \ \neg\varphi(x, \overline{s}, r) \tag{4}$$

Note that Claim 1 establishes that *most* sequences $\overline{s}$ satisfy $\forall r \ \varphi(x, \overline{s}, r)$, whereas Eq. (3) only requires the existence of *at least one* such $\overline{s}$. Similarly, Claim 2 establishes that for every $\overline{s}$ *most* choices of r violate $\varphi(x, \overline{s}, r)$, whereas Eq. (4) only requires that for every $\overline{s}$ there exists *at least one* such r.

7 Using Promise Problems to Define Modified Complexity Classes

In continuation to Section 6, we survey a recent suggestion of Vadhan for defining "modified" complexity classes in terms of promise problems [57]. We refer to language classes such as BPP/ log and io-BPP (i.e., "computations that take advice" and "infinitely often" classes). We comment that such classes are typically defined by modification to the operation of the computing device (or the conditions applied to its computations). Vadhan's suggestion is to define such classes by modification to the class itself, provided that the (resulting) class is understood as a class of promise problems. Indeed, this approach is sometimes taken with respect to language classes like P/poly but the extension to BPP and other probabilistic classes seems to require the use of promise problems. (Indeed, in view of the fact that BPP/poly = P/poly (cf. [1]), we demonstrate the approach with respect to BPP/ log (which is the focus of some recent studies [6, 16]).)

7.1 Probabilistic Machines That Take Advice

Nonuniform "advice" versions of standard complexity classes are typified by the following two equivalent definitions of the language class P/ log. The first states that $L \in$ P/ log if there exists a deterministic polynomial-time machine M and sequence $a_1, a_2, \ldots$ such that $|a_n| = \log n$ and $M(x, a_{|x|}) = \chi_L(x)$ for every x, where $\chi_L(x) = 1$ if and only if $x \in L$. The second states that $L \in \mathrm{P}/\log$ if there exists a language $L' \in \mathrm{P}$ and a sequence $a_1, a_2, \ldots$ such that $|a_n| = \log n$ and $x \in L$ if and only if $(x, a_{|x|}) \in L'$. Indeed, L' can be defined as the language accepted by the aforementioned machine M.

For BPP in place of P, however, the analogous first formulation does not seem to imply the second one: Consider a probabilistic polynomial-time machine M and sequence $a_1, a_2, \ldots$ such that $|a_n| = \log n$ and $\Pr[M(x, a_{|x|}) = \chi_L(x)] \geq 2/3$ for every x. Then, it is unclear which pair language (extending $L \in \mathrm{BPP}/\log$) is in BPP; for example, $\{(x, a) : \Pr[M(x, a) = 1] \geq 2/3\}$ is not necessarily a BPP-set.

However, as suggested by Vadhan [57], we can define an adequate promise problem that is in the (promise problem) class $\mathcal{BPP}$. Specifically, for L and a_n's as above, consider $\Pi' = (\Pi'_{\mathrm{YES}}, \Pi'_{\mathrm{NO}})$ such that $\Pi'_{\mathrm{YES}} = \{(x, a_{|x|}) : x \in L\}$ and $\Pi'_{\mathrm{NO}} = \{(x, a_{|x|}) : x \notin L\}$. Thus, we obtain a definition of $\mathrm{BPP}/\log$ in terms of promise problems in $\mathcal{BPP}$ that extend the original languages. Similarly, for a promise problem $\Pi = (\Pi_{\mathrm{YES}}, \Pi_{\mathrm{NO}}) \in \mathcal{BPP}/\log$ (under the first definition), we consider $\Pi' = (\Pi'_{\mathrm{YES}}, \Pi'_{\mathrm{NO}})$ such that $\Pi'_{\mathrm{YES}} = \{(x, a_{|x|}) : x \in \Pi_{\mathrm{YES}}\}$ and $\Pi'_{\mathrm{NO}} = \{(x, a_{|x|}) : x \in \Pi_{\mathrm{NO}}\}$, where the a_n's are as in the first definition. Thus, we may say that a promise problem $\Pi = (\Pi_{\mathrm{YES}}, \Pi_{\mathrm{NO}})$ is in the (promise problem) class $\mathcal{BPP}/\log$ if there exists a promise problem $\Pi' = (\Pi'_{\mathrm{YES}}, \Pi'_{\mathrm{NO}})$ in $\mathcal{BPP}$ and a sequence $a_1, a_2, \ldots$ such that $|a_n| = \log n$ and $x \in \Pi_{\mathrm{YES}}$ implies $(x, a_{|x|}) \in \Pi'_{\mathrm{YES}}$ while $x \in \Pi_{\mathrm{NO}}$ implies $(x, a_{|x|}) \in \Pi'_{\mathrm{NO}}$.

Needless to say, the same approach can be applied to other probabilistic classes (e.g., $\mathcal{RP}$ and $\mathcal{AM}$) and to any bound on the advice length. We note that the need for promise problems arises only in case of probabilistic classes (and not in case of deterministic or non-deterministic classes). The issue at hand is related to the difficulties regarding complete problems and hierarchy theorems (cf. Section 4.1); that is, not every advice (or machine) induces a *bounded-error* probabilistic computation, and focusing on the advice (or machines) that do induce such a computation is done by introducing a promise.

7.2   Infinitely Often Probabilistic Classes

Another modification with similar issues is the case of "infinitely often" classes. The standard definition of io-BPP is that a problem is in this class if there exists a probabilistic polynomial-time algorithm that solves it correctly for infinitely many input lengths. The alternative formulation would say that a promise problem $\Pi = (\Pi_{\mathrm{YES}}, \Pi_{\mathrm{NO}})$ is in the class io-$\mathcal{BPP}$ if there exists a promise problem $\Pi' = (\Pi'_{\mathrm{YES}}, \Pi'_{\mathrm{NO}})$ in $\mathcal{BPP}$ such that for infinitely many values of n it holds that $\Pi'_{\mathrm{YES}} \cap \{0, 1\}^n = \Pi_{\mathrm{YES}} \cap \{0, 1\}^n$ and $\Pi'_{\mathrm{NO}} \cap \{0, 1\}^n = \Pi_{\mathrm{NO}} \cap \{0, 1\}^n$.

8   Concluding Comments

We conclude with a couple of comments of "opposite nature": The first comment highlights the relevance of promise problems to the restricted study of ("traditional") complexity classes that refer to language recognition problems. The second comment asserts the applicability of the concept of promise problems to a wider scope of complexity questions, including the study of search (rather than decision) problems.

8.1   Implications on the Study of Classes of Languages

We have argued that promise problems are at least as natural as traditional language recognition problems, and that the former offer many advantages. Still tradition and simplicity (offered by language classes) have their appeal. We thus mention that in some cases, *the study of promise problems yields results about language classes.* The best example is the study of Statistical Zero-Knowledge (SZK), which is surveyed in Section 4: Using promise problems it is possible to present clear proofs of certain properties of the promise problem class SZK (e.g., that SZK is closed under complementation [51] and that any problem in SZK has a public-coin statistical zero-knowledge proof system [31]). But, then, it follows that the same properties hold for the *class of languages* having Statistical Zero-Knowledge proofs.

8.2   Applicability to Search Problems

As surveyed above, promise problems are a generalization of language recognition problems, and thus constitute a general form of decision problems. However, one may apply the concept of promise problems also in the context of search problems, and indeed such an application is at least as natural. For example, it is most natural to state search problems in terms of "promise problems" (rather than requiring their solver also to handle instances that have no solution (and hence also solve the corresponding decision problem)). That is, for a polynomially bounded relation R, the search (promise) problem is given x that has a solution to find such a solution (i.e., find a y such that $(x,y) \in R$). Hence, the promise is that x has a solution (i.e., y such that $(x,y) \in R$), and nothing is required in case x has no solution. (Note that the promise is important in case R is not an NP-relation.)

As in case of decision problems, search (promise) problems offer a formalism for the intuitive notion of special cases (i.e., problem restriction). In addition to the natural appeal of the promise problem formulation of search problems, such promise search problems offer a few fundamental advantages over traditional search problems. The best example is the connection between (bounded fan-in) circuit depth and communication complexity, established by Karchmer and Wigderson [38]. Specifically, the (bounded fan-in) circuit depth of a function $f : \{0,1\}^n \rightarrow \{0,1\}$ is shown to equal the communication complexity of the *search promise-problem* in which one party is given $x = x_1 \cdots x_n \in f^{-1}(1)$, the other party is given $y = y_1 \cdots y_n \in f^{-1}(0)$, and the task is to determine an $i \in [n]$ such that $x_i \neq y_i$. Furthermore, the monotone depth of f equals the communication complexity of a search problem with the same promise, where the task is to find an $i \in [n]$ such that $x_i = 1$ and $y_i = 0$. We stress that removing the promise yields a trivial communication complexity lower bound of n (e.g., by reduction from the communication complexity of the identity function), which of course has no relevance to the circuit depth of the function f.

8.3   The Bottom-Line

Our summary is that promise problems are a natural generalization of traditional language-recognition problems, and often convey both the original intent of the problem's framer and more information about the problem's complexity. Despite a needed qualification for Turing reductions (Section 5.1), most results for language classes carry over naturally and easily. In many cases, promise problems enable to represent natural concepts (e.g., problem restriction, unique solutions, and approximation) that cannot be represented in terms of language-recognition problems. In other cases, the generalization to promise problems allows to derive appealing results that are not known for the corresponding language classes. Shimon Even, together with Selman and Yacobi, gave first voice to the technical formulation of this natural outlook on the computational world.

Acknowledgments

I am grateful to Salil Vadhan for various insightful comments and suggestions. I also wish to thank Ran Raz and Avi Wigderson for answering my questions and making additional suggestions. Finally, many thanks to the anonymous referee for his/her numerous comments as well as for the courageous attempt to improve my writing style.

References

1. L. Adleman. Two theorems on random polynomial-time. In *19th FOCS*, pages 75–83, 1978.
2. D. Aharonov and O. Regev. Lattice problems in $\mathcal{NP} \cap \text{co}\mathcal{NP}$. In *45th FOCS*, pages 362–371, 2004.
3. W. Aiello and J. Håstad. Statistical zero-knowledge languages can be recognized in two rounds. *JCSS*, Vol. 42 (3), pages 327–345, 1991. Preliminary version in *28th FOCS*, 1987.
4. N. Alon and R. Boppana. The monotone circuit complexity of Boolean functions. *Combinatorica*, Vol. 7 (1), pages 1–22, 1987.
5. N. Alon and A. Shapira. A Separation Theorem in Property Testing. Unpublished manuscript, 2004.
6. B. Barak. A Probabilistic-Time Hierarchy Theorem for "Slightly Nonuniform" Algorithms. In *Random'02*, LNCS 2483, pages 194–208, 2002.
7. M. Bellare, O. Goldreich and M. Sudan. Free Bits, PCPs and Non-Approximability – Towards Tight Results. *SICOMP*, Vol. 27, No. 3, pages 804–915, 1998. Extended abstract in *36th FOCS*, pages 422–431, 1995.

8. C.H. Bennett, G. Brassard and J.M. Robert. Privacy Amplification by Public Discussion. *SICOMP*, Vol. 17, pages 210–229, 1988. Preliminary version in *CRYPTO'85*, Springer-Verlag LNCS Vol. 218, pages 468–476 (titled "How to Reduce your Enemy's Information").

9. E. Ben-Sasson, M. Sudan, S. Vadhan and A. Wigderson. Randomness-efficient low degree tests and short PCPs via epsilon-biased sets. In *35th STOC*, pages 612–621, 2003.

10. H. Buhrman and L. Fortnow. One-sided versus two-sided randomness. In Proceedings of the *16th Symposium on Theoretical Aspects of Computer Science*. LNCS, Springer, Berlin, 1999.

11. J. Cai and A. Nerurkar. A note on the non-NP-hardness of approximate lattice problems under general Cook reductions. *IPL*, Vol. 76, pages 61-66, 2000. See comment in [21].

12. L. Carter and M. Wegman. Universal Hash Functions. *JCSS*, Vol. 18, 1979, pages 143–154.

13. S. Even, A.L. Selman, and Y. Yacobi. The Complexity of Promise Problems with Applications to Public-Key Cryptography. *Inform. and Control*, Vol. 61, pages 159–173, 1984.

14. S. Even and Y. Yacobi. Cryptography and NP-Completeness. In proceedings of *7th ICALP*, Springer-Verlag, LNCS Vol. 85, pages 195–207, 1980. See [13].

15. L. Fortnow, The complexity of perfect zero-knowledge. In *Advances in Computing Research*, Vol. 5 (S. Micali, ed.), JAC Press Inc., pages 327–343, 1989. Preliminary version in *19th STOC*, 1987.

16. L. Fortnow and R. Santhanam. Hierarchy theorems for probabilistic polynomial time. In *45th FOCS*, pages 316–324, 2004.

17. M.R. Garey and D.S. Johnson. *Computers and Intractability: A Guide to the Theory of NP-Completeness*. W.H. Freeman and Company, New York, 1979.

18. C. Glasser, A.L. Selman, and L. Zhang. Survey of Disjoint NP-Pairs and Relations to Propositional Proof Systems. This volume.

19. O. Goldreich. *Foundation of Cryptography – Basic Tools*. Cambridge University Press, 2001.

20. O. Goldreich. Zero-Knowledge twenty years after its invention. *Quaderni di Matematica*, Vol. 13 (Complexity of Computations and Proofs, ed. J. Krajicek), pages 249–304, 2004. Available from `http://www.wisdom.weizmann.ac.il/~oded/zk-tut02.html`.

21. O. Goldreich. Comments regarding [22] at web-page `http://www.wisdom.weizmann.ac.il/~oded/p_lp.html`.

22. O. Goldreich and S. Goldwasser, On the Limits of Non-Approximability of Lattice Problems. *JCSS*, Vol. 60, pages 540–563, 2000. Extended abstract in *30th STOC*, 1998.

23. O. Goldreich, S. Goldwasser, and D. Ron. Property testing and its connection to learning and approximation. *Journal of the ACM*, pages 653–750, July 1998. Extended abstract in *37th FOCS*, 1996.

24. O. Goldreich and E. Kushilevitz. A Perfect Zero-Knowledge Proof for a Decision Problem Equivalent to Discrete Logarithm. *Jour. of Crypto.*, Vol. 6 (2), pages 97–116, 1993. Extended abstract in proceedings of *Crypto'88*.

25. O. Goldreich, S. Micali and A. Wigderson. Proofs that Yield Nothing but their Validity or All Languages in NP Have Zero-Knowledge Proof Systems. *JACM*, Vol. 38, No. 1, pages 691–729, 1991. Preliminary version in *27th FOCS*, 1986.

26. O. Goldreich and D. Ron. Property testing in bounded degree graphs. *Algorithmica*, pages 302–343, 2002. Extended abstract in *29th STOC*, 1997.

27. O. Goldreich and D. Ron. A sublinear bipartite tester for bounded degree graphs. *Combinatorica*, 19(3):335–373, 1999. Extended abstract in *30th STOC*, 1998.

28. O. Goldreich, A. Sahai, and S. Vadhan. Honest-Verifier Statistical Zero-Knowledge equals general Statistical Zero-Knowledge. In *30th STOC*, pages 399–408, 1998.

29. O. Goldreich, A. Sahai, and S. Vadhan. Can Statistical Zero-Knowledge be Made Non-Interactive? or On the Relationship of SZK and NISZK. In *Proceedings of Crypto99*, Springer LNCS Vol. 1666, pages 467–484.

30. O. Goldreich and M. Sudan. Locally Testable Codes and PCPs of Almost-Linear Length. Preliminary version in *43rd FOCS*, 2002. ECCC Report TR02-050, 2002.

31. O. Goldreich and S. Vadhan. Comparing Entropies in Statistical Zero-Knowledge with Applications to the Structure of SZK. In *14th IEEE Conference on Computational Complexity*, pages 54–73, 1999.

32. O. Goldreich and D. Zuckerman. Another proof that BPP subseteq PH (and more). ECCC Report TR97-045, 1997.

33. S. Goldwasser, S. Micali and C. Rackoff. The Knowledge Complexity of Interactive Proof Systems. *SIAM Journal on Computing*, Vol. 18, pages 186–208, 1989. Preliminary version in *17th STOC*, 1985.

34. J. Grollmann and A.L. Selman. Complexity Measures for Public-Key Cryptosystems. *SIAM Jour. on Comput.*, Vol. 17 (2), pages 309–335, 1988.

35. J. Hastad. Clique is hard to approximate within $n^{1-\epsilon}$. *Acta Mathematica*, Vol. 182, pages 105–142, 1999. Preliminary versions in *28th STOC* (1996) and *37th FOCS* (1996).

36. J. Hastad. Getting optimal in-approximability results. In *29th STOC*, pages 1–10, 1997.

37. R. Impagliazzo, L.A. Levin and M. Luby. Pseudorandom Generation from One-Way Functions. In *21st STOC*, pages 12–24, 1989.

38. M. Karchmer and A. Wigderson. Monotone Circuits for Connectivity Require Super-logarithmic Depth. *SIAM J. Discrete Math.*, Vol. 3 (2), pages 255–265, 1990. Preliminary version in *20th STOC*, 1988.

39. E. Knill. Quantum randomness and nondeterminism. `quant-ph/9610012`, 1996.

40. A.Yu. Kitaev, A.H. Shen and M.N. Vyalyi. Classical and Quantum Computation. Vol. 47 of *Graduate Studies in Mathematics*, AMS, 2002.

41. T. Kuhn. *The Structure of Scientific Revolution*. University of Chicago, 1962.

42. C. Lautemann. BPP and the Polynomial Hierarchy. *IPL*, 17, pages 215–217, 1983.

43. A. Lempel. Cryptography in Transition. *Computing Surveys*, Dec. 1979.

44. N. Nisan and A. Wigderson. A note on Rank versus Communication Complexity. *Combinatorica*, Vol. 15 (4), pages 557–566, 1995.

45. T. Okamoto. On relationships between statistical zero-knowledge proofs. *JCSS*, Vol. 60 (1), pages 47–108, 2000. Preliminary version in *28th STOC*, 1996.

46. R. Raz. Exponential Separation of Quantum and Classical Communication Complexity. In *31st STOC*, pages 358–367, 1999.

47. A. Razborov. Lower bounds for the monotone complexity of some Boolean functions. In *Doklady Akademii Nauk SSSR*, Vol. 281, No. 4, 1985, pages 798-801. English translation in *Soviet Math. Doklady*, 31:354-357, 1985.

48. A. Razborov. Lower bounds of monotone complexity of the logical permanent function. *Matematicheskie Zametki*, Vol. 37, No. 6, 1985, pages 887-900. English translation in *Mathematical Notes of the Academy of Sci. of the USSR*, 37:485-493, 1985.

49. A. Razborov. On the Distributional Complexity of Disjointness. *Theoretical Computer Science*, Vol. 106, pages 385–390, 1992.

50. R. Rubinfeld and M. Sudan. Robust characterization of polynomials with applications to program testing. *SIAM Journal on Computing*, 25(2), pages 252–271, 1996.

51. A. Sahai and S. Vadhan. A complete problem for Statistical Zero-Knowledge. *Journal of the ACM*, 50(2):196–249, March 2003. Preliminary version in *38th FOCS*, 1997.

52. M. Sipser. A Complexity Theoretic Approach to Randomness. In *15th STOC*, pages 330–335, 1983.

53. L. Stockmeyer. On Approximation Algorithms for #P. *SIAM Journal on Computing*, Vol. 14 (4), pages 849–861, 1985. Preliminary version in *15th STOC*, pages 118–126, 1983.

54. E. Tardos. The gap between monotone and non-monotone circuit complexity is exponential. *Combinatorica*, Vol. 8 (1), pages 141–142, 1988.

55. S. Vadhan. A Study of Statistical Zero-Knowledge Proofs. PhD Thesis, Department of Mathematics, MIT, 1999. Available from http://www.eecs.harvard.edu/~salil/papers/phdthesis-abs.html.

56. S. Vadhan. An Unconditional Study of Computational Zero Knowledge. In *45th FOCS*, pages 176–185, 2004.

57. S. Vadhan. Using promise problems to define modified complexity classes. Email to the author (Jan. 24, 2005).

58. L.G. Valiant and V.V. Vazirani. NP is as Easy as Detecting Unique Solutions. *Theoretical Computer Science*, Vol. 47 (1), pages 85–93, 1986. Preliminary version in *17th STOC*, pages 458–463, 1985.

Appendix A: Proof Sketches for Theorems 4 and 5

We prove Theorem 5 first, and establish Theorem 4 later while using similar techniques. We start by observing that solving the counting problem $\#R_{\mathsf{SAT}}$ for very narrow margins of error is reducible to solving it for very large margins of error. That is, for $f(n) = 1 + (1/\mathrm{poly}(n))$ and $g(n) < \exp(n^c)$ for any $c \in (0,1)$, it holds that $\#R^f_{\mathsf{SAT}}$ is Karp-reducible to $\#R^g_{\mathsf{SAT}}$. The reduction is based on the observation that, for formulae $\phi_1, ..., \phi_t$ over disjoint sets of variables, it holds that $R_{\mathsf{SAT}}(\wedge^t_{i=1}\phi_i) = \{\langle \tau_1, ..., \tau_t \rangle : (\forall i)\, \tau_i \in R_{\mathsf{SAT}}(\phi_i)\}$. Thus, $|R_{\mathsf{SAT}}(\phi^t)| = |R_{\mathsf{SAT}}(\phi)|^t$, where ϕ^t is the formula obtained by concatenating t copies of the formula ϕ (while using different variables in each copy). It follows that, for any polynomially bounded t, the problem $\#R^f_{\mathsf{SAT}}$ is Karp-reducible to $\#R^g_{\mathsf{SAT}}$, where $g(t(n) \cdot n) = f(n)^{t(n)}$, by mapping (ϕ, N) to $(\phi^{t(|\phi|)}, N^{t(|\phi|)})$.

Reducing $\#R^g_{\mathsf{SAT}}$ to SAT, *for sufficiently large g.* In view of the foregoing, we may focus on randomly reducing $\#R^g_{\mathsf{SAT}}$ to SAT, for $g(n) = n^2$. Given an instance (ϕ, N), with $1 \leq N < g(|\phi|)$, we reduce ϕ to itself, and notice that YES-instances are certainly satisfiable (because $N \geq 1$), whereas NO-instances are not satisfiable (because they have less than $N/g(|\phi|) < 1$ satisfying assignments).[12] Thus, in this case the reduction is valid. However, the interesting case is when $N \geq g(|\phi|)$, which in particular implies $N > |\phi|$.

Given an instance (ϕ, N), with $N > |\phi|$, our goal is to create a random formula ϕ' such that the *expected* cardinality of $R_{\mathrm{SAT}}(\phi')$ equals $|R_{\mathrm{SAT}}(\phi)|/2^k$, where $k \overset{\mathrm{def}}{=} \log_2 N - \log_2 |\phi| \geq 0$. Furthermore, with very high probability, if $|R_{\mathrm{SAT}}(\phi)| \geq N$ then $|R_{\mathrm{SAT}}(\phi')| > N/2^{k+1} > 1$ and if $|R_{\mathrm{SAT}}(\phi)| < N/g(|\phi|)$ then $R_{\mathrm{SAT}}(\phi') = \emptyset$ (because $2^{-k} \cdot N/g(|\phi|) \ll 1$).

We create the formula ϕ' as the conjunction of ϕ and ϕ_h, where $h : \{0,1\}^n \to \{0,1\}^k$ is a randomly chosen (Universal-2 [12]) hashing function and $\phi_h(x_1, ..., x_n) = 1$ if and only if $h(x_1, ..., x_n) = 0^k$. We stress that ϕ and ϕ_h use the same variables $x_1, ..., x_n$, and that ϕ_h can be obtained by a parsimonious reduction of the computation of h (i.e., verifying that $h(x_1, ..., x_n) = 0^k$) to SAT. That is, we consider the randomized mapping

$$(\phi, N) \to \phi' \text{ where } \phi'(x) \overset{\mathrm{def}}{=} \phi(x) \wedge (h(x) = 0^{\log_2(N/|\phi|)}) \tag{5}$$
$$\text{and } h : \{0,1\}^{|x|} \to \{0,1\}^{\log_2(N/|\phi|)} \text{ is a random hash function.}$$

Using the "Leftover Hashing Lemma" [52, 8, 37] it follows that, with very high probability, if $|R_{\mathrm{SAT}}(\phi)| \geq N$ then $|R_{\mathrm{SAT}}(\phi')| > N/2^{k+1} > 1$ and if $|R_{\mathrm{SAT}}(\phi)| < N/g(|\phi|)$ then $R_{\mathrm{SAT}}(\phi') = \emptyset$. Thus, we randomly reduced the instance (ϕ, N) of $\#R_{\mathrm{SAT}}^g$ to deciding whether or not ϕ' is satisfiable. That is, *the randomized mapping $(\phi, N) \mapsto \phi'$ of Eq. (5) is a randomized Karp-reduction of $\#R_{\mathrm{SAT}}^g$ to* SAT. Combined with the reduction of $\#R_{\mathrm{SAT}}^f$ to $\#R_{\mathrm{SAT}}^g$, this completes the proof of Theorem 5.

Proof of Theorem 4: To prove Theorem 4 we combine the foregoing ideas with two additional observations. The first observation is that if an n-variable formula ϕ is unsatisfiable then, for every $i \in \{0, 1, ..., n\}$, the pair $(\phi, 2^i)$ is a NO-instance of $\#R_{\mathrm{SAT}}^g$, whereas in case ϕ is satisfiable then, for $i = \lfloor \log_2 |R_{\mathrm{SAT}}(\phi)| \rfloor$, the pair $(\phi, 2^i)$ is a YES-instance of $\#R_{\mathrm{SAT}}^g$. Furthermore, in the latter case, $2^i \leq |R_{\mathrm{SAT}}(\phi)| < 2^{i+1}$. For sake of simplicity, we assume below that $i \geq \log |\phi|$. The second observation is that in case $2^i \leq |R_{\mathrm{SAT}}(\phi)| < 2^{i+1}$, with very high probability, the formula ϕ_i' (randomly constructed as in Eq. (5) using $N = 2^i$), has at least $m \overset{\mathrm{def}}{=} 2^i/2^{(i-\log|\phi|)+1} = |\phi|/2$ satisfying assignments and at most $8m$ satisfying assignments. Our goal is to reduce the problem of counting the number of satisfying assignments of ϕ_i' to uSAT. We consider a Cook-reduction that, for every possible value $j \in \{m, ..., 8m\}$, constructs a formula $\phi_{i,j}''$ that is satisfiable if and only if ϕ_i' has at least j satisfying assignments; for example, we may use

---

$$\phi''_{i,j}(x_1^{(1)}, ..., x_n^{(1)}, ..., x_1^{(j)}, ..., x_n^{(j)})$$

$$= \left(\bigwedge_{\ell=1}^{j} \phi'_i(x_1^{(\ell)}, ..., x_n^{(\ell)}) \right) \wedge \left(\bigwedge_{\ell=1}^{j-1} \left((x_1^{(\ell)}, ..., x_n^{(\ell)}) < (x_1^{(\ell+1)}, ..., x_n^{(\ell+1)}) \right) \right) \quad (6)$$

where $(x_1^{(\ell)}, ..., x_n^{(\ell)}) < (x_1^{(\ell+1)}, ..., x_n^{(\ell+1)})$ if and only if $x_q^{(\ell)} < x_q^{(\ell+1)}$ for some q and $x_q^{(\ell)} \leq x_q^{(\ell+1)}$ for every q. Furthermore, note that if ϕ'_i has exactly j satisfying assignments then $\phi''_{i,j}$ has a unique satisfying assignment. This suggests the following randomized Cook-reduction from SAT to uSAT:

1. On input an n-variable formula ϕ, the oracle machine constructs the formulae $\phi''_{i,j}$, for every $i \in \{\log |\phi|, ..., |\phi|\}$ and $j \in \{1, ..., 8m\}$, where ϕ'_i is obtained by applying Eq. (5) to the pair $(\phi, 2^i)$, and $\phi''_{i,j}$ is obtained by applying Eq. (6) to ϕ'_i.
   (The case that ϕ has less than $|\phi|$ satisfying assignments is covered by $i = \log |\phi|$, where effectively no hashing takes place, and thus $\phi'_i = \phi$. For this reason, we have let j range in $\{1, ..., 8m\}$ rather than in $\{m, ..., 8m\}$, a change that causes no harm to larger values of i.)
2. The oracle machine queries the oracle on each of the formulae $\phi''_{i,j}$ and accepts if and only if at least one answer is positive.

Note that if ϕ is satisfiable then, with very high probability, at least one of the formulae $\phi''_{i,j}$ has a unique satisfying assignment (and thus the corresponding query will be answered positively). On the other hand, if ϕ is unsatisfiable then all the formulae $\phi''_{i,j}$ are unsatisfiable (and thus all queries will be answered negatively). This completes the proof of Theorem 4.

Appendix B: More Details Regarding the Study of SZK

In this appendix, we provide a more detailed presentation of the material surveyed in Section 4.2, and describe some of the ideas underlying the proof of central results. We start by recalling a few underlying notions.

The statistical difference (or variation distance) between the distributions (or the random variables) X and Y is defined as

$$\Delta(X, Y) \overset{\text{def}}{=} \frac{1}{2} \cdot \sum_e |\Pr[X=e] - \Pr[Y=e]| = \max_S \{\Pr[X \in S] - \Pr[Y \in S]\} \quad (7)$$

We say that X and Y are δ-close if $\Delta(X, Y) \leq \delta$ and that they are δ-far if $\Delta(X, Y) \geq \delta$. Note that X and Y are identical if and only if they are 0-close, and are disjoint (or have disjoint support) if and only if they are 1-far. The entropy of a distribution (or random variables) X is defined as

$$H(X) \overset{\text{def}}{=} \sum_e \Pr[X=e] \cdot \log_2(1/\Pr[X=e]). \quad (8)$$

The entropy of a distribution is always non-negative and is zero if and only if the distribution is concentrated on a single element. In general, if a distribution that has support size N then its entropy is at most $\log_2 N$.

The distribution represented (or generated) by a circuit $C : \{0,1\}^n \to \{0,1\}^m$ assigns each string $\alpha \in \{0,1\}^m$ probability $|\{s : C(s) = \alpha\}|/2^n$. The corresponding random variable is $C(U_n)$, where U_n denotes a random variable uniformly distributed over $\{0,1\}^n$. A function $\mu : \mathsf{N} \to [0,1]$ is called negligible if it decreases faster than any polynomial fraction; that is, for every positive polynomial p and all sufficiently large n it holds that $\mu(n) < 1/p(n)$. A function $\nu : \mathsf{N} \to [0,1]$ is called noticeable if $\nu(n) > 1/p(n)$ for some positive polynomial p and all sufficiently large n.

B.1   The Class SZK and Its Complete Problems

The class $\mathcal{SZK}$ consists of promise problems that have an interactive proof system that is "statistically zero-knowledge" (with respect to the honest verifier). Recall that interactive proof systems are two-party protocols in which a computationally unbounded prover may convince a probabilistic polynomial-time verifier to accept YES-instances, whereas no prover can fool the verifier into accepting NO-instances. Both assertions hold with high probability, which can be amplified by repetitions.

Definition 11 ([15, 24], following [33]) *The two-party protocol (P,V) is called an* interactive proof system *for the promise problem $\Pi = (\Pi_{\mathrm{YES}}, \Pi_{\mathrm{NO}})$ if V is a probabilistic polynomial-time interactive machine and the following two conditions hold*

1. Completeness: *For any $x \in \Pi_{\mathrm{YES}}$, with probability at least $2/3$, the verifier V accepts after interacting with the prover P on common input x.*
2. Soundness: *For any $x \in \Pi_{\mathrm{NO}}$, with probability at least $2/3$, the verifier V rejects after interacting with* any strategy *on common input x.*

We denote by $\langle P,V \rangle(x)$ the local view of V when interacting with P on common input x, where the local view consists of x, the internal coin tosses of V, and the sequence of messages it has received from P. The proof system (P,V) is said to be statistical zero-knowledge *if there exists a probabilistic polynomial-time machine S, called a* simulator, *such that for $x \in \Pi_{\mathrm{YES}}$ the statistical difference between $\langle P,V \rangle(x)$ and $S(x)$ is negligible as a function of $|x|$.*

We stress that the completeness and zero-knowledge conditions refer only to YES-instances, whereas the soundness condition refers only to NO-instances. We mention that Definition 11 refers only to honest-verifiers, but it is known that any problem that has an interactive proof satisfying Definition 11 also has one that is statistical zero-knowledge in general (i.e., with respect to arbitrary verifiers); see [28, 31].

Definition 12 *The class $\mathcal{SZK}$ consists of all promise problems that have a statistical zero-knowledge interactive proof system.*

The class $\mathcal{SZK}$ contains some promise problems that are widely believed not to be in $\mathcal{BPP}$ (e.g., it contains a promise problem that is computationally equivalent to the Discrete Logarithm Problem, cf. [24]). On the other hand, $\mathcal{SZK} \subseteq \mathcal{AM} \cap$ co$\mathcal{AM}$ (cf. [15, 3]), which in turn lies quite low in the Polynomial-Time Hierarchy.

Approximating the distance between distributions. We consider promise problems that take as input a pair of circuits and refer to the statistical difference between the two corresponding distributions (generated by the two circuits). For (threshold) functions $c, f : \mathsf{N} \to [0, 1]$, where $c \leq f$, the promise problem $\mathtt{GapSD}^{c,f} = (\mathtt{Close}^c, \mathtt{Far}^f)$ is defined such that $(C_1, C_2) \in \mathtt{Close}^c$ if $\Delta(C_1, C_2) \leq c(|C_1| + |C_2|)$ and $(C_1, C_2) \in \mathtt{Far}^f$ if $\Delta(C_1, C_2) > f(|C_1| + |C_2|)$. In particular, we focus on promise problem $\mathtt{GapSD} \stackrel{\text{def}}{=} \mathtt{GapSD}^{\frac{1}{3}, \frac{2}{3}}$. Interestingly, the complexity of $\mathtt{GapSD}$, which captures quite a good approximation requirement, is computationally equivalent to a very crude approximation requirement (e.g., $\mathtt{GapSD}^{0.01, 0.99}$). That is, the former problem is Karp-reducible to the latter:

Theorem 13 [51]: *For some $\alpha > 0$, there exists a Karp-reduction of $\mathtt{GapSD}^{\frac{1}{3}, \frac{2}{3}}$ to $\mathtt{GapSD}^{\epsilon, 1-\epsilon}$, where $\epsilon(n) = 2^{-n^\alpha}$. More generally, for every polynomial-time computable $c, f : \mathsf{N} \to [0, 1]$ such that $c(n) < f(n)^2 - (1/\mathrm{poly}(n))$ it holds that $\mathtt{GapSD}^{c,f}$ is Karp-reducible to $\mathtt{GapSD}^{\epsilon, 1-\epsilon}$.*

Using a trivial reduction in the other direction, we conclude that *for every $c, f :$ $\mathsf{N} \to [0, 1]$ such that $c(n) \geq 2^{-n^\alpha}$, $c(n) < f(n)^2 - (1/\mathrm{poly}(n))$ and $f(n) \geq$ $1 - 2^{-n^\alpha}$, the problems $\mathtt{GapSD}^{c,f}$ and $\mathtt{GapSD} = \mathtt{GapSD}^{\frac{1}{3}, \frac{2}{3}}$ are computationally equivalent* (under Karp reductions). This equivalence is useful in determining the complexity of $\mathtt{GapSD}$ (as well as all these $\mathtt{GapSD}^{c,f}$'s). Specifically, in order to show that $\mathcal{SZK}$ is Karp-reducible to $\mathtt{GapSD}$, it is shown that $\mathcal{SZK}$ is Karp-reducible to $\mathtt{GapSD}^{\frac{1}{2p^2}, \frac{1}{p}}$, for some polynomial p. On the other hand, in order to show that $\mathtt{GapSD}$ is in $\mathcal{SZK}$, it is shown that for $\epsilon(n) = 2^{-n^\alpha}$ the problem $\mathtt{GapSD}^{\epsilon, 1-\epsilon}$ is in $\mathcal{SZK}$. Thus, one gets

Theorem 14 [51]: *The promise problem $\mathtt{GapSD}$ is $\mathcal{SZK}$-complete* (under Karp reductions).

We stress that the promise problem nature of $\mathtt{GapSD}$ seems essential for showing that $\mathtt{GapSD} \in \mathcal{SZK}$. On the other hand, the class $\mathcal{SZK}$ reduces naturally to a promise problem with a non-trivial promise. For details, see Section B.2.

Approximating the entropy of a distribution. We consider two types of computational problems related to approximating the entropy of a distribution. The first type consists of promise problems that take as input a circuit and a value and refers to the relation between the entropy of (the distribution generated by) the circuit and the given value. The second type of promise problems take as input a pair of circuits and refer to the relation between the entropies of the corresponding distributions (generated by the two circuits). Note that the two types of problems are computationally equivalent (i.e., each is Cook-reducible to the other). We focus on the second type of problems, because (unlike the first type) they are known to be complete for $\mathcal{SZK}$ under Karp-reductions. Specifically, for a positive (slackness) function $s : \mathsf{N} \to \mathsf{R}^+$, the promise problem $\mathtt{GapENT}^s = (\mathtt{Smaller}^s, \mathtt{Larger}^s)$ is defined such that $(C_1, C_2) \in \mathtt{Smaller}^s$ if $\mathrm{H}(C_1) \leq \mathrm{H}(C_2) - s(|C_1| + |C_2|)$ and $(C_1, C_2) \in \mathtt{Larger}^f$ if $\mathrm{H}(C_1) \geq \mathrm{H}(C_2) + s(|C_1| + |C_2|)$. We focus on promise problem $\mathtt{GapENT} \overset{\text{def}}{=} \mathtt{GapENT}^1$, and mention the following two simple facts:

Fact 1: For every positive polynomial p and $\ell_\epsilon(n) = n^{1-\epsilon}$ for any $\epsilon > 0$, it holds that the problems $\mathtt{GapENT}^{1/p}$, $\mathtt{GapENT}$ and $\mathtt{GapENT}^{\ell_\epsilon}$ are computationally equivalent (under Karp reductions).

Fact 2: The problem $\mathtt{GapENT}$ is Karp-reducible to its complement by the reduction that maps (C_1, C_2) to (C_2, C_1).

It turns out that the computational problems regarding entropy are computationally equivalent to the computational problems regarding statistical distance:

Theorem 15 [31]: *The promise problems* $\mathtt{GapENT}$ *and* $\mathtt{GapSD}$ *are computationally equivalent under Karp reductions.*

Combining Theorems 14 and 15, it follows that $\mathtt{GapENT}$ is $\mathcal{SZK}$-complete (under Karp-reductions). Using Fact 2, it follows that $\mathcal{SZK}$ is closed under complementation.

B.2   Comments Regarding the Proofs of Theorems 13–15

The proofs of Theorems 13 and 15 rely on sophisticated manipulations of distributions (or rather the corresponding sampling circuits). Although these proofs are quite interesting, we focus on the proof of Theorem 14, which provides the bridge between the aforementioned specific computational problems and the class $\mathcal{SZK}$. Indeed, the proof of Theorem 14 highlights the role of promise problems (with non-trivial promises) in the study of $\mathcal{SZK}$, whereas the proofs of Theorems 13 and 15 merely translate one promise problem (with a non-trivial promise) to another.

Theorem 14 was proven by Sahai and Vadhan [51], and here we sketch the ideas underlying their proof. Their proof consists of two parts: (1) showing that $\mathtt{GapSD}$ has a statistical zero-knowledge proof system, and (2) showing that any problem in $\mathcal{SZK}$ is Karp-reducible to $\mathtt{GapSD}$.

The problem $\mathtt{GapSD}$ *has a statistical zero-knowledge proof system:* Using Theorem 13, it suffices to show such a proof system for $\mathtt{GapSD}^{\epsilon, 1-\epsilon}$, where $\epsilon : \mathsf{N} \to [0, 1]$ is a negligible function (e.g., $\epsilon(n) = 2^{-n^\alpha}$ for some $\alpha > 0$). Actually, we present such a proof system for the complement problem (i.e., $(\mathtt{Far}^{1-\epsilon}, \mathtt{Close}^\epsilon)$), and rely on the (highly non-trivial) fact that $\mathtt{GapSD}$ is reducible to its complement.[13]

---

[13] This fact follows by combining Theorem 15 an Fact 2. An alternative proof of the fact that $\mathtt{GapSD}$ is reducible to its complement was given in [51], before Theorem 15 was stated (let alone proved). Another alternative is to rely on an even earlier result of Okamoto by which $\mathcal{SZK}$ is closed under complementation [45].

Following an idea that originates in [33, 25], the protocol proceeds as follows, with the aim of establishing that the two input distributions are far apart. The verifier selects one of the input distributions at random and presents the prover with a random sample generated according to this distribution. The verifier accepts if and only if the prover correctly identifies the distribution from which the sample was taken. Observe that if the input distributions are far apart then the prover can answer correctly with very high probability. On the other hand, if the input distributions are very close then the prover cannot guess the correct answer with probability significantly larger than 1/2. This establishes that the protocol is an interactive proof (and thus that GapSD is in co$\mathcal{AM}$). It can be shown that this protocol is actually statistical zero-knowledge, intuitively because the verifier learns nothing from the prover's correct answer which is a priori known to to the verifier (in case the two distributions are far apart).

Any problem in $\mathcal{SZK}$ is Karp-reducible to GapSD: We rely on Okamoto's Theorem by which any problem in $\mathcal{SZK}$ has a *public-coin*[14] statistical zero-knowledge proof system [45]. (We comment that an alternative proof of that theorem has subsequently appeared in [31], who showed that $\mathcal{SZK}$ is Karp-reducible to GapENT while the latter problem has a public-coin statistical zero-knowledge proof system.) We consider an arbitrary (*public-coin*) statistical zero-knowledge proof system. Following Fortnow [15], we observe a discrepancy between the behavior of the simulator on YES-instances versus NO-instances:

- In case the input is a YES-instance, the simulator outputs transcripts that are very similar to those in the real interaction. In particular, these transcripts are accepting and the verifier's behavior in them is as in a real interaction. Resorting to the public-coin condition, this means that the verifier's messages in the simulation are (almost) uniformly distributed independently of prior messages.
- In case the input is a NO-instance, the simulator must output either rejecting transcripts or transcripts in which the verifier's behavior is significantly different from the verifier's behavior in a real interaction. In particular, the only way the simulator can produce accepting transcripts is by producing transcripts in which the verifier's messages are not "random enough" (i.e., they depend, in a noticeable way, on previous messages).

Thus assuming, without loss of generality, that the simulator only produces accepting transcripts, we consider two types of distributions. The first type of the distributions is obtained by truncating a random simulator-produced transcript at a random "location" (after some verifier message), whereas the second type is obtained by doing the same while replacing the last verifier message by a

---

[14] An interactive proof is said to be of the public-coin type if the verifier is required to send the outcome of any coin it tosses as soon as it sees it. In other words, the verifier's messages are uniformly distributed strings (of predetermined length), and the verifier's decision depends only on the messages exchanged (rather than on some secret random choices made by the verifier).

random one. Note that both distributions can be implemented by polynomial-size circuits that depend on the input to the proof system being analyzed (and that these two circuits can be constructed in polynomial-time given the said input). The key observation is that if the input is a YES-instance then the two corresponding distributions will be very close, whereas if the input is a NO-instance then there will be a noticeable distance between the two corresponding distributions. Thus, we reduced any problem having a (public-coin) statistical zero-knowledge proof system to $\mathsf{GapSD}^{\mu,\nu}$, where μ is a negligible function and ν is a noticeable function. The proof is completed by using Theorem 13 (while noting that $\mu(n) < \nu(n)^2 - (1/\mathrm{poly}(n))$).

Alternative proofs of Theorems 14 and 15: In sketching the proof of Theorem 14, we relied on two theorems of Okamoto [45]: The closure of $\mathcal{SZK}$ under complementation and the existence of public-coin statistical zero-knowledge proof systems for any problem in $\mathcal{SZK}$. Since Okamoto's arguments are hard to follow, it is worthwhile noting that an alternative route does exist. In [31] it is proved that GapENT is $\mathcal{SZK}$-complete (under Karp-reductions), without relying on Okamoto's results (but while using some of his ideas). Furthermore, the statistical zero-knowledge proof system presented for GapENT is of the public-coin type. Thus, the two aforementioned theorems of Okamoto follow (using the fact that GapENT is easily reducible to its complement). Consequently, the proof of Theorem 14 need not refer to Okamoto's paper [45]. (Theorem 15 follows immediately from the fact that both GapENT and GapSD are $\mathcal{SZK}$-complete, but a direct proof is possible by employing the ideas underlying [31, 51].)

Appendix C: On the Derandomization of BPP Versus MA

The following presentation is adapted from [32, Sec. 5.4]. We denote by $\mathcal{MA}$ the class of promise problems of the form $\Pi = (\Pi_{\mathrm{YES}}, \Pi_{\mathrm{NO}})$, where there exists a polynomial p and a polynomial-time (verifier) V such that

$$x \in \Pi_{\mathrm{YES}} \implies \exists w \in \{0,1\}^{p(|x|)} \quad \Pr_{r \in \{0,1\}^{p(|x|)}}[V(x,w,r) = 1] = 1$$

$$x \in \Pi_{\mathrm{NO}} \implies \forall w \in \{0,1\}^{p(|x|)} \quad \Pr_{r \in \{0,1\}^{p(|x|)}}[V(x,w,r) = 1] \leq \frac{1}{2}$$

(All other complexity classes used below also refer to promise problems.)

Proposition (folklore): *Suppose that $\mathcal{RP} \subseteq \mathcal{DTIME}(t)$, for some monotonically non-decreasing and time-constructible function $t : \mathbb{N} \to \mathbb{N}$. Then, $\mathcal{MA} \subseteq \cup_{i \in \mathbb{N}} \mathcal{NTIME}(t_i)$, where $t_i(n) = t(n^i)$.*

In particular, $\mathcal{RP} = \mathcal{P}$ implies $\mathcal{MA} = \mathcal{NP}$.

Proof: Each promise problem $\Pi = (\Pi_{\mathrm{YES}}, \Pi_{\mathrm{NO}})$ in $\mathcal{MA}$ gives rise to a promise problem $\Pi' = (\Pi'_{\mathrm{YES}}, \Pi'_{\mathrm{NO}})$, where

$$\Pi'_{\mathrm{YES}} \stackrel{\text{def}}{=} \{(x,w) : \forall r \in \{0,1\}^{p(|x|)} \ V(x,w,r) = 1\}$$
$$\Pi'_{\mathrm{NO}} \stackrel{\text{def}}{=} \{(x,w) : x \in \Pi_{\mathrm{NO}}\}.$$

where p and V are as above. Note that, for every $(x, w) \in \Pi'_{\text{YES}}$ it holds that $\text{Pr}_{r \in \{0,1\}^{p(|x|)}}[V(x, w, r) = 1] = 1$, whereas for every $(x, w) \in \Pi_{\text{NO}}$ it holds that $\text{Pr}_{r \in \{0,1\}^{p(|x|)}}[V(x, w, r) = 1] \leq 1/2$. Thus, $\Pi' \in \text{co}\mathcal{RP}$ (i.e., $(\Pi'_{\text{NO}}, \Pi'_{\text{YES}}) \in \mathcal{RP}$). Using the hypothesis (and the closure of $\mathcal{DTIME}$ under complementation), we have $\Pi' \in \mathcal{DTIME}(t)$. On the other hand, note that for every $x \in \Pi_{\text{YES}}$ there exists $w \in \{0, 1\}^{p(|x|)}$ such that $(x, w) \in \Pi'_{\text{YES}}$, whereas for every $x \in \Pi_{\text{NO}}$ and every $w \in \{0, 1\}^{p(|x|)}$ it holds that $(x, w) \in \Pi'_{\text{NO}}$. Thus, Π is "non-deterministically reducible" to Π' (i.e., by a "reduction" that maps x to (x, w), where $w \in \{0, 1\}^{p(|x|)}$, such that $x \in \Pi_{\text{YES}}$ is mapped to $(x, w) \in \Pi'_{\text{YES}}$), and $\Pi \in \mathcal{NTIME}(t')$ follows, where $t'(n) = t(n + p(n)) < t(n^i)$ for some $i \in \mathbb{N}$. The proposition follows. ∎

A Pebble Game for Internet-Based Computing

Grzegorz Malewicz[1] and Arnold L. Rosenberg[2]

[1] Dept. of Computer Science, University of Alabama, Tuscaloosa, AL 35487, USA
greg@cs.ua.edu
[2] Dept. of Computer Science, University of Massachusetts, Amherst, MA 01003, USA
rsnbrg@cs.umass.edu

Abstract. Advances in technology have rendered the Internet a viable medium for employing multiple independent computers collaboratively in the solution of a single computational problem, leading to the new genre of collaborative computing that we term *Internet-based computing* (*IC*). Scheduling a computation for IC presents challenges that were not encountered with earlier modalities of collaborative computing, especially when the computation's constituent tasks have interdependencies that constrain their order of execution. This paper surveys an ongoing study of (an abstraction of) the scheduling problem for such computations for IC. The work employs a "pebble game on computation-dags," that abstracts the process of allocating a computation's interdependent tasks to participating remote computers. The goal of a schedule, motivated by two related scheduling challenges, is to maximize the production rate of tasks that are eligible for execution. First, in many modalities of IC, remote computers become available at unpredictable times. Always having a maximal number of execution-eligible tasks enhances the utilization of available resources. Second, the fact that remote computers are often not dedicated to this IC computation, hence, may be more dilatory than anticipated, can lead to a type of "gridlock" that results when a computation stalls because (due to dependencies) all execution-eligible tasks are already allocated to remote computers. These motivating challenges raise the hope that the optimality results presented here within an abstract IC setting have the potential of improving efficiency and fault-tolerance in real IC settings.

1 Introduction

A variety of so-called *pebble games* on dags[3] (directed acyclic graphs) have been shown, over the course of several decades, to yield elegant formal analogues of a variety of problems related to scheduling the tasks/nodes of a computation-dag. The basic idea underlying such games is to use tokens (called "pebbles") to model the progress of a computation on a dag: the placement or removal of pebbles of various types—which is constrained by the dependencies modeled by the dag's arcs[4]—represents the changing (computational) status of the tasks represented

---

[3] Precise definitions of all required notions appear in Section 2.
[4] Pending the definitions of Section 2, one can refer to an algorithms text such as [5] for examples of task interdependencies and their representations via dags.

O. Goldreich et al. (Eds.): Shimon Even Festschrift, LNCS 3895, pp. 291–312, 2006.
© Springer-Verlag Berlin Heidelberg 2006

by the dag's nodes. Pebble games have been used to study problems as diverse as register allocation [17, 3], interprocessor communication in parallel computers [11], "out-of-core" memory accesses [10], and the bandwidth-minimization problem for sparse matrices (which can be formulated as a genre of scheduling problem) [20]. Additionally, pebble games have been shown to model many complexity-theoretic problems perspicaciously; see the survey [18]. The current paper is devoted to surveying ongoing joint work, [19, 21, 16], by the authors and M. Yurkewych (U. Massachusetts), which uses a new pebble game to study the problem of scheduling computation-dags for *Internet-based computing* (*IC*, for short). While this new game shares its basic structure with the "no recomputation allowed" pebble game of [20], it differs markedly from that game in the resource one strives to optimize.

A word about IC will explain the pebble game we study. Advancing technology has rendered the Internet a viable medium for employing multiple independent computers collaboratively in the solution of a single computational problem. A variety of mechanisms have been developed for IC, with "Web-based computing" [14], Peer-to-Peer computing (P2PC) [2, 23], and Grid computing [6, 7] being among the most popular.[5] Most forms of IC—including those just cited—lend themselves naturally to the master-slave computing metaphor, in which a master computer enlists the aid of remote "slave" (or, client) computers to collaborate in the computation of a massive collection of compute-intensive tasks. In rough terms, the differences among the listed modalities are as follows. In "Web-based computing," the remote clients are individuals who allow the master to download a program that will run in background on each client's pc; the clients are typically anonymous and, hence, untrusted; the project usually exists to perform a single computation. Grid computing—so named in analogy with a power grid—typically involves a fixed assemblage of computing sites that contract with one another to share computing resources (possibly, but not necessarily, including computing cycles); Grid members are usually mutually trusted. P2PC often shares with "Web-based computing" the anonymity of remote clients; it usually shares with Grid computing the revolving role of master and client, hence, a lifetime that goes beyond a single computation.

As with all new computing technologies, IC engenders novel scheduling challenges, even while enabling a large variety of computations that could not be handled efficiently by any fixed-size assemblage of dedicated computing agents (e.g., multiprocessors or clusters of workstations). Two related challenges that arise in IC motivate our study. First, in many modalities of IC, remote clients become available (to receive work) at unpredictable times. Second, the fact that remote clients are often not dedicated to the IC computation being performed raises the possibility that some may be slower than anticipated in returning the results from tasks allocated to them. (Indeed, in "Web-based computing," a client may

---

[5] Definitions and terminology in this fast-evolving field tend to vary from one researcher to another, but the definitions here should convey the essential nature of the three modalities of IC. We put "Web-based computing" in quotes because this specific modality has no generally accepted name.

never return its results.) When the tasks being computed are mutually independent, then (finite) delays by clients are just an annoyance; in particular, delays by "old" clients can never preclude having a new task available for allocation to a new client who becomes available. In contrast, when the tasks being computed have interdependencies that constrain their order of execution, dilatory clients may cause the supply of eligible tasks to be very small at certain times. Indeed, in the limit, an IC computation could occasionally encounter a type of "gridlock" wherein the computation stalls because (due to intertask dependencies) all tasks that are eligible for execution are already in the hands of remote clients. The dual scheduling challenges inherent in the preceding scenarios—to enhance the utilization of remote clients and to prevent "gridlock"—motivated the work we survey here.

As is common in the literature on scheduling (cf. [8, 9]), the studies we survey view the intertask dependencies of the computations being scheduled as having the structure of a dag. The goal of our schedules is to allocate the tasks of a given computation-dag to remote clients in a way that *always maximizes the number of tasks that are eligible for execution*. Although details must await further development and/or reference to [21], we can pictorially hint at the significance of the quality metric we are studying. Imagine that one wants to schedule a computation whose task-dependencies have the structure of the evolving mesh in (the upper left corner of) Fig. 1. If one schedules the dag along its "diagonal levels," as depicted in Fig. 2, then after having executed x tasks, one has roughly $\sqrt{x}$ tasks that are eligible to be the next executed task. In contrast, if one chooses to schedule the dag along its "square shells," as depicted in Fig. 3, then one never has access to more than three tasks that are eligible for execution. This example presents an atypically extreme contrast, but it should suggest that the rate of producing execution-eligible tasks may vary significantly for a given dag depending on the schedule used to execute the dag.

Of course, even if one were able to schedule all dags optimally within our idealized setting, one may not always eliminate the two motivating challenges. However, our scheduling strategies would provide guidelines that would provably improve utilization of remote clients and decrease the likelihood of gridlock— when tasks are executed in the order in which they are assigned to the clients. (One avenue toward achieving the desired order is to monitor the behavior and performance of remote clients, as mandated in [1, 13, 22].) And, importantly, the guidelines we derive prescribe actions that are under the control of the IC master and are independent of the behavior of the remote clients!

Our presentation centers on three topics. In Section 2.2, we define the IC Pebble Game that underlies the theory we are developing. Section 3 presents several results that suggest the range of ways that a given family of dags can fit into our embryonic theory—from not admitting an optimal schedule, at one extreme, to admitting infinitely many such schedules, at the other. Section 4 sketches some of the analyses used to derive optimal schedules for certain very uniform families of dags, notably, those in Fig. 1. Section 5 describes our latest, most exciting work, which establishes a foundation for a decomposition-based

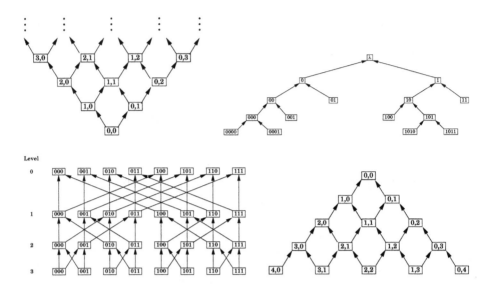

Fig. 1. *Clockwise from upper left: the (2-dimensional) evolving mesh, a (binary) reduction-tree, a (2-dimensional) reduction-mesh (or, pyramid dag), an FFT-dag.*

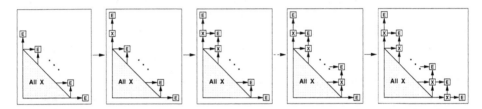

Fig. 2. *Computing a typical diagonal level of the evolving mesh.* "X" *denotes an* EXECUTED *node;* "E" *denotes an* ELIGIBLE *node.*

procedure that derives optimal schedules for a broad range of dags of quite complex, nonuniform structures. Finally, Section 6 describes our response to the fact that some dags admit no optimal schedules within the theory discussed thus far: a batched-scheduling analogue of our theory, within which optimal schedules exist for all families of dags. The major issue in the batched-scheduling setting is how complex (near-)optimal schedules are to derive.

2   A Formal Model for Scheduling Dags for IC

2.1   Computation-Dags

A *directed graph* (*digraph*, for short) $\mathcal{G}$ is given by: a set of *nodes* $N_{\mathcal{G}}$ and a set of *arcs* (or, *directed edges*) $A_{\mathcal{G}}$, each having the form $(u \rightarrow v)$, where $u, v \in N_{\mathcal{G}}$. A *path* in $\mathcal{G}$ is a sequence of arcs that share adjacent endpoints, as in the following path from node u_1 to node u_n:

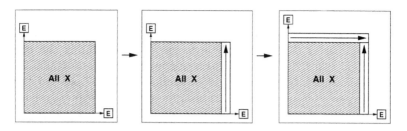

Fig. 3. *A schedule for* $\mathcal{M}_2$ *that traverses* square *levels.* "X" *denotes an* EXECUTED *node;* "E" *denotes an* ELIGIBLE *node. The long arrows indicate sequences of node-executions.*

$$(u_1 \rightarrow u_2), \ (u_2 \rightarrow u_3), \ \ldots, \ (u_{n-2} \rightarrow u_{n-1}), \ (u_{n-1} \rightarrow u_n) \tag{1}$$

A *dag* (short for *directed acyclic graph*) $\mathcal{G}$ is a digraph that has no cycles; i.e., $\mathcal{G}$ cannot contain a path of the form (1) wherein $u_1 = u_n$. When a dag $\mathcal{G}$ is used to model a computation, i.e., is a *computation-dag*:

- each node $v \in N_{\mathcal{G}}$ represents a task of the computation;
- an arc $(u \rightarrow v) \in A_{\mathcal{G}}$ represents the dependence of task v on task u: v cannot be executed until u is.

For each arc $(u \rightarrow v) \in A_{\mathcal{G}}$, we call u a *parent* of v and v a *child* of u in $\mathcal{G}$. The transitive extensions of the parent and child relations are, respectively, the *ancestor* and *descendant* relations. Excepting the degenerate dag that has no nodes: every dag has at least one parentless node (which is called a *source*); every finite dag has at least one childless node (which is called a *sink*). The *outdegree* of a node is its number of children. A dag $\mathcal{G}$ is *bipartite* if:

1. $N_{\mathcal{G}}$ can be partitioned into subsets X and Y such that, for every arc $(u \rightarrow v) \in A_{\mathcal{G}}$, $u \in X$ and $v \in Y$;
2. each $v \in N_{\mathcal{G}}$ is *incident* to some arc of $\mathcal{G}$, i.e., is either the node u or the node w of some arc $(u \rightarrow w) \in A_{\mathcal{G}}$. (Prohibiting "isolated" nodes avoids degeneracies.)

$\mathcal{G}$ is *connected* if, when one ignores the orientation of $\mathcal{G}$'s arcs, there is a path connecting every pair of distinct nodes. A *connected bipartite* dag $\mathcal{H}$ is a **constituent** of $\mathcal{G}$ just when:

1. $\mathcal{H}$ is an *induced subdag* of $\mathcal{G}$: $N_{\mathcal{H}} \subseteq N_{\mathcal{G}}$, and $A_{\mathcal{H}}$ comprises all arcs $(u \rightarrow v) \in A_{\mathcal{G}}$ such that $\{u, v\} \subseteq N_{\mathcal{H}}$.
2. $\mathcal{H}$ is *maximal*: the induced subdag of $\mathcal{G}$ on any *superset* of $\mathcal{H}$'s nodes—i.e., any set S such that $N_{\mathcal{H}} \subset S \subseteq N_{\mathcal{G}}$—is not connected and bipartite.

Let $\mathcal{G}_1, \mathcal{G}_2, \ldots, \mathcal{G}_n$ be connected bipartite dags that are pairwise disjoint, in the sense that $N_{\mathcal{G}_i} \cap N_{\mathcal{G}_j} = \emptyset$ for all distinct indices i and j. The *sum* of these dags, denoted $\mathcal{G}_1 + \mathcal{G}_2 + \cdots + \mathcal{G}_n$, is the bipartite dag whose node-set and arc-set are, respectively, the unions of the corresponding sets of $\mathcal{G}_1, \mathcal{G}_2, \ldots, \mathcal{G}_n$.

We have not posited the *finiteness* of computation-dags. While the inter-task dependencies in nontrivial computations usually have cycles—caused, say, by loops—it is useful to "unroll" these loops when scheduling the computation's individual tasks. This converts the computation's (possibly modest-size) computation-digraph into a sequence of expanding "prefixes" of what "evolves" into an enormous—often infinite—computation-dag. One typically has better algorithmic control over the "steady-state" scheduling of such computations if one expands these computation-dags to their infinite limits and concentrates on scheduling tasks in a way that leads to a computationally expedient sequence of evolving prefixes.

Fig. 1 displays four dags that are studied in [19, 21]: the mesh-dag in the upper left is an infinite dag (which has no sinks); the other three dags are finite. In Section 5, we outline the (de)composition-based theory of [16], which shows how to construct these four dag-families from bipartite building blocks.

2.2   The Internet-Computing Pebble Game

For brevity, we describe the Internet-Computing (IC) Pebble Game within a "pull"-based scheduling framework, in which remote clients approach the server seeking work; the reader can easily adapt our description to a "push"-based framework, in which the server polls remote clients for availability.

The Idealized IC Pebble Game   The IC Pebble Game on a computation-dag $\mathcal{G}$ involves one player S, the *Server*, who has access to unlimited supplies of two types of pebbles: ELIGIBLE pebbles, whose presence indicates a task's eligibility for execution, and EXECUTED pebbles, whose presence indicates a task's having been executed. We now present the rules of the Game, which simplify those of the original IC Pebble Game of [19, 21].

> Our simplification resides in the assumption that by monitoring remote clients (as mandated in, say, [1, 13, 22]) the Server can enhance the likeli-hood, if not the certainty, that remotely allocated tasks will be executed in order of their allocation. We idealize by assuming that the Server can ensure this ordering exactly.

Fig. 4 presents the rules of the IC Pebble Game; Fig. 2 illustrates the rules via a succession of moves of the Game on the 2-dimensional evolving mesh.

For each step t of a play of the IC Pebble Game on a dag $\mathcal{G}$, let $X^{(t)}$ denote the number of EXECUTED pebbles on $\mathcal{G}$'s nodes at step t, and let $E^{(t)}$ denote the analogous number of ELIGIBLE pebbles. Of course, $X^{(t)} = t$ in our idealized version of the Game, although this is not true in the original version of [19]

> *We measure the quality of a play of the IC Pebble Game on a dag $\mathcal{G}$ by the size of $E^{(t)}$ at each step t of the play—the bigger $E^{(t)}$ is, the better. Our goal is an **IC optimal** schedule, in which, for all steps t, $E^{(t)}$ is as big as possible.*

— S begins by placing an ELIGIBLE pebble on each unpebbled source of $\mathcal{G}$.
/*Unexecuted sources are always eligible for execution, having no parents
whose prior execution they depend on.*/
— At each step, S
  • selects a node that contains an ELIGIBLE pebble,
  • replaces that pebble by an EXECUTED pebble,
  • places an ELIGIBLE pebble on each unpebbled node of $\mathcal{G}$ all of whose
    parents contain EXECUTED pebbles.
— S's goal is to allocate nodes in such a way that every node v of $\mathcal{G}$ *eventually*
contains an EXECUTED pebble.
/*This modest goal is necessitated by the possibility that $\mathcal{G}$ is infinite.*/

Fig. 4. *The Rules of the IC Pebble Game*

The significance of IC quality—hence of IC optimality—stems from the following
intuitive scenarios. (1) Schedules that produce ELIGIBLE nodes maximally fast
may reduce the chance of a computation's "stalling" because no new tasks can
be allocated pending the return of already assigned ones (the "gridlock" of the
Introduction). (2) If the Server receives a batch of requests for nodes at (roughly)
the same time, then an IC-optimal schedule allows maximally many requests to
be satisfied, thereby enhancing the exploitation of clients' available resources.

3   The Boundaries of the Playing Field

The property of IC optimality is so demanding that it is not *a priori* clear that
such schedules ever exist! The property demands that there be a *single* schedule
Σ for a dag $\mathcal{G}$ such that, at *every* step of the computation, Σ maximizes the
number of ELIGIBLE nodes across *all* schedules for $\mathcal{G}$. In principle, it could be
that every schedule that maximizes the number of ELIGIBLE nodes at some step t
requires that a certain set of t nodes is EXECUTED, while every analogous schedule
for step $t + 1$ requires that a disjoint set of $t + 1$ nodes is EXECUTED. Indeed,
there exist (simple) dags that do preclude IC-optimal scheduling for precisely
this reason. However, there is a large class of computationally significant dags
that can be scheduled IC optimally. In this section, we exhibit, in turn:

  – simple dags that admit no IC-optimal schedule;
  – a familiar family of dags (evolving meshes), each of which admits a unique
    strategy for producing IC-optimal schedules; we also show that a natural
    alternative to this schedule is actually pessimal in IC quality;
  – a familiar family of dags (evolving trees) all of whose schedules are IC opti-
    mal.

Regrettably, we do not yet know the complexity of determining whether or not a
given dag admits any IC-optimal schedule; this is an inviting research challenge.

Dags that admit no IC-optimal schedule. We begin with two recalcitrant dags; the reader can easily produce others.

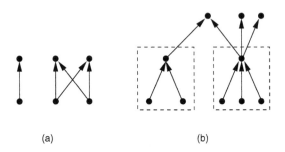

(a)   (b)

Fig. 5. *Two simple dags that admit no IC-optimal schedule.*

Fig. 5 contains two simple dags that do not admit any IC-optimal schedule — for precisely the reason mentioned in the opening paragraph of the section. For the 2-component dag of Fig. 5(a): in order to maximize the number of ELIGIBLE nodes at time $t = 1$, after one node is EXECUTED, one must begin executing the dag with the (unique) outdegree-1 source; in order to maximize the number of ELIGIBLE nodes at time $t = 2$, after two nodes are EXECUTED, one must begin executing the dag with the two outdegree-2 sources. For the tree-dag[6] of Fig. 5(b): in order to maximize the number of ELIGIBLE nodes at time $t = 2$, after two nodes are EXECUTED, one must begin executing the dag with the subtree in the lefthand dashed box; in order to maximize the number of ELIGIBLE nodes at time $t = 4$, after four nodes are EXECUTED, one must begin executing the dag with the subtree in the righthand dashed box.

Dags with a unique IC-optimal scheduling strategy. Fig. 1 (upper left) depicts the first four levels of the *evolving two-dimensional mesh-dag* $\mathcal{M}_2$. The nodes of $\mathcal{M}_2$ are all ordered pairs of nonnegative integers; its arcs connect each node $\langle v_1, v_2 \rangle$ to its two children $\langle v_1+1, v_2 \rangle$ and $\langle v_1, v_2+1 \rangle$. Node $\langle 0, 0 \rangle$ is $\mathcal{M}_2$'s unique source (often called its *origin*). The kth *diagonal level* of $\mathcal{M}_2$, denoted L_k, is the set of nodes whose coordinates sum to k. While $\mathcal{M}_2$ admits infinitely many IC-optimal schedules, all of them implement the strategy of proceeding along successive diagonal levels, from one end to the other.

Theorem 1 ([19]) (a) *For any schedule that allocates nodes sequentially along successive diagonal levels of* $\mathcal{M}_2$, $E^{(t)} = n$ *whenever* $\binom{n}{2} \leq t < \binom{n+1}{2}$.
(b) *For any schedule for* $\mathcal{M}_2$, *if t lies in the preceding range, then* $E^{(t)}$ *can be as large as n, but no larger.*

---

[6] A *tree-dag* $\mathcal{T}$ is any dag such that, if one ignores the orientations of $\mathcal{T}$'s arcs, then the resulting graph is a tree (in the graph-theoretic sense).

It follows that $\mathcal{M}_2$'s IC-optimal schedules are precisely the diagonal-threading schedules.

The intuition underlying Theorem 1 resides in the following facts.

- Each row or column of $\mathcal{M}_2$ contains at most one ELIGIBLE node.
- All ancestors (parents, parents of parents, ...) of each ELIGIBLE node of $\mathcal{M}_2$ are EXECUTED.

Theorem 1 asserts that a *lazy* regimen for executing $\mathcal{M}_2$—i.e., one that always executes the *oldest* ELIGIBLE node, say, by proceeding up each diagonal level of $\mathcal{M}_2$—is IC optimal (albeit not uniquely so). In contrast, an *eager* regimen—i.e., one that always executes the *newest* ELIGIBLE node—is actually *pessimal* in IC quality. One implementation of the eager regimen is the "square-shell" schedule depicted schematically in Fig. 3. By fleshing out this schematic depiction to a level of detail commensurate with that of the lazy, "diagonal-level," schedule of Fig. 2, the reader will find that, under the "square-shell" schedule, no more than *three* nodes of $\mathcal{M}_2$ are ever simultaneously ELIGIBLE, in contrast with the ever-growing number of ELIGIBLE nodes promised by Theorem 1 for any lazy schedule.

Dags for which any *schedule is IC optimal.* Consider the *evolving binary out-tree* of Fig. 6. A simple argument shows that *every* valid schedule for the evolving

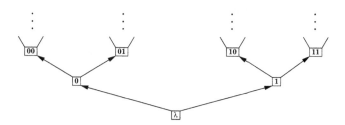

Fig. 6. *An evolving binary out-tree.*

binary out-tree $\mathcal{T}$ is IC optimal. To wit, at every moment during the execution of $\mathcal{T}$, the EXECUTED nodes are the internal nodes of a full binary sub-out-tree $\mathcal{T}'$ of $\mathcal{T}$, and the ELIGIBLE nodes are the leaves of $\mathcal{T}'$. It follows that at every step of any schedule for $\mathcal{T}$, the number of ELIGIBLE nodes is precisely one more than the number of EXECUTED nodes.

The messages of this section.

- There are significant families of dags that admit IC-optimal schedules.
- The disparity between the IC quality of an optimal schedule and that of a natural competitor can be very large.

– No scheduling strategy is going to guarantee IC-optimal schedules for all dags.

The remaining sections summarize our responses to these messages.

4   IC-optimal Schedules for Specific Families of Dags

In this section, we complete the scheduling story for the dags in Fig. 1 by dealing with the three "reductive" dags.

Theorem 2 ([21]) *A schedule for any reduction-mesh, reduction-tree, or FFT-dag is IC optimal if, and only if, it is parent-oriented—i.e., it executes all parents of a node in consecutive steps.*

Since the proofs for the three families of dags share are almost identical in structure (cf. [21]), we sketch the proof only for the reduction-mesh.

The nodes of the ℓ-*level reduction-mesh* M_ℓ comprise the set of ordered pairs of integers $\{\langle x, y \rangle \mid 0 \leq x + y < \ell\}$. M_ℓ's arcs connect each node $v = \langle x, y \rangle$ to node $\langle x - 1, y \rangle$ whenever $x > 0$ and to node $\langle x, y - 1 \rangle$ whenever $y > 0$. The integer $x + y$ is the *level* of node $\langle x, y \rangle$. M_ℓ's ℓ source nodes are the nodes at level $\ell - 1$; M_ℓ's unique sink node is node $\langle 0, 0 \rangle$, the sole occupant of level 0.

Focus on a play of the IC Pebble Game on M_ℓ. Say that at step t of the play, each level $l \in \{0, 1, \ldots, \ell - 1\}$ of M_ℓ has $E_l^{(t)}$ ELIGIBLE nodes and $X_l^{(t)}$ EXECUTED nodes. Let c be the smallest level-number for which $E_c^{(t)} + X_c^{(t)} > 0$.

Claim. *Given the current profile* $\langle X_l^{(t)} \mid 0 \leq l < \ell \rangle$ *of* EXECUTED *nodes:*

1. *The aggregate number of* ELIGIBLE *nodes at time* t, $E^{(t)} \stackrel{\text{def}}{=} \sum_{i=0}^{\ell-1} E_i^{(t)}$, *is maximized if all* EXECUTED *nodes on each level of* M_ℓ *are consecutive.*[7]
2. *Once* $E^{(t)}$ *is so maximized, we have* $c \leq E^{(t)} \leq c + 1$.

Each nonsource ELIGIBLE node of M_ℓ has two EXECUTED parents; any two consecutive nonsource ELIGIBLE nodes share an EXECUTED parent. We thus have the following system of inequalities.

$$
\begin{aligned}
E_l^{(t)} &\leq X_{l+1}^{(t)} - X_l^{(t)} - 1 \quad \text{for} \quad l \in \{c, c+1, \ldots, \ell - 2\}; \\
E_{\ell-1}^{(t)} &= \ell - X_{\ell-1}^{(t)}.
\end{aligned}
\tag{2}
$$

1. If all EXECUTED nodes occur consecutively along a level $l + 1$ of M_ℓ, then the inequality involving $E_l^{(t)}$ in (2) is an equality. Therefore, all inequalities in (2) are equalities when the EXECUTED nodes at every level occur consecutively. Further, such consecutiveness may decrease the value of c, by rendering new nodes ELIGIBLE at lower-numbered levels. Consequently, this arrangement of EXECUTED nodes maximizes the value of $E^{(t)}$.

---

[7] Nodes $u_0, u_1, \ldots, u_{k-1}$ are *consecutive* on level l of M_ℓ just when each $u_j = \langle m + j, l - m - j \rangle$ for some $0 \leq m \leq l - k$, $0 \leq j < k$.

2. Summing the (now) equalities in system (2) yields an explicit expression for the maximum value of $E^{(t)}$ in terms of $\sum_{i=0}^{\ell-1} X_i^{(t)} = t$, namely: $E^{(t)} = \sum_{i=c}^{\ell-1} E_i^{(t)} = c+1-X_c^{(t)}$. Part (2) of the claim now follows, because when the EXECUTED nodes at each level of M_ℓ occur consecutively, we must have $X_c^{(t)} \leq 1$: a larger value would imply that $X_{c-1}^{(t)} + E_{c-1}^{(t)} > 0$.

For reduction-meshes, parent-orientation means "level-by-level" execution. For reduction-trees, the phrase means that each tree-node u and its "sibling" (i.e., the node that shares a child with u) must be executed in consecutive steps. For FFT-dags, the phrase means that each node u and its "butterfly partner" (i.e., the node that shares two children with u) must be executed in consecutive steps.

5   Toward a Theory of Scheduling *Composite* Dags

The similarities in the structures of the proofs of Theorem 2 for its three families of dags led us to seek a structure-based explanation of the similarities. We now describe the results of this quest, which has gone far beyond just the motivating explanation.

A hallmark of the nascent scheduling theory of [16] is that it seeks explicit IC-optimal schedules only for connected bipartite dags (which experience has shown is already often quite a challenge). It then uses these bipartite dags as building blocks for constructing complex dags that inherit their IC-optimal schedules from those of the bipartite dags. We outline this development in this section.

The following simple result is quite useful in analyzing scheduling strategies for possible IC optimality. It should allow the reader to intuit the proofs for several of the results that we present.

Lemma 3 ([16]) *If a schedule Σ for a dag $\mathcal{G}$ is altered to execute all of $\mathcal{G}$'s nonsinks before any of its sinks, then the IC quality of the resulting schedule is no less than Σ's.*

5.1   A Sampler of Bipartite Building Blocks

Our study applies to any repertoire of *connected bipartite building-block dags* that one chooses to build complex dags from. For illustration, though, we focus on the following specific building blocks. The following descriptions proceed left to right along successive rows of Fig. 7. For all descriptions, we use the drawings in Fig. 7 to refer to "left" and "right."

The first three dags are named for the letters suggested by their topologies.

W-dags. For each integer $d > 1$, the $(1,d)$-*W-dag* $\mathcal{W}_{1,d}$ has one source and d sinks; its d arcs connect the source to each sink. Inductively, for positive integers a, b, the $(a+b, d)$-W-dag $\mathcal{W}_{a+b,d}$ is obtained from the (a, d)-W-dag $\mathcal{W}_{a,d}$ and the (b, d)-W-dag $\mathcal{W}_{b,d}$ by identifying (or, merging) the rightmost sink of the former dag with the leftmost sink of the latter. W-dags epitomize "expansive" computations.

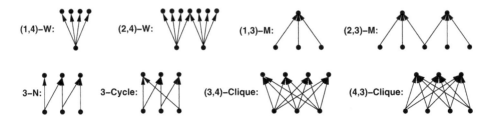

Fig. 7. *The building blocks of semi-uniform dags.*

M-dags. For each integer $d > 1$, the $(1, d)$-*M-dag* $\mathcal{M}_{1,d}$ has d sources and one sink; its d arcs connect each source to the sink. Inductively, for positive integers a, b, the $(a + b, d)$-M-dag $\mathcal{M}_{a+b,d}$ is obtained from the (a, d)-M-dag $\mathcal{M}_{a,d}$ and the (b, d)-M-dag $\mathcal{M}_{b,d}$ by identifying (or, merging) the rightmost source of the former dag with the leftmost source of the latter. M-dags epitomize "contractive" (or, "reductive") computations.

N-dags. For each integer $s > 0$, the s-*N-dag* $\mathcal{N}_s$ has s sources and s sinks; its $2s - 1$ arcs connect each source v to sink v and to sink $v + 1$ if the latter exists. Specifically, $\mathcal{N}_s$ is obtained from $\mathcal{W}_{s-1,2}$ by adding a new source on the right whose sole arc goes to the rightmost sink. The leftmost source of $\mathcal{N}_s$ has a child that has no other parents; we call this source the *anchor* of $\mathcal{N}_s$.

(Bipartite) Cycle-dags. For each integer $s > 1$, the s-*(Bipartite) Cycle-dag* $\mathcal{C}_s$ is obtained from $\mathcal{N}_s$ by adding a new arc from the rightmost source to the leftmost sink—so that each source v has arcs to sinks v and $v + 1 \bmod s$.

(Bipartite) Clique-dags. For integers $s, s' > 1$, the (s, s')-*(Bipartite) Clique-dag* $\mathcal{Q}_{s,s'}$ has s sources, s' sinks, and an arc from each source to each sink. (We actually deal only with (s, s)-Cliques, which we henceforth denote $\mathcal{Q}_s$; we present general (s, s')-Cliques as an invitation to the reader.)

5.2   Building Dags Via Composition

Our basic technique for constructing complex dags is the following inductively defined operation of *composition*.

- We start with any set **B** of connected bipartite dags; these will serve as our base set.
- Given dags $\mathcal{G}_1, \mathcal{G}_2 \in \mathbf{B}$—which could be copies of the same dag with nodes renamed to achieve disjointness—we obtain a composite dag $\mathcal{G}$ as follows.
  - Let the composite dag $\mathcal{G}$ begin as the sum, $\mathcal{G}_1 + \mathcal{G}_2$, of the dags $\mathcal{G}_1, \mathcal{G}_2$. We rename nodes to ensure that $N_{\mathcal{G}}$ is disjoint from $N_{\mathcal{G}_1}$ and $N_{\mathcal{G}_2}$.
  - We select some set S_1 of sinks from the copy of $\mathcal{G}_1$ in the sum $\mathcal{G}_1 + \mathcal{G}_2$, and an equal-size set S_2 of sources from the copy of $\mathcal{G}_2$ in the sum.
  - We pairwise identify (i.e., merge) the nodes in the sets S_1 and S_2 in some way. The resulting set of nodes is $\mathcal{G}$'s node-set; the induced set of arcs is $\mathcal{G}$'s arc-set.
- We add the dag $\mathcal{G}$ thus obtained to the base set **B**.

We denote the composition operation by ⇑ and refer to the resulting dag $\mathcal{G}$ as a *composite dag of type* $[\mathcal{G}_1 \Uparrow \mathcal{G}_2]$. (Note that the structure of $\mathcal{G}$ is not identified uniquely by its type. Our theory does not require knowledge of this detailed structure.) The roles of $\mathcal{G}_1$ and $\mathcal{G}_2$ in creating $\mathcal{G}$ are asymmetric: $\mathcal{G}_1$ contributes sinks, while $\mathcal{G}_2$ contributes sources.

We can now simply illustrate the natural correspondence between the node-set of a composite dag and those of its constituents, via Fig. 1:

- The evolving mesh $\mathcal{M}_2$ is composite of type $\mathcal{W}_{1,2} \Uparrow \mathcal{W}_{2,2} \Uparrow \mathcal{W}_{3,2} \Uparrow \cdots$.
- A binary reduction-tree is obtained by pairwise composing many instances of $\mathcal{M}_{1,2}$ (seven instances in the figure).
- The reduction-mesh $\mathcal{M}_5$ is composite of type $\mathcal{M}_{5,2} \Uparrow \mathcal{M}_{4,2} \Uparrow \mathcal{M}_{3,2} \Uparrow \mathcal{M}_{2,2} \Uparrow \mathcal{M}_{1,2}$.
- The FFT dag is obtained by pairwise composing many instances of $\mathcal{C}_2 = \mathcal{Q}_2$ (twelve instances in the figure).

As hinted at in the preceding description, the composition operation is associative, so we do not have to keep track of the order in which constituent dags are incorporated into a composite dag.

Lemma 4 ([16]) *The composition operation on dags is associative. That is, for all dags $\mathcal{G}_1$, $\mathcal{G}_2$, $\mathcal{G}_3$, a dag is composite of type $[[\mathcal{G}_1 \Uparrow \mathcal{G}_2] \Uparrow \mathcal{G}_3]$ if, and only if, it is composite of type $[\mathcal{G}_1 \Uparrow [\mathcal{G}_2 \Uparrow \mathcal{G}_3]]$.*

One can garner intuition for the proof of Lemma 4 from the dags on Fig. 8.

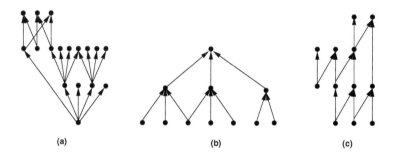

(a)   (b)   (c)

Fig. 8. *A sampler of composite dags, each of which admits an IC-optimal schedule.*

5.3   The Priority Relation $\succeq$

The next ingredient in our scheduling theory is the following relation on bipartite dags. This relation is the mechanism that we can often use to "inherit" an IC-optimal schedule for a composite dag from IC-optimal schedules for its constituents.

Let the disjoint bipartite dags G_1 and G_2, having s_1 and s_2 sources, respectively, admit the IC-optimal schedules Σ_1 and Σ_2, respectively. Say that the following inequalities hold.[8]

$$(\forall x \in [0, s_1]) \, (\forall y \in [0, s_2]) : \\ E_{\Sigma_1}(x) + E_{\Sigma_2}(y) \; \leq \; E_{\Sigma_1}(\min\{s_1, x + y\}) + E_{\Sigma_2}(\max\{0, x + y - s_1\}). \tag{3}$$

Then we say that G_1 *has priority over* G_2, denoted $G_1 \succeq G_2$.

By Lemma 3, the inequalities in (3) basically say that one never decreases IC quality by executing a source of G_1, in preference to a source of G_2, whenever possible.

It is important, both conceptually and algorithmically, that the relation $\succeq$ is transitive. This fact is a bit trickier to prove than one might think at first blush.

Lemma 5 ([16]) *The relation $\succeq$ on bipartite dags is transitive.*

One simple, but consequential application of Lemma 5 is:

Corollary 6 ([16]) *Let $G_1, G_2, \ldots, G_n$ be pairwise disjoint bipartite dags. If $G_1 \succeq G_2 \succeq \cdots \succeq G_n$, then $G_1 \succeq (G_2 + G_3 + \cdots + G_n)$.*

5.4   Scheduling $\succeq$-Linear Compositions of Composite Dags

We arrive finally at the first major result of the theory, which provides the sought explanation for the structures of the proofs of Theorem 2 for the three families of dags. More importantly, this result gives structure to our quest for a decomposition-based scheduling theory.

Say that dag G is a $\succeq$-*linear composition* of the connected bipartite dags $G_1, G_2, \ldots, G_n$ if:

1. G is composite of type $G_1 \Uparrow G_2 \Uparrow \cdots \Uparrow G_n$;
2. each $G_i \succeq G_{i+1}$, for all $i \in [1, n-1]$.

Dags that are $\succeq$-linear compositions inherit IC-optimal schedules from their constituents.

Theorem 7 ([16]) *Let G be a $\succeq$-linear composition of $G_1, G_2, \ldots, G_n$, where each G_i admits an IC-optimal schedule Σ_i. The schedule Σ for G that proceeds as follows is IC optimal.*

1. *Σ executes the nodes of G that correspond to sources of G_1, in the order mandated by Σ_1, then the nodes that correspond to sources of G_2, in the order mandated by Σ_2, and so on, for all $i \in [1, n]$.*
2. *Σ finally executes all sinks of G in any order.*

The proof of Theorem 7 essentially demonstrates that when a dag G is a $\succeq$-*linear composition*, then the priority relation $\succeq$ on G's bipartite constituents is compatible with the executional priorities that are inherent in G's being a dag.

---

[8] $[a, b]$ denotes the set of integers $\{a, a+1, \ldots, b\}$.

5.5   Scheduling Composite Dags Via Decomposition

The framework developed thus far in this section is descriptive rather than pre-scriptive. *If* a computation-dag $\mathcal{G}$ is constructed from bipartite building blocks via composition, and *if* we can identify the "blueprint" used to construct $\mathcal{G}$, and *if* the underlying building blocks are interrelated in a certain way, *then* the strategy prescribed in Theorem 7 produces an optimal schedule for $\mathcal{G}$. We now describe how the algorithmic challenge hidden in the preceding *if*s is addressed in [16]: given a computation-dag $\mathcal{G}$, how does one apply the preceding frame-work to it? The algorithms we describe now attempt to decompose $\mathcal{G}$ in order to expose the structure needed to apply Theorem 7. We thereby derive IC-optimal schedules for a large variety of dags.

We describe a suite of algorithms that: (a) reduce any computation-dag $\mathcal{G}$ to its "transitive skeleton" $\mathcal{G}'$, a simplified version of $\mathcal{G}$ that shares the same set of optimal schedules; (b) decompose $\mathcal{G}'$ to determine whether or not it is constructed from bipartite building blocks via composition, thereby exposing a "blueprint" for $\mathcal{G}'$; (c) specify an optimal schedule for any such $\mathcal{G}'$ that is built from building blocks that are interrelated under $\succeq$.

Skeletonizing Input Dags For any dag $\mathcal{G}$ and nodes $u, v \in N_\mathcal{G}$, we write $u \Rightarrow_\mathcal{G} v$ to indicate that there is a path from u to v in $\mathcal{G}$. An arc $(u \rightarrow v) \in A_\mathcal{G}$ is a *shortcut* if there is a path $u \Rightarrow_\mathcal{G} v$ that does not include the arc. Of course, removing shortcuts from a dag does not alter internode connectivities. By removing all shortcuts from a dag $\mathcal{G}$, one obtains $\mathcal{G}$'s *(transitive) skeleton* (or, *transitive reduction*). This dag, which is unique, is the smallest subdag of $\mathcal{G}$ that shares $\mathcal{G}$'s transitive closure [5]; we call this dag $\sigma(\mathcal{G})$. One finds in [12] a polynomial-time algorithm that generates $\sigma(\mathcal{G})$ from $\mathcal{G}$. (In fact, a very simple algorithm suffices, that just removes, in turn, each arc $(u \rightarrow v)$ from $\mathcal{G}$ and tests if v is still accessible from u.)

Eliminating shortcuts is a critical first step in our decompsitional scheduling strategy, because dags that are compositions of bipartite dags have no shortcuts.

Since $\mathcal{G}$ shares its node-set with $\sigma(\mathcal{G})$, any schedule that executes one dag also executes the other. This is important because any schedule executes $\mathcal{G}$ as efficiently (in IC quality) as it executes $\sigma(\mathcal{G})$. A special case of this result appears in [19].

Theorem 8 ([16]) *A schedule Σ has the same IC quality when executing a dag $\mathcal{G}$ as when executing $\sigma(\mathcal{G})$. In particular, if Σ is IC optimal for $\sigma(\mathcal{G})$, then it is IC optimal for $\mathcal{G}$.*

Decomposing a Composite Dag Once we have a shortcut-free dag $\mathcal{G}$, we can start trying to decompose it, to find subdags whose composition yields $\mathcal{G}$. We now describe this process.

Selecting a constituent. We begin by selecting any constituent[9] of $\mathcal{G}$ all of whose sources are also sources of $\mathcal{G}$; call the selected constituent $\mathcal{B}_1$ (the notation emphasizing that $\mathcal{B}_1$ is *bipartite*).

In Fig. 1: Every candidate $\mathcal{B}_1$ for the FFT dag comprises a copy of $\mathcal{C}_2 = \mathcal{Q}_2$ included in levels 2 and 3; every candidate for the reduction-tree comprises a copy of $\mathcal{M}_{1,2}$; the unique candidate for the reduction-mesh comprises $\mathcal{M}_{4,2}$.

Detaching a constituent. We "detach" $\mathcal{B}_1$ from $\mathcal{G}$ by deleting the nodes of $\mathcal{G}$ that correspond to sources of $\mathcal{B}_1$, all incident arcs, and all resulting isolated sinks. We thereby replace $\mathcal{G}$ with a pair of dags $\langle \mathcal{B}_1, \mathcal{G}' \rangle$, where $\mathcal{G}'$ is the remnant of $\mathcal{G}$ remaining after $\mathcal{B}_1$ is detached.

If the remnant $\mathcal{G}'$ is not empty, then we continue the process of selection and detachment. If $\mathcal{G}$ was a composition of bipartite dags, then we produce a sequence of the form

$$\mathcal{G} \implies \langle \mathcal{B}_1, \mathcal{G}' \rangle \implies \langle \mathcal{B}_1, \langle \mathcal{B}_2, \mathcal{G}'' \rangle \rangle \implies \langle \mathcal{B}_1, \langle \mathcal{B}_2, \langle \mathcal{B}_3, \mathcal{G}''' \rangle \rangle \rangle \implies \cdots,$$

that leads ultimately to a complete decomposition of $\mathcal{G}$ into a sequence comprising all of its constituents: $\mathcal{B}_1, \mathcal{B}_2, \ldots, \mathcal{B}_n$.

We claim that the described process does, indeed, recognize whether or not $\mathcal{G}$ is a composite dag, and, if so, it produces the constituents from which $\mathcal{G}$ is composed (possibly, of course, in an order that differs from their original order of composition).

Theorem 9 ([16]) *Let the dag* $\mathcal{G}$ *be composite of type* $\mathcal{G}_1 \Uparrow \mathcal{G}_2 \Uparrow \cdots \Uparrow \mathcal{G}_n$. *The decomposition process produces a sequence* $\mathcal{B}_1, \mathcal{B}_2, \ldots, \mathcal{B}_n$ *of constituents of* $\mathcal{G}$ *such that:*

- $\mathcal{G}$ *is composite of type* $\mathcal{B}_1 \Uparrow \mathcal{B}_2 \Uparrow \cdots \Uparrow \mathcal{B}_n$;
- $\{\mathcal{B}_1, \mathcal{B}_2, \ldots, \mathcal{B}_n\} = \{\mathcal{G}_1, \mathcal{G}_2, \ldots, \mathcal{G}_n\}$.

It is fruitful to construct a *super-dag* that abstracts a dag $\mathcal{G}$'s structure, as exposed by the decomposition process. This super-dag, which we denote $\mathcal{S}(\mathcal{B}_1 \Uparrow \cdots \Uparrow \mathcal{B}_n)$, has the constituents $\mathcal{B}_1, \mathcal{B}_2, \ldots, \mathcal{B}_n$ as its nodes and has an arc from each constituent $\mathcal{B}_i$ to each of the constituents that it is detached from during the decomposition. Fig. 9 depicts the super-dag obtained from decomposing the 3-dimensional FFT dag. Easily that the linearization $\mathcal{B}_1, \ldots, \mathcal{B}_n$ produced by the described decomposition process is a *topological sort* [5] of the super-dag $\mathcal{S}(\mathcal{B}_1 \Uparrow \cdots \Uparrow \mathcal{B}_n)$.

Scheduling a Composite Dag Via Its Super-Dag Our remaining challenge is to determine, given a super-dag $\mathcal{S}(\mathcal{B}_1 \Uparrow \cdots \Uparrow \mathcal{B}_n)$ that is produced by our decomposition process, whether or not there is a topological sort of the super-dag that linearizes the supernodes in an order that honors relation $\succeq$. We now present sufficient conditions for this to occur, verified via a linearization algorithm.

---

[9] Recall the technical definition of "constituent" from Section 2.1.

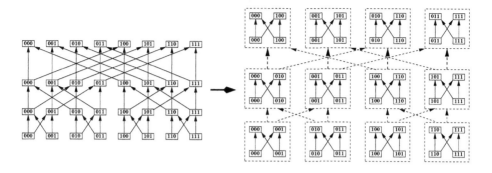

Fig. 9. *The 3-dimensional FFT dag and its associated super-dag.*

Theorem 10 ([16]) *Say that the dag $\mathcal{G}$ is composite of type $\mathcal{B}_1 \Uparrow \cdots \Uparrow \mathcal{B}_n$ and that, for each pair of constituents, $\mathcal{B}_i$, $\mathcal{B}_j$ with $i \neq j$, either $\mathcal{B}_i \succeq \mathcal{B}_j$ or $\mathcal{B}_j \succeq \mathcal{B}_i$. Then $\mathcal{G}$ is a $\succeq$-linear composition whenever the following holds.*

Whenever $\mathcal{B}_j$ is a child of $\mathcal{B}_i$ in $\mathcal{S}(\mathcal{B}_1 \Uparrow \cdots \Uparrow \mathcal{B}_n)$, we have $\mathcal{B}_i \succeq \mathcal{B}_j$.

Theorem 10 is proved via the following algorithm that determines whether or not $\mathcal{G}$ is a $\succeq$-linear composition of the $\mathcal{B}_i$.

1. We begin with a topological sort, $\widehat{\mathcal{B}} \stackrel{\text{def}}{=} \mathcal{B}_{\alpha(1)}, \ldots, \mathcal{B}_{\alpha(n)}$ of $\mathcal{S}_{\mathcal{G}} \stackrel{\text{def}}{=} \mathcal{S}(\mathcal{B}_1 \Uparrow \cdots \Uparrow \mathcal{B}_n)$.

2. We invoke the hypothesis that $\succeq$ is a (weak) order on the $\mathcal{B}_i$'s to reorder $\widehat{\mathcal{B}}$ according to $\succeq$, using a *stable*[10] comparison sort.

Let $\mathcal{B} \stackrel{\text{def}}{=} \mathcal{B}_{\beta(1)} \succeq \cdots \succeq \mathcal{B}_{\beta(n)}$ be the linearization of $\mathcal{S}_{\mathcal{G}}$ produced by the sort. We claim that $\mathcal{B}$ is also a topological sort of $\mathcal{S}_{\mathcal{G}}$. This follows easily because we start with a topological sort of $\mathcal{S}_{\mathcal{G}}$ and employ a stable sort on relation $\succeq$. We conclude that $\mathcal{G}$ is composite of type $\mathcal{B}_{\beta(1)} \Uparrow \cdots \Uparrow \mathcal{B}_{\beta(n)}$. In other words, $\mathcal{B}$ is the desired $\succeq$-linearization of $\mathcal{G}$.

Once we have the decomposition $\mathcal{B}$, we can invoke Theorem 7 to obtain an IC-optimal schedule for $\mathcal{G}$.

6   A Batched Approach to Scheduling

Our development of a dag-scheduling theory for IC is still ongoing: we are making steady progress in both refining and extending the work described in Section 5. Yet, the stringent demands of IC optimality, as reflected in the requirement that a schedule maximize the number of ELIGIBLE nodes *at every step* of a computation, guarantees the existence of simple dags that admit no IC-optimal schedule (cf. Section 3); hence, they preclude this theory from ever being comprehensive.

---

[10] Stability means that if $\mathcal{B}_i \succeq \mathcal{B}_j$ and $\mathcal{B}_j \succeq \mathcal{B}_i$, then the sort maintains the original positions of $\mathcal{B}_i$ and $\mathcal{B}_j$.

Responding to this fact, we are investigating alternative formulations of the IC-scheduling problem. We have developed one such in [15], and we describe its rudiments in this section.

In the *batched* version of the IC Pebble Game, which abstracts the batched version of the IC-scheduling problem, the Server does not respond to individual requests by Clients as they come in. Instead, it services requests at fixed intervals, hence responds to batches of requests rather than individual ones. This formulation of IC scheduling simplifies the scheduling problem along one axis, while complicating it along another. We now focus on optimizing the production of ELIGIBLE nodes:

1. for a single step of the computation, rather than uniformly for all steps;
2. while executing r (perforce, ELIGIBLE) nodes as a batch, rather than a single node.

The (algorithmic) greed built into this version of IC-scheduling—by the first condition—ensures that there is an optimal solution to every instance of the problem. The complication built into this version—by the second condition—turns out to endow the challenge of finding this optimal solution with the likely computational intractability of NP-hardness. (The solution is easy to find when $r = 1$.) Fig. 10 presents the rules of the game.

— S begins by placing an ELIGIBLE pebble on each unpebbled source node of $\mathcal{G}$.
  /*Unexecuted source nodes are always eligible for execution, having no parents whose prior execution they depend on.*/
— At each step t—when there is some number, say e_t, of ELIGIBLE pebbles on $\mathcal{G}$'s nodes—S is approached by some number, say r_t, of Clients, requesting tasks. In response, S:
  • selects $\min\{e_t, r_t\}$ tasks that contain ELIGIBLE pebbles,
  • replaces those pebbles by EXECUTED pebbles,
  • places ELIGIBLE pebbles on each unpebbled node of $\mathcal{G}$ all of whose parents contain EXECUTED pebbles.
— S's goal is to allocate nodes in such a way that every node v of $\mathcal{G}$ *eventually* contains an EXECUTED pebble.
  /*This modest goal is necessitated by the possibility that $\mathcal{G}$ may be infinite.*/

Fig. 10. *The Rules of the Batch-IC Pebble Game*

As in earlier sections, we call a node ELIGIBLE (resp., EXECUTED) when it contains an ELIGIBLE (resp., an EXECUTED) pebble, and we talk about executing nodes rather than tasks.

The Batch-IC scheduling problem (BICSO). Our goal is to play the Batch-IC Pebble Game in a way that maximizes the number of ELIGIBLE pebbles on $\mathcal{G}$ at every step of the Game. That is, for each step t of a play of the Game on a dag $\mathcal{G}$, if there are currently e_t ELIGIBLE tasks, and if r_t Clients request tasks, then we want the Server to execute a set of $\min\{e_t, r_t\}$ ELIGIBLE nodes that will result in the largest possible number of ELIGIBLE tasks at step $t + 1$. We thus arrive at the following optimization problem.

Batched IC-Scheduling (Optimization version) (**BICSO**)
Instance: $\imath = \langle \mathcal{G}, X, E; r \rangle$, where:
- $\mathcal{G}$ is a computation-dag;
- X and E are disjoint subsets of $N_{\mathcal{G}}$ that satisfy the following;
  There is a step of some play of the Batched IC Pebble Game on $\mathcal{G}$ in which X is the set of EXECUTED nodes and E the set of ELIGIBLE nodes on $\mathcal{G}$.
- r is in the set $[1, |E|]$.
Problem: Find a cardinality-r set $R \subseteq E$ that maximizes the number of ELIGIBLE nodes on $\mathcal{G}$ after executing the nodes in R, given that the nodes in X are already EXECUTED.

Note that the process of solving BICSO automatically carries with it a guarantee of optimality.

In contrast to the search for IC-optimal schedules for dags, every instance of BICSO can be solved! The only question is how hard it is computationally to find a solution. Unfortunately, solving BICSO is likely computationally intractable, even for dags of quite restricted structure.

Theorem 11 ([15]) *BICSO is NP-hard, even when restricted to bipartite dags.*

Of course, results such as Theorem 11 automatically trigger a search for special classes of dags that can be scheduled optimally in polynomial time. Not surprisingly, bipartite tree-dags—*and compositions thereof*—are the first such class that we discovered. The algorithm guaranteed by the following theorem contains a dynamic program as a central component.

Theorem 12 ([15]) *There is a polynomial-time algorithm Σ_{tree} that solves BICSO for any bipartite tree-dag $\mathcal{T}$.*

Theorem 12 is actually more textured than it seems to be at first. On the optimistic side: The theorem gives us more scheduling power than is immediately apparent. Specifically, we show in [15] how to build upon the theorem to solve BICSO for any *composition* of bipartite tree-dags. This is important, since compositions of such tree-dags need not be either leveled or (in their undirected incarnations) cycle-free. On the less-optimistic side: Algorithm Σ_{tree} is computationally rather inefficient: its timing polynomial has high degree. In response, we have sought nontrivial classes of dags for which we could solve BICSO *efficiently*, even if the solution was only *approximate*. We use the word "approximate" here in its usual technical sense: we insist that the number of ELIGIBLE nodes produced by the scheduling algorithm in response to r requests be within a predictable factor of the maximum possible number, given the then-current number of ELIGIBLE nodes.

Our initial success in this quest involved the family **E** of *bipartite expansive-dags*. Each such dag $\mathcal{E}$ is a bipartite dag wherein each source v has an associated number $\varphi_v \geq 2$ such that: v has φ_v children that have no parent other than v and at most φ_v other children. Expansive-dags epitomize computations that are "expansive" but may have complex interdependencies. A simple algorithm that we call Algorithm $\Sigma_{\exp}$ approximates a solution to BICSO for the family **E**.

Algorithm $\Sigma_{\exp}$ implements the following natural, fast heuristic for scheduling a dag $\mathcal{E} \in \mathbf{E}$. For each source v of $\mathcal{E}$, say that there are φ_v nodes that have v as their sole parent and ψ_v nodes that have other parents also. If there are r requests for ELIGIBLE nodes at time t, then $\Sigma_{\exp}$ selects the r ELIGIBLE nodes that have the largest associated integers φ_v. Of course, this greedy heuristic may deviate from optimality because it ignores the "bonuses" that may arise from executing ELIGIBLE nodes that are siblings in $\mathcal{E}$, but it does come close to optimality.

Theorem 13 ([15]) *For any instance $\iota = \langle \mathcal{E}, X, E; r \rangle$ of BICSO, where $\mathcal{E} \in \mathbf{E}$, Algorithm $\Sigma_{\exp}$ will, in time $O(|E|)$, find a solution to BICSO, whose increase in the number of* ELIGIBLE *nodes is at least one-fourth the optimal increase.*

Work continues in trying to extend both Theorems 12 and 13, by expanding the classes of dags for which we can tractably solve, or quickly approximate a solution to BICSO.

7   Conclusions and Projections

We have described two related pebble games that abstract the problem of scheduling computation-dags for Internet-based computing. Both games place an ELIGIBLE pebble (which represents a task's being eligible for execution) on every node all of whose parents contain EXECUTED pebbles (which represents a task's having been executed). At each step: one game selects a single ELIGIBLE pebble to replace with an EXECUTED pebble; the other selects a variable number of ELIGIBLE pebbles to replace (based on the input). With both games, the placement of a new EXECUTED pebble may cause the placement of new ELIGIBLE pebbles. Both games strive, under somewhat different rules, to maximize the number of nodes that contain ELIGIBLE pebbles.

The **IC Pebble Game** takes as input a dag $\mathcal{G}$. It seeks an execution schedule for $\mathcal{G}$ that maximizes the number of nodes that hold ELIGIBLE pebbles *at every step of the game*. We have described the underpinnings of a theory of scheduling under the IC Pebble Game, which builds on the decomposition of an input dag $\mathcal{G}$ into bipartite "building-block" dags. When the decomposition exposes $\mathcal{G}$ to be a *composition of building blocks* that are suitably iterrelated under the *priority relation*, then the theory generates a schedule for $\mathcal{G}$ that is optimal. Ongoing work, some in collaboration with G. Cordasco (U. Salerno), seeks to expand the range of dags that the theory can schedule optimally, both by expanding the repertoire of building blocks that it can deal with [4] and by extending the scope of the priority relation. Other work, some in collaboration with I. Foster and

M. Wilde of Argonne National Laboratory, seeks to assess the impact of the emerging theory on a real IC project.

The **Batched-IC Pebble Game** takes as input a dag $\mathcal{G}$, some e of whose nodes hold ELIGIBLE pebbles, and an integer $r \leq e$ of "requests." It seeks to find a set of r nodes currently holding ELIGIBLE pebbles such that executing those nodes will allow the placement of maximally many new ELIGIBLE pebbles. Results obtained thus far have shown the problem of solving instances of this problem optimally to be NP-hard (with the decision version being NP-complete). The problem is solvable in polynomial time for composite tree-dags, yet not efficiently. The problem is efficiently approximable for certain special classes of dags. Ongoing work here is delving further into the search for efficiently schedulable classes of dags and efficiently approximable classes.

For both pebble games, attempts are also being made to assess the quality of schedules produced by simple heuristics.

Acknowledgment. The research of A. Rosenberg was supported in part by NSF Grant CCF-0342417. It is a pleasure to acknowledge the many contributions of our collaborator, Matt Yurkewych of the University of Massachusetts Amherst.

References

1. R. Buyya, D. Abramson, J. Giddy (2001): A case for economy Grid architecture for service oriented Grid computing. *10th Heterogeneous Computing Wkshp.*
2. W. Cirne and K. Marzullo (1999): The Computational Co-Op: gathering clusters into a metacomputer. *13th Intl. Parallel Processing Symp.*, 160–166.
3. S.A. Cook (1974): An observation on time-storage tradeoff. *J. Comp. Syst. Scis. 9*, 308–316.
4. G. Cordasco, G. Malewicz, A.L. Rosenberg (2005): Optimal schedules for expansive and reductive dags in Internet-based computing (tentative title). In process, Univ. Massachusetts Amherst.
5. T.H. Cormen, C.E. Leiserson, R.L. Rivest, C. Stein (2001): *Introduction to Algorithms (2nd ed.)*. MIT Press, Cambridge, MA.
6. I. Foster and C. Kesselman [eds.] (2004): *The Grid: Blueprint for a New Computing Infrastructure* (2nd ed.) Morgan-Kaufmann, San Francisco, CA.
7. I. Foster, C. Kesselman, S. Tuecke (2001): The anatomy of the Grid: enabling scalable virtual organizations. *Intl. J. High Performance Computing Applications 15*, 200–222.
8. A. Gerasoulis and T. Yang (1992): A comparison of clustering heuristics for scheduling dags on multiprocessors. *J. Parallel Distr. Comput. 16*, 276–291.
9. L. He, Z. Han, H. Jin, L. Pan, S. Li (2000): DAG-based parallel real time task scheduling algorithm on a cluster. *Intl. Conf. on Parallel and Distr. Processing Techniques and Applications*, 437–443.
10. J.-W. Hong and H.T. Kung (1981): I/O complexity: the red-blue pebble game. *13th ACM Symp. on Theory of Computing*, 326–333.
11. J.E. Hopcroft, W. Paul, L.G. Valiant (1977): On time versus space. *J. ACM 24*, 332–337.
12. H.T. Hsu (1975): An algorithm for finding a minimal equivalent graph of a digraph. *J. ACM 22*, 11–16.

13. D. Kondo, H. Casanova, E. Wing, F. Berman (2002): Models and scheduling guidelines for global computing applications. *Intl. Parallel and Distr. Processing Symp.*, 79.
14. E. Korpela, D. Werthimer, D. Anderson, J. Cobb, M. Lebofsky (2000): SETI@home: massively distributed computing for SETI. In *Computing in Science and Engineering* (P.F. Dubois, Ed.) IEEE Computer Soc. Press, Los Alamitos, CA.
15. G. Malewicz and A.L. Rosenberg (2005): On batch-scheduling dags for Internet-based computing. *Euro-Par 2005*, to appear
16. G. Malewicz, A.L. Rosenberg, M. Yurkewych (2005): Toward a theory for scheduling dags in Internet-based computing. *IEEE Trans. Comput.*, to appear. See also: On scheduling complex dags for Internet-based computing. *Intl. Parallel and Distributed Processing Symp.*, 2005.
17. M.S. Paterson, C.E. Hewitt (1970): Comparative schematology. *Project MAC Conf. on Concurrent Systems and Parallel Computation*, ACM Press, 119–127.
18. N.J. Pippenger (1980): Pebbling. In *5th IBM Symp. on Math. Foundations of Computer Science*.
19. A.L. Rosenberg (2004): On scheduling mesh-structured computations for Internet-based computing. *IEEE Trans. Comput. 53*, 1176–1186.
20. A.L. Rosenberg and I.H. Sudborough (1983): Bandwidth and pebbling. *Computing 31*, 115–139.
21. A.L. Rosenberg and M. Yurkewych (2005): Guidelines for scheduling some common computation-dags for Internet-based computing. *IEEE Trans. Comput. 54*, 428–438.
22. X.-H. Sun and M. Wu (2003): Grid Harvest Service: a system for long-term, application-level task scheduling. *IEEE Intl. Parallel and Distributed Processing Symp.*, 25.
23. S.W. White and D.C. Torney (1993): Use of a workstation cluster for the physical mapping of chromosomes. *SIAM NEWS*, March, 1993, 14–17.

On Teaching Fast Adder Designs:
Revisiting Ladner & Fischer

Guy Even

School of Electrical Engineering, Tel-Aviv University, Israel
`guy@eng.tau.ac.il`

Abstract. We present a self-contained and detailed description of the parallel-prefix adder of Ladner and Fischer. Very little background is assumed in digital hardware design. The goal is to understand the rational behind the design of this adder and view the parallel-prefix adder as an outcome of a general method.

1 Introduction

This essay is about how to teach adder designs for undergraduate Computer Science (CS) and Electrical Engineering (EE) students. For the past eight years I have been teaching the second hardware course in Tel-Aviv University's EE school. Although the goal is to teach how to build a simple computer from basic gates, the part I enjoy teaching the most is about addition. At first I thought I felt so comfortable with teaching about adders because it is a well defined, very basic question, and the solutions are elegant and can be proved rigorously without exhausting the students. After a few years, I was able to summarize all these nice properties simply by saying that this is the most algorithmic topic in my course. Teaching has helped me realize that appreciation of algorithms is not straightforward; I was lucky to have been influenced by my father. In fact, while writing this essay, I constantly asked myself how he would have presented this topic.

When writing this essay I had three types of readers in mind.

- Lecturers of undergraduate CS hardware courses. Typically, CS students have a good background in discrete math, data structures, algorithms, and finite automata. However, CS students often lack enthusiasm for hardware, and the only concrete machine they are comfortable with is a Turing machine. To make teaching easier, I added a rather long preliminaries section that defines the hardware model and presents some hardware terminology. Even combinational gates and flip-flops are briefly described.
- Lecturers of undergraduate EE hardware students. In contrast to CS students, EE students are often enthusiastic about hardware (including devices, components, and commercial products), but are usually indifferent to formal specification, proofs, and asymptotic bounds. These students are eager

O. Goldreich et al. (Eds.): Shimon Even Festschrift, LNCS 3895, pp. 313–347, 2006.

to learn about the latest buzzwords in VLSI and microprocessors. My challenge, when I teach about adders, is to convince the students that learning about an old and solved topic is useful.

— General curious readers (especially students). Course material is often presented in the shortest possible way that is still clear enough to follow. I am not aware of a text that tells a story that sacrifices conciseness for insight about hardware design. I hope that this essay could provide such insight to students interested in learning more about what happens "behind the screen".

1.1 On Teaching Hardware

Most hardware textbooks avoid abstractions, definitions, and formal claims and proofs (Müller and Paul [MüllerPaul00] is an exception). Since I regard hardware design as a subbranch of algorithms, I think it is a disadvantage not to follow the format of algorithm books. I tried to follow this rule when I prepared lecture notes for my course [Even04]. However, in this essay I am lax in following this rule. First, I assumed that the readers could easily fill in the missing formalities. Second, my impression of my father's teaching was that he preferred clarity over formality. On the other hand, I tried to present the development of a fast adder in a systematic fashion and avoid ad-hoc solutions. In particular, a distinction is made between concepts and representation (e.g., we interpret the celebrated "generate-carry" and "propagate-carry" bits as a representation of functions, and introduce them rather late in a specific design in Sec. 6.7).

I believe the communication skills of most computer engineers would greatly benefit if they acquired a richer language. Perhaps one should start by modeling circuits by graphs and using graph terminology (e.g., out-degree vs. fanout, depth vs. delay). I decided to stick to hardware terminology since I suspect that people with a background in graphs are more flexible. Nevertheless, whenever possible, I tried to use graphs to model circuits (e.g., netlists, communication graphs).

1.2 A Brief Summary

This essay mainly deals with the presentation of one of the parallel-prefix adders of Ladner and Fischer [LadnerFischer80]. This adder was popularized by Brent and Kung [BrentKung82] who presented a regular layout for it as well as a reduction of its fanout. In fact, it is often referred to as the "Brent-Kung" adder. Our focus is on a detailed and self-contained explanation of this adder.

The presentation of the parallel-prefix adder in many texts is short but lacks intuition (see [BrentKung82, MüllerPaul00, ErceLang04]). For example, the carry-generate (g_i) and carry-propagate (p_i) signals are introduced as a way to compute the carry bits without explaining their origin[1]. In addition, an associative operator is defined over pairs (g_i, p_i) as a way to reduce the task of

---

[1] The signals g_i and p_i are defined in Sec. 6.7

computing the carry-bits to a prefix problem. However, this operator is introduced without explaining how it is derived.

Ladner and Fischer's presentation does not suffer from these drawbacks. The parallel-prefix adder is systematically obtained by "parallelizing" the "bit-serial adder" (i.e., the trivial finite state machine with two states, see Sec. 4.1). According to this explanation the pair of carry-generate and carry-propagate signals represent three functions defined over the two states of the bit-serial adder. The associative operator is simply a composition of these functions. The mystery is unravelled and one can see the role of each part.

Ladner and Fischer's explanation is not long or complicated, yet it does not appear in textbooks. Perhaps a detailed and self-contained presentation of the parallel-prefix adder will influence the way parallel-prefix adders are taught. I believe that students can gain much more by understanding the rational behind such an important design. Topics taught so that the students can add some items to their "bag of tricks" often end up in the "bag of obscure and forgotten tricks". I believe that the parallel-prefix adder belongs to the collection of fundamental algorithmic paradigms and can be presented as such.

1.3   Confusing Terminology (A Note for Experts)

The terms "parallel-prefix adder" and "carry-lookahead adders" are used inconsistently in the literature. Our usage of these terms refers to the algorithmic method employed in obtaining the design rather than the specifics of each adder. We use the term "parallel-prefix adder" to refer to an adder that is based on a reduction of the task of computing the carry-bits to a prefix problem (defined in Sec. 6.3). In particular, parallel-prefix adders in this essay are based on the parallel prefix circuits of Ladner and Fischer. The term "carry-lookahead adder" refers to an adder in which special gates (called carry-lookahead gates) are organized in a tree-like structure. The topology is not precisely a tree for two reasons. First, often connections are made between nodes in the same layer. Second, information flows both up and down the tree

One can argue justifiably that, according to this definition, a carry-lookahead is a special case of a parallel-prefix adder. To help the readers, we prefer to make the distinction between the two types of adders.

1.4   Questions

Questions for the students appear in the text. The purpose of these questions is to help students check their understanding, consider alternatives to the text, or just think about related issues that we do not focus on. Before presenting a topic, I usually try to convince the students that they have something to learn. The next question is a good example for such an attempt.

Question 1.   1. What is the definition of an adder? (Note that this is a question about hardware design, not about Zoology.)

2. Can you prove the correctness of the addition algorithm taught in elementary school?
3. (Assuming students are familiar with the definition of the delay (i.e., depth) of a combinational circuit) What is the smallest possible delay of an adder? Do you know of an adder that achieves this delay?
4. Suppose you are given the task of adding very long numbers. Could you share this work with friends so that you could work on it simultaneously to speed up the computation?

1.5   Organization

We begin with a rather long preliminaries in Section 2. This section is a brief review of digital hardware design. In Section 3, we define two types of binary adders: a combinational adder and a bit-serial adder. In Section 4, we present trivial designs for each type of adder. The synchronous adder is an implementation of a finite state machine with two states. The combinational adder is a "ripple-carry adder". In Section 5, we prove lower bounds on the cost and delay of a combinational adder. Section 6 is the heart of the essay. In it, we present the parallel-prefix circuit of Ladner and Fischer as well as the parallel-prefix adder. In Section 7, we discuss various issues related to adders and their implementation. In Section 8, we briefly outline the history of adder designs. We close with a discussion that attempts to speculate why the insightful explanation in [LadnerFischer80] has not made it into textbooks.

2   Preliminaries

2.1   Digital Operation

We assume that inputs and outputs of devices are always either zero or one. This assumption is unrealistic due to the fact that the digital value is obtained by rounding an analog value that changes continuously (e.g., voltage). There is a gap between analog values that are rounded to zero and analog values that are rounded to one. When the analog value is in this gap, its digital value is neither zero or one.

The advantage of this assumption is that it simplifies the task of designing digital hardware. We do need to take precautions to make sure that this unrealistic assumption will not render our designs useless. We set strict design rules for designing circuits to guarantee well defined functionality.

We often use the term *signal*. A signal is simply zero or one value that is output by a gate, input to a gate, or delivered by a wire.

2.2   Building Blocks

The first issue we need to address is what are our building blocks? The building blocks are combinational gates, flip-flops, and wires. We briefly describe these objects.

Combinational gates. A *combinational gate* (or gate, in short) is a device that implements a Boolean function. What does this mean? Consider a Boolean function $f : \{0,1\}^k \to \{0,1\}^\ell$. Now consider a device G with k inputs and ℓ outputs. We say that G *implements* f if the outputs of G equal $f(\alpha) \in \{0,1\}^\ell$ when the input equals $\alpha \in \{0,1\}^\ell$. Of course, the evaluation of $f(\alpha)$ requires time and cannot occur instantaneously. This is formalized by requiring that the inputs of G remain stable with the value α for at least d units of time. After d units of time elapse, the output of G stabilizes on $f(\alpha)$. The amount of time d that is required for the output of G to stabilize on the correct value (assuming that the inputs are stable during this period) is called the *delay* of a gate.

Typical gates are inverters and gates that compute the Boolean OR/AND/XOR of two bits. We depict gates by boxes; the functionality is written in the box. We use the convention that information flows rightwards or downwards. Namely, The inputs of a box are on the right side and the outputs are on the left side (or inputs on the top side and outputs on the bottom side).

We also consider a particular gate, called a *full-adder*, that is useful for addition, defined as follows.

Definition 1 (Full-Adder). *A* full-adder *is a combinational gate with* 3 *inputs* $x, y, z \in \{0,1\}$ *and* 2 *outputs* $c, s \in \{0,1\}$ *that satisfies:* $2c + s = x + y + z$. *(Note that each bit is viewed as a zero/one integer.)*

The output s of a full-adder is called the *sum output*, while the output c of a full-adder is called the *carry-out output*. We denote a full-adder by FA. A *half-adder* is a degenerate full-adder with only two inputs (i.e., the third input z of the full-adder is always input a zero).

We do not discuss here how to build a full-adder from basic gates. Since a full-adder has a constant number of inputs and outputs, every (reasonable) implementation has constant cost and delay.

Flip-Flops. A flip-flop is a memory device. Here we use only a special type of flip-flops called *edge triggered D-flip-flops*. What does this mean? We assume that time is divided into intervals, each interval is called a clock cycle. Namely, the ith clock cycle is the interval $(t_i, t_{i+1}]$. A flip-flop has one input (denoted by D), and one output (denoted by Q). The output Q during clock cycle $i + 1$ equals the value of the input D at time t_i (end of clock cycle i). We denote a flip-flop by FF.

This functionality is considered as memory because the input D is sampled at the end of clock cycle i. The sampled value is stored and output during the next clock cycle (i.e., clock cycle $i + 1$). Note that D may change during clock cycle $i + 1$, but the output Q must stay fixed.

We remark that three issues are ignored in this description: (i) Initialization. What does a flip-flop output during the first clock cycle (i.e., clock cycle zero)? We assume that it outs a zero. (ii) Timing - we ignore timing issues such as setup time and hold time. (iii) The clock signal is missing. (The role of the clock signal is to mark the beginning of each clock cycle.) In fact, every flip-flop has an additional input for a global signal called the clock signal. We assume that the clock signal is used only for feeding these special inputs, and that all the clock inputs of all the flip-flops are fed by the same clock signal. Hence, we may ignore the clock signal.

Wires. The idea behind connecting a wire between two components is to take the output of one component and use it as input to another component. We refer to an input or an output of a component as a *port*. In this essay, a wire can connect exactly two ports; one port is an output port and the other port is an input port. We assume also that a wire can deliver only a single bit.

We will later see that there are strict rules regarding wires. To give an idea of these rules, note that it makes little sense to connect two outputs ports to each other. However, it does make sense to feed multiple input ports by the same output port. We, therefore, allow different wires to be connected to the same output port. However, we do not allow more than one wire to be connected to the same input port. The reason is that multiple wires feeding the same input port could cause an ambiguity in the definition of the input value.

The number of inputs ports that are fed by the same output port is called the *fanout* of the output. In the hardware design community, the fanout is often defined as the number of inputs minus one. We prefer to define the fanout as the out-degree.

We remark that very often connections are depicted by *nets*. A net is a set of input and output ports that are connected by wires. In graph terminology, a net is a hyperedge. Since we are not concerned in this essay with the detailed physical design, and do not consider complicated connections such as buses, we consider only point-to-point connections. An output port that feeds multiple input ports can be modelled by a star of directed edges that emanate from the output port. Hence, one can avoid using nets. Another option is to add trivial gates in branching points, as described in the next section.

2.3   Combinational Circuits

One can build complex circuits from gates by connecting wires between gates. However, only a small subset of such circuits are combinational. (In the theory community combinational circuits are known as Boolean circuits.) To guarantee well defined functionality, strict design rules are defined regarding the connections between gates. Only circuits that abide these rules are called combinational circuits. We now describe these rules.

The first rule is the "output-to-input" rule which says: *the starting point of every connection must be an output port and the end point must be an input port.* There is a problem with this rule; namely, how do we feed the external inputs to input ports? We encounter a similar problem with external outputs. Instead of setting exceptions for external inputs and outputs, perhaps the best way to solve this problem is by defining special gates for external inputs and outputs. We define an input gate as a gate with a single output port and no input port. An input gate simply models an external signal that is fed to the circuit. Similarly, we define an output gate as a gate with a single input port and no output port.

The second rule says: *feed every input port exactly once*. This means that there must be exactly one connection to every input port of every gate. We do not allow input ports that are not fed by some signal. The reason is that an unconnected input may violate the digital abstraction (namely, we cannot decide whether it feeds a zero or a one). We do not allow multiple connections to feed the same input port. The reason is that different connections to th same input port may deliver different values, and then the input value is not well defined. (Note that we do allow the same output port to be connected to multiple inputs and we also allow unconnected output ports.)

The third rule is the "no-cycles" rule which says: *a connection is forbidden if it closes a directed cycle*. The idea is that we can model a circuit C using a directed graph $G(C)$. This graph is called the *netlist* of C. We assign a vertex for every gate and a directed edge for every wire. The orientation of the edge is from the output port to the input port. (Note that we can always orient the edges thanks to the output-to-input rule). The no-cycles rule simply does not allow directed cycles in the directed graph $G(C)$.

Finally, we add a word about nets. Very often connections in circuits are not depicted only by point-to-point wires. Instead, one often draws nets that form a connection between more than two ports (see, for example, the schematic on the right side of Fig. 1). A net is depicted as a tree whose leaves are the ports of the net. We refer to the interior vertices of these trees as branching points. We interpret every branching point in a drawing of a net as a trivial combinational gate with one input port and one output port (the output port may be connected to many input ports). In this trivial gate, the output value simply equals the input value. Hence one should feel comfortable with branching points as long as nets contain exactly one output port.

Question 2. Consider the circuits depicted in Figure 1. Can you explain why these are not valid combinational circuits?

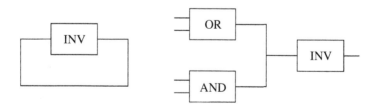

Fig. 1. Two examples of non-combinational circuits.

The definition we use for combinational circuits is syntactic; namely, we only require that the connections between the components follow some simple rules.

Our focus was syntax and we did not say anything about functionality. This, of course, does not mean that we are not interested in functionality! In natural languages (like English), the meaning of a sentence with correct syntax may not be well defined. The syntactic definition of combinational circuits has two main advantages: (i) It is easy to check if a circuit is indeed combinational. (ii) The functionality of every combinational circuit is well defined. The task of determining the output values given the input values is referred to as *logical simulation*; it can performed as follows. Given the input values of the circuit, one can scan the circuit starting from the inputs and determine all the values of the inputs and outputs of gates. When this scan ends, the output values of the circuit are known. The order in which the gates should be scanned is called *topological order*, and this order can can be computed in linear time [Even79, Sec. 6.5]. It follows that combinational circuits implement Boolean functions just as gates do. The difference is that functions implemented by combinational circuits are bigger (i.e., have more inputs/outputs).

2.4   Synchronous Circuits

Synchronous circuits are built from gates and flip-flops. Obviously, not every collection of gates and flip-flops connected by wires constitutes a "legal" synchronous circuit. Perhaps the simplest way to define a synchronous circuit is by a reduction that maps synchronous circuits to combinational circuits.

Consider a circuit C that is simply a set of gates and flip-flops connected by wires. We assume that the first two rules of combinational circuit are satisfied: Namely, wires connect outputs ports to inputs ports and every input port is fed exactly once.

We are now ready to decide whether C is a synchronous circuit. The decision is based on a reduction that replaces every flip-flops by fictive input and output gates as follows. For every flip-flop in C, we remove the flip-flop and add an output-gate instead of the input D of the flip-flop. Similarly, we add an input-gate instead of the output Q of the flip-flop. Now, we say that C is a synchronous circuit if C' is a combinational circuit. Figure 2 depicts a circuit C and the circuit C' obtained by removing the flip-flops in C.

Question 3. Prove that every directed cycle in a synchronous circuit contains at least one flip-flop. (By cycle we mean a closed walk that obeys the output-to-input orientation.)

As in the case of combinational circuits, the definition of synchronous circuits is syntactic. The functionality of a synchronous circuit can be modeled by a finite state machine, defined below.

2.5   Finite State Machines

A finite state machine (also known as a finite automaton with outputs or a transducer) is an abstract machine that is described by a 6-tuple: $\langle Q, q_0, \Sigma, \Delta, \delta, \gamma \rangle$

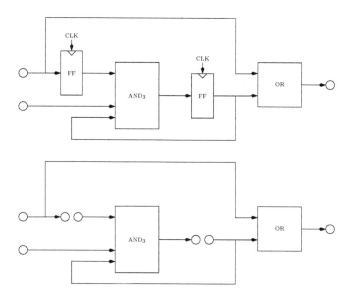

Fig. 2. A synchronous circuit C and the corresponding combinational circuit C' after the flip-flops are removed.

as follows. (1) Q is a finite set of states. (2) $q_0 \in Q$ is an initial state, namely, we assume that in clock cycle 0 the machine is in state q_0. (3) Σ is the input alphabet and Δ is the output alphabet. In each cycle, a symbol $\sigma \in \Sigma$ is fed as input to the machine, and the machine outputs a symbol $y \in \Delta$. (4) The *transition function* $\delta : Q \times \Sigma \to Q$ specifies the next state: If in cycle i the machine is in state q and the input equals σ, then, in cycle $i+1$, the state of the machine equals $\delta(q, \sigma)$. (5) The *output function* $\gamma : Q \times \Sigma \to \Delta$ specifies the output symbol as follows: when the state is q and the input is σ, then the machine outputs the symbol $\gamma(q, \sigma)$.

One often depicts a finite state machine using a directed graph called a *state diagram*. The vertices of the state diagrams are the set of states Q. We draw a directed edge (q', q'') and label the edge with a pair of symbols (σ, y), if $\delta(q', \sigma) = q''$ and $\gamma(q', \sigma) = y$. Note that every vertex in the state diagram has $|\Sigma|$ edges emanating from it.

We now explain how the functionality of a synchronous circuit C can be modeled by a finite state machine. Let F denote the set of flip-flops in C. For convenience, we index the flip-flops in F, so $F = \{f_1, \ldots, f_k\}$, where $k = |F|$. Let $f_j(i)$ denote the bit that is output by flip-flop $f_j \in F$ during the ith clock cycle. The set of states is $Q = \{0, 1\}^k$. The state $q \in Q$ during cycle i is simply the binary string $f_1(i) \cdots f_k(i)$. The initial state q_0 is the binary string whose jth bit equals the value of $f_j(0)$ (recall that we assumed that each flip-flop is initialized so that it outputs a predetermined value in cycle 0). The input alphabet Σ is $\{0, 1\}^{|I|}$, where I denotes the set of input gates in C. The jth bit of $\sigma \in \Sigma$ is the value of fed by the jth input gate. Similarly, the output alphabet Δ is $\{0, 1\}^{|Y|}$, where Y denotes the set of output gates in C. The transition function is uniquely determined by C since C is a synchronous circuit. Namely, given a state $q \in Q$ and an input $\sigma \in \Sigma$, we apply logical simulation to the reduced combinational circuit C' to determine the values fed to each flip-flop. These values are well defined since C' is combinational. The vector of values input to the flip-flops by the end of clock cycle i is the state $q' \in Q$ during the next clock cycle. The same argument determines the output function.

We note that the definition given above for a finite state machine is often called a *Mealy machine*. There is another definition, called a *Moore machine*, that is a slightly more restricted [HopUll79]. In a Moore machine, the domain of the output function γ is Q rather than $Q \times \Sigma$. Namely, the output is determined by the current state, regardless of the current input symbol.

2.6   Cost and Delay

Every combinational circuit has a cost and a delay. The cost of a combinational circuit is the sum of the costs of the gates in the circuit. We use only gates with a constant number of inputs and outputs, and such gates have a constant cost. Since we are not interested in the constants, we simply attach a unit cost to every gate, and the cost of a combinational circuit equals the number of gates in the circuit.

The delay of a combinational circuit is defined similarly to the delay of a gate. Namely, the delay is the smallest amount of time required for the outputs to stabilize, assuming that the inputs are stable.

Let C denote a combinational circuit. The delay of a path p in the netlist $G(C)$ is the sum of the delays of the gates in p.

Theorem 1. *The delay of combinational circuit C is not greater than the maximum delay of a path in $G(C)$.*

Proof. Focus on a single gate g in C. Let $d(g)$ denote the delay of g. If all the inputs of g stabilize at time t, then we are guaranteed that g's outputs are stable at time $t + d(g)$. Note that $t + d(g)$ is an upper bound; in reality, g's output may stabilize much sooner.

We assume that all the inputs of C are stable at time $t = 0$. We now describe an algorithm that labels each net with a delay t that specifies when we are guaranteed that the signal on the net stabilizes. We do so by topologically sorting the gates in the netlist $G(C)$ (this is possible since $G(C)$ is acyclic). Now, we visit the gates according to the topological order. If g is an input gate, then we label the net that it feeds by zero. If g is not an input gate, then the topological order implies that all the nets that feed g have been already labeled. We compute the maximum label $t_{\max}$ appearing on nets that feed g (so that we are guaranteed that all the inputs of g stabilize by t_{max}). We now attach the label $t_{\max} + d(g)$ to all the nets that are fed by g. The statement in the beginning of the proof assures us that every output of g is indeed stable by time $t_{\max} + d(g)$.

It is easy to show by induction on the topological order that the label attached to nets fed by gate g equals the maximum delay of a path from an input gate to g. Hence, the theorem follows.

To simplify the discussion, we attach a unit delay to every gate. With this simplification, the delay of a circuit C is the length of the longest path in the netlist graph $G(C)$.

We note that a synchronous circuit also has a cost that is the sum of the gate costs and flip-flop costs. Instead of a delay, a synchronous circuit has a minimum clock period. The minimum clock period equals the delay of the combinational circuit obtained after the flip-flops are removed (in practice, one should also add the propagation delay and the hold time of the flip-flops).

2.7   Notation

A binary string with n bits is often represented by an indexed vector $X[n-1:0]$. Note that an uppercase letter is used to denote the bit-vector. In this essay we use indexed vectors only with combinational circuits. So in a combinational circuit $X[i]$ always means the ith bit of the vector $X[n-1:0]$.

Notation of signals (i.e., external inputs and outputs, and interior values output by gates) in synchronous circuits requires referring to the clock cycle. We denote the value of a signal x during clock cycle t in a synchronous circuit by $x[t]$. Note that a lowercase letter is used to denote a signal in a synchronous circuit.

We use lowercase letter for synchronous circuits and uppercase letters for combinational circuits. Since we deal with very simple synchronous circuits, this distinction suffices to avoid confusion between $x[i]$ and $X[i]$: The symbol $x[i]$ means the value of the signal x in the ith clock cycle. The symbol $X[i]$ means the ith bit in the vector $X[n-1:0]$.

Referring to the value of a signal (especially in a synchronous circuit) introduces some confusion due to the fact that signals change all the time from zero to one and vice-versa. Moreover, during these transitions, there is a (short) period of time during which the value of the signal is neither zero or one. The guiding principle is that we are interested in the stable value of a signal. In the case of a combinational circuit this means that we wait sufficiently long after the inputs are stable. In the case of a synchronous circuit, the functionality of flip-flops implies that the outputs of each flip-flop are stable shortly after a clock cycle begins and remain stable till the end of the clock cycle. Since the clock period is sufficiently long, all other nets stabilize before the end of the clock cycle.

2.8   Representation of Numbers

Our goal is to design fast adders. For this purpose we must agree on how nonnegative integers are represented. Throughout this essay we use binary representation. Namely, a binary string $X[n-1:0] \in \{0,1\}^n$ represents the number $\sum_{i=0}^{n-1} X[i] \cdot 2^i$.

In a synchronous circuit a single bit signal $x[t]$ can be also used to represent a nonnegative integer. The ith bit equals $x[i]$, and therefore, the number represented by $x[t]$ equals $\sum_{t=0}^{n-1} x[t]$.

Although students are accustomed to binary representation, it is useful to bear in mind that this is not the only useful representation. Negative numbers require a special representation, the most common is known as two's complement. In two's complement representation, the binary string $X[n-1:0]$ represents the integer $-2^{n-1} \cdot X[n-1] + \sum_{i=0}^{n-2} X[i] \cdot 2^i$.

There are representations in which the same number may be have more than one representation. Such representations are called *redundant representations*. Interestingly, redundant representations are very useful. One important example is carry-save representation. In carry-save representation, a nonnegative integer is represented by two binary strings (i.e., two bits are used for each digit). Each binary string represents an integer in binary representation, and the number represented by two such binary strings is the sum of the numbers represented by the two strings. An important property of carry-save representation is that addition in carry-save representation can be computed with constant delay (this can be done by using only full-adders). Addition with constant delay is vital for the design of fast multipliers.

3   Definition of a Binary Adder

Everybody knows that computers compute arithmetic operations; even a calculator can do it! So it is hardly a surprise that every computer contains a hardware device that adds numbers. We refer to such a device as an *adder*. Suppose we wish to design an adder. Before we start discussing how to design an adder, it is useful to specify or define exactly what we mean by this term.

3.1   Importance of Specification

Unfortunately, the importance of a formal specification is not immediately understood by many students. This is especially true when it comes to seemingly obvious tasks such as addition. However, there are a few issues that the specification of an adder must address.

Representation: How are addends represented? The designer must know how numbers are represented. For example, an adder of numbers represented in unary representation is completely different than an adder of numbers represented in binary representation. The issue of representation is much more important when we consider representations of signed numbers (e.g., two's complement and one's complement) or redundant representations (e.g., carry-save representation).

Another important issue is how to represent the computed sum? After all, the addends already represent the sum, but this is usually not satisfactory. A reasonable and useful assumption is to require the same representation for the addends and the sum (i.e., binary representation in our case). The main advantage of this assumption is that one could later use the sum as an addend for subsequent additions.

In this essay we consider only binary representation.

Model: How are the inputs fed to the circuit and how is the output obtained? We consider two extremes: a combinational circuit and a synchronous circuit. In a combinational circuit, we assume that there is a dedicated port for every bit of the addends and the sum. For example, if we are adding two 32-bit numbers, then there are 32×3 ports; 32×2 ports are input ports and 32 ports are output ports. (We are ignoring in this example the carry-in and the carry-out ports.)

In a synchronous circuit, we consider the bit-serial model in which there are exactly two input ports and one output port. Namely, there are exactly three ports regardless of the length of the addends. The synchronous circuit is very easy to design and will serve as a starting point for combinational designs. In Section 3.3 we present a bit-serial adder.

We are now ready to specify (or define) a combinational binary adder and a serial binary adder. We specify the combinational adder first, but design a serial adder first. The reason is that we will use the implementation of the serial adder to design a simple combinational adder.

3.2   Combinational Adder

Definition 2. *A* combinational binary adder *with input length n is a combinational circuit specified as follows.*

Input: $A[n-1:0], B[n-1:0] \in \{0,1\}^n$.
Output: $S[n-1:0] \in \{0,1\}^n$ *and* $C[n] \in \{0,1\}$.
Functionality:

$$C[n] \cdot 2^n + \sum_{i=0}^{n-1} S[i] \cdot 2^i = \sum_{i=0}^{n-1} A[i] \cdot 2^i + \sum_{i=0}^{n-1} B[i] \cdot 2^i. \tag{1}$$

We denote a combinational binary adder with input length n by ADDER(n). The inputs $A[n-1:0]$ and $B[n-1:0]$ are the binary representations of the addends. Often an additional input $C[0]$, called the *carry-in bit*, is used. To simplify the presentation we omit this bit at this stage (we return to it in Section 7.1). The output $S[n-1:0]$ is the binary representation of the sum modulo 2^n. The output $C[n]$ is called the *carry-out bit* and is set to 1 if the sum is at least 2^n.

Question 4. Verify that the functionality of ADDER(n) is well defined. Show that, for every $A[n-1:0]$ and $B[n-1:0]$ there exist unique $S[n-1:0]$ and $C[n]$ that satisfy Equation 1.

Hint: Show that the set of integers that can be represented by sums $A[n-1:0] + B[n-1:0]$ is contained in the set of integers that can be represented by sums $S[n-1:0] + 2^n \cdot C[n]$.

There are many ways to implement an ADDER(n). Our goal is to present a design of an ADDER(n) with optimal asymptotic delay and cost. In Sec. 5 we prove that every design of an ADDER(n) must have at least logarithmic delay and linear cost.

3.3   Bit-Serial Adder

We now define a synchronous adder that has two inputs and a single output.

Definition 3. *A* bit-serial binary adder *is a synchronous circuit specified as follows.*

Input ports: $a, b \in \{0, 1\}$.
Output ports: $s \in \{0, 1\}$.
Functionality: *For every clock cycle $n \geq 0$,*

$$\sum_{i=0}^{n} s[i] \cdot 2^i = \sum_{i=0}^{n} (a[i] + b[i]) \cdot 2^i \pmod{2^n}. \tag{2}$$

We refer to a bit-serial binary adder by s-adder. One can easily see the relation between $a[i]$ (e.g., the bit input in clock cycle i via port a) and $A[i]$ (e.g., the ith bit of the addend A). Note the lack of a carry-out in the specification of a s-adder.

4   Trivial Designs

In this section we present trivial designs for an s-adder and an ADDER(n). The combinational adder is obtained from the synchronous one by applying a time-space transformation.

4.1   A Bit-Serial Adder

In this section we present a design of a bit-serial adder. The design performs addition in the same way we are taught to add in school (i.e., from the least significant digit to the most significant digit). Figure 3 depicts a design that uses one flip-flop and one full-adder. The output of the flip-flop that is input to the full-adder is called the carry-in signal and is denoted by c_{in}. The carry output of the full-adder is input to the flip-flop, and is called the carry-out signal. It is denoted by c_{out}. We ignore the issue of initialization and assume that, in clock cycle zero, $c_{in}[0] = 0$.

We now present a correctness proof of the s-adder design. The proof is by induction, but is not totally straightforward. (Definitely not for undergraduate hardware design students!) The problem is that we need to strengthen Eq. 2 (we point out the difficulty within the proof). Another reason to insist on a complete proof is that, for most students, this is the first time they prove the correctness of the addition algorithm taught in school.

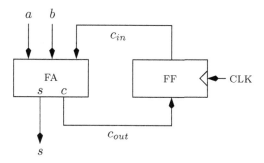

Fig. 3. A schematic of a serial adder.

Claim. The circuit depicted in Fig. 3 is a correct implementation of a s-adder.

Proof. We prove that the circuit satisfies Eq. 2. The proof is by induction on the clock cycle. The induction basis, for the clock cycle zero (i.e., $n = 0$), is easy to prove. In clock cycle zero, the inputs are $a[0]$ and $b[0]$. In addition $c_{in}[0] = 0$. By the definition of a full-adder, the output $s[0]$ equals XOR($a[0], b[0], c_{in}[0]$). It follows that $s[0] = \mathrm{mod}(a[0] + b[0], 2)$, as required.

We now try to prove the induction step for $n + 1$. Surprisingly, we are stuck. If we try to apply the induction hypothesis to the first n cycles, then we cannot claim anything about $c_{in}[n + 1]$. (We do know that $c_{in}[n + 1] = c_{out}[n]$, but Eq. 2 does not tell us anything about the value of $c_{out}[n]$.) On the other hand we cannot apply the induction hypothesis to the last n cycles (by decreasing the indexes of clock cycles by one) because $c_{in}[1]$ might equal 1.

The way to overcome this difficulty is to strengthen the statement we are trying to prove. Instead of Eq. 2, we prove the following stronger statement: For every clock cycle $n \geq 0$,

$$2^{n+1} \cdot c_{out}[n] + \sum_{i=0}^{n} s[i] \cdot 2^i = \sum_{i=0}^{n} (a[i] + b[i]) \cdot 2^i. \tag{3}$$

We prove Equation 3 by induction. When $n = 0$, Equation 3 follows from the functionality of a full-adder and the assumption that $c_{in}[0] = 0$. So now we turn again to the induction step, namely, we prove that Eq. 3 holds for $n + 1$.

The functionality of a full-adder states that

$$2 \cdot c_{out}[n + 1] + s[n + 1] = a[n + 1] + b[n + 1] + c_{in}[n + 1]. \tag{4}$$

We multiply both sides of Eq. 4 by 2^{n+1} and add it to Eq. 3 to obtain

$$2^{n+2} \cdot c_{out}[n+1] + 2^{n+1} \cdot c_{out}[n] + \sum_{i=0}^{n+1} s[i] \cdot 2^i = 2^{n+1} \cdot c_{in}[n+1] + \sum_{i=0}^{n+1} (a[i] + b[i]) \cdot 2^i. \tag{5}$$

The functionality of the flip-flop implies that $c_{out}[n] = c_{in}[n + 1]$, and hence, Eq. 3 holds also for $n + 1$, and the claim follows.

4.2   Ripple-Carry Adder

In this section we present a design of a combinational adder ADDER(n). The design we present is called a *ripple-carry adder*. We abbreviate and refer to a ripple-carry adder for binary numbers of length n as RCA(n). Although designing an RCA(n) from scratch is easy, we obtain it by applying a transformation, called a *time-space transformation*, to the bit-serial adder.

Ever since Ami Litman and my father introduced me to retiming [LeiSaxe81, LeiSaxe91, EvenLitman91, EvenLitman94], I thought it is best to describe designs by functionality preserving transformations. Namely, instead of obtaining a new design from scratch, obtain it by transforming a known design. Correctness of the new design follows immediately if the transformation preserves functionality. In this way a simple design evolves into a sophisticated design with much better performance. Students are rarely exposed to this concept, so I chose to present the ripple-carry adder as the outcome of a time-space transformation applied to the serial adder.

time-space transformation. We apply a transformation called a *time-space transformation*. This is a transformation that maps a directed graph (possibly with cycles) to an acyclic directed graph. In the language of circuits, this transformation maps synchronous circuits to combinational circuits.

Given a synchronous circuit C, construct a directed multi-graph $G = (V, E)$ with non-negative integer weights $w(e)$ defined over the edges as follows. This graph is called the *communication graph* of C [LeiSaxe91, EvenLitman94]. The vertices are the combinational gates in C (including branching points). An edge (u, v) in a communication graph models a p path in the netlist, all the interior nodes of which correspond to flip-flops. Namely, we draw an edge from u to v if there is a path in C from an output of u to an input v that traverses only flip-flops. Note that a direct wire from u to v also counts as a path with zero flip-flops. The weight of the edge (u, v) is set to the number of flip-flops along the path. Note that there might be several paths from u to v, each traversing a different number of flip-flops. For each such path, we add a parallel edge with the correct weight. Finally, recall that branching points are considered to be combinational gates, so such a path may not traverse a branching point. In Figure 4 a synchronous circuit and its communication graph are depicted. Following [LeiSaxe81], we depict the weight of an edge by segments across an edge (e.g., two segments mean that the weight is two).

The weight of a path in the communication graph is the sum of the edge weights along the path. The following claim follows from the definition of a synchronous circuit.

Claim. The weight of every cycle in the communication graph of a synchronous circuit is greater than zero.

Question 5. Prove Claim 4.2

Let n denote a parameter that specifies the number cycles used in the time-space transformation. The time-space transformation of a communication graph

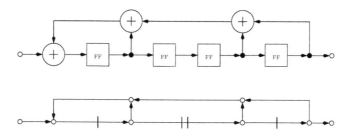

Fig. 4. A synchronous circuits and the corresponding communication graph (the gates are XOR gates).

$G = (V, E)$ is the directed graph $st_n(G) = (V \times \{0, \ldots, n-1\}, E')$ defined as follows. The vertex set is the Cartesian product of V and $\{0, \ldots, n-1\}$. Vertex (v, i) is a copy of v that corresponds to clock cycle i. There is an edge from (u, i) to (v, j) if: (i) $(u, v) \in E$, and (ii) $w(u, v) = j - i$. The edge $((u, i), (v, j))$ corresponds to the data transmitted from u to v in clock cycle i. Since there are $w(u, v)$ flip-flops along the path, the data arrives to v only in clock cycle $j = i + w(u, v)$.

Since the weight of every directed cycle in G is greater than zero, it follows that $st_n(G)$ is acyclic. We now build a combinational circuit C_n whose communication graph is $st_n(G)$. This is done by placing a gate of type v for each vertex (v, i). The connections use the same ports as they do in C.

There is a boundary problem that we should address. Namely, what feeds input ports of vertices (v, j) that "should" be fed by a vertex (u, i) with a negative index i? We encounter a similar problem with output ports of a vertex (u, i) that should feed a vertex (v, j) with an index $j \geq n$. We solve this problem by adding input gates that feed vertices (v, j) where $j < w(u, v)$. Similarly, we add output gates that are fed by vertices (u, i) where $i + w(u, v) > n$.

We are now ready to apply the time-space transformation to the bit-serial adder. We apply it for n clock cycles. The input gate a has now n instances that feed the signals $A[0], \ldots, A[n-1]$. The same holds for the other input b and the output s. The full-adder has now n instances denoted by $\text{FA}_0, \ldots, \text{FA}_{n-1}$. We are now ready to describe the connections. Since the input a feeds the full-adder in the bit-serial adder, we now use input $A[i]$ to feed the full-adder FA_i. Similarly, the input $B[i]$ feeds the full-adder FA_i. The carry-out signal c_{out} of the full-adder in the bit-serial adder is connected via a flip-flop to one of the inputs of the full-adder. Hence, we connect the carry-out port of full-adder FA_i to one of the inputs of FA_{i+1}. Finally, the output $S[i]$ is fed by the sum output of full-adder FA_i. Note that full-adder FA_0 is also fed by a "new" input gate that carries the carry-in signal $C[0]$. The carry-in signal is the initial state of the serial adder. The full-adder FA_{n-1} feeds a "new" output gate with the carry-out signal $C[n]$. Figure 5 depicts the resulting combinational circuit known as a ripple-carry adder. The netlist of the ripple-carry adder is simple and regular, and therefore, one can easily see that the outputs of FA_i depend only on the inputs $A[i:0]$ and $B[i:0]$.

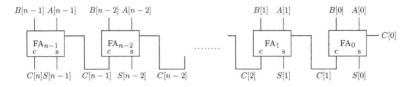

Fig. 5. A ripple-carry adder.

The main advantage of using the time-space transformation is that this transformation preserves functionality. Namely, there is a value-preserving mapping between the value of a signal $x[i]$ in clock cycle i in the synchronous circuit and the value of the corresponding signal in the combinational circuit. The main consequence of preserving functionality is that the correctness of $RCA(n)$ follows.

Question 6. Write a direct proof of the correctness of a ripple-carry adder $RCA(n)$. (That is, do not rely on the time-space transformation and the correctness of the bit-serial adder.)

4.3   Cost and Delay

The bit-serial adder consists of a single full-adder and a single bit flip-flop. It follows that the cost of the bit-serial adder is constant. The addition of n-bit numbers requires n clock cycles.

The ripple-carry adder $RCA(n)$ consists of n full-adders. Hence, its cost is linear in n. The delay is also linear since the path from $A[0]$ to $S[n-1]$ traverses all the n full-adders.

5   Lower Bounds

We saw that $RCA(n)$ has a linear cost and a linear delay. Before we look for better adder designs, we address the question of whether there exist cheaper or faster adders. For this purpose we need to prove lower bounds.

Lower bounds are rather mysterious to many students. One reason is mathematical in nature; students are not used to arguing about unknown abstract objects. The only lower bound that they probably encountered so far is the lower bound on the number of comparisons needed for sorting. Another reason is that they have been noticing a continuous improvement in computer performance and an ongoing reduction in computer costs. So many students are under the impression that there is no limit to faster and cheaper circuits. This impression might be even "well founded" if they heard about "Moore's Law". Finally, they are accustomed to thinking that better ways are awaiting the ingenious or lucky inventor.

We first state the lower bounds for the delay and cost of binary adders.

Theorem 2. *Assume that the number of inputs of every combinational gate is bounded by c. Then, for every combinational circuit G that implements an* ADDER(n), *the following hold:* $c(G) \geq n/c$ *and* $d(G) \geq \log_c n$.

How does one prove this theorem? The main difficulty is that we are trying prove something about an unknown circuit. We have no idea whether there exist better ways to add. Perhaps some strange yet very simple Boolean function of certain bits of the addends can help us compute the sum faster or cheaper? Instead of trying to consider all possible methods for designing adders, we rely on the simplest properties that every adder must have. In fact, the proof is based on topological properties common to every adder. There are two topological properties that we use: (i) There must be a path from every input to the output $S[n-1]$. (ii) The number of inputs of every combinational gate is bounded by c. Hence the proof of Claim 2 reveals an inherent limitation of combinational circuits rather than incompetence of designers.

Question 7. Prove Theorem 2.

Hint: Show that the output bit $S[n]$ depends on all the inputs. This means that one cannot determine the value of $S[n]$ without knowing the values of all the input bits. Prove that in every combinational circuit in which the output depends on all the inputs the delay is at least logarithmic and the cost is at least linear in the number of inputs. Rely on the fact that the number of inputs of each gate is at most c.

Returning the ripple-carry adder RCA(n), we see that its cost is optimal (upto a constant). However, its delay is linear. The lower bound is logarithmic, so much faster adders might exist. We point out that, in commercial microprocessors, 32-bit numbers are easily added within a single clock cycle. (In fact, in some floating point units numbers over 100 bits long are added within a clock cycle.) Clock periods in contemporary microprocessors are rather short; they are shorter than 10 times the delay of a full-adder. This means that even the addition of 32-bit numbers within one clock cycle requires faster adders.

6   The Adder of Ladner and Fischer

In this section we present an adder design whose delay and cost are asymptotically optimal (i.e., logarithmic delay and linear cost).

6.1   Motivation

Let us return to the bit-serial adder. The bit-serial adder is fed one bit of each addend in each clock cycle. Consider the finite-state diagram of the bit-serial adder depicted in Fig. 6. In this diagram there are two states: 0 and 1 (the state is simply the value stored in the flip-flop, namely, the carry-bit from the previous position). The computation of the bit-serial adder can be described as a walk in this diagram that starts in the initial state 0. If we know the sequence of the states in this walk, then we can easily compute the sum bits. The reason for this is that the output from state $q \in \{0,1\}$ when fed inputs $A[i]$ and $B[i]$ is $\text{XOR}(q, A[i], B[i])$.

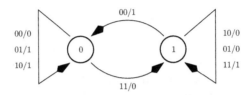

Fig. 6. The state diagram of the bit-serial adder.

We now consider the goal of *parallelizing* the bit-serial adder. By paralleling we mean the following. We know how to compute the sum bits if the addends are input one bit at a time. Could we compute the sum faster if we knew all the bits of the addends from the beginning? Another way to state this question is to ask whether we could quickly reconstruct the sequence of states that are traversed.

Perhaps it is easier to explain our goal if we consider a state-diagram with several states (rather than just two states). Let $\sigma_i \in \Sigma$ denote the symbol input in cycle i. We assume that in cycle zero the machine is in state q_0. When fed by the input sequence $\sigma_0, \ldots, \sigma_{n-1}$, the machine walks through the sequence of states $q_0, \ldots, q_n$ defined by transition function δ, as follows: $q_{i+1} = \delta(q_i, \sigma_i)$. The output sequence $y_0, y_1, \ldots, y_{n-1}$ is defined by output function γ, as follows: $y_i = \gamma(q_i, \sigma_i)$. We can interpret the inputs as "driving instructions" in the state diagram. Namely, in each cycle, the inputs instruct us how to proceed in the walk (i.e., "turn left", "turn right", or "keep straight"). Now, we wish to quickly reconstruct the walk from the sequence of n instructions.

Suppose we try to split this task among n players. The ith player is given the input symbol σ_i and performs some computation based on σ_i. After this, the players meet, and try to quickly glue together their pieces. Each player is confronted with the problem of simplifying the task of gluing. The main obstacle is that each player does not know the state q_i of the machine when the input symbol σ_i is input. At first, it seems that knowing q_i is vital if, for example, players i and $i + 1$ want to combine forces.

Ladner and Fisher proposed the following approach. Each player i computes (or, more precisely chooses) a restricted transition function $\delta_i : S \to S$, defined by $\delta_i(q) = \delta(q, \sigma_i)$. One can obtain a "graph" of δ_i if one considers only the edges in the state diagram that are labeled with the input symbol σ_i. By definition, if the initial state is q_0, then $q_1 = \delta_0(q_0)$. In general, $q_i = \delta_{i-1}(q_{i-1})$, and hence, $q_i = \delta_{i-1}(\delta_{i-2}(\cdots(\delta_0(q_0))\cdots))$. It follows that player i is satisfied if she computes the composition of the functions $\delta_{i-1}, \ldots, \delta_0$. Note that the function δ_i is determined by the input symbol σ_i, and the functions δ_i are selected from a fixed collection of $|\Sigma|$ functions.

Before we proceed, there is a somewhat confusing issue that we try to clarify. We usually think of an input σ_i as a parameter that is given to a function (say, f), and the goal is to evaluate $f(\sigma)$. Here, the input σ_i is used to select a function δ_i. We do not evaluate the function δ_i; instead, we compose it with other functions.

We denote the composition of two function f and g by $f \circ g$. Note that $(f \circ g)(q) \triangleq f(g(q))$. Namely, g is applied first, and f is applied second.

We denote the composition of the functions $\delta_1, \ldots, \delta_0$ by the function π_i. Namely, $\pi_0 = \delta_0$, and $\pi_{i+1}(q) = \delta_{i+1}(\pi_i(q))$. Assume that, given representations of two functions f and g, the representation of the composition $f \circ g$ can be computed in constant time. This implies that the function π_{n-1} can be, of course, computed with linear delay. The goal in parallelizing the computation is to compute π_{n-1} with logarithmic delay. Moreover, we wish to compute all the functions $\pi_0, \ldots, \pi_{n-1}$ with logarithmic delay and with overall linear cost.

Recall that our goal is to compute the output sequence rather than the sequence of states traversed by the state machine. Obviously, if we compute the walk $q_0, \ldots, q_{n-1}$, then we can easily compute the output sequence simply by $y_i = \gamma(q_i, \sigma_i)$. Hence, each output symbol y_i can be computed with constant delay and cost after q_i is computed. This means that we have reduced the problem of parallelizing the computation of a finite state machine to the problem of parallelizing the computation of compositions of functions.

Remember that our goal is to design an optimal adder. So the finite state machine we are interested in is the bit-serial adder. There are four possible input symbols corresponding to the values of the bits $a[i]$ and $b[i]$. The definition of full-adder (and the state diagram), imply that the function δ_i satisfies the following condition for $q \in \{0, 1\}$:

$$\delta_i(q) = \begin{cases} 0 & \text{if } q + a[i] + b[i] < 2 \\ 1 & \text{if } q + a[i] + b[i] \geq 2. \end{cases}$$

$$= \begin{cases} 0 & \text{if } a[i] + b[i] = 0 \\ q & \text{if } a[i] + b[i] = 1. \\ 1 & \text{if } a[i] + b[i] = 2. \end{cases} \tag{6}$$

In the literature the sum $a[i] + b[i]$ is often represented using the carry "kill/propagate/generate" signals [CLR90]. This terminology is justified by the following explanation.

- When $a[i] + b[i] = 0$, it is called "kill", because the carry-out is zero. In Eq. 6, we see that when $a[i] + b[i] = 0$, the value of the function δ_i is always zero.

- When $a[i] + b[i] = 1$, it is called "propagate", because the carry-out equals the carry-in. In Eq. 6, we see that when $a[i] + b[i] = 1$, the function δ_i is the identity function.
- When $a[i] + b[i] = 2$, it is called "generate", because the carry-out is one. In Eq. 6, we see that when $a[i] + b[i] = 2$, the value of the function δ_i is always one.

From this discussion, it follows that the "kill/propagate/generate" jargon (also known as k, p, g) is simply a representation of $a[i] + b[i]$. We prefer to abandon this tradition and use a different notation described below.

Equation 6 implies that, in a bit-serial adder, δ_i can be one of three functions: the zero function (i.e., value is always zero), the one function (i.e., value is always one), and the identity function (i.e., value equals the parameter). We denote the zero function by f_0, the one function by f_1, and the identity function by f_{id}.

The fact that these three functions are closed under composition can be easily verified since for every $x \in \{0, id, 1\}$:

$$f_0 \circ f_x = f_0$$
$$f_{id} \circ f_x = f_x$$
$$f_1 \circ f_x = f_1.$$

Finally, we point out that the "multiplication" table presented for the operator defined over the alphabet $\{k, p, g\}$ is simply the table of composing the functions f_0, f_{id}, f_1 (see, for example, [CLR90]).

6.2   Associativity of Composition

Before we continue, we point out an important property of compositions of functions whose domain and range are identical (e.g., Q is both the domain and the range of all the functions δ_i).

Consider a set Q and the set $\mathcal{F}$ of all functions from Q to Q. An *operator* is a function $\star : \mathcal{F} \times \mathcal{F} \to \mathcal{F}$. (One could define operators $\star : A \times A \to A$ with respect to any set A. Here we need only operators with respect to $\mathcal{F}$.) Since $\star$ is a dyadic function, we denote by $f \star g$ the image of $\star$ when applied to f and g.

Composition of functions is perhaps the most natural operator. Given two functions, $f, g \in \mathcal{F}$, the composition of f and g is the function h defined by $h(q) = f(g(q))$. We denote the composition of f and g by $f \circ g$.

Definition 4. *An operator* $\star : \mathcal{F} \times \mathcal{F} \to \mathcal{F}$ *is associative if, for every three functions* $f, g, h \in \mathcal{F}$, *the following holds:*

$$(f \star g) \star h = f \star (g \star h).$$

We note that associativity is usually defined for dyadic functions, namely, $f(a, f(b, c)) = f(f(a, b), c)$. Here, we are interested in operators (i.e., functions of functions), so we consider associativity of operators. Of course, associativity means the same in both cases, and one should not be confused by the fact that the value of an operator is a function.

We wish to prove that composition is an associative operator. Although this is an easy claim, it is often hard to convince the students that it is true. One could prove this by a reduction to multiplication of zero-one matrices. Instead, we provide a "proof by diagram".

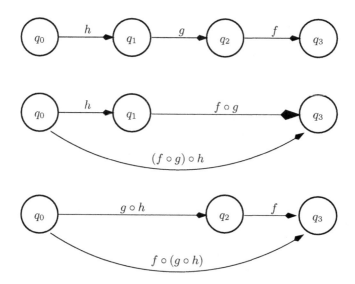

Fig. 7. Composition of functions is associative.

Claim. Composition of functions is associative.

Proof. Consider three functions $f, g, h \in \mathcal{F}$ and an element $q_0 \in Q$. Let $q_1 = h(q_0)$, $q_2 = g(q_1)$, and $q_3 = f(q_2)$. In Figure 7 we depict the compositions $(f \circ g) \circ h$ and $f \circ (g \circ h)$. In the second line of the figure we depict the following.

$$((f \circ g) \circ h)(q_0) = (f \circ g)(h(q_0))$$
$$= f(g(h(q))).$$

In third line of the figure we depict the following.

$$f \circ (g \circ h)(q) = f((g \circ h)(q))$$
$$= f(g(h(q))).$$

Hence ,the claim follows.

The associativity of an operator allows us to write expressions without parenthesis. Namely, $f_1 \circ f_2 \circ \cdots \circ f_n$ is well defined.

Question 8. Let $Q = \{0, 1\}$. Find an operator in $\mathcal{F}$ that is not associative.

Interestingly, in most textbooks that describe parallel-prefix adders, an associative dyadic function is defined over the alphabet $\{k, p, g\}$ (or over its representation by the two bits p and g). This associative function is usually presented without any justification or motivation. As mentioned at the end of Sec. 6.1, this function is simply the composition of f_0, f_{id}, f_1. The associativity of this operator is usually presented as a special property of addition. In this section we showed this is not the case. The special property of addition is that the functions $\delta(\cdot, \sigma)$ are closed under composition. Associativity, on the other hand, is simply the associativity of composition.

6.3   The Parallel Prefix Problem

In Section 6.1, we reduced the problem of designing a fast adder to the problem of computing compositions of functions. In Section 6.2, we showed that composition of functions is an associative operator. This motivates the definition of the prefix problem.

Definition 5. *Consider a set $\mathcal{A}$ and an associative operator $\star : \mathcal{A} \times \mathcal{A} \to \mathcal{A}$. The prefix problem is defined as follows.*

Input: $\delta_0, \ldots, \delta_{n-1} \in \mathcal{A}$.
Output: $\pi_0, \ldots, \pi_{n-1} \in \mathcal{A}$ *defined by* $\pi_0 = \delta_0$ *and* $\pi_i = \delta_i \star \cdots \star \delta_0$, *for* $0 < i < n$.
   (Note that $\pi_{i+1} = \delta_{i+1} \star \pi_i$.*)*

Assume that we have picked a (binary) representation for elements in $\mathcal{A}$. Moreover, assume that we have an implementation of the associative operator $\star$ with respect to this implementation. Namely, a $\star$-gate is a combinational gate that when fed two representations of elements $f, g \in \mathcal{A}$, outputs a representation of $f \star g$. Our goal now is to design a fast circuit for solving the prefix problem using $\star$-gates. Namely, our goal is to design a parallel prefix circuit.

Note that the operator $\star$ need not be commutative, so the inputs of the $\star$-gate cannot be interchanged. This means that there is a difference between the left input and the right input of a $\star$-gate.

6.4   The Parallel Prefix Circuit

We are now ready to describe a combinational circuit for the prefix problem. We will use only one building block, namely, a $\star$-gate. We assume that the cost and the delay of a $\star$-gate are constant (i.e., do not depend on n).

We begin by considering two circuits; one with optimal cost and the second with optimal delay. We then present a circuit with optimal cost and delay.

Linear cost but linear delay. Figure 8 depicts a circuit for the prefix problem with linear cost. The circuit contains $(n - 1)$ copies of $\star$-gates, but its delay is also linear. In fact, this circuit is very similar to the ripple carry adder.

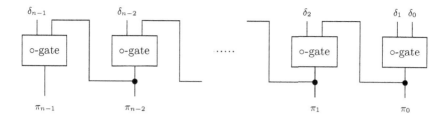

Fig. 8. A prefix computation circuit with linear delay.

Logarithmic delay but quadratic cost. The ith output π_i can be computed by circuit with the topology of a balanced binary tree, where the inputs are fed from the leaves, a $\star$-gate is placed in each node, and π_i is output by the root. The circuit contains $(i-1)$ copies of $\star$-gates and its delay is logarithmic in i. We could construct a separate tree for each π_i to obtain n circuits with logarithmic delay but quadratic cost.

Our goal now is to design a circuit with logarithmic delay and linear cost. Intuitively, the design based on n separate trees is wasteful because the same computations are repeated in different trees. Hence, we need to find an efficient way to "combine" the trees so that computations are not repeated.

Parallel prefix computation. We now present a circuit called PPC(n) for the prefix problem. The design we present is a recursive design. For simplicity, we assume that n is a power of 2. The design for $n = 2$ simply outputs $\pi_0 \leftarrow \delta_0$ and $\pi_1 \leftarrow \delta_1 \star \delta_0$. The recursion step is depicted in Figure 9. Adjacent inputs are paired and fed to a $\star$-gate. The $n/2$ outputs of the $\star$-gates are fed to a PPC($n/2$) circuit. The outputs $\pi_0', \ldots, \pi_{n/2-1}'$ of the PPC($n/2$) circuit are directly connected to the odd indexed outputs, namely, $\pi_{2i+1} \leftarrow \pi_i'$. Observe that wires carrying the inputs with even indexes are drawn (or routed) over the PPC($n/2$) box; these "even indexed" wires are not part of the PPC($n/2$) design. The even indexed outputs (for $i > 0$) are obtained as follows: $\pi_{2i} \leftarrow \delta_{2i} \star \pi_{i-1}'$.

6.5   Correctness

Claim. The design depicted in Fig. 9 is correct.

Proof. The proof of the claim is by induction. The induction basis holds trivially for $n = 2$. We now prove the induction step. Consider the PPC($n/2$) used in a PPC(n). Let δ_i' and π_i' denote the inputs and outputs of the PPC($n/2$), respectively. The ith input $\delta'[i]$ equals $\delta_{2i+1} \star \delta_{2i}$. By associativity and the induction hypothesis, the ith output π_i' satisfies:

$$\pi_i' = \delta_i' \star \cdots \delta_0'$$
$$= (\delta_{2i+1} \star \delta_{2i}) \star \cdots \star (\delta_1 \star \delta_0)$$

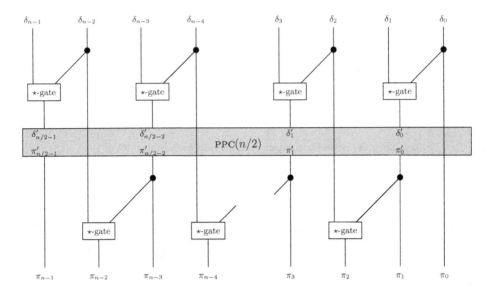

Fig. 9. A recursive design of $\text{PPC}(n)$. (The wires that pass over the $\text{PPC}(n/2)$ box carry the even indexed inputs $\delta_0, \delta_2, \ldots, \delta_{n-2}$. These signals are not part of the $\text{PPC}(n/2)$ circuit. The wires are drawn in this fashion only to simplify the drawing.)

It follows that the output π_{2i+1} equals the composition $\delta_{2i+1} \star \cdots \star \delta_0$, as required. Hence, the odd indexed outputs $\pi_1, \pi_3, \ldots, \pi_{n-1}$ are correct.

Finally, output in position $2i$ equals $\delta_{2i} \star \pi'_{i-1} = \delta_{2i} \star \pi_{2i-1} = \delta_{2i} \star \cdots \star \delta_0$. It follows that the even indexed outputs are also correct, and the claim follows.

6.6 Delay and Cost Analysis

The delay of the $\text{PPC}(n)$ circuit satisfies the following recurrence:

$$d(\text{PPC}(n)) = \begin{cases} d(\star\text{-gate}) & \text{if } n = 2 \\ d(\text{PPC}(n/2)) + 2 \cdot d(\star\text{-gate}) & \text{otherwise.} \end{cases}$$

If follows that

$$d(\text{PPC}(n)) = (2\log n - 1) \cdot d(\star\text{-gate}).$$

The cost of the $\text{PPC}(n)$ circuit satisfies the following recurrence:

$$c(\text{PPC}(n)) = \begin{cases} c(\star\text{-gate}) & \text{if } n = 2 \\ c(\text{PPC}(n/2)) + (n-1) \cdot c(\star\text{-gate}) & \text{otherwise.} \end{cases}$$

Let $n = 2^k$, it follows that

$$c(\text{PPC}(n)) = \sum_{i=2}^{k} (2^i - 1) \cdot c(\star\text{-gate}) + c(\star\text{-gate})$$

$$= (2n - \log n - 1) \cdot c(\star\text{-gate}).$$

We conclude with the following corollary.

Corollary 1. *If the delay and cost of an $\star$-gate is constant, then*

$$d(\text{PPC}(n)) = \Theta(\log n)$$
$$c(\text{PPC}(n)) = \Theta(n).$$

Question 9. This question deals with the implementation of $\circ$-gates for general finite state machines. (Recall that a $\circ$-gate implements composition of restricted transition functions.)

1. Suggest a representation (using bits) for the functions δ_i.
2. Design a $\circ$-gate with respect to your representation, namely, explain how to compute composition of functions in this representation.
3. What is the size and delay of the $\circ$-circuit with this representation? How does it depend on Q and Σ?

6.7 The Parallel-Prefix Adder

So far, the description of the parallel-prefix adder has been rather abstract. We started with the state diagram of the serial adder, attached a function δ_i to each input symbol σ_i, and computed the composition of the functions. In essence, this leads to a fast and cheap adder design. In this section we present a concrete design based on this construction. For this design we choose a specific representation of the functions f_0, f_1, f_{id}. This representation appears in Ladner and Fischer [LadnerFischer80], and in many subsequent descriptions (see [BrentKung82, ErceLang04, MüllerPaul00]). To facilitate comparison with these descriptions, we follow the notation of [BrentKung82]. In Fig. 10, a block diagram of this parallel-prefix adder is presented.

The carry-generate and carry-propagate signals. We decide to represent the functions $\delta_i \in \{f_0, f_1, f_{id}\}$ by two bits: g_i - the carry-generate signal, and p_i - the carry propagate signal. Recall that the ith input symbol σ_i is the pair of bits $A[i], B[i]$. We simply use the binary representation of $A[i] + B[i]$. The binary representation of the sum $A[i] + B[i]$ requires two bits: one for units and the second for twos. We denote the units bit by p_i and the twos bit by g_i. The computation of p_i and g_i is done by a half-adder that is input $A[i]$ and $B[i]$.

Implementation of the $\circ$-gate. Now that we have selected a representation of the functions f_0, f_1, f_{id}, we need to design the $\circ$-gate that implements composition. This is an easy task: if $(g, p) = (g_1, p_1) \circ (g_2, p_2)$, then $g = g_1 \text{ OR } (p_1 \text{ AND } g_2)$ and $p = p_1 \text{ AND } p_2$.

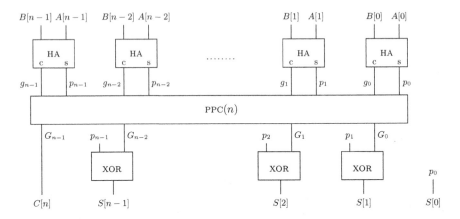

Fig. 10. A parallel-prefix adder. (HA denotes a half-adder and the PPC(n) circuit consists of ∘-gates described below.)

Putting things together. The pairs of signals $(g_0, p_0), (g_1, p_1), \ldots (g_{n-1}, p_{n-1})$ are input to a PPC(n) circuit that contains only ∘-gates. The output of the PPC(n) is a representation of the functions π_i, for $0 \le i < n$. We denote the pair of bits used to represent π_i by (G_i, P_i). We do not need the outputs $P_0, \ldots, P_{n-1}$, so we only depict the outputs $G_0, \ldots, G_{n-1}$.

Recall that state q_{i+1} equals $\pi_i(q_0)$. Since $q_0 = 0$, it follows that $\pi_i(q_0) = 1$ if and only if $\pi_i = f_1$, namely, if and only if $G_i = 1$. Hence $q_{i+1} = G_i$.

We are now ready to compute the sum bits: Since $S[0] = \text{XOR}(A[0], B[0])$, we may reuse $p[0]$, and output $S[0] = p[0]$. For $0 < i < n$, $S[i] = \text{XOR}(A[i], B[i], q_i) = \text{XOR}(p_i, G_{i-1})$. Finally, for those interested in the carry-out bit $C[n]$, simply note that $C[n] = q_n = G_{n-1}$.

Question 10. Compute the number of gates of each type in the parallel prefix adder presented in this chapter as a function of n.

In fact it is possible to save some hardware by using degenerate ∘-gates when only the carry-generate bit of the output of a ∘-gate is used. This case occurs for the $n/2 - 1$ ∘-gates whose outputs only feed outputs of the PPC(n) circuit. More generally, one could define such degenerate ∘-gates recursively, as follows. A ∘-gate is degenerate if: (i) Its output feeds only an output of the PPC(n) circuit, or (ii) Its output is connected only to ports that are the right input ports of degenerate ∘-gates or an output port of the PPC(n)-circuit.

Question 11. How many degenerate ∘-gates are there?

7   Further Topics

In this section we discuss various topics related to adders.

7.1   The Carry-In Bit

The role of the carry-in bit is somewhat mysterious at this stage. To my knowledge, no programming language contains an instruction that uses a carry-in bit. For example, we write $S := A + B$, and we do not have an extra single bit variable for the carry-in. So why is the carry-in included in the specification of an adder?

There are two justifications for the carry-in bit. The first justification is that one can build adders of numbers of length $k + \ell$ by serially connecting an adder of length k and an adder of length ℓ. The carry-out output of the adder of the lower bits is fed to the carry-in inputs of the adder of the higher bits.

A more important justification for the carry-in bit is that it is needed for a constant time reduction from subtraction of two's complement numbers to binary addition. A definition of the two's complement representation and a proof of the reduction appear in [MüllerPaul00, Even04].

In this essay, we do not define two's complement representation or deal with subtraction. However, most hardware design courses deal with these issues, and hence, they need the carry-in input. The carry-in input creates very little complications, so we do not mind considering it, even if its usefulness is not clear at this point.

Formally, the carry-in bit is part of the input and is denoted by $C[0]$. The goal is to compute $A + B + C[0]$.

There are three main ways to compute $A + B + C[0]$ within the framework of this essay:

1. The carry-in bit $C[0]$ can be viewed as a setting of the initial state q_0. Namely, $q_0 = C[0]$. This change has two effects on the parallel-prefix adder design: (i) The sum bit $S[0]$ equals $\gamma(q_0, \sigma_0)$. Hence $S[0] = \text{XOR}(C[0], p[0])$. (ii) For $i > 0$, the sum bit $S[i]$ equals $\text{XOR}(\pi_{i-1}(q_0), p_i)$. Hence $S[i]$ can be computed from 4 bits: p_i, $C[0]$, and G_{i-1}, P_{i-1}. The disadvantage of this solution is that we need an additional gate for the computation of each sum bit. The advantage of this solution is that it suggests a simple way to design a compound adder (see Sec. 7.2).

2. The carry-in bit can be viewed as two additional bits of inputs, namely $A[-1] = B[-1] = C[0]$, and then $C[0] = 2^{-1} \cdot (A[-1] + B[-1])$. This means that we reduce the task of computing $A + B + C[0]$ to the task of adding numbers that are longer by one bit. The disadvantage of this solution is that the $\text{PPC}(n)$ design is particularly suited for n that is a power of two. Increasing n (that is a power of two) by one incurs a high overhead in cost.

3. Use δ_0' instead of δ_0, where δ_0' is defined as follows:

$$\delta_0' \triangleq \begin{cases} f_0 & \text{if } A[0] + B[0] + C[0] < 2 \\ f_1 & \text{if } A[0] + B[0] + C[0] \geq 2. \end{cases}$$

This setting avoids the assignment $\delta_0' = f_{id}$, but still satisfies: $\delta_0'(q_0) = q_1$. Hence, the $\text{PPC}(n)$ circuit outputs the correct functions even when δ_0 is replaced by δ_0'. The sum bit $S[0]$ is computed directly by $\text{XOR}(A[0], B[0], C[0])$.

Note that the implementation simply replaces the half-adder used to compute p_0 and g_0 by a full-adder that is also input $C[0]$. Hence, the overhead for dealing with the carry-in bit $C[0]$ is constant.

7.2   Compound Adder

A compound adder is an adder that computes both $A + B$ and $A + B + 1$. Compound adders are used in floating point units to implement rounding. Interestingly, one does not need two complete adders to implement a compound adder since hardware can be shared. On the other hand, this method does not allow using a PPC(n) circuit with degenerate $\circ$-gates.

The idea behind sharing is that, in the first method way for computing $A + B + C[0]$, we do not rely on the carry-in $C[0]$ to compute the functions π_i. Only after the functions π_i are computed, the carry-in bit $C[0]$ is used to determine the initial state q_0. The sum bits then satisfy $S[i] = \text{XOR}(\pi_{i-1}(q_0), p_i)$, for $i > 0$. This means that we can share the circuitry that computes π_i for the computation of the sum $A + B$ and the incremented sum $A + B + 1$.

7.3   Fanout

The fanout of a circuit is the maximum fanout of an output port in the circuit. Recall that the fanout of an output port is the number of input ports that are connected to it. In reality, a large fanout slows down the circuit. The main reason for this in CMOS technology is that each input port has a capacity. An output port has to charge (or discharge) all the capacitors corresponding to the input ports that it feeds. As the capacitance increases linearly with the fanout, the delay associated with stabilizing the output signal also increases linearly with the fanout. So it is desirable to keep the fanout low. (Delay grows even faster if resistance is taken into account.)

Ladner and Fischer provided a complete analysis of the fanout. In the PPC(n) circuit, the only cause for fanout greater than 2 are the branchings after the output of PPC($n/2$). Namely, only nets that feed outputs have a large fanout. Let $fo(i, n)$ denote the fanout of the net that feeds π_i. It follows that $fo(2i, 2n) = 1$ and $fo(2i + 1, 2n) = 1 + fo(i, n)$. Hence, the fanout is logarithmic. Brent and Kung [BrentKung82] reduced the fanout to two but increased the cost to $O(n \log n)$. One could achieve the same fanout while keeping the cost linear (see question below for details); however, the focus in [BrentKung82] was area and a regular layout, not cost.

Question 12. In this question we consider fanout in the PPC(n) design and suggest a way to reduce the fanout so that it is at most two.

-   What is the maximum fanout in the PPC(n) design?
-   Find the output port with the largest fanout in the PPC(n) circuit.
-   The fanout can be made constant if buffers are inserted according to the following recursive rule. Insert a buffer in every branching point of the PPC(n)

that is fed by an output of the PPC($n/2$) design (such branching points are depicted in Fig. 9 by filled circles below the PPC($n/2$) circuit). (A buffer is a combinational circuit that implements the identity function. A buffer can be implemented by cascading two inverters.)

- By how much does the insertion of buffers increase the cost and delay?
- The rule for inserting buffers to reduce the fanout can be further refined to save hardware without increasing the fanout. Can you suggest how?

7.4   Tradeoffs Between Cost and Delay

Ladner and Fischer [LadnerFischer80] present an additional type of "recursive step" for constructing a parallel prefix circuit. This additional recursive step reduces the delay at the price of increasing the cost and the fanout. For input length n, Ladner and Fischer suggested $\log n$ different circuits. Loosely speaking, in the kth circuit, one performs k recursive steps of the type we have presented and then $\log n - k$ recursive steps of the second type. We refer the reader to [LadnerFischer80] for the details and the analysis.

In the computer arithmetic literature several circuits have been suggested for fast adders. For example, Kogge and Stone suggested a circuit with logarithmic delay but its cost is $O(n \log n)$. Its asymptotic layout area is the same as the area of the layout suggested by Brent and Kung. More details about other variations and hybrids can be found in [Knowles99, Zimmerman98].

7.5   VLSI Area

This essay is about hardware designs of adders. Since such adders are implemented as VLSI circuits, it makes sense to consider criteria that are relevant to VLSI circuits. With all other factors remaining the same, area is more important in VLSI than counting the number of gates. By area one means the area required to draw the circuit on a chip. (Note, however, that more gates usually consume more power.)

More about the model for VLSI area can be found in [Shiloach76, Thompson80]. In fact, the first formal research on the area of hardware circuits was done by Yossi Shiloach under the supervision of my father during the mid 70's in the Weizmann institute. This research was motivated by discussions with the people who were building the second Golem computer using printed circuit boards.

In addition to reducing fanout, Brent and Kung "unrolled" the recursive construction depicted in Fig. 9 and presented a regular layout for adders. The area of this layout is $O(n \log n)$.

8   An Opinionated History of Adder Designs

The search for hardware algorithms for addition with short delay started in the fifties. Ercegovac and Lang [ErceLang04] attribute the earliest reference for an adder with a logarithmic delay to Weinberger and Smith [WeinSmith58]. This adder is often referred to as a *carry-lookahead adder*. On the other hand, Ladner and Fischer cite Ofman [Ofman63] for a carry-lookahead adder. In Cormen et. al., it is said that the circuit of Weinberger and Smith required large gates as opposed to Ofman's design that required constant size gates. So it seems that in modern terms, the delay of the design of Weinberger and Smith was not logarithmic.

I failed to locate these original papers. However, one can find descriptions of carry-lookahead adders in [CLR90, ErceLang04, Koren93, Kornerup97]). The topology of a carry-lookahead adder is a complete tree of degree 4 with $\log_4 n$ levels (any constant is fine, and 4 is the common choice). The data traverses the tree from the leaves to the root and then back to the leaves. The carry-lookahead adder is rather complicated and seems to be very specific to addition. Even though the delay is logarithmic and the cost is linear, this adder was not the "final word" on adders. In fact, due to its complexity, only the first two layers of the carry-lookahead adder are described in detail in most textbooks (see [Koren93, ErceLang04]).

The conditional-sum adder [Sklansky60] is a simple adder with logarithmic delay. However, its cost is $O(n \log n)$. Even worse, it has a linear fanout, so in practice the delay is actually $O(\log^2 n)$. The main justification for the conditional-sum adder is that it is simple.

There was not much progress in the area of adder design till Ladner and Fischer [LadnerFischer80] presented the *parallel-prefix adder*. Their idea was to design small combinational circuits that simulate finite-state machines (i.e., transducers). They reduced this task to a *prefix problem* over an associative operator. They call a circuit that computes the prefix problem a *parallel prefix circuit*. When applied to the trivial bit-serial adder (implemented by a two-state finite state machine), their method yielded a combinational adder with logarithmic delay, linear size, and logarithmic fanout.

In fact, for every n, Ladner and Fischer presented $\log n$ different parallel prefix circuits; these designs are denoted by $\mathcal{P}_k(n)$, for $k = 0, \ldots, \log n$. The description of the circuits uses recursion. There are two types of recursion steps, the combination of which yields $\log n$ different circuits. (For simplicity, we assume throughout that n is a power of 2.) All these designs share a logarithmic delay and a linear cost, however, the constants vary; reduced delay results in increased size. Ladner and Fischer saw a similarity between $\mathcal{P}_{\log n}(n)$ and the carry-lookahead adder. However, they did not formalize this similarity and resorted to the statement that 'we believe that our circuit is essentially the same as the "carry-lookahead" adder'.

Theory of VLSI was at a peak when Ladner and Fischer published their paper. Brent and Kung [BrentKung82] popularized the parallel-prefix adder by presenting a regular layout for the parallel-prefix adder $\mathcal{P}_{\log n}(n)$. The layout was obtained by "unrolling" the recursion in [LadnerFischer80]. The area of the layout is $O(n \log n)$. In addition, Brent and Kung reduced the fanout of $\mathcal{P}_{\log n}(n)$ to two by inserting buffers. Since they were concerned with area and were not sensitive to cost, they actually suggested increasing the cost to $O(n \log n)$ (see

Question 12 for guidelines regarding the reduction of the fanout while keeping the cost linear.)

The adder presented in Brent and Kung's paper is specific for addition. The ideas of parallelizing the computation of a finite-state machine and the prefix problem are not mentioned in [BrentKung82]. This meant that when teaching the Brent-Kung adder, one introduces the carry-generate and carry-propagate signals (i.e., g_i, p_i signals) without any motivation (i.e., it is a way to compute the carry bits, but where does this originate from?). Even worse, the associative operator is magically pulled out of the hat. Using simple and short algebraic arguments, it is shown that applying this operator to a prefix problem leads to the computation of the carry bits. Indeed, the proofs are short, but most students find this explanation hard to follow. I suspect that the cause of this difficulty is that no meaning is attached to the associative operator and to the g_i, p_i signals.

The prefix problem gained a lot of success in the area of parallel computation. Leighton [Leighton91] presented a parallel prefix algorithm on a binary tree. The presentation is for a synchronous parallel architecture, and hence the computation proceeds in steps. I think that the simplest way to explain the parallel prefix algorithm on a binary tree is as a simulation of Ladner and Fischer's $\mathcal{P}_{\log n}(n)$ circuit on a binary tree. In this simulation, every node in the binary tree is in charge of simulating two nodes in $\mathcal{P}_{\log n}$ (one before the recursive call and one after the recursive call).

In Cormen et. al. [CLR90], a carry-lookahead adder is presented. This adder is a hybrid design that combines the presentation of Brent and Kung and the presentation of Leighton. The main problem with the presentation in [CLR90] is that the topology is a binary tree and data traverses the tree in both directions (i.e., from the leaves to the root and back to the leaves). Hence, it is not obvious that the design is combinational (although it is). So the presentation in Cormen et. al. introduces an additional cause for confusion. On the other hand, this might be the key to explaining Ofman's adder. Namely, perhaps Ofman's adder can be viewed as a simulation of $\mathcal{P}_{\log n}(n)$ on a binary tree.

9   Discussion

When I wrote this essay, I was surprised to find that in many senses the presentation in [LadnerFischer80] is better than subsequent textbooks on adders (including my own class notes). Perhaps the main reason for not adopting the presentation in [LadnerFischer80] is that Ladner and Fischer presented two recursion steps. The combination of these steps led to $\log n$ designs for parallel prefix circuits with n arguments (and hence for n-bit adders). Only one of these designs has gained much attention (i.e., the circuit they called $\mathcal{P}_{\log n}(n)$). There are probably two reasons for the success of this circuit: fanout and area. Brent and Kung were successful in reducing the fanout and presenting a regular layout only for $\mathcal{P}_{\log n}(n)$. It seems that these multiple circuits confused many who preferred then only to consider the presentation of Brent and Kung. In this essay only one type of recursion is described for constructing only one parallel prefix circuit, namely, the $\mathcal{P}_{\log n}(n)$ circuit.

I can speculate that the arithmetic community was accustomed to the carry-propagate and carry-generate signals from the carry-lookahead adder. They did not need motivation for it and the explanation of Ladner and Fischer was forgotten in favor of the presentation of Brent and Kung.

Another speculation is that the parallel computation community was interested in describing parallel algorithms on "canonic" graphs such as binary trees and hypercubes (see [Leighton91]). The special graph of Ladner and Fischer did not belong to this family. In hardware design, however, one has usually freedom in selecting the topology of the circuit, so canonic graphs are not important.

Acknowledgments

I thank Oded Goldreich for motivating me to write this essay. Thanks to his continuous encouragement, this essay was finally completed. Helpful remarks were provided by Oded Goldreich, Peter-Michael Seidel, and Ami Litman. I am grateful for their positive feedback and constructive suggestions.

References

[BrentKung82] R. P. Brent and H. T. Kung, "Regular Layout for Parallel Adders", IEEE Trans. Comp., Vol. C-31, no. 3, pp. 260-264. 1982. Available online at http://web.comlab.ox.ac.uk/oucl/work/richard.brent/pd/rpb060.pdf

[CLR90] T. H. Cormen, C. E. Leiserson, R. L. Rivest, Introduction to Algorithms, The MIT Press, 1990.

[ErceLang04] M. D. Ercegovac and T. Lang, Digital Arithmetic, Morgan Kaufmann, 2004.

[Even79] Shimon Even, Graph Algorithms, Computer Science Press, 1979.

[Even04] G. Even, "Lecture Notes in Computer Structure", manuscript, 2004.

[EvenLitman91] Shimon Even and Ami Litman, "A systematic design and explanation of the Atrubin multiplier", Sequences II, Methods in Communication, Security and Computer Sciences, R. Capocelli et.al. (Editors), pp. 189-202, Springer-Verlag, 1993.

[EvenLitman94] Shimon Even and Ami Litman. "On the capabilities of systolic systems", Mathematical Systems Theory, 27:3-28, 1994.

[HopUll79] John E. Hopcroft and Jeffrey D. Ullman, Introduction to automata theory, languages, and computation, Addison Wesley, 1979.

[Knowles99] S. Knowles, "A Family of Adders" Proceedings of the 14th IEEE Symposium on Computer Arithmetic, pp 30-34, 1999. (Note that the figures in the published proceedings are wrong.)

[Koren93] Israel Koren, Computer Arithmetic Algorithms, Prentice Hall, 1993.

[Kornerup97] Peter Kornerup, "Chapter 2 on Radix Integer Addition", manuscript, 1997.

[LadnerFischer80] R. Ladner and M. Fischer, "Parallel prefix computation", J. Assoc. Comput. Mach., 27, pp. 831–838, 1980.

[Leighton91] F. Thomson Leighton, *Introduction to parallel algorithms and architectures: array, trees, hypercubes*, Morgan Kaufmann, 1991.

[LeiSaxe81] C. E. Leiserson and J. B. Saxe, "Optimizing synchronous systems", *Journal of VLSI and Computer Systems*, 1:41-67, 1983. (Also appeared in Twenty-Second Annual Symposium on Foundations of Computer Science, pp. 23-36, 1981.)

[LeiSaxe91] C. E. Leiserson and J. B. Saxe. "Retiming synchronous circuitry". *Algorithmica*, 6(1)5-35, 1991.

[MüllerPaul00] S. M. Müller and W. J. Paul, *Computer Architecture: complexity and correctness*, Springer, 2000.

[Ofman63] Yu Ofman, "On the algorithmic complexity of discrete functions", *Sov Phys Dokl*, 7, pp. 589-591, 1963.

[Shiloach76] Yossi Shiloach, *Linear and planar arrangements of graphs*, Ph.D. thesis, Dept. of Applied Mathematics, Weizmann Institute, 1976.

[Sklansky60] J. Sklansky, "An evaluation of several two-summand binary adders", *IRE Trans. on Electronic Computers*, EC-9:213-26, 1960.

[Thompson80] Clark David Thompson, *A Complexity Theory For VLSI*, PhD Thesis, Carnegie Mellon University, 1980.

[WeinSmith58] A. Weinberger and J. L. Smith, "A logic for high-speed addition", *Nat. Bur. Stand. Circ.*, 591:3-12, 1958.

[Zimmerman98] R. Zimmermann, *Binary Adder Architectures for Cell-Based VLSI and their Synthesis*, PhD thesis, Swiss Federal Institute of Technology (ETH) Zurich, Hartung-Gorre Verlag, 1998. Available online at `http://www.iis.ee.ethz.ch/~zimmi/`

On Teaching the Basics of Complexity Theory⋆

Oded Goldreich

Department of Computer Science and Applied Mathematics,
Weizmann Institute of Science, ISRAEL
oded.goldreich@weizmann.ac.il

Abstract. We outline a conceptual framework for teaching the basic notions and results of complexity theory. Our focus is on using definitions and on organizing the presentation in a way that reflects the fundamental nature of the material. We do not attempt to provide a self-contained presentation of the material itself, but rather outline our suggestions regarding how this material should be presented in class. In addition, we express our opinions on numerous related issues.

We focus on the P-vs-NP Question, the general notion of a reduction, and the theory of NP-completeness. In particular, we suggest presenting the P-vs-NP Question both in terms of search problems and in terms of decision problems (where NP is viewed as a class of proof systems). As for the theory of NP-completeness, we suggest highlighting the mere existence of NP-complete sets.

1 Introduction

This is a highly opinionated essay that advocates a concept-oriented approach towards teaching technical material such as the basics of complexity theory. In addition to making various suggestions, I express my opinion on a variety of related issues. I do hope that this essay will stir discussion and maybe even affect the way some courses are being taught.

1.1 Teaching and Current Student Perception of Complexity Theory

Shimon Even had a passion for good teaching, and so writing this essay in his memory seems most appropriate. In my opinion, good teaching is an art (and, needless to say, Shimon was one of its top masters). It is hard (if at all possible) to cultivate artistic talents, but there are certain basic principles that underly each art form, and these can be discussed.

One central aspect of good teaching is putting things in the right perspective; that is, a perspective that clarifies the motivation for the various definitions and results. Nothing should be easier when it comes to complexity theory: It is easy

---

⋆ This essay was written for the current volume. The technical presentation was adapted from earlier lecture notes (e.g., [4]).

O. Goldreich et al. (Eds.): Shimon Even Festschrift, LNCS 3895, pp. 348–374, 2006.

to provide a good perspective on the basic notions and results of complexity theory, because these are of fundamental nature and of great intuitive appeal. Unfortunately, often this is not the way this material is taught. The annoying (and quite amazing) consequences are students that have only a vague understanding of the *conceptual meaning* of these fundamental notions and results.

1.2 The Source of Trouble and Eliminating It

In my opinion, it all boils down to taking the time to explicitly discuss the conceptual meaning of definitions and results. After all, the most important aspects of a scientific discovery are the intuitive question that it addresses, the reason that it addresses this question, the way it phrases the question, the approach that underlies its answer, and the ideas that are embedded in the answer. All these have to be reflected in the way the discovery is presented. In particular, one should use the "right" definitions (i.e., those that reflect better the fundamental nature of the notion being defined), and proceed in the (conceptually) "right" order. Two concrete examples follow.

Typically[1], NP is defined as the class of languages recognized by nondeterministic polynomial-time machines. Even bright students may have a hard time figuring out (by themselves) why one should care about such a class. On the other hand, when defining NP as the class of assertions that have easily verifiable proofs, each student is likely to understand its fundamental nature. Furthermore, the message becomes even more clear when discussing the search version analogue.

Similarly, one typically[1] takes the students throughout the detailed proof of Cook's Theorem before communicating to them the striking message (i.e., that "universal" problems exist at all, let alone that many natural problems like SAT are universal). Furthermore, in some cases, this message is not communicated explicitly at all.

1.3 Concrete Suggestions

The rest of this essay provides concrete suggestions for teaching the basics of complexity theory, where by the basics I mean the P-vs-NP Question and the theory of NP-completeness. This material is typically taught as part of an undergraduate course on computability and complexity theory, and my suggestions are targeted primarily at computer scientists teaching such a course. However, I believe that my suggestions are valid regardless of the context in which this material is being taught.

I assume that the basic material itself is well-known to the reader. Thus, *my focus is not on the material itself, but rather on how it should be presented in class.* The two most important suggestions were already mentioned above:

---

[1] However, exceptions do exists: There are teachers and even textbooks that deviate from the standard practice being bashed here.

1. The teacher should communicate the fundamental nature of the P-vs-NP Question while referring to definitions that (clearly) reflect this nature. In particular, I suggest explicitly presenting the implication of the P-vs-NP Question on the complexity of search problems, in addition to presenting the implication to decision problems.
2. The teacher should communicate the striking significance of the mere existence of NP-complete problems (let alone natural ones), before exhausting the students with complicated reductions.

Additional suggestions include providing a general perspective on the concept of a reduction, establishing tight relations between the complexity of search and decision problems, decoupling the proof of NP-hardness of SAT by using Circuit-SAT as an intermediate problem, and mentioning some additional topics (e.g., NP-sets that are neither in P nor NP-complete) rather than a host of NP-completeness results.

I advocate a model-independent presentation of the questions and results of complexity theory. I claim that most questions and results in complexity theory (like all results of computability theory) hold for any reasonable model of computation and can be presented with minimal reference to the specifics of the model.[2] In fact, in most cases, the specific model of computation is irrelevant. Typically, the presentation needs to refer to the specifics of the model of computation only when encoding the relation between consecutive instantaneous configurations of computation (see Section 4.3). However, such an encoding is possible for any reasonable model of computation, and this fact should be stressed.

It is also important to start a course (or series of lectures) by providing a wide perspective on its subject matter, which in this case is complexity theory. I would say that complexity theory is a central field of (Theoretical) Computer Science, concerned with the study of the *intrinsic* complexity of computational tasks, where this study tend to aim at *generality*: The field focuses on natural computational resources (most notably time), and the effect of limiting these resources on the *class of problems* that can be solved. Put in other words, complexity theory aims at understanding the *nature of efficient computation*. I suggest re-iterating the wider goals of complexity theory at the end of the course (or series of lectures), and illustrating them at that point by sketching a few of the active research directions and the results obtained in them. My own suggestion for such a brief overview is presented in Section 6.

Finally, until we reach the day in which every student can be assumed to have understood the meaning of the P-vs-NP Question and of NP-completeness, I suggest not to assume such an understanding when teaching an advanced complexity

---

[2] The specifics of the (reasonable) model are irrelevant for all questions and results mentioned in this essay, except for Theorem 6 where the model is important only for the exact bound on the slow-down of the optimal algorithm. Similarly, the specifics of the model effect the exact quantitative form of hierarchy theorems, but not their mere existence. Finally, in contrary to some beliefs, the specifics of the model are irrelevant also for most results regarding space complexity, provided that reasonable accounting of work-space is applied.

theory course. Instead, I suggest starting such a course with a fast discussion of the P-vs-NP Question and NP-completeness, making sure that the students understand the conceptual meaning of these basics.[3] (Needless to say, the rest of the course should also clarify the conceptual meaning of the material being taught.)

1.4 A Parenthetical Comment on Computability Versus Complexity

This essay refers to the current situation in many schools, where the basics of complexity theory are taught within a course in which material entitled "computability" plays at least an equal role. The essay is confined to the "complexity" part of such a course, and takes the "computability" part for granted.

Let me seize the opportunity and express my opinion on this combined course on computability and complexity theory. In my opinion, complexity theory should play the main role in this course, whereas the basic concepts and results of computability theory should be regarded as an important preliminary material. That is, I view computability theory as setting the stage for the study of the complexity of the computational tasks that can be automated at all. Thus, the computability aspects of such a course should be confined to establishing that the intuitive notion of an algorithm can be rigorously defined, and to emphasizing the uncomputability of most functions and of some natural functions (e.g., the Halting predicate). This includes introducing the idea of a universal algorithm, but does not included extensive programming with Turing machines or extensive study of (complexity-free) Turing reductions. Needless to say, I oppose the teaching of finite automata (let alone context-free grammars) within such a course.

Expanding upon the opinions expressed in the last paragraph is beyond the scope of the current essay. On the other hand, the rest of this essay is independent of the foregoing remarks. That is, it refers to the basic material of complexity theory, regardless of the question within which course this material is taught and what role does it play in such a course.

1.5 Organization

Section 2 contains a presentation of the P-vs-NP Question both in terms of search problems and in terms of decision problems. Section 3 contains a general treatment of reductions as well as a subsection on "self-reducibility" (of search problems). Section 4 contains a presentation of the basic definitions and results of the theory of NP-completeness (as well as a mention of the existence of NP-sets that are neither in P nor NP-complete). Section 5 mentions three additional topics that are typically not taught in a basic course on computability and complexity theory. These topics include the conjectured non-existence of coNP-sets that are NP-complete, the existence of optimal search algorithms for NP-relations, and the notion of promise problems.

---

[3] In fact, this essay is based on my notes for three lectures (covering the basic material), which were given in a graduate course on complexity theory (see [4]).

As a general rule, the more standard the material is, the less detail we provide about is actual technical contents. Our focus is on the conceptual contents of the material, and technical details are given merely for illustration. We stress again that this essay is not supposed to serve as a textbook, but rather as a conceptual framework.

The essay is augmented by a brief overview of complexity theory. Unlike the rest of this essay, which assumes familiarity with the material, this overview (Section 6) is supposed to be accessible to the novice (or an "outsider"), and may be used accordingly. One possible use is as a base for introductory comments on complexity theory to be made either at the beginning of a graduate course on the topic or at the end of the (currently prevailing) undergraduate course on computability and complexity theory.

2   P Versus NP

Most students have heard of P and NP before, but we suspect that many have not obtained a good explanation of what the P-vs-NP Question actually represents. This unfortunate situation is due to using the standard technical definition of NP (which refers to nondeterministic polynomial-time) rather than using (somehat more cumbersome) definitions that clearly capture the fundamental nature of NP. Below, we take the alternative approach. In fact, we present two fundamental formulations of the P-vs-NP Question, one in terms of search problems and the other in terms of decision problems.

Efficient computation. The teacher should discuss the association of efficient computation with polynomial-time algorithms, stressing that this association merely provides a convenient way of addressing fundamental issues.[4] In particular, polynomials are merely a "closed" set of moderately growing functions, where "closure" means closure under addition, multiplication and functional composition. These closure properties guarantee the closure of the class of efficient algorithms under natural algorithmic composition operations such as sequential execution and subroutine calls. (The specifics of the model of computation are also immaterial, as long as the model is "reasonable"; this strengthening of the Church–Turing Thesis is called the Cobham–Edmonds Thesis.)

2.1   The Search Version: Finding Versus Checking

In the eyes of non-experts, search problems are more natural than decision problems: typically, people seeks solutions more than they stop to wonder whether or not solutions exist. Thus, we recommend starting with a formulation of the

---

[4] Indeed, we claim that these fundamental issues are actually independent of the aforementioned association. For example, the question of whether finding a solution is harder than verifying its validity makes sense under any reasonable notion of "hardness". Similarly, the claim that factoring (or any other "NP problem") is "easily reducible" to SAT holds for many reasonable notions of "easy to compute" mappings.

P-vs-NP Question in terms of search problems. Admittingly, the cost is more cumbersome formulations (presented in Figure 1), but it is more than worthwhile. Furthermore, the equivalence to the decision problem formulation gives rise to conceptually appealing exercises.

We focus on polynomially-bounded relations, where a relation $R \subseteq \{0,1\}^* \times \{0,1\}^*$ is polynomially-bounded if there exists a polynomial p such that for every $(x, y) \in R$ it holds that $|y| \leq p(|x|)$. For such a relation it makes sense to ask whether, given an "instance" x, one can efficiently find a "solution" y such that $(x, y) \in R$. The polynomial bound on the length of the solution (i.e., y) guarantees that the intrinsic complexity of outputting a solution may not be due to the length (or mere typing) of the required solution.

The class P as a natural class of search problems. With each polynomially-bounded relation R, we associate the following search problem: *given x find y such that $(x, y) \in R$ or state that no such y exists.* The class $\mathcal{P}$ corresponds[5] to the class of search problems that are solvable in polynomial-time; that is, a relation R (or rather the search problem of R) is polynomial-time solvable if there exists a polynomial-time algorithm that given x find y such that $(x, y) \in R$ or state that no such y exists.

The class NP as another natural class of search problems. A polynomially-bounded relation R is called an NP-relation if, given an alleged instance-solution pair, one can efficiently check whether or not the pair is valid; that is, there exists a polynomial-time algorithm that given x and y determines whether or not $(x, y) \in R$. The class $\mathcal{NP}$ corresponds[5] to the class of search problems for NP-relations (and contains a host of natural search problems). It is reasonable to focus on search problems for NP-relations, because the ability to efficiently recognize a valid solution seems to be a natural prerequisite for a discussion regarding the complexity of finding such solutions. (Indeed, one can introduce (unnatural) non-NP-relations for which the search problem is solvable in polynomial-time; still the restriction to NP-relations is very natural.)

The P versus NP question in terms of search problems: *Is it the case that the search problem of any NP-relation can be solved in polynomial-time?* In other words, if it is easy to check whether or not a given solution for a given instance is correct, then is it also easy to find a solution to a given instance?

If $\mathcal{P} = \mathcal{NP}$ (in terms of search problems) then this would mean that whenever solutions to given instances can be efficiently verified for correctness it is also the case that these solutions can be efficiently found (when given only the instance). This would mean that all reasonable search problems (i.e., all NP-relations) are easy to solve. Needless to say, such a situation would contradict the intuitive

---

[5] We leave it to the teacher whether to actually define $\mathcal{P}$ (resp., $\mathcal{NP}$) as a class of search problems or to reserve this notion for the relevant class of decision problems (and merely talk about a "correspondence" between the search and decision problem classes). Our own preference is to introduce different notations for the search problem classes (see Figure 1).

feeling (and daily experience) that some reasonable search problems are hard to solve. On the other hand, if $\mathcal{P} \neq \mathcal{NP}$ then there exist reasonable search problems (i.e., some NP-relations) that are hard to solve. This conforms with our daily experience by which some reasonable problems are easy to solve whereas others are hard to solve.

Recall that search problems refer to binary relations. For such a relation R, the corresponding search problem is given x to find y such that $(x, y) \in R$ (or assert that no such y exists). We suggest defining two classes of search problems.

- $\mathcal{PF}$ (standing for "Poly-Find") denotes the class of search problems that are solvable in polynomial-time. That is, $R \in \mathcal{PF}$ if there exists a polynomial time algorithm that given x finds y such that $(x, y) \in R$ (or assert that no such y exists).
- $\mathcal{PC}$ (standing for "Poly-Check") denotes the class of search problems that correspond to polynomially-bounded binary relations that are "checkable" in polynomial-time. That is, $R \in \mathcal{PC}$ if the following two conditions hold
  1. For some polynomial p, if $(x, y) \in R$ then $|y| \leq p(|x|)$.
  2. There exists a polynomial-time algorithm that given (x, y) determines whether or not $(x, y) \in R$.

In terms of search problems the P-vs-NP Question consists of asking whether or not $\mathcal{PC}$ is contained in $\mathcal{PF}$. The conjectured inequality $\mathcal{P} \neq \mathcal{NP}$ implies that $\mathcal{PC} \setminus \mathcal{PF} \neq \emptyset$.

Fig. 1. P-vs-NP in terms of search problems: notational suggestions.

2.2   The Decision Version: Proving Versus Verifying

We suggest starting by asserting the natural stature of decision problems (beyond their role in the study of search problems). After all, some people do care about the truth, and so determining whether a given object has some claimed property is an appealing problem. The P-vs-NP Question refers to the complexity of answering such questions for a wide and natural class of properties associated with the class $\mathcal{NP}$. The latter class refers to properties that have efficient proof systems allowing for the verification of the claim that a given object has a predetermined property (i.e., is a member of a predetermined set).

For an NP-relation R, we denote the set of instances having a solution by L_R; that is, $L_R = \{x : \exists y \, (x, y) \in R\}$. Such a set is called an NP-set, and $\mathcal{NP}$ denotes the class of all NP-sets. Intuitively, an NP-set is a set of valid statements (i.e., statements of membership of a given x in L_R) that can be efficiently verified when given adequate proofs (i.e., a corresponding NP-witness y such that $(x, y) \in R$). This leads to viewing NP-sets as proof systems.

NP-proof systems. Proof systems are defined in terms of their verification procedures. Here we focus on the natural class of efficient verification procedures, where efficiency is represented by polynomial-time computations. (We should either require that the time is polynomial in terms of the statement or confine ourselves to "short proofs" – that is, proofs of length that is bounded by a polynomial in the length of the statement.) NP-relations correspond to proof systems with efficient verification procedures. Specifically, the NP-relation R corresponds to the (proof system with a) verification procedure that checks whether or not the alleged statement-proof pair is in R. This proof system satisfies the natural completeness and soundness conditions: every true statement (i.e., $x \in L_R$) has a valid proof (i.e., an NP-witness y such that $(x, y) \in R$), whereas false statements (i.e., $x \notin L_R$) have no valid proofs (i.e., $(x, y) \notin R$ for all y's).

Recall that decision problems refer to membership in sets. We suggest defining two classes of decision problems, which indeed coincide with the standard definitions of $\mathcal{P}$ and $\mathcal{NP}$.

- $\mathcal{P}$ denotes the class of decision problems that are solvable in polynomial-time. That is, $S \in \mathcal{P}$ if there exists a polynomial time algorithm that given x determines whether or not $x \in S$.
- $\mathcal{NP}$ denotes the class of decision problems that have NP-proof systems. The latter are defined in terms of a (deterministic) polynomial-time verification algorithm. That is, $S \in \mathcal{NP}$ if there exists a polynomial p and a polynomial-time algorithm V that satisfy the following completeness and soundness conditions:
  1. Completeness: if $x \in S$ then there exists y of length at most $p(|x|)$ such that $V(x, y) = 1$.
     (Such a string y is called an NP-witness for $x \in S$.)
  2. Soundness: if $x \notin S$ then for every y it holds that $V(x, y) = 0$.
  Indeed, the point is defining $\mathcal{NP}$ as a class of sets of assertions having efficient verification procedures.

In terms of decision problems the P-vs-NP Question consists of asking whether or not $\mathcal{NP}$ is contained in $\mathcal{P}$. Since $\mathcal{P} \subseteq \mathcal{NP}$, the question is phrased as whether or not $\mathcal{NP}$ equals $\mathcal{P}$.

Fig. 2. P-vs-NP in terms of decision problems: notational suggestions.

The P versus NP question in terms of decision problems: *Is it the case that NP-proofs are useless?* That is, is it the case that for every efficiently verifiable proof system one can easily determine the validity of assertions (without being given suitable proofs)? If that were the case, then proofs would be meaningless, because they would have no fundamental advantage over directly determining the validity of the assertion. Denoting by $\mathcal{P}$ the class of sets that can be decided efficiently (i.e., by a polynomial-time algorithm), the conjecture

$\mathcal{P} \neq \mathcal{NP}$ asserts that proofs are useful: there exists NP-sets that cannot be decided by a polynomial-time algorithm, and so for these sets obtaining a proof of membership (for some instances) is useful (because we cannot efficiently determine membership by ourselves).

2.3   Equivalence of the Two Formulations

We strongly recommend proving that *the two formulations of the P-vs-NP Questions are equivalent*. That is, the search problem of every NP-relation is solvable in polynomial time if and only if membership in any NP-set can be decided in polynomial time (see Figure 3). This justifies the focus on the latter (simpler) formulation.

Referring the notations of Figures 1 and 2, we prove that $\mathcal{PC} \subseteq \mathcal{PF}$ if and only if $\mathcal{NP} = \mathcal{P}$.

- Suppose that the inclusion holds for the search version (i.e., $\mathcal{PC} \subseteq \mathcal{PF}$). Let L be an arbitrary NP-set and R_L be the corresponding witness relation. Then R_L is a NP-relation, and by the hypothesis its search problem is solvable in polynomial time (i.e., $R_L \in \mathcal{PC} \subseteq \mathcal{PF}$). This yields a polynomial-time decision procedure for L; i.e., given x try to find y such that $(x, y) \in R_L$ (and output "yes" iff such a y was found). Thus, $\mathcal{NP} = \mathcal{P}$ follows.
- Suppose that $\mathcal{NP} = \mathcal{P}$ (as classes of sets), and let R be an arbitrary NP-relation. Then the set $S_R \stackrel{\text{def}}{=} \{(x, y') : \exists y'' \text{ s.t. } (x, y'y'') \in R\}$ (where $y'y''$ denotes the concatenation of y' and y'') is in $\mathcal{NP}$ and hence in $\mathcal{P}$. This yields a polynomial-time algorithm for solving the search problem of R, by extending a prefix of a potential solution bit by bit (while using the decision procedure to determine whether or not the current prefix is valid). Thus, $\mathcal{PC} \subseteq \mathcal{PF}$ follows.

Fig. 3. A proof that $\mathcal{PC} \subseteq \mathcal{PF}$ if and only if $\mathcal{NP} = \mathcal{P}$.

We also suggest mentioning that $\mathcal{NP}$ is sometimes defined as the class of sets that can be decided by a *fictitious* device called a nondeterministic polynomial-time machine (and that this is the source of the notation NP). The reason that this class of fictitious devices is important is because it captures (indirectly) the definition of NP-proof systems. We suggest proving that indeed the definition of $\mathcal{NP}$ in terms of nondeterministic polynomial-time machine equals our definition of $\mathcal{NP}$ (in terms of the class of sets having NP-proof systems).

3   Reductions and Self-Reducibility

We assume that many students have heard of reductions, but again we fear that most of them have obtained a conceptually poor view of their nature. We believe that this is due to expositions that start with a technical definition of many-to-one (polynomial-time) reductions (i.e., Karp-reductions), rather than with a motivational discussion. Below, we take an the alternative approach, presenting first the general notion of (polynomial-time) reductions among computational problems, and viewing the notion of a Karp-reduction as an important special case that suffices (and is more convenient) in many cases.

3.1   The General Notion of a Reduction

Reductions are procedures that use "functionally specified" subroutines. That is, the functionality of the subroutine is specified, but its operation remains unspecified and its running-time is counted at unit cost. Analogously to algorithms, which are modeled by Turing machines, reductions can be modeled as *oracle* (Turing) machines. A reduction solves one computational problem (which may be either a search or decision problem) by using oracle (or subroutine) calls to another computational problem (which again may be either a search or decision problem). We focus on efficient (i.e., polynomial-time) reductions, which are often called Cook reductions.

The key property of reductions is that they translate efficient procedures for the subroutine into efficient procedures for the invoking machine. That is, if one problem is Cook-reducible to another problem and the latter is polynomial-time solvable then so is the former.

The most popular case is of reducing decision problems to decision problems, but we will also consider reducing search problems to search problems or reducing search problems to decision problems. Indeed, a good exercise is showing that the search problem of any NP-relation can be reduced to deciding membership in some NP-set (which is the actual contents of the second item of Figure 3).

A Karp-reduction is a special case of a reduction (from a decision problem to a decision problem). Specifically, for decision problems L and L', we say that L is Karp-reducible to L' if there is a reduction of L to L' *that operates as follows*: On input x (an instance for L), the reduction computes x', makes query x' to the oracle L' (i.e., invokes the subroutine for L' on input x'), and answers whatever the latter returns. This Karp-reduction is often represented by the polynomial-time computable mapping of x to x'; that is, a polynomial-time computable f is called a Karp-reduction of L to L' if for every x it holds that $x \in L$ iff $f(x) \in L'$.

Indeed, a Karp-reduction is a syntactically restricted notion of a reduction. This restricted case suffices for many cases (e.g., most importantly for the theory of NP-completeness (when developed for decision problems)), but not in case we want to reduce a search problem to a decision problem. Furthermore, whereas each decision problem is reducible to its complement, some decision problems are not Karp-reducible to their complement (e.g., the trivial decision problem).[6] Likewise, each decision problem in $\mathcal{P}$ is reducible to any computational problem by a reduction that does not use the subroutine at all, whereas such a trivial

---

[6] We call a decision problem trivial if it refers to either the empty set or the set of all strings.

reduction is disallowed by the syntax of Karp-reductions. (Nevertheless, a popular exercise of dubious nature is to show that any decision problem in $\mathcal{P}$ is Karp-reducible to any *non-trivial* decision problem.)

We comment that Karp-reductions may (and should) be augmented in order to handle reductions of search problems to search problems. Such an augmented Karp-reduction of the search problem of R to the search problem of R' operates as follows: On input x (an instance for R), the reduction computes x', makes query x' to the oracle R' (i.e., invokes the subroutine for searching R' on input x') obtaining y' such that $(x', y') \in R'$, and uses y' to compute a solution y to x (i.e., $(x, y) \in R$). Thus, such a reduction can be represented by two polynomial-time computable mappings, f and g, such that $(x, g(x, y')) \in R$ for any y' that solves $f(x)$ (i.e., y' that satisfies $(f(x), y') \in R'$). (Indeed, in general, unlike in the case of decision problems, the reduction cannot just return y' as an answer to x.)

We say that two problems are computationally equivalent if they are reducible to one another. This means that the two problems are essentially equally hard (or equally easy).

3.2   Self-Reducibility of Search Problems

We suggest introducing the notion of self-reducibility[7] for several reasons. Most importantly, it further justifies the focus on decision problems (see discussion following Proposition 1). In addition, it illustrates the general notion of a reduction, and asserts its relevance beyond the theory of NP-completeness.

The search problem of R is called self-reducible if it can be reduced to the decision problem of $L_R = \{x : \exists y \ (x, y) \in R\}$ (rather than to the set S_R as in Figure 3). Note that the decision problem of L_R is always reducible to the search problem for R (e.g., invoke the search subroutine and answer "yes" if and only if it returns some string (rather than the "no solution" symbol)).

We will see that all NP-relations that correspond to NP-complete sets are self-reducible, mostly via "natural reductions". We start with SAT, the set of satisfiable Boolean formulae (in CNF). Let R_{SAT} be the set of pairs (ϕ, τ) such that τ is a satisfying assignment to the formulae ϕ. Note that R_{SAT} is an NP-relation (i.e., it is polynomially-bounded and easy to decide (by evaluating a Boolean expression)).

Proposition 1 (R_{SAT} is self-reducible): *The search problem of R_{SAT} is reducible to SAT.*

---

[7] Our usage of this term differs from the traditional one. Traditionally, a decision problem is called self-reducible if it is Cook-reducible to itself via a reduction that on input x only makes queries that are smaller than x (according to some appropriate measure on the size of strings). Under some natural restrictions (i.e., the reduction takes the disjunction of the oracle answers) such reductions yield reductions of search to decision (as discussed in the main text).

Thus, the search problem of R_{SAT} is computationally equivalent to deciding membership in SAT. Hence, in studying the complexity of SAT, we also address the complexity of the search problem of R_{SAT}. This justifies the relevance of decision problems to search problems in a stronger sense than established in Section 2.3: The study of decision problems determines not only the complexity of the class of "NP-search" problems but rather determines the complexity of each individual search problem that is self-reducible.

Proof: Given a formula ϕ, we use a subroutine for SAT in order to find a satisfying assignment to ϕ (in case such an assignment exists). First, we query SAT on ϕ itself, and return "no solution" if the answer we get is 'false'. Otherwise, we let τ, initiated to the empty string, denote a prefix of a satisfying assignment of ϕ. We proceed in iterations, where in each iteration we extend τ by one bit. This is done as follows: First we derive a formula, denoted ϕ', by setting the first $|\tau|+1$ variables of ϕ according to the values $\tau 0$. Next we query SAT on ϕ' (which means that we ask whether or not $\tau 0$ is a prefix of a satisfying assignment of ϕ). If the answer is positive then we set $\tau \leftarrow \tau 0$ else we set $\tau \leftarrow \tau 1$ (because if τ is a prefix of a satisfying assignment of ϕ and $\tau 0$ is not a prefix of a satisfying assignment of ϕ then $\tau 1$ must be a prefix of a satisfying assignment of ϕ).

A key point is that each formula ϕ' (which contains Boolean variables as well as constants) can be simplified to contain no constants (in order to fit the canonical definition of SAT, which disallows Boolean constants). That is, after replacing some variables by constants, we should simplify clauses according to the straightforward boolean rules (e.g., a false literal can be omitted from a clause and a true literal appearing in a clause allows omitting the entire clause). ∎

Advanced comment: A reduction analogous to the one used in the proof of Proposition 1 can be presented also for other NP-search problems (and not only for NP-complete ones).[8] Consider, for example, the problem Graph 3-Colorability and prefixes of a 3-coloring of the input graph. Note, however, that in this case the process of getting rid of constants (representing partial solutions) is more involved.[9] In general, if you don't see a "natural" self-reducibility process for some NP-complete relation, you should know that a self-reduction process does exist (alas it maybe not be a natural one).

Theorem 2 *The search problem of the NP-relation of any NP-complete set is self-reducible.*

---

[8] We assume that the students have heard of NP-completeness. If this assumption does not hold for your class, then the presentation of the following material should be postponed (to Section 4.1 or to an even later stage).

[9] Details can left as an exercise to the student. You may hint that a partial 3-coloring can be hard-wired into the graph by augmenting the graph with adequate gadgets that force equality (or inequality) between the colors of two vertices (of our choice).

Proof: Let R be an NP-relation of the NP-complete set L_R. In order to reduce the search problem of R to deciding L_R, we compose the three reductions mentioned next:

1. The search problem of R is reducible to the search problem of R_{SAT} (by the NP-completeness of the latter).
2. The search problem of R_{SAT} is reducible to SAT (by Proposition 1).
3. The decision problem SAT is reducible to the decision problem L_R (by the NP-completeness of the latter).

The theorem follows.   ∎

4   NP-completeness

Some (or most) students have heard of NP-completeness before, but we suspect that many have missed important conceptual points. Specifically, we stress that the mere existence of NP-complete sets (regardless of whether this is SAT or some other set) is amazing.

4.1   Definitions

The standard definition is that a set is NP-complete if it is in $\mathcal{NP}$ and every set in $\mathcal{NP}$ is reducible to it via a Karp-reduction. Indeed, there is no reason to insist on Karp-reductions (rather than using arbitrary reductions), except that the restricted notion suffices for all positive results and is easier to work with.

We will also refer to the search version of NP-completeness. We say that a binary relation is NP-complete if it is an NP-relation and every NP-relation is reducible to it.

We stress that the mere fact that we have defined something (i.e., NP-completeness) does not mean that this thing exists (i.e., that there exist objects that satisfy the definition). *It is indeed remarkable that NP-complete problems do exist.* Such problems are "universal" in the sense that solving them allows solving any other (reasonable) problem.

4.2   The Existence of NP-complete Problems

We suggest not to confuse the mere existence of NP-complete problems, which is remarkable by itself, with the even more remarkable existence of "natural" NP-complete problems. We believe that the following proof facilitates the delivery of this message as well as focusing on the essence of NP-completeness, rather than on more complicated technical details.

Theorem 3 *There exist NP-complete relations and sets.*

Proof: The proof (as well as any other NP-completeness proof) is based on the observation that some NP-relations (resp., NP-sets) are "rich enough" to encode all NP-relations (resp., NP-sets). This is most obvious for the "generic" NP-relation, denoted R_U (and defined below), which is used to derive the simplest proof of the current theorem.

The relation R_U consists of pairs $(\langle M, x, 1^t\rangle, y)$ such that M is a description of a (deterministic) Turing machine that accepts the pair (x, y) within t steps, where $|y| \leq t$. (Instead of requiring that $|y| \leq t$, one may require that M is canonical in the sense that it reads its entire input before halting.) It is easy to see that R_U is an NP-relation, and thus $L_U \overset{\text{def}}{=} \{\overline{x} : \exists y \ (\overline{x}, y) \in R_U\}$ is an NP-set. Indeed, R_U is recognizable by a universal Turing machine, which on input $(\langle M, x, 1^t\rangle, y)$ emulates (t steps of) the computation of M on (x, y), and U indeed stands for universal (machine). (Thus, the proof extends to any reasonable model of computation, which has adequate universal machines.)

We now turn to showing that any NP-relation is reducible to R_U. As a warm-up, let us first show that any NP-set is Karp-reducible to L_U. Let R be an NP-relation, and $L_R = \{x : \exists y \ (x, y) \in R\}$ be the corresponding NP-set. Let p_R be a polynomial bounding the length of solutions in R (i.e., $|y| \leq p_R(|x|)$ for every $(x, y) \in R$), let M_R be a polynomial-time machine deciding membership (of alleged (x, y) pairs) in R, and let t_R be a polynomial bounding its running-time. Then, the Karp-reduction maps an instance x (for L) to the instance $\langle M_R, x, 1^{t_R(|x|+p_R(|y|))}\rangle$.

Note that this mapping can be computed in polynomial-time, and that $x \in L$ if and only if $\langle M_R, x, 1^{t_R(|x|+p_R(|y|))}\rangle \in L_U$.

To reduce the search problem of R to the search problem of R_U, we use essentially the same reduction. On input an instance x (for R), we make the query $\langle M_R, x, 1^{t_R(|x|+p_R(|y|))}\rangle$ to the search problem of R_U and return whatever the latter returns. Note that if $x \notin L_R$ then the answer will be "no solution", whereas for every x and y it holds that $(x, y) \in R$ if and only if $(\langle M_R, x, 1^{t_R(|x|+p_R(|y|))}\rangle, y) \in R_U$. ∎

Advanced comment. Note that the role of 1^t in the definition of R_U is to make R_U an NP-relation. In contrast, consider the relation $R_H \overset{\text{def}}{=} \{(\langle M, x\rangle, y) : M(xy) = 1\}$ (which corresponds to the halting problem). Indeed, the search problem of any relation (an in particular of any NP-relation) is Karp-reducible to the search problem of R_H, but the latter is not solvable at all (i.e., there exists no algorithm that halts on every input and on input $\overline{x} = \langle M, x\rangle$ outputs y such that $(\overline{x}, y) \in R_H$ iff such a y exists).

4.3   CSAT, SAT, and Other NP-complete Problems

Once the mere existence of NP-complete problems has been established, we suggest informing the students of the fact that many natural problems are NP-complete, and demonstrating this fact with a few examples. Indeed, SAT is a good first example, both because the reduction to it is instructive and because it is a convenient starting point to further reductions. As a second example, we suggest various variants of the Set Cover problem. Additional reductions may be deferred to homework assignments, and presenting them in class seems inadequate in the context of a course on complexity theory.

We suggest establishing the NP-completeness of SAT by a reduction from the circuit satisfaction problem (CSAT), after establishing the NP-completeness of the latter. Doing so allows decoupling two important issues in the proof of the NP-completeness of SAT: (1) the emulation of Turing machines by circuits, and (2) the encoding of circuits by formulae with auxiliary variables. Following is a rough outline, which focuses on the decision version.

CSAT. Define Boolean circuits as directed acyclic graphs with internal vertices, called gates, labeled by Boolean operations (of arity either 2 or 1), and external vertices called terminals that are associated with either inputs or outputs. When setting the inputs of such a circuit, all internal nodes are assigned values in the natural way, and this yields a value to the output(s), called an evaluation of the circuit on the given input. Define the satisfiability problem of such circuits as determining, for a given circuit, whether there exists a setting to its inputs that makes its (first) output evaluate to 1. Prove the NP-completeness of the circuit satisfaction problem (CSAT), by reducing any NP-set to it (where the set is represented by the machine that recognizes the corresponding NP-relation). The reduction boils down to encoding possible computations of a Turing machine by a corresponding layered circuit, where each layer represents an instantaneous configuration of the machine, and the relation between consecutive configurations is captured by ("uniform") local gadgets in the circuit. For further details, see Figure 4. (The proof extends to any other "reasonable" model of efficient computation.)

The above reduction is called "generic" because it (explicitly) refers to any (generic) NP-set. However, the common practice is to establish NP-completeness by a reduction from some NP-complete set (i.e., a set already shown to be NP-complete). This practice is based on the fact that if an NP-complete problem Π is reducible to some problem Π' in NP then Π' is NP-complete. The proof of this fact boils down to asserting the transitivity of reductions.

SAT. Define Boolean formulae, which may be viewed as Boolean circuits with a tree structure. Prove the NP-completeness of the formula satisfaction problem (SAT), even when the formula is given in a nice form (i.e., CNF). The proof is by a reduction from CSAT, which in turn boils down to introducing auxiliary variables in order to cut the computation of a deep circuit into a conjunction of related computations of shallow (i.e., depth-2) circuits (which may be presented as CNF formulae). The aforementioned auxiliary variables hold the possible values of the internal wires of the circuit.

3SAT. Note that the formulae resulting from the latter reduction are in conjunctive normal form (CNF) with each clause referring to three variables (i.e., two corresponding to the input wires of a node/gate and one to its output wire).

Following are some additional comments on the proof of the NP-completeness of CSAT. These comments refer to the high-level structure of the reduction, and do not provide a full (low-level) description of it.

For a machine M_R (as in the proof of Theorem 3), we will represent the computation of M_R on input (x, y), where x is the input to the reduction and y is undetermined, by a circuit C_x that takes such a string y as input. Thus, $C_x(y) = 1$ if and only if M_R accepts (x, y), and so C_x is satisfiable if and only if $x \in L_R$. The reduction maps x to a circuit C_x as follows.

The circuit C_x consists of layers such that the ith layers of wires (connecting the $i - 1^{\text{st}}$ and i^{th} layers of vertices) represents the instantaneous configuration of $M_R(x, y)$ just before the ith step. In particular, the gates of the $i + 1^{\text{st}}$ layer are designed to guaranteed that the instantaneous configuration of $M_R(x, y)$ just before the ith step is transformed to the instantaneous configuration of $M_R(x, y)$ just before the $i + 1^{\text{st}}$ step. Only the first layer of C_x depends on x itself (which is "hard-wired" into the circuit). The rest of the construction depends only on $|x|$ and M_R.

Fig. 4. Encoding computations of a Turing machine in a Boolean circuit.

Thus, the above reduction actually establishes the NP-completeness of 3SAT (i.e., SAT restricted to CNF formula with up to three variables per clause). Alternatively, reduce SAT (for CNF formula) to 3SAT (i.e., satisfiability of 3CNF formula) by replacing long clauses with conjunctions of three-variable clauses using auxiliary variables.

In order to establish the NP-completeness of the search version of the aforementioned problems we need to present a polynomial-time mapping of solutions for the target problem (e.g., SAT) to solutions for the origin problem (e.g., CSAT). Note that such a mapping is typically explicit in the argument establishing the validity of the Karp-reduction.

Set Cover and other problems. If time permits, one may want to present another class of NP-complete problems, and our choice is of Set Cover. There is a simple reduction from SAT to Set Cover (with the sets corresponding to the sets of clauses that are satisfied when assigning a specific Boolean variable a specific Boolean value). When applied to a restricted version of SAT in which each variable appears in at most three clauses, the same reduction implies the NP-completeness of a version of Set Cover in which each set contains at most three elements. (Indeed, one should first establish the NP-completeness of the aforementioned restricted version of SAT.) Using the restricted version of Set Cover one may establish the NP-completeness of Exact Cover (even when restricted to 3-element sets). The latter problem is a convenient starting point for further reductions.

4.4   NP Sets That Are Neither in P nor NP-complete

Many (to say the least) other NP-sets have been shown to be NP-complete. A very partial list includes Graph 3-Colorability, Subset Sum, (Exact) Set Cover, and the Traveling Salesman Problem. (Hundreds of other natural problems can be found in [3].) Things reach a situation in which some computer scientists seem to expect any NP-set to be either NP-complete or in $\mathcal{P}$. This naive view is wrong:

Theorem 4 *Assuming $\mathcal{NP} \neq \mathcal{P}$, there exist NP-sets that are neither NP-complete nor in $\mathcal{P}$.*

We mention that some natural problems (e.g., factoring) are conjectured to be neither solvable in polynomial-time nor NP-hard, where a problem Π is NP-hard if any NP-set is reducible to solving Π. See discussion following Theorem 5. We recommend to either state Theorem 4 without a proof or merely provide the proof idea (which is sketched next).

Proof idea. The proof is by modifying a set in $\mathcal{NP} \setminus \mathcal{P}$ such that to fail all possible reductions (to this set) and all possible polynomial-time decision procedures (for this set). Specifically, we start with some $L \in \mathcal{NP} \setminus \mathcal{P}$ and derive $L' \subset L$ (which is also in $\mathcal{NP} \setminus \mathcal{P}$) by making each reduction (say of L) to L' fail by dropping finitely many elements from L (until the reduction fails), whereas all possible polynomial-time fail to decide L' (which differ from L only on a finite number of inputs). We use the fact that any reduction (of some set in $\mathcal{NP} \setminus \mathcal{P}$) to a finite set (i.e., a finite subset of L) must fail (and this failure is due to a finite set of queries), whereas any efficient decision procedure for L (or L modified on finitely many inputs) must fail on some finite portion of all possible inputs (of L). The process of modifying L into L' proceeds in iterations, alternatively failing a potential reduction (by dropping sufficiently many strings from the rest of L) and failing a potential decision procedure (by including sufficiently many strings from the rest of L). This can be done efficiently because it is inessential to determine the optimal points of alternation (where sufficiently many strings were dropped (resp., included) to fail a potential reduction (resp., decision procedure)). Thus, L' is the intersection of L and some set in $\mathcal{P}$, which implies that $L' \in \mathcal{NP} \setminus \mathcal{P}$.

5   Three Additional Topics

The following topics are typically not mentioned in a basic course on complexity. Still, pending on time constraints, we suggest covering them at some minimal level.

5.1   The Class coNP and NP-completeness

By prepending the name of a complexity class (of decision problems) with the prefix "co" we mean the class of complement sets; that is,

$$\mathrm{co}\mathcal{C} \overset{\text{def}}{=} \{\{0,1\}^* \setminus L : L \in \mathcal{C}\}$$

Specifically, $\text{co}\mathcal{NP} = \{\{0,1\}^* \setminus L : L \in \mathcal{NP}\}$ is the class of sets that are complements of NP-sets. That is, if R is an NP-relation and $L_R = \{x : \exists y \ (x,y) \in R\}$ is the associated NP-set then $\{0,1\}^* \setminus L_R = \{x : \forall y \ (x,y) \notin R\}$ is the corresponding coNP-set.

It is widely believed that $\mathcal{NP}$ is not closed under complementation (i.e., $\mathcal{NP} \neq \text{co}\mathcal{NP}$). Indeed, this conjecture implies $\mathcal{P} \neq \mathcal{NP}$ (because $\mathcal{P}$ is closed under complementation). The conjecture $\mathcal{NP} \neq \text{co}\mathcal{NP}$ means that some coNP-sets (e.g., the complements of NP-complete sets) do not have NP-proof systems; that is, there is no NP-proof system for proving that a given formula is not satisfiable.

If indeed $\mathcal{NP} \neq \text{co}\mathcal{NP}$ then some NP-sets cannot be Karp-reducible to any coNP-set.[10] However, each NP-set is reducible to some coNP-set (by a straightforward Cook-reduction that just flips the answer), and so the non-existence of such Karp-reduction does not seem to represent anything really fundamental. In contrast, we mention that $\mathcal{NP} \neq \text{co}\mathcal{NP}$ implies that some NP-sets cannot be reduced to sets in the intersection $\mathcal{NP} \cap \text{co}\mathcal{NP}$ (even under general (i.e., Cook) reductions). Specifically,

Theorem 5 *If $\mathcal{NP} \cap \text{co}\mathcal{NP}$ contains an NP-hard set then $\mathcal{NP} = \text{co}\mathcal{NP}$.*

Recall that a set is NP-hard if every NP-set is reducible to it (possibly via a general reduction). Since $\mathcal{NP} \cap \text{co}\mathcal{NP}$ is conjectured to be a proper superset of $\mathcal{P}$, it follows (using the conjecture $\mathcal{NP} \neq \text{co}\mathcal{NP}$) that there are NP-sets that are neither in $\mathcal{P}$ nor NP-hard (specifically, the sets in $(\mathcal{NP} \cap \text{co}\mathcal{NP}) \setminus \mathcal{P}$ are neither in $\mathcal{P}$ nor NP-hard). Notable candidates are sets related to the integer factorization problem (e.g., the set of pairs (N, s) such that s has a square root modulo N that is a quadratic residue modulo N and the least significant bit of s equals 1).

Proof: Suppose that $L \in \mathcal{NP} \cap \text{co}\mathcal{NP}$ is NP-hard. Given any $L' \in \text{co}\mathcal{NP}$, we will show that $L' \in \mathcal{NP}$. We will merely use the fact that L' reduces to L (which is in $\mathcal{NP} \cap \text{co}\mathcal{NP}$). Such a reduction exists because L' is reducible $\overline{L}' \stackrel{\text{def}}{=} \{0,1\}^* \setminus L'$ (via a general reduction), whereas $\overline{L}' \in \mathcal{NP}$ and thus is reducible to L (which is NP-hard).

To show that $L' \in \mathcal{NP}$, we will present an NP-relation, R', that characterizes L' (i.e., $L' = \{x : \exists y \ (x,y) \in R'\}$). The relation R' consists of pairs of the form $(x, ((z_1, \sigma_1, w_1), ..., (z_t, \sigma_t, w_t)))$, where on input x the reduction of L' to L accepts after making the queries $z_1, ..., z_t$, obtaining the corresponding answers $\sigma_1, ..., \sigma_t$, and for every i it holds that if $\sigma_i = 1$ then w_i is an NP-witness for $z_i \in L$, whereas if $\sigma_i = 0$ then w_i is an NP-witness for $z_i \in \{0,1\}^* \setminus L$.

---

[10] Specifically, we claim that sets in $\mathcal{NP} \setminus \text{co}\mathcal{NP}$ cannot be Karp-reducible to sets in $\text{co}\mathcal{NP}$. In fact, we prove that only sets in $\text{co}\mathcal{NP}$ are Karp-reducible to sets in $\text{co}\mathcal{NP}$. Equivalently, let us prove that only sets in $\mathcal{NP}$ are Karp-reducible to sets in $\mathcal{NP}$, where the equivalence follows by noting that a reduction of L to L' is also a reduction of $\{0,1\}^* \setminus L$ to $\{0,1\}^* \setminus L'$. Indeed, suppose that L Karp-reduces to $L' \in \mathcal{NP}$. Then $L \in \mathcal{NP}$ by virtue of the NP-relation $\{(x,y) : (f(x),y) \in R'\}$, where R' is the witness relation of L'.

We stress that we use the fact that both L and $\overline{L} \overset{\text{def}}{=} \{0,1\}^* \setminus L$ are NP-sets, and refer to the corresponding NP-witnesses. Note that R' is indeed an NP-relation: The length of solutions is bounded by the running-time of the reduction (and the corresponding NP-witnesses). Membership in R' is decided by checking that the sequence of (z_i, σ_i)'s matches a possible query-answer sequence in an accepting execution of the reduction[11] (ignoring the correctness of the answers), and that all answers (i.e., σ_i's) are correct. The latter condition is easily verified by use of the corresponding NP-witnesses. ∎

5.2   Optimal Search Algorithms for NP-relations

The title of this section sounds very promising, but our guess is that the students will be less excited once they see the proof. We claim the existence of an *optimal search algorithm for any NP-relation*. Furthermore, we will explicitly present such an algorithm, and prove that it is optimal (without knowing its running time).

Theorem 6 *For every NP-relation R there exists an algorithm A that satisfies the following:*

1. *A correctly solves the search problem of R.*
2. *There exists a polynomial p such that for every algorithm A' that correctly solves the search problem of R and for every $x \in L_R = \{z : \exists y \, (z, y) \in R\}$ it holds that $t_A(x) = O(t_{A'}(x) + p(|x|))$, where t_A (resp., $t_{A'}$) denotes the number of steps taken by A (resp., A') on input x.*

We stress that the hidden constant in the O-notation depends only on A', but in the following proof the dependence is exponential in the length of the description of algorithm A' (and it is not known whether a better dependence can be achieved). Optimality holds in a "point-wise" manner (i.e., for every input), and the additive polynomial term (i.e., $p(|x|)$) is insignificant in case the NP-problem is not solvable in polynomial-time. On the other hand, the optimality of algorithm A refers only to inputs that have a solution (i.e., $x \in L_R$). Interestingly, we establish the optimality of A without knowing what its (optimal) running-time is. Furthermore, *the P-sv-NP Question boils down to determining the running time of a single explicitly presented algorithm* (i.e., the optimal algorithm A). Finally, we note that the theorem as stated refers only to models of computation that have machines that can emulate a given number of steps of other machines with a constant overhead. We mention that in most natural models the overhead of such emulation is at most poly-logarithmic in the number of steps, in which case it holds that $t_A(x) = \widetilde{O}(t_{A'}(x) + p(|x|))$.

Proof sketch: Fixing R, we let M be a polynomial-time algorithm that decides membership in R, and let p be a polynomial bounding the running-time of M. We

---

[11] That is, we need to verify that on input x, after obtaining the answers $\sigma_1, ..., \sigma_{i-1}$ to the first $i - 1$ queries, the ith query made by the reduction equals z_i.

present the following algorithm A that merely runs all possible search algorithms "in parallel" and checks the results provided by each of them (using M), halting whenever it obtains a correct solution.

Since there are infinitely many possible algorithms, we should clarify what we mean by "running them all in parallel". What we mean is to run them at different rates such that the infinite sum of rates converges to 1 (or any other constant). Specifically, we will run the ith possible algorithm at rate $1/(i+1)^2$. Note that a straightforward implementation of this idea may create a significant overhead, involved in switching frequently from the computation of one machine to another. Instead we present an alternative implementation that proceeds in iterations. In the jth iteration, for $i = 1, ..., 2^{j/2}$, we emulate $2^j/(i+1)^2$ steps of the ith machine. Each of these emulations is conducted in one chunk, and thus the overhead of switching between the various emulations is insignificant (in comparison to the total number of steps being emulated). We stress that in case some of these emulations halts with output y, algorithm A invokes M on input (x, y) and output y if and only if $M(x, y) = 1$. Furthermore, the verification of a solution provided by a candidate algorithm is also emulated at the expense of its step-count. (Put in other words, we augment each algorithm with a canonical procedure (i.e., M) that checks the validity of the solution offered by the algorithm.)

In order to guarantee that A also halts on $x \notin L_R$, we let it run an exhaustive search for a solution, in parallel to all searches, and halt with output $\perp$ in case this exhaustive search fails.

Clearly, whenever $A(x)$ outputs y (i.e., $y \neq \perp$) it must hold that $(x, y) \in R$. To show the optimality of A, we consider an arbitrary algorithm A' that solves the search problem of R. Our aim is to show that A is not much slower than A'. Intuitively, this is the case because the overhead of A results from emulating other algorithms (in addition to A'), but the total number of emulation steps wasted (due to these algorithms) is inversely proportional to the rate of algorithm A', which in turn is exponentially related to the length of the description of A'. The punch-line is that since A' is fixed, the length of its description is a constant. ∎

5.3   Promise Problems

Promise problems are a natural generalization of decision problems (and search problems can be generalized in a similar manner). In fact, in many cases, promise problems provide the more natural formulation of a decision problem. Formally, promise problems refer to a three-way partition of the set of all strings into yes-instances, no-instances, and instances that violate the promise. A potential decider is only required to distinguish yes-instances from no-instances, and is allowed arbitrary behavior on inputs that violate the promise. Standard decision problems are obtained as a special case by postulating that all inputs are allowed (i.e., the promise is trivial).

In contrary to the common perception, promise problems are no offshoot for *abnormal* situations, but *are rather the norm*: Indeed, the standard and natural presentation of natural decision problems is actually in terms of promise problems, although the presentation rarely refers explicitly to the terminology of promise problems. Consider a standard entry in [3] (or any similar compendium) reading something like "given a planar graph, determine whether or not ...". A more formal statement will refer to strings that represent planar graphs. Either way, the natural formulation actually refers to a promise problem (where the promise in this case is that the input is a planar graph).

We comment that the discrepancy between the intuitive promise problem formulation and the standard formulation in terms of decision problems can be easily bridged in the case that there exists an efficient algorithm for determining membership in the "promise set" (i.e., the set of instances that satisfy the promise). In this case, the promise problem is computationally equivalent to deciding membership in the set of yes-instances. However, in case the promise set is not tractable, the terminology of promise problems is unavoidable. Examples include the notion of "unique solutions", the formulation of "gap problems" that capture various approximation tasks, and complete problems for various probabilistic complexity classes. For a recent survey on promise problems and their applications, the interested reader is referred to [5].

6   A Brief Overview of Complexity Theory

(The following text was originally written as a brief overview of complexity theory, intended for the novice. It can also be used as a basis for communicating the essence of complexity theory to the outside (i.e., to scientists in other disciplines and even to the general interested public). Thus, unlike the rest of this essay, which is intended for the teacher, this section is intended for the student (or for other "outsiders" that the teacher may wish to address). The text starts with an overview of the P-vs-NP Question and the theory of NP-completeness, repeating themes that were expressed in the previous sections. Still, in light of the different potential uses of this text, I preferred not to eliminate this part of the overview.)

Complexity Theory is concerned with the study of the *intrinsic complexity* of computational tasks. Its "final" goals include the determination of the complexity of any well-defined task. Additional "final" goals include obtaining an understanding of the relations between various computational phenomena (e.g., relating one fact regarding computational complexity to another). Indeed, we may say that the former type of goals is concerned with *absolute* answers regarding specific computational phenomena, whereas the latter type is concerned with questions regarding the *relation* between computational phenomena.

Interestingly, the current success of Complexity Theory in coping with the latter type of goals has been more significant. In fact, the failure to resolve questions of the "absolute" type, led to the flourishing of methods for coping with questions of the "relative" type. Putting aside for a moment the frustration caused by the failure, we must admit that there is something fascinating in the success: in some sense, establishing relations between phenomena is more revealing than making statements about each phenomenon. Indeed, the first example that comes to mind is the theory of NP-completeness. Let us consider this theory, for a moment, from the perspective of these two types of goals.

Complexity Theory has failed to determine the intrinsic complexity of tasks such as finding a satisfying assignment to a given (satisfiable) propositional formula or finding a 3-coloring of a given (3-colorable) graph. But it has established that these two seemingly different computational tasks are in some sense the same (or, more precisely, are computationally equivalent). We find this success amazing and exciting, and hope that the reader shares our feeling. The same feeling of wonder and excitement is generated by many of the other discoveries of Complexity Theory. Indeed, the reader is invited to join a fast tour of some of the other questions and answers that make up the field of Complexity Theory.

We will indeed start with the "P versus NP Question". Our daily experience is that it is harder to solve a problem than it is to check the correctness of a solution (e.g., think of either a puzzle or a research problem). Is this experience merely a coincidence or does it represent a fundamental fact of life (or a property of the world)? Could you imagine a world in which solving any problem is not significantly harder than checking a solution to it? Would the term "solving a problem" not lose its meaning in such a hypothetical (and impossible in our opinion) world? The denial of the plausibility of such a hypothetical world (in which "solving" is not harder than "checking") is what "P different than NP" actually means, where P represents tasks that are efficiently solvable and NP represents tasks for which solutions can be efficiently checked.

The mathematically (or theoretically) inclined reader may also consider the task of proving theorems versus the task of verifying the validity of proofs. Indeed, finding proofs is a special type of the aforementioned task of "solving a problem" (and verifying the validity of proofs is a corresponding case of checking correctness). Again, "P different than NP" means that there are theorems that are harder to prove than to be convinced of correctness when presented with a proof. This means that the notion of a proof is meaningful (i.e., that proofs do help when trying to be convinced of the correctness of assertions). Here NP represents sets of assertions that can be efficiently verified with the help of adequate proofs, and P represents sets of assertions that can be efficiently verified from scratch (i.e., without proofs).

In light of the foregoing discussion it is clear that the P-versus-NP Question is a fundamental scientific question of far-reaching consequences. The fact that this question seems beyond our current reach led to the development of the theory of NP-completeness. Loosely speaking, this theory identifies a set of computational problems that are as hard as NP. That is, the fate of the P-versus-NP Question lies with each of these problems: if any of these problems is easy to solve then so are all problems in NP. Thus, showing that a problem is NP-complete provides evidence to its intractability (assuming, of course, "P different than NP"). Indeed, demonstrating NP-completeness of computational tasks is a central tool in indicating hardness of natural computational problems, and it has been used extensively both in computer science and in other disciplines. NP-completeness indicates not only the conjectured intractability of a problem but rather also its "richness" in the sense that the problem is rich enough to "encode" any other problem in NP. The use of the term "encoding" is justified by the exact meaning of NP-completeness, which in turn is based on establishing relations between different computational problems (without referring to their "absolute" complexity).

The foregoing discussion of the P-versus-NP Question also hints to *the importance of representation*, a phenomenon that is central to complexity theory. In general, complexity theory is concerned with problems the solutions of which are implicit in the problem's statement. That is, the problem contains all necessary information, and one merely needs to process this information in order to supply the answer.[12] Thus, complexity theory is concerned with manipulation of information, and its transformation from one representation (in which the information is given) to another representation (which is the one desired). Indeed, a solution to a computational problem is merely a different representation of the information given; that is, a representation in which the answer is explicit rather than implicit. For example, the answer to the question of whether or not a given Boolean formula is satisfiable is implicit in the formula itself (but the task is to make the answer explicit). Thus, complexity theory clarifies a central issue regarding representation; that is, the distinction between what is explicit and what is implicit in a representation. Furthermore, it even suggests a quantification of the level of non-explicitness.

In general, complexity theory provides new viewpoints on various phenomena that were considered also by past thinkers. Examples include the aforementioned concepts of proofs and representation as well as concepts like randomness, knowledge, interaction, secrecy and learning. We next discuss some of these concepts and the perspective offered by complexity theory.

The concept of *randomness* has puzzled thinkers for ages. Their perspective can be described as ontological: they asked "what is randomness" and wondered whether it exist at all (or is the world deterministic). The perspective of complexity theory is behavioristic: it is based on defining objects as equivalent if they cannot be told apart by any efficient procedure. That is, a coin toss is (defined to be) "random" (even if one believes that the universe is deterministic) if it is infeasible to predict the coin's outcome. Likewise, a string (or a distribution of strings) is "random" if it is infeasible to distinguish it from the uniform distribution (regardless of whether or not one can generate the latter). Interestingly, randomness (or rather pseudorandomness) defined this way is efficiently expandable; that is, under a reasonable complexity assumption (to be discussed next), short pseudorandom strings can be deterministically expanded into long pseudorandom strings. Indeed, it turns out that randomness is intimately related to intractability. Firstly, note that the very definition of pseudorandomness refers to

---

[12] In contrast, in other disciplines, solving a problem may require gathering information that is not available in the problem's statement. This information may either be available from auxiliary (past) records or be obtained by conducting new experiments.

intractability (i.e., the infeasibility of distinguishing a pseudorandomness object from a uniformly distributed object). Secondly, as hinted above, a complexity assumption that refers to the existence of functions that are easy to evaluate but hard to invert (called *one-way functions*) imply the existence of deterministic programs (called *pseudorandom generators*) that stretch short random seeds into long pseudorandom sequences. In fact, it turns out that the existence of pseudorandom generators is equivalent to the existence of one-way functions.

Complexity Theory offers its own perspective on the concept of *knowledge* (and distinguishes it from information). It views knowledge as the result of a hard computation. Thus, whatever can be efficiently done by anyone is not considered knowledge. In particular, the result of an easy computation applied to publicly available information is not considered knowledge. In contrast, the value of a hard to compute function applied to publicly available information is knowledge, and if somebody provides you with such a value then it has provided you with knowledge. This discussion is related to the notion of *zero-knowledge* interactions, which are interactions in which no knowledge is gained. Such interactions may still be useful, because they may assert the *correctness* of specific knowledge that was provided beforehand.

The foregoing paragraph has explicitly referred to *interaction*. It has pointed one possible motivation for interaction: gaining knowledge. It turns out that interaction may help in a variety of other contexts. For example, it may be easier to verify an assertion when allowed to interact with a prover rather than when reading a proof. Put differently, interaction with some teacher may be more beneficial than reading any book. We comment that the added power of such *interactive proofs* is rooted in their being randomized (i.e., the verification procedure is randomized), because if the verifier's questions can be determined beforehand then the prover may just provide the transcript of the interaction as a traditional written proof.

Another concept related to knowledge is that of *secrecy*: knowledge is something that one party has while another party does not have (and cannot feasibly obtain by itself) – thus, in some sense knowledge is a secret. In general, complexity theory is related to *Cryptography*, where the latter is broadly defined as the study of systems that are easy to use but hard to abuse. Typically, such systems involve secrets, randomness and interaction as well as a complexity gap between the ease of proper usage and the infeasibility of causing the system to deviate from its prescribed behavior. Thus, much of Cryptography is based on complexity theoretic assumptions and its results are typically transformations of relatively simple computational primitives (e.g., one-way functions) into more complex cryptographic applications (e.g., a secure encryption scheme).

We have already mentioned the context of *learning* when referring to learning from a teacher versus learning from a book. Recall that complexity theory provides evidence to the advantage of the former. This is in the context of gaining knowledge about publicly available information. In contrast, computational learning theory is concerned with learning objects that are only partially available to the learner (i.e., learning a function based on its value at a few random locations or even at locations chosen by the learner). Complexity Theory sheds light on the intrinsic limitations of learning (in this sense).

Complexity Theory deals with a variety of computational tasks. We have already mentioned two fundamental types of tasks: *searching for solutions* (or "finding solutions") and *making decisions* (e.g., regarding the validity of assertion). We have also hinted that in some cases these two types of tasks can be related. Now we consider two additional types of tasks: *counting the number of solutions* and *generating random solutions*. Clearly, both the latter tasks are at least as hard as finding arbitrary solutions to the corresponding problem, but it turns out that for some natural problems they are not significantly harder. Specifically, under some natural conditions on the problem, approximately counting the number of solutions and generating an approximately random solution is not significantly harder than finding an arbitrary solution.

Having mentioned the notion of *approximation*, we mention that the study of the complexity of finding approximate solutions has also received a lot of attention. One type of approximation problems refers to an objective function defined on the set of potential solutions. Rather than finding a solution that attains the optimal value, the approximation task consists of finding a solution that obtains an "almost optimal" value, where the notion of "almost optimal" may be understood in different ways giving rise to different levels of approximation. Interestingly, in many cases even a very relaxed level of approximation is as difficult to achieve as the original (exact) search problem (i.e., finding an approximate solution is as hard as finding an optimal solution). Surprisingly, these hardness of approximation results are related to the study of *probabilistically checkable proofs*, which are proofs that allow for ultra-fast probabilistic verification. Amazingly, every proof can be efficiently transformed into one that allows for probabilistic verification based on probing a *constant* number of bits (in the alleged proof). Turning back to approximation problems, we note that in other cases a reasonable level of approximation is easier to achieve than solving the original (exact) search problem.

Approximation is a natural relaxation of various computational problems. Another natural relaxation is the study of *average-case complexity*, where the "average" is taken over some "simple" distributions (representing a model of the problem's instances that may occur in practice). We stress that, although it was not stated explicitly, the entire discussion so far has referred to "worst-case" analysis of algorithms. We mention that worst-case complexity is a more robust notion than average-case complexity. For starters, one avoids the controversial question of what are the instances that are "important in practice" and correspondingly the selection of the class of distributions for which average-case analysis is to be conducted. Nevertheless, a relatively robust theory of average-case complexity has been suggested, albeit it is far less developed than the theory of worst-case complexity.

In view of the central role of randomness in complexity theory (as evident, say, in the study of pseudorandomness, probabilistic proof systems, and cryptography), one may wonder as to whether the randomness needed for the various applications can be obtained in real-life. One specific question, which received a lot of attention, is the possibility of "purifying" randomness (or "extracting good randomness from bad sources"). That is, can we use "defected" sources of randomness in order to implement almost perfect sources of randomness. The answer depends, of course, on the model of such defected sources. This study turned out to be related to complexity theory, where the most tight connection is between some type of *randomness extractors* and some type of pseudorandom generators.

So far we have focused on the time complexity of computational tasks, while relying on the natural association of efficiency with time. However, time is not the only resource one should care about. Another important resource is *space*: the amount of (temporary) memory consumed by the computation. The study of space complexity has uncovered several fascinating phenomena, which seem to indicate a fundamental difference between space complexity and time complexity. For example, in the context of space complexity, verifying proofs of validity of assertions (of any specific type) has the same complexity as verifying proofs of invalidity for the same type of assertions.

In case the reader feels dizzy, it is no wonder. We took an ultra-fast air-tour of some mountain tops, and dizziness is to be expected. Needless to say, a good graduate course in complexity theory should consist of climbing some of these mountains by foot, step by step, and stopping to look around and reflect.

Absolute Results (a.k.a. Lower-Bounds). As stated up-front, absolute results are not known for many of the "big questions" of complexity theory (most notably the P-versus-NP Question). However, several highly non-trivial absolute results have been proved. For example, it was shown that using negation can speed-up the computation of monotone functions (which do not require negation for their mere computation). In addition, many promising techniques were introduced and employed with the aim of providing a "low-level" analysis of the progress of computation. However, the focus of this overview was on the connections among various computational problems and notions, which may be viewed as a "high-level" study of computation.

Historical Notes

Many sources provide historical accounts of the developments that led to the formulation of the *P vs NP Problem* and the development of the theory of NP-completeness (see, e.g., [3]). We thus refrain from attempting to provide such an account.

One technical point that we mention is that the three "founding papers" of the theory of NP-completeness (i.e., [1, 6, 8]) refer to the three different terms of reductions used above. Specifically, Cook used the general notion of polynomial-time reduction [1], often referred to as Cook-reductions. The notion of Karp-reductions originates from Karp's paper [6], whereas its augmentation to search problems originates from Levin's paper [8]. It is worth noting that unlike Cook and Karp's works, which treat decision problems, Levin's work is stated in terms of search problems.

The existence of NP-sets that are neither in P nor NP-complete (i.e., Theorem 4) was proven by Ladner [7], Theorem 5 was proven by Selman [9], and the existence of optimal search algorithms for NP-relations (i.e., Theorem 6) was proven by Levin [8]. (Interestingly, the latter result was proved in the same paper in which Levin presented the discovery of NP-completeness, independently of Cook and Karp.) Promise problems were explicitly introduced by Even, Selman and Yacobi [2].

Acknowledgments

I am grateful to Arny Rosenberg, Alan Selman, Salil Vadhan, and an anonymous referee for their useful comments and suggestions.

References

1. S.A. Cook. The Complexity of Theorem Proving Procedures. In *3rd STOC*, pages 151–158, 1971.
2. S. Even, A.L. Selman, and Y. Yacobi. The Complexity of Promise Problems with Applications to Public-Key Cryptography. *Inform. and Control*, Vol. 61, pages 159–173, 1984.
3. M.R. Garey and D.S. Johnson. *Computers and Intractability: A Guide to the Theory of NP-Completeness*. W.H. Freeman and Company, New York, 1979.
4. O. Goldreich. *Introduction to Complexity Theory – notes for a one-semester course*. Weizmann Institute of Science, Spring 2002. Available from http://www.wisdom.weizmann.ac.il/~oded/cc.html
5. O. Goldreich. On Promise Problems: A Survey. This volume.
6. R.M. Karp. Reducibility among Combinatorial Problems. In *Complexity of Computer Computations*, R.E. Miller and J.W. Thatcher (eds.), Plenum Press, pages 85–103, 1972.
7. R.E. Ladner. On the Structure of Polynomial Time Reducibility. *Jour. of the ACM*, 22, 1975, pages 155–171.
8. L.A. Levin. Universal Search Problems. *Problemy Peredaci Informacii 9*, pages 115–116, 1973. Translated in *problems of Information Transmission 9*, pages 265–266.
9. A. Selman. On the structure of NP. *Notices Amer. Math. Soc.*, Vol 21 (6), page 310, 1974.

State

Arnold L. Rosenberg

Dept. of Computer Science, University of Massachusetts, Amherst, MA 01003, USA
rsnbrg@cs.umass.edu

Abstract. The notion of state is fundamental to the design and analysis of virtually all computational systems. The Myhill-Nerode Theorem of Finite Automata theory—and the concepts underlying the Theorem— is a source of sophisticated fundamental insights about a large class of state-based systems, both finite-state and infinite-state systems. The Theorem's elegant algebraic characterization of the notion of state often allows one to analyze the behaviors and resource requirements of such systems. This paper reviews the Theorem and illustrates its application to a variety of formal computational systems and problems, ranging from the design of circuits, to the analysis of data structures, to the study of state-based formalisms for machine-learning systems. It is hoped that this survey will awaken many to, and remind others of, the importance of the Theorem and its fundamental insights.

A dedication. I decided to contribute this piece to this volume because Shimon Even is largely—albeit indirectly—responsible for the piece. I learned about Finite Automata Theory and the Myhill-Nerode Theorem in a course taught by Shimon at Harvard during his last year of graduate school and my first. I further learned from associating with Shimon, during a friendship of more than 42 years, a commitment to effective teaching and the importance of defending strongly held positions, even when they run counter to prevailing trends.

1   Introduction

A paean to the Myhill-Nerode Theorem. The notion of state is fundamental to the design and analysis of virtually all computational systems, from the sequential circuits that underlie sophisticated hardware, to the semantic models that enable optimizing compilers, to leading-edge machine-learning concepts, to the models used in discrete-event simulation. Decades of experience with state-based systems have taught that all but the simplest display a level of complexity that makes them hard—conceptually and/or computationally—to design and analyze. One brilliant candle in this gloomy scenario is the Myhill-Nerode Theorem, which supplies a *rigorous, mathematical,* analogue of the following informal characterization of the notion "state."

> The state of a system comprises that fragment of its history that allows it to behave correctly in the future.

O. Goldreich et al. (Eds.): Shimon Even Festschrift, LNCS 3895, pp. 375–398, 2006.
© Springer-Verlag Berlin Heidelberg 2006

Superficially, it may appear that this definition of "state" is of no greater *operational* significance than is the foundational identification of the number *eight* with the infinitude of sets that contain eight elements. This appearance is illusory. The Myhill-Nerode Theorem turns out to be a conceptual and technical powerhouse when analyzing a surprising range of problems concerning the state-transition systems that occur in so many guises within the field of computation. Indeed, although the Theorem resides most naturally within the theory of Finite Automata—it first appeared in [13]; an earlier, weaker version appeared in [12]; the most accessible presentation appeared in [15]—it has manifold lessons for the analysis of many problems associated with any state-transition system, even those having infinitely many states.

It is my goal to back up the preceding praise for the Myhill-Nerode Theorem by reviewing both the Theorem and a sampler of its applications. In subsequent sections, I review the work of several researchers from the 1960's, whose work on a variety of problems relating to state-transition systems can be viewed as applying the fundamental insights that underlie the Theorem. While the Theorem originated as a cornerstone of the theory of Finite Automata,[1] several of the systems we consider here are quite removed from the standard Finite Automaton model.

A pedagogical ax to grind. Permit me now to step away from technical matters to pedagogical ones. I argue here via case studies that the Myhill-Nerode Theorem, in the insights that it supplies and the formal settings that it suggests, is one of the real gems of the foundational branch of theoretical computer science. To the extent that this evaluation is accurate, it is regrettable that the Theorem, and its algebraic message and insights, have disappeared from virtually all modern introductory texts on "computation theory," despite that fact that all of these begin with a section on finite automata. For illustration, as I was examining texts for my introductory course in this area, I perused [3, 4, 8, 9, 11, 18] and found the Theorem only in the first edition of [4]; its second edition, [3], no longer presents it! While it is not my intention to speculate at length on why the Theorem has been systematically excluded from the aforementioned texts, I suspect that it is due to a narrowing of attitudes over the years/decades about what constitutes the *foundational* branch of theoretical computer science. Whereas earlier attitudes identified "computation theory" with all approaches to a mathematical foundation—as defined in texts by some compendium of loosely related material from the theories of automata, formal languages, computability, and complexity—modern attitudes seem to posit the overriding importance of complexity theory (even while texts continue to include a smattering of material from the three other theories). Thus, understanding the essential nature of computation, as manifest in the resources required to compute various functions, has largely displaced (in the introductory course, at least) the attempt to develop mathematical tools for understanding the structures that underlie the hardware

---

[1] In my opinion, only the Kleene-Myhill Theorem, which establishes the equivalence between Finite Automata and "Regular Expressions," rivals the Myhill-Nerode Theorem for importance in the theory of Finite Automata.

and software systems that we build and use. I believe that this trend is unsound, both technically and pedagogically. We present embryonic computer scientists with abstract models that we do little to motivate, and we largely deprive them of exposure to foundational material that is likely to be at least as meaningful to them in their professional lives as much of the esoterica that they are exposed to in what for many is their one and only course on "computation theory."

It is incumbent on me to justify my claims about the Myhill-Nerode Theorem—and, thereby, the more general claims I have just made. I do this by presenting the Theorem and a sampler of its applications. I acknowledge freely that my choice of material—as, perhaps, my position—is personal and eccentric. That said, I hope that the reader will at least find this essay provocative.

A roadmap. We begin by introducing, in Section 2, a very general, unstructured, model of state-transition system, that we call the *Online Automaton*. This model is intended to capture those aspects of a state-transition system that are captured by the Myhill-Nerode Theorem and its underlying concepts. We next turn in Section 3 to Finite Automata, and we develop the Myhill-Nerode Theorem (and its proof) in this, its "natural domain." Section 4 presents two applications of the Theorem to finite-state systems. In Section 4.1, we describe how to use the Theorem to prove that a language is not *regular*—i.e., is not acceptable by a Finite Automaton. We further argue in Section 4.1.2 that the proofs of nonregularity that emerge from the technique proposed in all modern texts—which use the so-called Pumping Lemma for regular languages—are never shorter and are seldom as perspicuous as the proofs advocated in Section 4.1. We invoke Occam's Razor[2] to argue for the reinstatement of the Myhill-Nerode Theorem as *the* fundamental technique for proofs of nonregularity. In Section 4.2, we describe how the Theorem supplies the foundation for the fundamental operation of "minimizing" a Finite Automaton, by coalescing states that are "equivalent" with respect to the language that the automaton accepts. Importantly for applications of the theory, such state minimization is purely algorithmic and requires no understanding of what the automaton does. We next leave the "natural domain" of the Theorem and describe three of its conceptual applications to state-transition systems that are not Finite Automata. Our first "indirect" application, in Section 5.1, describes a result from [6] that, informally, applies the Theorem to Online Automata that accept nonregular languages. This result quantitatively sharpens the Theorem's characterization of nonregular languages as those having infinite "memory requirements," by supplying a lower bound on these "requirements." The next study we review, in Section 5.2, lends structure to the infinitely many states of an Online Automaton, by specifying the organization of the memory that the states control. We arrive, thereby, at the notion of a *multi-tape multi-dimensional online Turing Machine*. Now, such models are not in vogue today, largely because they do not faithfully model the structure of digital computers and their peripherals. However, if one views such "machines" as stylized programs that manipulate multiple data structures—a linear "tape" is a linear list, a two-dimensional "tape" is an orthogonal list, etc.—then

---

[2] *"Entia non sunt multiplicanda praeter necessitatem"* (William of Occam, 14th cent.)

one can use such a model to advantage to prove nontrivial facts about data structures. The result that we adapt from [2] exposes the impact of memory structure on computational efficiency (specifically, time complexity), within the context of a simple data-retrieval problem. Our final "indirect" application of the Theorem, in Section 6, has implications for some of the voluminous work on probabilistic state-transition systems, such as are quite popular within the artificial-intelligence community. We present one of the most striking results from [14]: Even if Finite Automata are modified to make their state transitions probabilistic, the resulting model still accepts only regular sets when the probability that an input is accepted is always bounded away from the threshold required for acceptance. I close in Section 7 with a closing polemic advocating reinstituting the Myhill-Nerode Theorem within our theoretical computer science curricula.

The thread that connects all of the work we survey is the Myhill-Nerode Theorem and its underlying concepts. We hope that we have done justice to this work and that, after reading this piece, the reader will understand—and, hopefully, sympathize with—our claim that the Myhill-Nerode Theorem is a treasure that should be passed on to subsequent generations.

2   Online Automata and Their Languages

Languages. Let Σ be a finite set of (atomic) symbols (or, an **alphabet**). We denote by $\Sigma^\star$ the set of all finite-length strings of elements of Σ—including the *null string* ε, which is the unique string of length 0. A **word over** Σ is any element of $\Sigma^\star$; a **language over** Σ is any subset $L \subseteq \Sigma^\star$.

Equivalence relations on $\Sigma^\star$, specifically, *right-invariant* ones, cast a broad shadow in the theory, hence, in our survey.

An equivalence relation $\equiv$ on $\Sigma^\star$ is **right-invariant** if, for all $z \in \Sigma^\star$, $xz \equiv yz$ whenever $x \equiv y$.

Our particular focus will be on the following specific (right-invariant) equivalence relation on $\Sigma^\star$, which is defined in terms of a given language $L \subseteq \Sigma^\star$.

$$\text{For all } x, y \in \Sigma^\star : \quad [x \equiv_L y] \text{ iff } (\forall z \in \Sigma^\star)[[xz \in L] \Leftrightarrow [yz \in L]]. \quad (1)$$

The following important result is a simple exercise.

Lemma 1 *For all alphabets Σ and all languages $L \subseteq \Sigma^\star$, the equivalence relation $\equiv_L$ is right-invariant.*

Automata. An **online automaton** M is specified as follows:
$M = (Q, \Sigma, \delta, q_0, F)$, where

- Q is a (finite or infinite) set of *states*;
- Σ is a finite *alphabet*;
- δ is the *state-transition function*: $\delta : Q \times \Sigma \longrightarrow Q$;
- q_0 is M's *initial state*; it is the state M is in when you first "switch it on;"

- $F \subseteq Q$ is the set of *final* (or, *accepting*) states; these are the states that specify M's "response to" each input string $x \in \Sigma^\star$.

In order to make the OA model *dynamic* (so that it can "accept" a language), we need to talk about how an OA M responds to strings, not just to single symbols. We therefore *extend* the state-transition function δ to operate on $Q \times \Sigma^\star$, rather than just on $Q \times \Sigma$. It is crucial that our extension truly *extend* δ, i.e., that it agree with δ on strings of length 1 (which can, of course, be viewed as symbols). We call our extended function $\widehat{\delta}$ and define it via the following induction. For all $q \in Q$:

$$\widehat{\delta}(q, \varepsilon) = q$$
$$(\forall \sigma \in \Sigma, \ \forall x \in \Sigma^\star) \ \widehat{\delta}(q, \sigma x) = \widehat{\delta}(\delta(q, \sigma), x).$$

The first equation asserts that M responds only to the stimuli embodied by non-null strings. In the second equation, the unadorned "δ" highlights the fact that $\widehat{\delta}$ is an *extension* of δ.

We can finally define the **language accepted** (or, *recognized*) by M (sometimes called the "behavior" of M):

$$L(M) \overset{\text{def}}{=} \{x \in \Sigma^\star \mid \widehat{\delta}(q_0, x) \in F\}.$$

Since it can cause no confusion to "overload" the semantics of δ, we stop embellishing the extended δ with a hat and just write $\delta : Q \times \Sigma^\star \longrightarrow Q$.

In analogy with the equivalence relation $\equiv_L$ of Eq. 1, which is associated with a language L, we associate with each OA M the following equivalence relation on $\Sigma^\star$.

$$\text{For all } x, y \in \Sigma^\star : \quad [x \equiv_M y] \text{ iff } [\delta(q_0, x) = \delta(q_0, y)]. \tag{2}$$

The following is an immediate consequence of how we extended the state-transition function δ to $Q \times \Sigma^\star$, in particular, the fact that $\delta(q_0, xz) = \delta(\delta(q_0, x), z)$.

Lemma 2 *For each OA* $M = (Q, \Sigma, \delta, q_0, F)$:
  (a) *the equivalence relation* $\equiv_M$ *is right-invariant;*
  (b) $(\forall x, y \in \Sigma^\star) \ [x \equiv_M y] \ \text{iff} \ [x \equiv_{L(M)} y].$

3 Finite Automata and the Myhill-Nerode Theorem

A **finite automaton** (**FA**, for short) is an OA, $M = (Q, \Sigma, \delta, q_0, F)$, whose state-set Q is finite. A language L is **regular** iff there is an FA M such that $L = L(M)$.

We now prepare for our presentation of the Myhill-Nerode Theorem, which supplies a rigorous mathematical correspondent of the notion of "state." We begin with some basic definitions, facts, and notation. Let $\equiv$ be any equivalence relation on $\Sigma^\star$.

- For each $x \in \Sigma^*$, the $\equiv$-*class* that x belongs to is $[x]_\equiv \overset{\text{def}}{=} \{y \in \Sigma^* \mid x \equiv y\}$. (When the subject relation $\equiv$ is clear from context, we simplify notation by writing $[x]$ for $[x]_\equiv$.)
- The *classes* of $\equiv$ *partition* Σ^*.
- The *index* of $\equiv$ is the number of *classes* that it partitions Σ^* into.

Theorem 3 ([12, 13, 15]). (The Myhill-Nerode Theorem)
The following statements about a language $L \subseteq \Sigma^*$ are equivalent.

1. *L is regular.*
2. *L is the union of some of the equivalence classes of a right-invariant equivalence relation over Σ^* of finite index.*
3. *The right-invariant equivalence relation, $\equiv_L$ of Eq. 1 has finite index.*

Note. *The earliest version of the Theorem, in [12], uses congruences—i.e., equivalence relations that are both right- and left-invariant.*

Proof. We prove the (logical) equivalence of the Theorem's three statements by verifying the three cyclic implications: statement 1 implies statement 2, which implies statement 3, which implies statement 1.

(1) $\Rightarrow$ (2). Say that the language L is regular. There is, then, a FA $M = (Q, \Sigma, \delta, q_0, F)$ such that $L = L(M)$. Then the right-invariant equivalence relation $\equiv_M$ of Eq. 2 clearly has index no greater than $|Q|$. Moreover, L is the union of some of the classes of relation $\equiv_M$:

$$L = \{x \in \Sigma^* \mid \delta(q_0, x) \in F\} = \bigcup_{f \in F} \{x \in \Sigma^* \mid \delta(q_0, x) = f\}.$$

(2) $\Rightarrow$ (3). We claim that if L is "defined" via some (any) finite-index right-invariant equivalence relation, $\equiv$, on Σ^*, in the sense of statement 2, then the specific right-invariant equivalence relation $\equiv_L$ has finite index. We verify the claim by showing that the relation $\equiv$ must *refine* relation $\equiv_L$, in the sense that every equivalence class of $\equiv$ is totally contained in some equivalence class of $\equiv_L$. To see this, consider any strings $x, y \in \Sigma^*$ such that $x \equiv y$. By right invariance, then, for all $z \in \Sigma^*$, we have $xz \equiv yz$. Since L is, by assumption, the union of entire classes of relation $\equiv$, we must have

$$[xz \in L] \quad \text{if, and only if,} \quad [yz \in L].$$

We thus have

$$[x \equiv y] \Rightarrow [x \equiv_L y].$$

Since relation $\equiv$ has only finitely many classes, and since each class of relation $\equiv$ is a subset of some class of relation $\equiv_L$, it follows that relation $\equiv_L$ has finite index.

(3) $\Rightarrow$ (1). Say that L is the union of some of the classes of the finite-index right-invariant equivalence relation $\equiv_L$ on Σ^*. Let the distinct classes of $\equiv_L$ be $[x_1], [x_2], \ldots, [x_n]$, for some n strings $x_i \in \Sigma^*$. (Note that, because of the transitivity of relation $\equiv_L$, we can identify a class uniquely via any one of its constituent strings. This works, of course, for any equivalence relation.) We claim that these classes form the states of an FA $M = (Q, \Sigma, \delta, q_0, F)$ that accepts L. To wit:

1. $Q = \{[x_1], [x_2], \ldots, [x_n]\}$.
   This set is finite because $\equiv_L$ has finite index.
2. For all $x \in \Sigma^*$ and all $\sigma \in \Sigma$, define $\delta([x], \sigma) = [x\sigma]$.
   The right-invariance of relation $\equiv_L$ guarantees that δ is a well-defined function.
3. $q_0 = [\varepsilon]$.
   M's start state corresponds to its having read nothing.
4. $F = \{[x] \mid x \in L\}$

One verifies by an easy induction that M is a well-defined FA that accepts L.

4 Applying Myhill-Nerode Concepts to FA's

4.1 Proving that Languages Are Nonregular

FA's are very limited in their computing power due to the finiteness of their memories, i.e., of their sets of states. Indeed, the standard way to expose the limitations of FA's—by proving that a language L is not regular—is to establish somehow that the structure of L requires distinguishing among infinitely many mutually distinct situations.

4.1.1. The Continuation Lemma and Fooling Sets. Given the conceptual parsimony and power of Theorem 3, it is not surprising that the Theorem affords one a simple, yet powerful tool for proving that a language is not regular. This tool is encapsulated in the following corollary, which follows immediately from the equivalence of statements (1) and (3) in the Theorem. For reasons that we hope will become suggestive imminently, we refer to the corollary as "The Continuation Lemma." We maintain that the ensuing development should be viewed as *the primary tool* for proving that a language is not regular.

Lemma 4 (The Continuation Lemma)
Let $L \subseteq \Sigma^*$ be an infinite regular language. Every sufficiently large set of words over Σ contains at least two words x, y such that $x \equiv_L y$.

The Continuation Lemma has a natural interpretation in terms of FA's, namely, that an FA M has no "memory of the past" other than its current state. Specifically, if strings x and y lead M to the same state (from its initial state)—i.e., if $x \equiv_M y$—then no continuation/extension of the input string will ever allow M to determine which of x and y it actually read.

One applies the Continuation Lemma to the problem of showing that an infinite[3] language $L \subseteq \Sigma^\star$ is not regular by constructing a *fooling set for L*, i.e., an infinite set of words no two of which are equivalent with respect to L. In other words, an infinite set $S \subseteq \Sigma^\star$ is a fooling set for L if for every pair of words $x, y \in S$, there exists a word $z \in \Sigma^\star$ such that precisely one of xz and yz belongs to L.

> This technique has a natural interpretation in terms of FA's. Since any FA M has only finitely many states, any infinite set of words must (by the pigeonhole principle) always contain two, x and y, that are indistinguishable to M, in the sense that $x \equiv_M y$ (so that $x \equiv_{L(M)} y$; cf. Lemma 2). By the FA version of the Continuation Lemma, no continuation z can ever cause M to distinguish between x and y.

We now consider a few sample proofs of the nonregularity of languages, which suggest how direct and simple such proofs can be when they are based on the Continuation Lemma and fooling sets.

Application 1. *The language*[4] $L_1 = \{a^n b^n \mid n \in \mathbb{N}\} \subset \{a, b\}^\star$ *is not regular.*

We claim that the set $S_1 = \{a^k \mid k \in \mathbb{N}\}$ is a fooling set for L_1. To see this, note that, for any distinct words $a^i, a^j \in S_1$, we have $a^i b^i \in L_1$ while $a^j b^i \notin L_1$; hence, $a^i \not\equiv_{L_1} a^j$. By Lemma 4, L_1 is not regular.   □

Application 2. *The language* $L_2 = \{a^k \mid k$ *is a perfect square*$\}$ *is not regular.*

This application requires a bit of subtlety. We claim that L_2 is a fooling set for itself! To see this, consider any distinct words $a^{i^2}, a^{j^2} \in L_2$, where $j > i$. On the one hand, $a^{i^2} a^{2i+1} = a^{i^2+2i+1} = a^{(i+1)^2} \in L_2$; on the other hand, $a^{j^2} a^{2i+1} = a^{j^2+2i+1} \notin L_2$, because $j^2 < j^2 + 2i + 1 < (j+1)^2$; hence, $a^{i^2} \not\equiv_{L_2} a^{j^2}$. By Lemma 4, L_2 is not regular.   □

Applications 3 and 4. *The language*[5]

$$L_3 = \{x \in \{0,1\}^\star \mid x \text{ reads the same forwards and backwards;}$$
$$\text{symbolically, } x = x^R\}$$

(whose words are often called "palindromes"), and the language

$$L_4 = \{x \in \{0,1\}^\star \mid (\exists y \in \{0,1\}^\star)[x = yy]\}$$

(whose words are often called "squares"), are not regular.

We claim that the set $S_3 = \{10^k 1 \mid k \in \mathbb{N}\}$ is a fooling set for both L_3 and L_4. To see this, consider any pair of distinct words, $10^i 1$ and $10^j 1$, from S_3. On the one hand, $10^i 110^i 1 \in L_3 \cap L_4$; on the other hand, $10^j 110^i 1 \notin L_3 \cup L_4$; hence, $10^i 1 \not\equiv_{L_3} 10^j 1$, and $10^i 1 \not\equiv_{L_4} 10^j 1$. By Lemma 4, neither L_3 nor L_4 is regular.   □

---

[3] Easily, every finite language is regular; cf. [4].

[4] $\mathbb{N}$ denotes the positive integers. a^n denotes a string of n occurrences of string (or symbol) a.

[5] x^R denotes string x written backwards; e.g., $(\sigma_1 \sigma_2 \cdots \sigma_{n-1} \sigma_n)^R = \sigma_n \sigma_{n-1} \cdots \sigma_2 \sigma_1$.

4.1.2. The Pumping Lemma for Regular Languages. Inexplicably to me, most texts shun the proof strategy of Section 4.1.1, in favor of the more cumbersome—or, at least, never less cumbersome—use of the so-called Pumping Lemma for Regular Languages.

The phenomenon of "pumping" that underlies the Pumping Lemma is a characteristic of any finite closed system. Consider, for instance, any finite semigroup[6], $S = \{\alpha_1, \alpha_2, \ldots, \alpha_n\}$. Since there are only finitely many *distinct* products in any sequence of the form $\alpha_{i_1}, \alpha_{i_1}\alpha_{i_2}, \alpha_{i_1}\alpha_{i_2}\alpha_{i_3}, \ldots$, where each $\alpha_{i_j} \in S$, there must exist two products in the sequence, say $\alpha_{i_1}\alpha_{i_2}\cdots\alpha_{i_k}$ and $\alpha_{i_1}\alpha_{i_2}\cdots\alpha_{i_k}\alpha_{i_{k+1}}\cdots\alpha_{i_{k+\ell}}$ such that

$$\alpha_{i_1}\alpha_{i_2}\cdots\alpha_{i_k} = \alpha_{i_1}\alpha_{i_2}\cdots\alpha_{i_k}\alpha_{i_{k+1}}\cdots\alpha_{i_{k+\ell}}$$

within the semigroup. By associativity, then, for all $h \in \mathsf{N}$,

$$\alpha_{i_1}\alpha_{i_2}\cdots\alpha_{i_k} = \alpha_{i_1}\alpha_{i_2}\cdots\alpha_{i_k}\left(\alpha_{i_{k+1}}\cdots\alpha_{i_{k+\ell}}\right)^h,$$

where the power notation implies iterated multiplication within the semigroup.

Within the context of FA's, the phenomenon of "pumping" manifests itself as follows. Focus on an arbitrary FA $M = (Q, \Sigma, \delta, q_0, F)$. Any word $w \in \Sigma^\star$ of length[7] $\ell(w) \geq |Q|$ can be parsed into the form $w = xy$, where $y \neq \varepsilon$,[8] in such a way that $\delta(q_0, x) = \delta(q_0, xy)$. Since M is deterministic—i.e., since δ is a function—for all $h \in \mathsf{N}$,

$$\delta(q_0, x) = \delta(q_0, xy^h), \tag{3}$$

where, as earlier, the power notation implies iterated concatenation. Since the "pumping" depicted in Eq. 3 occurs also with words $w \in \Sigma^\star$ that admit a continuation $z \in \Sigma^\star$ that places them in $L(M)$—i.e., $wz \in L(M)$—we arrive finally at the Pumping Lemma. (Note the implicit invocation of Lemma 4 in our argument.)

Lemma 5 (The Pumping Lemma for Regular Languages)
For every infinite regular language L, there exists an integer $n \in \mathsf{N}$ such that: Every word $w \in L$ of length $\ell(w) \geq n$ can be parsed into the form $w = xyz$, where $\ell(xy) \leq n$ and $\ell(y) > 0$, in such a way that, for all $h \in \mathsf{N}$, $xy^hz \in L$.

The reader should easily see how to use Lemma 5 to prove that sets are not regular. The technique differs from our fooling set/Continuation Lemma technique mainly in the new (and nonintrinsic!) requirement that one of the "fooling" words must be a prefix of the other. I view this extraneous restriction as a sufficient argument not to use Lemma 5 for proofs of nonregularity. However, a common way of using the Lemma actually mandates looking for undesired "pumping" activity, rather than just for a pair of "fooling" words. For instance, a common pumping-based proof of the nonregularity of the language L_1 of Application 1 notes that the "pumped" word y of Lemma 5:

---

[6] A *semigroup* is a set of elements that are closed under an associative binary multiplication (denoted here by juxtaposition).

[7] $\ell(w)$ denotes the *length* of the string w.

[8] Of course, we could have $x = \varepsilon$.

1. cannot consist solely of a's, or else the block of a's becomes longer than the block of b's;
2. cannot consist solely of b's, or else the block of b's becomes longer than the block of a's;
3. cannot contain both an a and a b, or else the pumped word no longer has the form "a block of a's followed by a block of b's."

Even when one judiciously avoids this three-case argument by invoking the Lemma's length limit on the prefix xy, one is inviting/risking excessive complication by seeking a string that pumps. For instance, when proving the non-regularity of the language L_3 of palindromes, one must cope with the fact that any palindrome *does* pump about its center. (That is, for any palindrome w and any integer ℓ, if one parses w into $w = xyz$, where x and z both have length ℓ, then, indeed, for all $h \in \mathsf{N}$, the word $xy^h z$ is a palindrome.) Note that we are not suggesting that any of the problems we raise is insuperable, only that they unnecessarily complicate the proof process, hence violate Occam's Razor. The danger inherent in using Lemma 5 to prove that a language is not regular is mentioned explicitly in [9]:

> The pumping lemma is difficult for several reasons. Its statement is complicated, and it is easy to go astray in applying it.

We show now that the condition for a language to be regular that is provided in Lemma 5 is *necessary* but not sufficient. This contrasts with the *necessary and sufficient* condition provided by Theorem 3.

Lemma 6 ([20]) *Every string of length > 4 in the nonregular language*

$$L_5 = \{uu^R v \mid u, v \in \{0,1\}^\star;\ \ell(u), \ell(v) \geq 1\}$$

pumps in the sense of Lemma 5.

Proof. We paraphrase from [20]. Each string in L_5 consists of a nonempty even palindrome followed by another nonempty string. Say first that $w = uu^R v$ and that $\ell(w) \geq 4$. If $\ell(u) = 1$, then we can choose the first character of v as the nonnull "pumping" substring of Lemma 5. (Of course, the "pumped" strings are uninteresting in this case.) Alternatively, if $\ell(u) > 1$, then, since a^k is a palindrome for every $k > 1$, where a is the first character of u, we can let this first letter be the nonnull "pumping" substring of Lemma 5. In either case, the lemma holds.

Notably, the discussion in [20] ends with the following comment.

> For a practical test that exactly characterizes regular languages, see the Myhill-Nerode theorem.

For the record, Theorem 3 provides a simple proof that L_5 is not regular. Let x and y be distinct strings from the infinite language $L = (01)(01)^\star$, with $\ell(y) > \ell(x)$. (Strings in L consist of a sequence of one or more instances of 01.) Easily,

xx^R is an even-length palindrome, hence belongs to L_5 (with $v = \varepsilon$). However, one verifies easily that yx^R does not begin with an even-length palindrome, so that $yx^R \notin L_5$. To wit, if one could write yx^R in the form uu^Rv, then:

- u could not end with a 0, since the "center" substring 00 does not occur in yx^R;
- u could not end with a 1, since the unique occurrence of 11 in yx^R occurs to the right of the center of the string.

By Lemma 4, L_5 is not regular.   □

For completeness, we end this section with a version of Lemma 5 that supplies a condition that is both necessary and sufficient for a language to be regular. This version is rather nonperspicuous and a bit cumbersome, hence, is infrequently taught.

Lemma 7 ([5]) (The Necessary-and-Sufficient Pumping Lemma)
A language $L \subseteq \Sigma^\star$ is regular if, and only if, there exists an integer $n \in \mathbb{N}$ such that: Every word $w \in \Sigma^\star$ of length $\ell(w) \geq n$ can be parsed into the form $w = xyz$, where $\ell(y) > 0$, in such a way that, for all $z \in \Sigma^\star$:

- *if $wz \in L$, then for all $h \in \mathbb{N}$, $xy^hz \in L$;*
- *if $wz \notin L$, then for all $h \in \mathbb{N}$, $xy^hz \notin L$;*

4.2   Minimizing Finite Automata

Theorem 3 and its proof tell us two important things.

1. The notion of "state" underlying the FA model is embodied in the relations $\equiv_M$. More precisely, a state of an FA is a set of input strings that the FA "identifies," because—and so that—any two strings in the set are indistinguishable with respect to the language the FA accepts.
2. The *coarsest*—i.e., smallest-index—equivalence relation that "works" is $\equiv_L$, so that this relation embodies the *smallest* FA that accepts language L.

We can turn the preceding intuition into an algorithm for minimizing the state-set of a given FA. You can look at this algorithm as starting with any given equivalence relation that "defines" L (e.g., with any FA that accepts L) and iteratively "coarsifying" the relation as far as we can, thereby "sneaking" up on the relation $\equiv_L$.

The resulting algorithm for minimizing a FA $M = (Q, \Sigma, \delta, q_0, F)$ essentially computes the following equivalence relation on M's state-set Q. For $p, q \in Q$,

$$[p \equiv_\delta q] \text{ if, and only if, } (\forall x \in \Sigma^\star)[[\delta(p, x) \in F] \Leftrightarrow [\delta(q, x) \in F]]$$

This relation says that no input string will allow one to distinguish M's being in state p from M's being in state q. One can, therefore, coalesce states p and q to obtain a smaller FA that accepts $L(M)$. The equivalence classes of $\equiv_\delta$, i.e., the set

$$\{[p]_{\equiv_\delta} \mid p \in Q\}$$

are, therefore, the states of the smallest FA—call it $\widehat{M}$—that accepts $L(M)$. The state-transition function $\widehat{\delta}$ of $\widehat{M}$ is given by

$$\widehat{\delta}([p]_{\equiv_\delta}, \sigma) \ = \ [\delta(p, \sigma)]_{\equiv_\delta}.$$

Finally, the initial state of $\widehat{M}$ is $[q_0]_{\equiv_\delta}$, and the accepting states are $\{[p]_{\equiv_\delta} \mid p \in F\}$. One shows easily that $\widehat{\delta}$ is well defined and that $L(\widehat{M}) = L(M)$.

We simplify our explanation of how to compute $\equiv_\delta$ by describing an example concurrently with our description of the algorithm. We start with a very coarse approximation to $\equiv_\delta$ and iteratively improve the approximation. Fig. 1 presents the FA

$$M = (\{a, b, c, d, f, g, h\}, \{0, 1\}, \delta, a, \{c\})$$

for our example, in tabular form.

M	q	$\delta(q, 0)$	$\delta(q, 1)$	$q \in F$?
(start state) $\rightarrow$	a	b	f	$\notin F$
	b	g	c	$\notin F$
(final state) $\rightarrow$	c	a	c	$\in F$
	d	c	g	$\notin F$
	e	h	f	$\notin F$
	f	c	g	$\notin F$
	g	g	e	$\notin F$
	h	g	c	$\notin F$

Fig. 1. The FA M that we minimize.

Our initial partition[9] of Q is $Q - F$, F, to indicate that the null string ε witnesses the fact that no accepting state is equivalent to any nonaccepting state. This yields the initial partition of M's states:

$$[a, b, d, e, f, g, h]_1, \ [c]_1$$

(The subscript "1" indicates that this is the first discriminatory step). State c, being the unique final state, is not equivalent to any other state.

Inductively, we now look at the current, time-t, partition and try to "break apart" time-t blocks. We do this by feeding pairs of states in the same block single input symbols. If any symbol leads states p and q to different blocks, then, by induction, we have found a string x that discriminates between them. In detail, say that $\delta(p, \sigma) = r$ and $\delta(q, \sigma) = s$. If there is a string x that discriminates between states r and s—by showing them not to be equivalent under $\equiv_\delta$—then

---

[9] Recalling that partitions and equivalence relations are equivalent notions, we continue to use notation "$[ab \cdots z]$" to denote the set $\{a, b, \ldots, z\}$ viewed as a block of a partition (= equivalence class).

the string σx discriminates between states p and q. In our example, we find that input "0" breaks the big time-1 block, so that we get the "time 1.5" partition

$$[a, b, e, g, h]_{1.5}, \ [d, f]_{1.5}, \ [c]$$

and input "1" further breaks the block down. We end up with the time-2 partition

$$[a, e]_2, \ [b, h]_2, \ [g]_2, \ [d, f]_2, \ [c]_2$$

Let's see how this happens. First, we find that $\delta(d, 0) = \delta(f, 0) = c \in F$, while $\delta(q, 0) \notin F$ for $q \in \{a, b, e, g, h\}$. This leads to the "time-1.5" partition (since we have thus far used only one of the two input symbols). At this point, input "1" leads states a and e to block $\{d, f\}$, and it leads states b and h to block $\{c\}$; it leaves state g in its present block. We thus end up with the indicated time-2 partition. Further single inputs leave this partition unchanged, so it must be the coarsest partition that preserves $L(M)$.

The preceding sentence invokes the fact that, by a simple induction, if a partition persists under (i.e., is unchanged by) all single inputs, then it persists under all input strings. We claim that such a stable partition embodies the relation $\equiv_M$, hence, by Lemma 2, the relation $\equiv_{L(M)}$. To see this, consider any two states, p and q, that belong to the same block of a partition that persists under all input strings. Stability ensures that, for all $z \in \Sigma^*$, the states $\delta(p, z)$ and $\delta(q, z)$ belong to the same block of the partition; hence, either both states belong to F or neither does. In other words: If $\delta(q_0, x) = p$ and $\delta(q_0, y) = q$, for $x, y \in \Sigma^*$, then for all $z \in \Sigma^*$, either $\{p, q\} \subseteq F$, in which case $\{xz, yz\} \subseteq L(M)$, or $\{p, q\} \subseteq Q - F$, in which case $\{xz, yz\} \subseteq \Sigma^* - L(M)$. By definition, then, $x \equiv_M y$.

Returning to the algorithm, we have ended up with the FA $\widehat{M}$ of Fig. 2 as the minimum-state version of M.

$\widehat{M}$	q	$\widehat{\delta}(q, 0)$	$\widehat{\delta}(q, 1)$	$q \in F$?
(start state) $\rightarrow$	$[ae]$	$[bh]$	$[df]$	$\notin F$
	$[bh]$	$[g]$	$[c]$	$\notin F$
(final state) $\rightarrow$	$[c]$	$[ae]$	$[c]$	$\in F$
	$[df]$	$[c]$	$[g]$	$\notin F$
	$[g]$	$[g]$	$[ae]$	$\notin F$

Fig. 2. The FA $\widehat{M}$ that minimizes the FA M of Fig.1.

5   Applying Myhill-Nerode Concepts to Non-FA's

We present three applications of the concepts underlying Theorem 3 to state-transition systems other than FA's. Although our primary motivation is to expose interesting applications of Myhill-Nerode-type characterizations of "state," for the sake of completeness, we sketch out the derivations of the results that the characterizations lead to. The first two applications involve OA's that are strictly more powerful than FA's.

5.1   Memory Bounds for Online Automata

By Theorem 3, any OA M that accepts a nonregular language must have infinitely many states. We now present a result from [6] that sharpens this statement via an "infinitely-often" lower bound on the number of states an FA $M^{(n)}$ must have in order to correctly mimic M's (word-acceptance) behavior on all words of length $\leq n$ (thereby providing an "order-n approximation" of M). This bound assumes nothing about M other than its accepting a nonregular language. (Indeed, M's state-transition function δ need not even be computable.) In the context of this survey, this result removes Theorem 3 from the confines of the theory of FA's, by adapting it to a broader class of state-transition systems. This adaptation is achieved by converting the *word-relating* equivalence relation $\equiv_M$ to an *automaton-relating* relation that asserts the equivalence of two OA's on all words that are no longer than a chosen parameter.

Let L be a nonregular language, and let M be an OA that accepts L: $L = L(M)$. For any $n \in \mathsf{N}$, an FA $M^{(n)}$ is an *order-n approximate acceptor* of L or, equivalently, an *order-n approximation* of M if

$$\{x \in L(M^{(n)}) \mid \ell(x) \leq n\} \; = \; \{x \in L \mid \ell(x) \leq n\} \; = \; \{x \in L(M) \mid \ell(x) \leq n\}.$$

We denote by $\mathcal{A}_L(n)$ the (obviously monotonic nondecreasing) number of states in the smallest order-n approximate acceptor of L, as a function of n. This quantity can be viewed as a measure of L's "space complexity," in the sense that one needs $\lceil \log_2 \mathcal{A}_L(n) \rceil$ bi-stable devices (say, transistors) in order to implement an order-n approximate acceptor of L in circuitry.

The conceptual framework of Theorem 3 affords one easy access to a nontrivial lower bound on the "infinitely-often" behavior of $\mathcal{A}_L(n)$, *for any* nonregular language L.

Theorem 8 ([6]). *If the language L is nonregular, then, for infinitely many n,*

$$\mathcal{A}_L(n) > \frac{1}{2}n + 1. \tag{4}$$

Proof. Let M_1 and M_2 be OA's. For any $n \in \mathsf{N}$, we say that M_1 and M_2 are *n-equivalent*, denoted $M_1 \equiv_n M_2$, just when

$$\{x \in L(M_1) \mid \ell(x) \leq n\} \; = \; \{x \in L(M_2) \mid \ell(x) \leq n\}.$$

This relation is, thus, a parameterized extension of the relation $\equiv_M$ that is central to Theorem 3.

Our analysis of approximate acceptors of L builds on the following bound on the "degree" of equivalence of pairs of FA's.

Lemma 9 ([10]) *Let M_1 and M_2 be FA's with s_1 and s_2 states, respectively, such that $L(M_1) \neq L(M_2)$. Then $M_1 \not\equiv_{s_1+s_2-2} M_2$.*

Proof (Proof of Lemma 9). We bound from above the number of partition-refinements that suffice for the state-minimization algorithm of Section 4.2 to distinguish the initial states of M_1 and M_2 (which, by hypothesis, are distinguishable).

Since the algorithm is actually a "state-equivalence tester," we can apply it to state-transition systems that are not legal FA's, as long as we are careful to keep final and nonfinal state segregated from one another. We therefore apply the algorithm to the following "disconnected" FA M. Say that, for $i = 1, 2$, $M_i = (Q_i, \Sigma, \delta_i, q_{i,0}, F_i)$, where $Q_1 \cap Q_2 = \emptyset$. Then $M = (Q, \Sigma, \delta, \{q_{1,0}, q_{2,0}\}, F)$, where

- $Q = Q_1 \cup Q_2$
- for $q \in Q$ and $\sigma \in \Sigma$: $\delta(q, \sigma) = \begin{cases} \delta_1(q, \sigma) \text{ if } q \in Q_1 \\ \delta_2(q, \sigma) \text{ if } q \in Q_2 \end{cases}$
- $F = F_1 \cup F_2$.

Now, the fact that $L(M_1) \neq L(M_2)$ implies: (a) that $q_{1,0} \not\equiv_M q_{2,0}$; (b) that neither $Q - F$ nor F is empty. How many stages of the algorithm would be required, in the worst case, to distinguish states $q_{1,0}$ and $q_{2,0}$ within M, when the algorithm starts with the initial partition $\{Q - F, F\}$? Well, each stage of the algorithm, save the last, must "split" some block of the partition into two nonempty subblocks. Since one "split," namely, the separation of $Q - F$ from F, occurs before the algorithm starts applying input symbols, and since $|Q| = s_1 + s_2$, the algorithm can proceed for no more than $s_1 + s_2 - 2$ stages; after that many stages, all blocks would be singletons! In other words, if $p \not\equiv_M q$, for states $p, q \in Q$, then there is a string of length $\leq s_1 + s_2 - 2$ that witnesses the nonequivalence. Since we know that $q_{1,0} \not\equiv_M q_{2,0}$, this completes the proof.

Back to the theorem. For each $k \in \mathsf{N}$, Theorem 3 guarantees that there is a smallest integer $n > k$ such that $\mathcal{A}_L(k) = \mathcal{A}_L(n-1) < \mathcal{A}_L(n)$. The preceding inequality implies the existence of FA's M_1 and M_2 such that:

1. M_1 has $\mathcal{A}_L(n-1)$ states and is an $(n-1)$-approximate acceptor of L;
2. M_2 has $\mathcal{A}_L(n)$ and is an n-approximate acceptor of L.

By statement 1, $M_1 \equiv_{n-1} M_2$; by statements 1 and 2, $M_1 \not\equiv_n M_2$. By Lemma 9, then, $M_1 \not\equiv_{\mathcal{A}_L(n-1)+\mathcal{A}_L(n)-2} M_2$. Since $M_1 \equiv_{n-1} M_2$, we therefore have $\mathcal{A}_L(n-1) + \mathcal{A}_L(n) > n+1$, which yields Ineq. 4, since $\mathcal{A}_L(n-1) \leq \mathcal{A}_L(n) - 1$.

It is shown in [6] that Theorem 8 is as strong as possible, in that: the constants $\frac{1}{2}$ and 1 in Ineq. 4 cannot be improved; the phrase "infinitely many" cannot be strengthened to "all but finitely many."

5.2   Online Automata with Structured States

The preceding section derives a lower bound on the size of any OA M that accepts a nonregular language L, by bounding the number of classes of $\equiv_L$. In this section, we present lower bounds from [2] on the *time* a specific genre of OA requires to accept a language L, based on the "structure" of the OA's infinitely many states. Specifically, we analyze the behavior of "online" Turing Machines (TM's) whose infinitely many states arise from a collection of read-write "work tapes" of unbounded capacities. As in Section 5.1, the desired bound is achieved by adapting Theorem 3 to a broad class of infinite-state OA's. This adaptation is achieved here by parameterizing the word-relating equivalence relation $\equiv_M$; for each integer $t > 0$, the parameter-t relation $\equiv_M^{(t)}$ behaves like $\equiv_M$, but exposes only discriminations that M can make in t or fewer steps.

A word about TM's is in order, to explain why the study in this section is relevant to computer scientists. The TM model originated in the monumental study [19] that planted the seeds of computability theory, hence, also, of complexity theory. Lacking real digital computers as exemplars of the genre, Turing devised a model that served his purposes but that would be hard to justify today as a way for thinking about either computers or algorithms. Seen in this light, one surmises that TM's persists in today's textbooks on computation theory only because of their mathematical simplicity. However, I believe there is an alternative role for the TM model, which justifies continued attention—in certain contexts. Specifically, one can often devise varieties of TM that allow one to expose the impact of data-structure topology on the efficiency of certain computations. These TM's abstract the control portion of an algorithm down to a finite state-transition system and use the TM's "tapes" to model access to data structures. The study in [2] uses TM's in this way, focusing on the impact of tape topology on efficiency of retrieving sets of words. As such, the bounds here can be viewed as an early contribution to the theory of data structures. This perspective underlay both my "data graph" model [16] and Schönhage's "storage modification machine" model [17]. The interesting features here are the formulation of an information-retrieval problem as a formal language, and the use of the concepts underlying Theorem 3 to analyze the problem.

5.2.1. The Online TM Model. A *d-dimensional tape* is a linked data structure with an array-like topology, termed an *orthogonal list* in [7]. A tape is accessed via a *read-write head*—the TM-oriented name for a pointer. Each cell of a tape holds one symbol from a finite set Γ that contains a designated "blank" symbol; e.g., in a 32-bit computer, Γ could be the set of 32-bit binary words, and the "blank" symbol could be the word of all 1's. Access to cells within a tape is sequential: one can move the head at most one cell in any of the $2d$ permissible directions in a step.

An **online TM** M with t d-dimensional "work tapes" can be viewed as an FA that has access to t d-dimensional tapes. As with any FA, M has an *input port* via which it scans symbols from its input alphabet Σ; it also has a designated initial state and a designated set of final states.

Let me explain the role of the input port in M's "online" computing, by analogy with FA's. One can view an FA as a device that is passive until a symbol $\sigma \in \Sigma$ is "dropped into" its input port. If the FA is in a stable configuration at that moment—meaning that all bi-stable devices in its circuitry have stabilized—then the FA responds to input σ. The most interesting aspect of this response is that the FA indicates whether the entire sequence of input symbols that it has been presented up to that point—i.e., up to and including the last instance of symbol σ— is accepted. Note that the FA responds to input symbols in an *online* manner, making acceptance/rejection decisions about each prefix of the input string as that prefix has been read. Of course, once the FA has "digested" the last instance of symbol σ, by again reaching a stable configuration, then it is ready to "digest" another input symbol, when and if one is "dropped into" its input port.

The TM M uses its input port in much the manner just described. There is, however, a fundamental difference between an FA and an online TM. During the "passive" periods in which an FA does not accept new input symbols at its input port, the FA is typically waiting for its logic to stabilize, hence is usually not considered to be doing valuable computation. In contrast, during the "passive" periods in which an online TM does not accept new input symbols at its input port, the TM may be doing quite valuable subcomputations using its work tapes. Indeed, the study in [2] can be viewed as bounding (from below) the cumulative time that must be devoted to these "introspective" subcomputations when performing certain computations. With this intuitive background in place, a computational step by M depends on:

- its current state,
- the current input symbol, *if M's program reads the input at this step*,
- the t symbols (elements of Γ) currently scanned by the pointers on the t tapes.

On the basis of these, M:

- enters a new state (which may be the same as the current one),
- independently rewrites the symbols currently scanned on the t tapes (possibly with the same symbol as the current one),
- independently moves the read-write head on each tape at most[10] one square in one of the $2d$ allowable directions.

Notes. (a) When $d = 1$, we have a TM with t linear (i.e., one-dimensional) tapes. **(b)** Our tapes have array-like topologies because of the focus in [2]. It is easy to specify tapes with other regular topologies, such as trees of various arities.

One extends M's one-step computation to a multistep computation (whose goal is language recognition, as usual) as follows. To determine if a word $w =$

---

[10] "At most" means that a read-write head is allowed to remain stationary.

$\sigma_1\sigma_2\cdots\sigma_n \in \Sigma^\star$ is accepted by M—i.e., is in the language $L(M)$—one makes w's n symbols available, in sequence, at M's input port. If M starts in its initial state with all cells of all tapes containing the "blank" symbol, and it proceeds through a sequence of N steps that:

- includes n steps during which M "reads" an input symbol,
- ends with a step in which M is programmed to "read" an input symbol,

then M is said to *decide* w *in* N *steps*; if, moreover, M's state at step N is an accepting state, then M is said to *accept* w *in* N *steps*.

The somewhat complicated double condition for acceptance ("includes ..." and "ends with ...") ensures that, if M accepts w, then it does so unambiguously. Specifically, after reading the last symbol of w, M does not "give its answer" until it prepares to read the next input symbol (if that ever happens). This means that M cannot oscillate between accepting and nonaccepting states after reading the last symbol of w.

5.2.2. The Impact of Tape Structure on Memory Locality.

The *configuration* of an online TM M having t d-dimensional tapes, at any step of a computation, is the $(t+1)$-tuple $\langle q, \tau_1, \tau_2, \ldots, \tau_t \rangle$ defined as follows.

- q is the state of M's finite-state control (its associated FA);
- each τ_i is the d-dimensional configuration of symbols from Γ that comprises the non-"blank" portion of tape i, with one symbol highlighted (in some way) to indicate the current position of M's read-write head on tape i.

(M's configuration is often called its "total state.") The importance of this concept resides in the following. Say that, for $i = 1, 2$, the database-string $x_i \in \Sigma^\star$ leads M to configuration $C_M(x_i) = \langle q_i, \tau_{i1}, \tau_{i2}, \ldots, \tau_{it} \rangle$. If:

- $q_1 = q_2$; i.e., the configurations share the same state;
- for some integer $r \geq 1$, and all $i \in \{1, 2, \ldots, t\}$, tape configurations τ_{1i} and τ_{2i} are identical within r symbols of their highlighted symbols (which indicate where M's read-write heads reside),

then we say that the databases specified by x_1 and x_2 are r-*indistinguishable* by M, denoted $x_1 \equiv_M^{(r)} x_2$. This relation is an important parameterization of the FA-oriented relation $\equiv_M$ that is central to Theorem 3. Specifically, by analyzing relation $\equiv_M^{(r)}$, one can sometimes bound the time-complexity of various subcomputations by M, in the following sense.

Lemma 10 *Say that* $x_1 \equiv_M^{(r)} x_2$. *If there exists a* $y \in \Sigma^\star$ *such that one of* $x_1 y, x_2 y$ *belongs to* $L(M)$, *while the other does not, then, having read either of* x_1 *or* x_2, M *must compute for more than* r *steps while reading* y.

5.2.3. An Information-Retrieval Problem Formulated as a Language.

The following problem is used in [2] to expose the potential effect of tape structure on computational efficiency. We feed an online TM M a set of equal-length binary words, which we term a *database*. We then feed M a sequence of binary words, each of which is termed a *query*. After reading each query, M must respond "YES" if the query word occurs in the database, and "NO" if not.

The *database language* $L_{DB} \subseteq \Sigma^*$, where, $\Sigma = \{0, 1, :\}$, and ":" is a symbol distinct from "0" and "1," formalizes the preceding problem. Each word in L_{DB} has the form

$$\xi_1 : \xi_2 : \cdots : \xi_m :: \eta_1 : \eta_2 : \cdots : \eta_n$$

where, for some $k \in \mathbb{N}$,

- each ξ_i $(1 \leq i \leq m)$ and each η_j $(1 \leq j \leq n)$ is a length-k binary string;
- $m = 2^k$;
- $\eta_n \in \{\xi_1, \xi_2, \ldots, \xi_m\}$.

Both the sequence of ξ_i's and the sequence of η_j's can contain repetitions. In particular, we are interested only in the *set* of words $\{\xi_1, \xi_2, \ldots, \xi_m\}$ (the database). The *database string* "$\xi_1 : \xi_2 : \cdots : \xi_m$" is just the mechanism we use to present the database to M. Each word η_j is a *query*. In each word $x \in L_{DB}$, the double colon "::" separates the database from the queries, while the single colon ":" separates consecutive binary words.

The fact that we are interested only in whether or not *the last* query appears in the database reflects the *online* nature of the computation: M must respond to each query as it appears, with no knowledge of which is the last, hence, the important one. (This is essentially the challenge faced by all online algorithms.)

5.2.4. Tape Dimensionality and the Time to Recognize L_{DB}.

For simplicity, we focus henceforth on the sublanguages of L_{DB} that are parameterized by the common lengths of their binary words. For each $k \in \mathbb{N}$, $L_{DB}^{(k)}$ denotes the sublanguage in which each ξ_i and each η_j has length k. Note that each database-string in $L_{DB}^{(k)}$ has length $(k+1)2^k - 1$.

Focus on any fixed $L_{DB}^{(k)}$. If the database-strings x_1 and x_2 specify distinct databases, then there exists a query η that appears in the database specified by one of the x_i but not the other—so, precisely one of $x_1 :: \eta$ and $x_2 :: \eta$ belongs to $L_{DB}^{(k)}$. Database-strings that specify distinct databases must, thus, lead M to distinct configurations.

How "big" must these configurations be? On the one hand, there are $2^{2^k} - 1$ distinct databases (corresponding to each nonempty set of length-k ξ_i's). On the other hand, for any M with t d-dimensional tapes, there is an $\alpha_M > 0$ that depends only on M's structure, such that M has $\leq \alpha_M^{dtr}$ distinct configurations of "radius" r—meaning that all non-"blank" symbols on all tapes reside within r cells of the read-write heads. Thus, in order for each database to get a distinct configuration (so that $\equiv_M^{(r)}$ has $\geq 2^{2^k} - 1$ equivalence classes), the "radius" r must exceed $\beta_M \cdot 2^{k/d}$, for some $\beta_M > 0$ that depends only on M's structure.

Combining this bound with Lemma 10, we arrive at the following time bound.

Lemma 11 *If $L(M) = L_{DB}^{(k)}$, then, for some length-k query η, M must take[11] $>$ $\beta_M \cdot (2^{1/d})^k$ steps while reading η, for some $\beta_M > 0$ that depends only on M's structure.*

The reasoning behind Lemma 11 is *information theoretic*, depending only on the fact that the number of databases in $L_{DB}^{(k)}$ is doubly exponential in k, while the number of bounded-"radius" TM configurations is singly exponential. Therefore, no matter how M reorganizes its tape contents while responding to one bad query, there must be a query that is bad for the new configuration! By focusing on strings with 2^k bad queries, we thus obtain:

Theorem 12 ([2]). *Any online TM M with d-dimensional tapes that recognizes the language L_{DB} must, for infinitely many N, take time $> \beta_M \cdot (N/\log N)^{1+1/d}$ to process inputs of length N, for some constant $\beta_M > 0$ that depends only on M's structure.*

One finds in [2] a companion upper bound of $O(N^{1+1/d})$ for the problem of recognizing L_{DB}. Hence, Theorem 12 does expose the potential of nontrivial impact of data-structure topology on computational efficiency. In its time, the theorem also exposed one of the earliest examples of the cost of the *online* requirement. Specifically, L_{DB} can clearly be accepted *in linear time* by a TM M that has just a single, linear work tape, but that operates in an *offline* manner— meaning that M gets to see the entire input string before it must give an answer (so that it knows which query is important before it starts computing).

6   Finite Automata with Probabilistic Transitions

We now consider a rather different genre of OA's, namely, FA's whose state-transitions are *probabilistic*, with acceptance decisions depending on the probability of ending up in an final state. This is a very timely model to consider since probabilistic state-transition systems are currently quite in vogue in several areas of artificial intelligence, notably the growing area of machine learning. The main result that we present comes from [14]; it exhibits a nontrivial, somewhat surprising situation in which probabilistic state-transitions add no power to the model: The restricted automata accept only regular sets.

6.1   PFA's and Their Languages

We start with an FA, $M = (Q, \Sigma, \delta, q_0, F)$, and make its state-transitions and acceptance criterion *probabilistic*. We call the resulting model a *Probabilistic Finite Automaton* (*PFA*, for short).

---

[11] We write $2^{k/d}$ in the unusual form $(2^{1/d})^k$ to emphasize that the dimensionality of M's tapes (which is a fixed constant) appears only in the base of the exponential.

States. We simplify the formal development by positing that the state-set of the PFA M is $Q = \{1, 2, \ldots, n\}$, with $q_0 = 1$, and $F = \{m, m+1, \ldots, n\}$ for some $m \in Q$.

State-transitions. We replace M's state-transition function δ with a set of tables, one for each symbol of Σ. The table associated with $\sigma \in \Sigma$ indicates, for each pair of states $q, q' \in Q$, the probability—call it $\rho_{q,q'}$—that M ends up in state q' when started in state q and "fed" input symbol σ. It is convenient to present the state-transition tables as matrices. The table associated with $\sigma \in \Sigma$ is:

$$\Delta_\sigma = \begin{pmatrix} \rho_{1,1} & \rho_{1,2} & \cdots & \rho_{1,n} \\ \rho_{2,1} & \rho_{2,2} & \cdots & \rho_{2,n} \\ \vdots & \vdots & \ddots & \vdots \\ \rho_{n,1} & \rho_{n,2} & \cdots & \rho_{n,n} \end{pmatrix}$$

where each[12] $\rho_{i,j} \in [0,1]$, and, for each i, $\sum_j \rho_{i,j} = 1$.

States, revisited. The probabilistic nature of M's state-transitions forces us to distinguish between M's set of states—the set Q—and the "state" that reflects M situation at any point of a computation, which is a probability distribution over Q. We therefore define the *state-distribution* of M to be a vector of probabilities $q = \langle \pi_1, \pi_2, \ldots, \pi_n \rangle$, where each π_i is the probability that M is in state i. The *initial state-distribution* is $q_0 = \langle 1, 0, \ldots, 0 \rangle$.

State-transitions, revisited. Under the preceding formalism, the PFA analogue of the FA single-symbol state-transition $\delta(q, \sigma)$ is the vector-matrix product: $\widehat{\Delta}(q, \sigma) = q \times \Delta_\sigma$. By extension, the PFA analogue of the FA string state-transition $\delta(q, \sigma_1 \sigma_2 \cdots \sigma_k)$, where each $\sigma_i \in \Sigma$, is

$$\widehat{\Delta}(q, \sigma_1 \sigma_2 \cdots \sigma_k) \stackrel{\text{def}}{=} q \times \Delta_{\sigma_1} \times \Delta_{\sigma_2} \times \cdots \times \Delta_{\sigma_n}. \tag{5}$$

The language accepted by a PFA. The probabilistic analogue of acceptance by final state builds on the notion of an *(acceptance) threshold* $\theta \in [0,1]$. The string $x \in \Sigma^\star$ is *accepted* by M iff

$$p_M(x) \stackrel{\text{def}}{=} \sum_{i=m}^{n} \widehat{\Delta}(q_0, x)_i > \theta,$$

where $\widehat{\Delta}(q, x)_i$ denotes the ith coordinate of the tuple $\widehat{\Delta}(q, x)$. (Recall that M's final states are those whose integer-names are $\geq m$.) Thus, x is accepted iff it leads M from its initial state to its set of final states with probability $> \theta$. As with all OA's, the *language accepted by* M is the set of all strings that M accepts. Acknowledging the crucial role of the acceptance threshold θ, we denote this language by

$$L(M, \theta) \stackrel{\text{def}}{=} \{x \in \Sigma^\star \mid p_M(x) > \theta\}.$$

---

[12] As usual, $[0,1]$ denotes the closed real interval $\{x \mid 0 \leq x \leq 1\}$.

6.2 $L(M, \theta)$ Is Regular when θ Is "Isolated"

It is noted in [14] that even simple—e.g., two-state—PFA's can accept nonregular languages, when accompanied by an "unfavorable" acceptance threshold. When thresholds are "favorable," though, all PFA's accept regular languages.

The threshold $\theta \in [0, 1]$ is *isolated* for the PFA M iff there exist a real *constant of isolation* $\varepsilon > 0$ such that, for all $x \in \Sigma^\star$, $|p_M(x) - \theta| \geq \varepsilon$.

Theorem 13 ([14]). *For any PFA M and associated* isolated *acceptance threshold θ, the language $L(M, \theta)$ is regular.*

Proof. We sketch the proof from [14], which is a direct application of Theorem 3. Say that M has n states, a of which are final, and let $\varepsilon > 0$ be the constant of isolation. We claim that the relation $\equiv_{L(M,\theta)}$ cannot have more than $\kappa \overset{\text{def}}{=} [1 + (a/\varepsilon)]^{n-1}$ classes.

This bound is established by considering a set of k words that are mutually inequivalent under $\equiv_{L(M,\theta)}$, with the aim of showing that k cannot exceed κ. This is accomplished by converting M's language-related problem to a geometric setting, by considering, for each $x \in \Sigma^\star$, the point in n-dimensional space given by $\widehat{\Delta}(\boldsymbol{q}_0, w)$ (cf. Eq. 5).

In the language-related setting, we consider an arbitrary pair of inequivalent words, $x_i, x_j \in \Sigma^\star$, and note that there must exist $y \in \Sigma^\star$ such that (w.l.o.g.) $x_i y \in L(M, \theta)$ while $x_j y \notin L(M, \theta)$. In the geometric setting, this translates into the existence of three points:

$$\langle \xi_1^{(i)}, \xi_2^{(i)}, \ldots, \xi_n^{(i)} \rangle \text{ corresponding to } x_i$$
$$\langle \xi_1^{(j)}, \xi_2^{(j)}, \ldots, \xi_n^{(j)} \rangle \text{ corresponding to } x_j$$
$$\langle \eta_1, \eta_2, \ldots, \eta_n \rangle \text{ corresponding to } y$$

such that (here are the acceptance conditions):

$$\theta + \varepsilon < \xi_1^{(i)} \eta_1 + \xi_2^{(i)} \eta_2 + \cdots + \xi_n^{(i)} \eta_n;$$
$$\theta - \varepsilon \geq \xi_1^{(j)} \eta_1 + \xi_2^{(j)} \eta_2 + \cdots + \xi_n^{(j)} \eta_n.$$

Elementary reasoning then allows us to infer that

$$2(\varepsilon/a) \leq |\xi_1^{(i)} - \xi_1^{(j)}| + |\xi_2^{(i)} - \xi_2^{(j)}| + \cdots + |\xi_n^{(i)} - \xi_n^{(j)}|.$$

We next consider, for each $i \in \{1, 2, \ldots, k\}$, the set Λ_i comprising all points $\langle \xi_1, \xi_2, \ldots \xi_n \rangle$ such that

- $\xi_l \geq \xi_l^{(i)}$ for all $l \in \{1, 2, \ldots, n\}$ $\bullet \sum_{l=1}^{n} (\xi_l - \xi_l^{(i)}) = (\varepsilon/a).$

By bounding the volumes of the sets Λ_i, and arguing that no two share an internal point, one arrives at the following bounds on the cumulative volumes of the sets.

$$kc(\varepsilon/a)^{n-1} = \sum_{l=1}^{n} \text{Vol}(\Lambda_l) = c(1 + (\varepsilon/a))^{n-1}.$$

We infer directly that $k \leq [1 + (a/\varepsilon)]^{n-1}$, as was claimed.

7 Conclusions

It has been my goal to present a technical argument for the importance of the Myhill-Nerode Theorem and the concepts it uses to characterize the notion of "state." I have attempted to do so by reviewing several applications of (the concepts underlying) the Theorem, to areas as diverse as Finite Automata theory (Section 4.1), logic design (Section 4.2), space complexity (Section 5.1), the theory of data structures (Section 5.2), and artificial intelligence/machine learning (Section 6). To the extent that the role of theoretical computer science is to provide nonobvious conceptual frameworks for thinking/reasoning about and analyzing "real" computational settings and systems—and no one can dispute that this is at least one of the roles of the theory—the Myhill-Nerode Theorem is a success story for the field, one that should be in the toolbox of every theoretical computer scientist.

In closing, I want to stress that the Myhill-Nerode Theorem and its associated concepts is just one of the treasures from the 1960's that have slipped from front stage as automata-like models have slipped from favor. I would mention the product-decomposition work in [1] as another topic in the study of state-transition systems whose significance surely transcends the study of Finite Automata in which the work originated.

Acknowledgment. It is a pleasure to acknowledge my debt to Oded Goldreich for many perceptive comments, criticisms, and suggestions. I am grateful also to several others, notably Micah Adler and Ami Litman, for sharing insights and posing technical challenges.

References

1. J. Hartmanis and R.E. Stearns (1966): *Algebraic Structure Theory of Sequential Machines.* Prentice Hall, Englewood Cliffs, NJ.
2. F.C. Hennie (1966): On-line Turing machine computations. *IEEE Trans. Electronic Computers, EC-15*, 35–44.
3. J.E. Hopcroft, R. Motwani, J.D. Ullman (2001): *Introduction to Automata Theory, Languages, and Computation* (2nd ed.). Addison-Wesley, Reading, MA.
4. J.E. Hopcroft and J.D. Ullman (1979): *Introduction to Automata Theory, Languages, and Computation* (1st ed.) Addison-Wesley, Reading, MA.
5. J. Jaffe (1978): A necessary and sufficient pumping lemma for regular languages. *SIGACT News*, 48–49.
6. R.M. Karp (1967): Some bounds on the storage requirements of sequential machines and Turing machines. *J. ACM 14*, 478–489.
7. D.E. Knuth (1973): *The Art of Computer Programming: Fundamental Algorithms* (2nd ed.) Addison-Wesley, Reading, MA.
8. H.R. Lewis and C.H. Papadimitriou (1981): *Elements of the Theory of Computation.* Prentice-Hall, Englewood Cliffs, NJ.
9. P. Linz (2001): *An Introduction to Formal Languages and Automata* (3rd ed.) Jones and Bartlett Publ., Sudbury, MA.

10. E.F. Moore (1956): Gendanken experiments on sequential machines. In *Automata Studies* (C.E. Shannon and J. McCarthy, Eds.) *[Ann. Math. Studies 34]*, Princeton Univ. Press, Princeton, NJ, pp. 129–153.

11. B.M. Moret (1997): *The Theory of Computation*. Addison-Wesley, Reading, MA.

12. J. Myhill (1957): Finite automata and the representation of events. WADD TR-57-624, Wright Patterson AFB, Ohio, pp. 112–137.

13. A. Nerode (1958): Linear automaton transformations. *Proc. AMS 9*, 541–544.

14. M.O. Rabin (1963): Probabilistic automata. *Inform. Control 6*, 230–245.

15. M.O. Rabin and D. Scott (1959): Finite automata and their decision problems. *IBM J. Res. Develop. 3*, 114–125.

16. A.L. Rosenberg (1971): Data graphs and addressing schemes. *J. CSS 5*, 193–238.

17. A. Schönhage (1980): Storage modification machines. *SIAM J. Computing 9*, 490–508.

18. M. Sipser (1997): *Introduction to the Theory of Computation*. PWS Publishing Co., Boston, MA.

19. A.M. Turing (1936): On computable numbers, with an application to the Entscheidungsproblem. *Proc. London Math. Soc.* (ser. 2, vol. 42) 230–265; Correction *ibid.* (vol. 43) 544–546.

20. Wikipedia: The Free Encyclopedia (2005): http://en.wikipedia.org/wiki/Pumping_lemma

Author Index

Lecture Notes in Computer Science

For information about Vols. 1–3804

please contact your bookseller or Springer

Vol. 3848: J.-F. Boulicaut, L. De Raedt, H. Mannila (Eds.), Constraint-Based Mining and Inductive Databases. X, 401 pages. 2006. (Sublibrary LNAI).

Vol. 3847: K.P. Jantke, A. Lunzer, N. Spyratos, Y. Tanaka (Eds.), Federation over the Web. X, 215 pages. 2006. (Sublibrary LNAI).

Vol. 3846: H. J. van den Herik, Y. Björnsson, N.S. Netanyahu (Eds.), Computers and Games. XIV, 333 pages. 2006.

Vol. 3845: J. Farré, I. Litovsky, S. Schmitz (Eds.), Implementation and Application of Automata. XIII, 360 pages. 2006.

Vol. 3844: J.-M. Bruel (Ed.), Satellite Events at the MoDELS 2005 Conference. XIII, 360 pages. 2006.

Vol. 3843: P. Healy, N.S. Nikolov (Eds.), Graph Drawing. XVII, 536 pages. 2006.

Vol. 3842: H.T. Shen, J. Li, M. Li, J. Ni, W. Wang (Eds.), Advanced Web and Network Technologies, and Applications. XXVII, 1057 pages. 2006.

Vol. 3841: X. Zhou, J. Li, H.T. Shen, M. Kitsuregawa, Y. Zhang (Eds.), Frontiers of WWW Research and Development - APWeb 2006. XXIV, 1223 pages. 2006.

Vol. 3840: M. Li, B. Boehm, L.J. Osterweil (Eds.), Unifying the Software Process Spectrum. XVI, 522 pages. 2006.

Vol. 3839: J.-C. Filliâtre, C. Paulin-Mohring, B. Werner (Eds.), Types for Proofs and Programs. VIII, 275 pages. 2006.

Vol. 3838: A. Middeldorp, V. van Oostrom, F. van Raamsdonk, R. de Vrijer (Eds.), Processes, Terms and Cycles: Steps on the Road to Infinity. XVIII, 639 pages. 2005.

Vol. 3837: K. Cho, P. Jacquet (Eds.), Technologies for Advanced Heterogeneous Networks. IX, 307 pages. 2005.

Vol. 3836: J.-M. Pierson (Ed.), Data Management in Grids. X, 143 pages. 2006.

Vol. 3835: G. Sutcliffe, A. Voronkov (Eds.), Logic for Programming, Artificial Intelligence, and Reasoning. XIV, 744 pages. 2005. (Sublibrary LNAI).

Vol. 3834: D.G. Feitelson, E. Frachtenberg, L. Rudolph, U. Schwiegelshohn (Eds.), Job Scheduling Strategies for Parallel Processing. VIII, 283 pages. 2005.

Vol. 3833: K.-J. Li, C. Vangenot (Eds.), Web and Wireless Geographical Information Systems. XI, 309 pages. 2005.

Vol. 3832: D. Zhang, A.K. Jain (Eds.), Advances in Biometrics. XX, 796 pages. 2005.

Vol. 3831: J. Wiedermann, G. Tel, J. Pokorný, M. Bieliková, J. Štuller (Eds.), SOFSEM 2006: Theory and Practice of Computer Science. XV, 576 pages. 2006.

Vol. 3830: D. Weyns, H. V.D. Parunak, F. Michel (Eds.), Environments for Multi-Agent Systems II. VIII, 291 pages. 2006. (Sublibrary LNAI).

Vol. 3829: P. Pettersson, W. Yi (Eds.), Formal Modeling and Analysis of Timed Systems. IX, 305 pages. 2005.

Vol. 3828: X. Deng, Y. Ye (Eds.), Internet and Network Economics. XVII, 1106 pages. 2005.

Vol. 3827: X. Deng, D.-Z. Du (Eds.), Algorithms and Computation. XX, 1190 pages. 2005.

Vol. 3826: B. Benatallah, F. Casati, P. Traverso (Eds.), Service-Oriented Computing - ICSOC 2005. XVIII, 597 pages. 2005.

Vol. 3824: L.T. Yang, M. Amamiya, Z. Liu, M. Guo, F.J. Rammig (Eds.), Embedded and Ubiquitous Computing - EUC 2005. XXIII, 1204 pages. 2005.

Vol. 3823: T. Enokido, L. Yan, B. Xiao, D. Kim, Y. Dai, L.T. Yang (Eds.), Embedded and Ubiquitous Computing - EUC 2005 Workshops. XXXII, 1317 pages. 2005.

Vol. 3822: D. Feng, D. Lin, M. Yung (Eds.), Information Security and Cryptology. XII, 420 pages. 2005.

Vol. 3821: R. Ramanujam, S. Sen (Eds.), FSTTCS 2005: Foundations of Software Technology and Theoretical Computer Science. XIV, 566 pages. 2005.

Vol. 3820: L.T. Yang, X.-s. Zhou, W. Zhao, Z. Wu, Y. Zhu, M. Lin (Eds.), Embedded Software and Systems. XXVIII, 779 pages. 2005.

Vol. 3819: P. Van Hentenryck (Ed.), Practical Aspects of Declarative Languages. X, 231 pages. 2005.

Vol. 3818: S. Grumbach, L. Sui, V. Vianu (Eds.), Advances in Computer Science - ASIAN 2005. XIII, 294 pages. 2005.

Vol. 3817: M. Faundez-Zanuy, L. Janer, A. Esposito, A. Satue-Villar, J. Roure, V. Espinosa-Duro (Eds.), Nonlinear Analyses and Algorithms for Speech Processing. XII, 380 pages. 2006. (Sublibrary LNAI).

Vol. 3816: G. Chakraborty (Ed.), Distributed Computing and Internet Technology. XXI, 606 pages. 2005.

Vol. 3815: E.A. Fox, E.J. Neuhold, P. Premsmit, V. Wuwongse (Eds.), Digital Libraries: Implementing Strategies and Sharing Experiences. XVII, 529 pages. 2005.

Vol. 3814: M. Maybury, O. Stock, W. Wahlster (Eds.), Intelligent Technologies for Interactive Entertainment. XV, 342 pages. 2005. (Sublibrary LNAI).

Vol. 3813: R. Molva, G. Tsudik, D. Westhoff (Eds.), Security and Privacy in Ad-hoc and Sensor Networks. VIII, 219 pages. 2005.

Vol. 3812: C. Bussler, A. Haller (Eds.), Business Process Management Workshops. XIII, 520 pages. 2006.

Vol. 3811: C. Bussler, M.-C. Shan (Eds.), Technologies for E-Services. VIII, 127 pages. 2006.

Vol. 3810: Y.G. Desmedt, H. Wang, Y. Mu, Y. Li (Eds.), Cryptology and Network Security. XI, 349 pages. 2005.

Vol. 3809: S. Zhang, R. Jarvis (Eds.), AI 2005: Advances in Artificial Intelligence. XXVII, 1344 pages. 2005. (Sublibrary LNAI).

Vol. 3808: C. Bento, A. Cardoso, G. Dias (Eds.), Progress in Artificial Intelligence. XVIII, 704 pages. 2005. (Sublibrary LNAI).

Vol. 3807: M. Dean, Y. Guo, W. Jun, R. Kaschek, S. Krishnaswamy, Z. Pan, Q.Z. Sheng (Eds.), Web Information Systems Engineering - WISE 2005 Workshops. XV, 275 pages. 2005.

Vol. 3806: A.H. H. Ngu, M. Kitsuregawa, E.J. Neuhold, J.-Y. Chung, Q.Z. Sheng (Eds.), Web Information Systems Engineering - WISE 2005. XXI, 771 pages. 2005.

Vol. 3805: G. Subsol (Ed.), Virtual Storytelling. XII, 289 pages. 2005.